# Pharmacology of Hormonal Polypeptides and Proteins

## ADVANCES IN EXPERIMENTAL MEDICINE AND BIOLOGY



Volume 1
THE RETICULOENDOTHELIAL SYSTEM AND ATHEROSCLEROSIS
Edited by N. R. Di Luzio and R. Paoletti • 1967

Volume 2
PHARMACOLOGY OF HORMONAL POLYPEPTIDES AND PROTEINS
Edited by N. Back, L. Martini, and R. Paoletti • 1968

# Pharmacology of Hormonal Polypeptides and Proteins

Proceedings of an International Symposium on the Pharmacology of Hormonal Polypeptides, held in Milan, Italy, September 14-16, 1967

Edited by
**Nathan Back**
*Department of Biochemical Pharmacology*
*School of Pharmacy*
*State University of New York*
*Buffalo, New York*

**Luciano Martini**
*Institute of Pharmacology*
*University of Milan*
*Milan, Italy*

and

**Rodolfo Paoletti**
*Chairman, Institute of Pharmacology*
*University of Cagliari*
*Cagliari, Italy*

PLENUM PRESS • NEW YORK • 1968

Library of Congress Catalog Card Number 68-19184


A Division of Plenum Publishing Corporation
227 West 17 Street, New York, N. Y. 10011



*Printed in the United States of America*

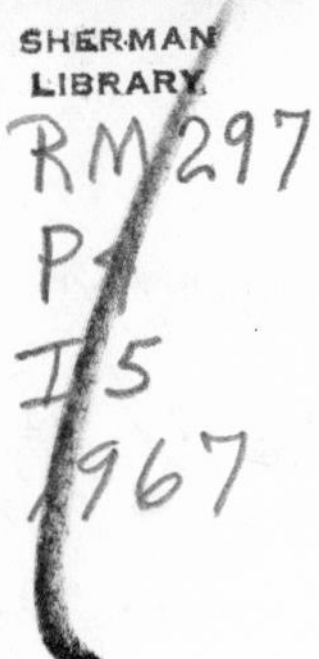



# Contents

PEPTIDE AND PROTEIN CHEMISTRY



POSTERIOR PITUITARY HORMONES

HYPOTHALAMIC NEUROHUMORAL PRINCIPLES

## ANTERIOR PITUITARY HORMONES

# SOME NEW METHODS IN PEPTIDE SYNTHESIS

Iphigenia Photaki

Laboratory of Organic Chemistry University of Athens

Athens, Greece

The aim of this lecture is to report the contribution of the Laboratory of Organic Chemistry, University of Athens, to the development of new methods of peptide synthesis.

The prerequisite for peptide synthesis is the temporary disappearance of the ampholytic properties of amino acids. The different methods and steps in the synthesis of peptides (1) are concerned with the following:

1) the mode of temporary protection of the $\alpha$-amino group
2) the mode of temporary protection of the carboxyl group
3) the mode of temporary protection of active groups of side chains (such as $\omega$-amino group, $\omega$-carboxyl group, sulfhydryl, etc) and
4) the methods of coupling amino acids under conditions which exclude racemization.

Let us begin with the protection of the $\alpha$-amino group.

The new era in the development of peptide synthesis began with the so-called carbobenzoxy method of Bergmann-Zervas (2). Thirty five years after its introduction the carbobenzoxy method is still the method of choice, especially since Bergmann and Zervas (3) have adapted it to almost all the known amino acids and du Vigneaud (4) to S-benzyl cysteine. These procedures made possible the synthesis (1) of thousands of peptides and of peptide hormones e.g. oxytocin (du Vigneaud), ACTH (Schwyzer), insulin (Katsoyannis, Zahn, Du).

However, it became clear later that as the peptide chain is lengthened, the removal of the carbobenzoxy group by catalytic hydrogenolysis becomes more difficult, and the use of HBr(5) as the splitting agent or the hydrogenolysis by reduction with sodium in liquid ammonia(4), have the disadvantage of possible secondary reactions(1). For such cases alternative N-protecting groups should be used, and many methods have been and are being worked out in our laboratory. Three such methods are the trityl method(6), which was developed simultaneously by Velluz(7) in France, the dibenzylphosphoryl method(8), and especially the nitrophenylsulphenyl method(9) which in our opinion constitutes, along with the carbobenzoxy method, the most significant advance in the problem of the protection of the α-amino group. After the formation of the peptide bond, the removal of the trityl(TRI) group is accomplished easily e.g. by alcoholysis in the presence of 1 equiv. of hydrogen chloride whereas the removal of the dibenzylphosphoryl group can be accomplished either by catalytic hydrogenolysis or with very dilute hydrogen bromide even in non-polar solvents. (The N-phosphorylpeptides formed as intermediates hydrolyze immediately due to the acidic environment). The nitrophenylsulphenyl (NPS)group is removed by the action of 2-3 equiv. of hydrogen chloride in non-polar solvents. Usually the N-protected peptide is dissolved in a solvent, or a mixture of solvents, from which the resulting hydrochloride salt precipitates as soon as it is formed, thus minimizing the possibility of side reactions(10). The removal of the NPS-group can be effected also

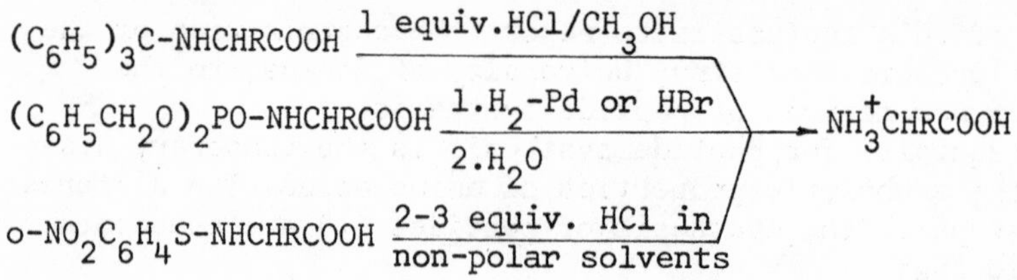

$(C_6H_5)_3C\text{-NHCHRCOOH} \xrightarrow{\text{1 equiv.HCl/}CH_3OH}$

$(C_6H_5CH_2O)_2PO\text{-NHCHRCOOH} \xrightarrow[\text{2.}H_2O]{\text{1.}H_2\text{-Pd or HBr}} NH_3^+CHRCOOH$

$o\text{-}NO_2C_6H_4S\text{-NHCHRCOOH} \xrightarrow[\text{non-polar solvents}]{\text{2-3 equiv. HCl in}}$

Fig. 1

with one equiv. of strong acids in polar solvents(9), and with various other reagents as has been reported by Kessler and Iselin(11), Poduška et al.(12), Fontana et al.(13), Bradenburg(14), Wünsch (15).

When the NPS-group is removed by acids or other reagents from N-NPS-peptides bearing a free thiol group, an N→S transfer of the NPS-group takes place(16). Therefore, in these cases the removal of the NPS-group must be done either in the S-protected cysteine peptides or in the oxidized -S-S-peptides, i.e. in the cystine peptides.

In many cases it is possible to use at first the carbobenzoxy method for the synthesis of a peptide, and then to exchange the amino-protecting group for the trityl or the NPS-group, so that the synthesis can be continued on a new basis. This possibility was pointed out in our laboratory long ago(6), and has in the meantime been applied successfully by Schwyzer(17) and by Zahn(18).

The second problem in peptide synthesis, i.e. the temporary masking of the carboxyl group is usually accomplished by esterification by methyl or ethyl alcohol. The ester group is later removed by alkaline saponification, which procedure in some cases is not without danger. Some progress was afforded by the use of the benzylesters introduced by Bergmann-Zervas(19,1). However the removal of the benzyl group by hydrogenolysis becomes more difficult with the growing length of the peptide chain, and is impossible in the presence of sulfur containing amino acids.

A significant progress was accomplished by the introduction of benzhydryl esters in our laboratory(20) and in the laboratories of Fruton(21) and of Hiskey(22). This group can be removed either by hydrogen chloride (or hydrogen bromide) in non-polar solvents or by hydrogenolysis. The full application of the use of diphenylmethyl esters was only made possible after the introduction of the NPS-method since the removal of the N-NPS-group can be accomplished under conditions which leave the benzhydryl groups intact(20).

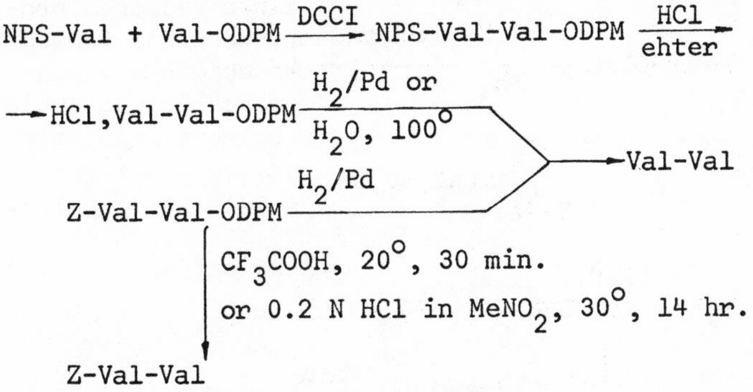

NPS-Val + Val-ODPM $\xrightarrow{\text{DCCI}}$ NPS-Val-Val-ODPM $\xrightarrow[\text{ehter}]{\text{HCl}}$

→ HCl,Val-Val-ODPM $\xrightarrow[H_2O,\ 100^o]{H_2/Pd\ or}$ → Val-Val

Z-Val-Val-ODPM $\xrightarrow{H_2/Pd}$ → Val-Val

Z-Val-Val-ODPM ↓ $CF_3COOH$, 20°, 30 min. or 0.2 N HCl in $MeNO_2$, 30°, 14 hr. → Z-Val-Val

NPS = $o\text{-}NO_2C_6H_4S$, Z = $C_6H_5CH_2OCO$, DPM = $Ph_2CH$

Fig. 2

Other valuable intermediates are the phenacyl esters of amino acids(23,20b). Lately a simple method for the preparation of the phenacyl esters of amino acids has been introduced in our laboratory(20b). The removal of the phenacyl group is illustrated in fig.3.

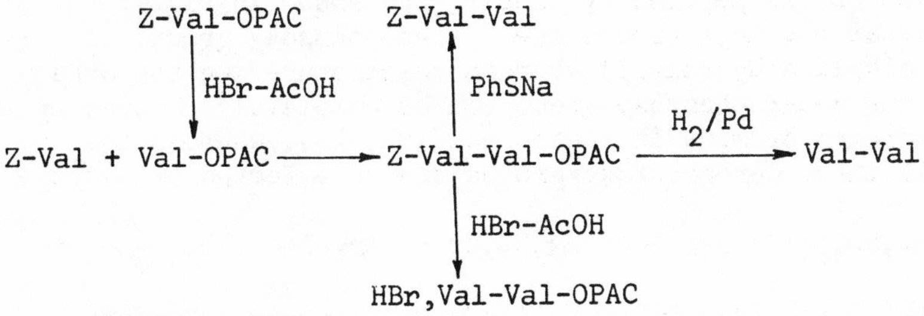

Z-Val-OPAC ↓ HBr-AcOH → Z-Val + Val-OPAC

Z-Val + Val-OPAC → Z-Val-Val-OPAC $\xrightarrow{H_2/Pd}$ Val-Val

Z-Val-Val-OPAC ↑ PhSNa → Z-Val-Val

Z-Val-Val-OPAC ↓ HBr-AcOH → HBr,Val-Val-OPAC

PAC = $PhCOCH_2$

Fig. 3

The carboxyl protection is of special interest in the case of dicarboxylic acids(24)because there must be a differentiation of the $\alpha$ and $\omega$ position. Figure 4 illustrates the preparation of diphenylmethyl and phenacyl mono-esters or mixed diesters of L-glutamic acid.

As regards the problem of the protection of the amino acid side chain, the way of protection of the guanido moiety of arginine by the carbobenzoxy group and the use of N$\alpha$, N$\omega$, N$\omega'$, -tricarboben-

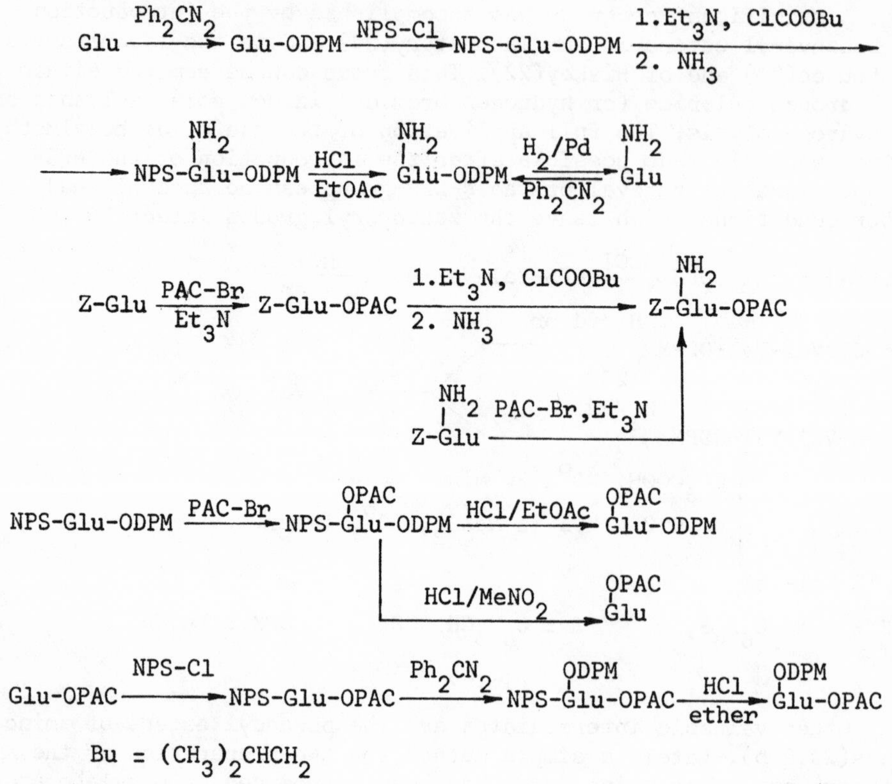

$$\text{Glu} \xrightarrow{Ph_2CN_2} \text{Glu-ODPM} \xrightarrow{\text{NPS-Cl}} \text{NPS-Glu-ODPM} \xrightarrow[\text{2. } NH_3]{\text{1.} Et_3N\text{, ClCOOBu}}$$

$$\longrightarrow \text{NPS-}\overset{NH_2}{\overset{|}{\text{Glu}}}\text{-ODPM} \xrightarrow[\text{EtOAc}]{\text{HCl}} \overset{NH_2}{\overset{|}{\text{Glu}}}\text{-ODPM} \underset{Ph_2CN_2}{\overset{H_2/Pd}{\rightleftarrows}} \overset{NH_2}{\overset{|}{\text{Glu}}}$$

$$\text{Z-Glu} \xrightarrow[Et_3N]{\text{PAC-Br}} \text{Z-Glu-OPAC} \xrightarrow[\text{2. } NH_3]{\text{1.} Et_3N\text{, ClCOOBu}} \text{Z-}\overset{NH_2}{\overset{|}{\text{Glu}}}\text{-OPAC}$$

$$\text{Z-}\overset{NH_2}{\overset{|}{\text{Glu}}} \xrightarrow{\text{PAC-Br}, Et_3N} \text{Z-}\overset{NH_2}{\overset{|}{\text{Glu}}}\text{-OPAC}$$

$$\text{NPS-Glu-ODPM} \xrightarrow{\text{PAC-Br}} \text{NPS-}\overset{\text{OPAC}}{\overset{|}{\text{Glu}}}\text{-ODPM} \xrightarrow{\text{HCl/EtOAc}} \overset{\text{OPAC}}{\overset{|}{\text{Glu}}}\text{-ODPM}$$

$$\text{NPS-}\overset{\text{OPAC}}{\overset{|}{\text{Glu}}}\text{-ODPM} \xrightarrow{\text{HCl/MeNO}_2} \overset{\text{OPAC}}{\overset{|}{\text{Glu}}}$$

$$\text{Glu-OPAC} \xrightarrow{\text{NPS-Cl}} \text{NPS-Glu-OPAC} \xrightarrow{Ph_2CN_2} \text{NPS-}\overset{\text{ODPM}}{\overset{|}{\text{Glu}}}\text{-OPAC} \xrightarrow[\text{ether}]{\text{HCl}} \overset{\text{ODPM}}{\overset{|}{\text{Glu}}}\text{-OPAC}$$

$\text{Bu} = (CH_3)_2CHCH_2$

Fig. 4

zoxy-arginine(3b) in peptide synthesis, are shown in fig. 5. It is remarkable that one acyl group, the $\omega'$-carbobenzoxy group, is removable by alkaline hydrolysis at room temperature. On the other hand, only the $\alpha$-carbobenzoxy group can be completely removed by means of hydrogen bromide in acetic acid(9b, 25), whereas the removal of all three carbobenzoxy groups can be effected by hydrogenolysis.

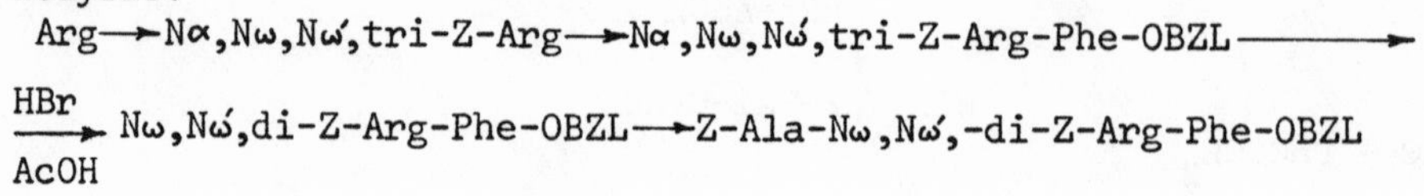

$$\text{Arg} \rightarrow N\alpha,N\omega,N\omega'\text{,tri-Z-Arg} \rightarrow N\alpha,N\omega,N\omega'\text{,tri-Z-Arg-Phe-OBZL} \longrightarrow$$

$$\xrightarrow[\text{AcOH}]{\text{HBr}} N\omega,N\omega'\text{,di-Z-Arg-Phe-OBZL} \rightarrow \text{Z-Ala-}N\omega,N\omega'\text{,-di-Z-Arg-Phe-OBZL}$$

Fig. 5

It is noteworthy that even tritylation of the guanido moiety constitutes an excellent protection(20a). The treatment described for the effective splitting off the $\alpha$-trityl group, has no action on the other two, thus offering the possibility for lengthening of the peptide chain from the amino end. As a matter of fact the TRI-group in the guanido moiety cannot be removed by hydrogenolysis or by hydrochloric acid but only by hydrogen bromide in acetic acid(26).

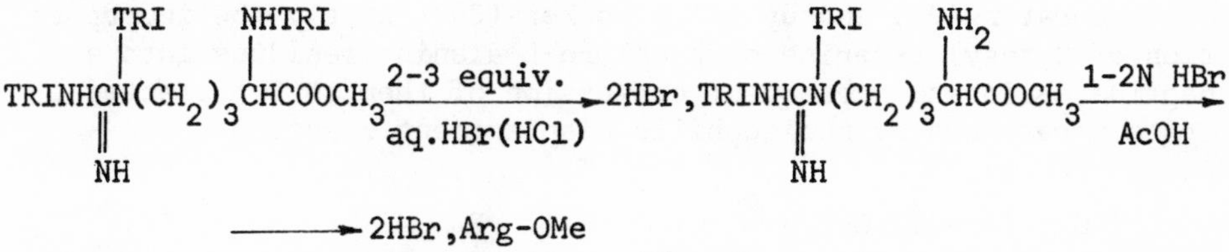

$$\underset{\underset{NH}{\|}}{TRINHCN}(CH_2)_3\overset{NHTRI}{CH}COOCH_3 \text{ (TRI on N)} \xrightarrow[\text{aq.HBr(HCl)}]{\text{2-3 equiv.}} 2HBr, \underset{\underset{NH}{\|}}{TRINHCN}(CH_2)_3\overset{NH_2}{CH}COOCH_3 \text{ (TRI on N)} \xrightarrow[\text{AcOH}]{\text{1-2N HBr}} 2HBr,Arg\text{-}OMe$$

Fig. 6

In addition to the well known du Vigneaud's S-benzyl group(27), which is split off only by sodium in liquid ammonia, other S-protecting groups for cysteine were necessary which could be split off differentially and under mild conditions. Thus more efficient methods for protection of SH groups are required, mainly for the solution of the problem of synthesis of unsymmetrical cystine peptides, especially those bearing two or more -S-S-bridges within the molecule(28). All the groups illustrated in fig. 7 can be selectively

$(C_6H_5)_3C$ = TRI  $C_6H_5CO$ = Bz  $C_6H_5CH_2OCO$ = Z  $(CH_3OC_6H_4)(C_6H_5)_2C$ = MMT

$(C_6H_5)_2CH$ = DPM  $CH_3CO$ = Ac  $CH_3OC_6H_4CH_2OCO$ = pMZ

Fig. 7

removed without affecting sensistive parts of the molecule, and especially any already existing -S-S-bridge.

Table 1

| GROUPS | REMOVED BY | % CLEAVAGE |
|---|---|---|
| 1. S-TRI(in N-protected pept.) | a. $AgNO_3$-Pyr. | 95-100(20°,1 min.) |
| | b. $HgCl_2$ | " (20°,5 min.) |
| S-TRI(in N-unprotected peptides) | c. N HBr-AcOH | 95 (20°,5 min.) |
| | d. 0.2 N " | 75-80 (20°,5 min.) |
| | e. TFA-phenol | 50(20°or70°,15 min) |
| 2. S-MMT | TFA-phenol | 95 (20°,15 min.) |
| 3. S-DPM | a. TFA-phenol | 98 (70°,15 min.) |
| | b. TFA-phenol | 75 (20°,18 hrs.) |
| 4. S-Bz | 1 eq. 0.1N MeONa-MeOH | 98 (20°,10 min.) |
| 5. S-Ac | a. 1 eq. 0.1N MeONa-MeOH | 98 (20°, 5 min.) |
| | b. 1 N HBr-AcOH | 98 (20°,30 min.) |
| 6. S-Z | a. 1 eq. 0.1N MeONa-MeOH | 90 (20°,30 min.) |
| | b. TFA-phenol | 90 (70°,30 min.) |
| 7. S-pMZ | a. 0.2N HBr-AcOH | 85 (20°, 5 min.) |
| | b. 1.5N HCl-AcOH | 75 (20°,15 min.) |
| | c. TFA-anisol | 50 (20°,30 min.) |
| | d. 5 eq. 0.4N MeONa-MeOH | 95 (20°,15 min.) |

Another approach to the problem of side protection and especially to the synthesis of cysteine peptides has been followed in our laboratory(29) and by other workers(30), namely the incorporation of O-tosyl-L-serine or β-chloro-L-alanine residues into a peptide chain and subsequent conversion of them to S-protected L-cysteine residues by nucleophilic displacement reactions.

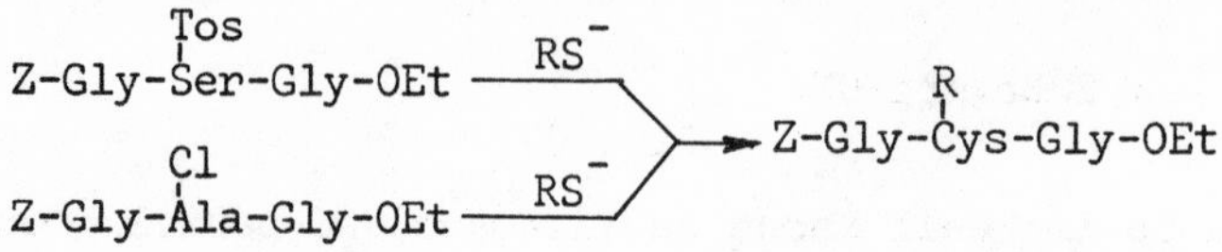

$$\text{Z-Gly-}\underset{}{\overset{\text{Tos}}{\text{Ser}}}\text{-Gly-OEt} \xrightarrow{RS^-} \text{Z-Gly-}\overset{\text{R}}{\text{Cys}}\text{-Gly-OEt}$$

$$\text{Z-Gly-}\overset{\text{Cl}}{\text{Ala}}\text{-Gly-OEt} \xrightarrow{RS^-} \text{Z-Gly-}\overset{\text{R}}{\text{Cys}}\text{-Gly-OEt}$$

Tos = $CH_3C_6H_4SO_2$,  R = $C_6H_5CO$,  $CH_3CO$

Fig. 8

In the remaining part of the talk I should like to give some examples which illustrate the importance of having S-protected cysteines such as those mentioned above. First, they can be used for the preparation of not only simple peptides of cystine and cysteine but also of cyclic peptides of the oxytocin (31) or the vasopressin type.

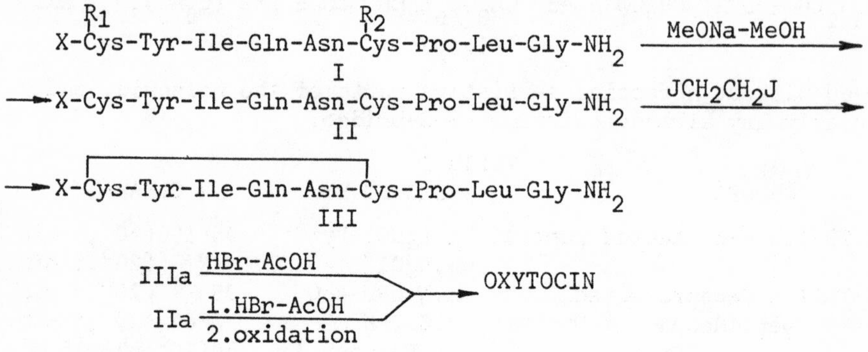

$$\text{X-}\overset{R_1}{\text{Cys}}\text{-Tyr-Ile-Gln-Asn-}\overset{R_2}{\text{Cys}}\text{-Pro-Leu-Gly-NH}_2 \text{ (I)} \xrightarrow{\text{MeONa-MeOH}}$$

$$\longrightarrow \text{X-Cys-Tyr-Ile-Gln-Asn-Cys-Pro-Leu-Gly-NH}_2 \text{ (II)} \xrightarrow{JCH_2CH_2J}$$

$$\longrightarrow \text{X-}\overline{\text{Cys-Tyr-Ile-Gln-Asn-Cys}}\text{-Pro-Leu-Gly-NH}_2 \text{ (III)}$$

$$\text{IIIa} \xrightarrow{\text{HBr-AcOH}} \text{OXYTOCIN}$$

$$\text{IIa} \xrightarrow[\text{2.oxidation}]{\text{1.HBr-AcOH}} \text{OXYTOCIN}$$

| I | | | II | III |
|---|---|---|---|---|
| a: X=Z, | $R_1$=Bz,Z, | $R_2$=Bz | a: X=Z | a: X=Z |
| b: X=BOC, | $R_1$,$R_2$=Bz | | b: X=BOC | b: X=BOC |
| c: X=NPS, | $R_1$,$R_2$=Bz | | c: X=NPS | |

Fig. 9

The peptides in figures 10(32) and 11(33) demonstrate the use of some of the new amino, carboxyl and side chain protecting groups developed in our laboratory. The removal of all the protecting groups can be effected selectively thus giving new possibilities in the process of the synthesis.

Another example of the usefullness of the new methods is the synthesis of the 6-11 fragment of insulin which bears a sulfhydryl group in addition to an -S-S-bridge (28b,34). This fragment (cf.fig.12) could not have been synthesized in another way.

In the meantime following the same principle and using the same intermediates Hiskey(35)has prepared many cystine peptides.

```
                                      OtBu           TRI
NPS-Leu-Tyr-Gln-Leu-N3 + HCl,Glu-Asn-Tyr-Cys-Asn-ODPM ——>
   m.p. 196-199°
                                OtBu           TRI
———> NPS-Leu-Tyr-Gln-Leu-Glu-Asn-Tyr-Cys-Asn-ODPM
           m.p. 206°, [α]D -29°(c 1, DMF)
```

Fig. 10

```
         Bz  TRI
BOC-Cys-Cys-Ala-Gly-OPAC ——PhSNa——>
   m.p.180°, [α]D -17°(c 2,DMF)

        Bz   TRI                        Bz  Ac
———> BOC-Cys-Cys-Ala-Gly + HBr,Val-Cys-Ser-OMe ———————>
m.p.132°, [α]D -21.6°(c 2,DMF), m.p.176°, [α]D -4°(c 5,DMF)
        Bz   TRI                Bz  Ac
———> BOC-Cys-Cys-Ala-Gly-Val-Cys-Ser-OMe ——MeONa-MeOH——>
                 m.p. 161-3°
             TRI
———> BOC-Cys-Cys-Ala-Gly-Val-Cys-Ser-OMe    (34)
```

Fig. 11

```
   TRI                 TRI        1.HgCl2 2.H2S 3.(CH2J)2
BOC-Cys-Cys-Ala-Gly-Val-Cys-Ser-OMe ——————————————————————>
        DPM                       4.Removal of the BOC-group

          ┌───────────────────────┐
———> HCl,Cys-Cys-Ala-Gly-Val-Cys-Ser-OMe
             DPM

   Bz                  Bz         1.Methanolysis 2.Oxidation
NPS-Cys-Cys-Ala-Gly-Val-Cys-Ser-OMe ——————————————————————>
        TRI                       3. Removal of the NPS-group
          ┌───────────────────────┐
———> HCl,Cys-Cys-Ala-Gly-Val-Cys-Ser-OMe
             TRI
```

Fig. 12

The progress achieved by the use of the various S-protecting groups which can be split off selectively is shown from their use in work done towards the synthesis of unsymmetrical cystine peptides bearing two -S-S-bridges. The following heptapeptides(36)have been synthesized in a manner proving the established structure illustrated in fig. 13. Upon treatment of the peptide (I) with mercury chloride the mercaptide III is obtained and from this, by treatment with $H_2S$, the heptapeptide IV is prepared which bears two protected and two unprotected sulfhydryl groups. Oxidation of the two free SH-groups, splitting offthe N-carbobenzoxy group affords the formation of a cyclic peptide which, unfortunately is a mixture of monomer and polymers.

```
    CO—Cys-Ala-Cys-OMe              CO—Cys-Ala-Cys-OMe
Z-NHCH  Bz       TRI            Z-NHCH  Bz       DPM
    CH2                             CH2
    CH2 Bz       TRI                CH2 Bz       DPM
    CO—Cys-Gly-Cys-OMe              CO—Cys-Gly-Cys-OMe
```

m.p.174°, $[\alpha]_D$-19.5°(c 2,DMF) m.p.192°, $[\alpha]_D$-40°(c 2,DMF)

I II

```
    CO—Cys-Ala-Cys-OMe                 CO—Cys-Ala-Cys-OMe
Z-NHCH  Bz      HgCl      H2S      Z-NHCH  Bz
    CH2                  ------>       CH2
    CH2 Bz      HgCl                   CH2 Bz
    CO—Cys-Gly-Cys-OMe                 CO—Cys-Gly-Cys-OMe
```

m.p.197-199° m.p.196°,$[\alpha]_D$-43.1°(c 1,DMF)

III Fig. 13 IV

It remains to say a few words about the methods of coupling in peptide synthesis. The contribution of our laboratory in this step is indeed very small. I would only like to report that our method of coupling using diphenylphosphorylchloride (9) has proved to be very effective, even in coupling N-trityl amino acids, which do not couple satisfactorily as other mixed anhydrides because of steric hindrance.

## References

1)J.P.Greenstein and M.Winitz,"Chemistry of the Amino Acids", J. Wiley,New York,1961;J.S.Fruton in"Advances in Protein Chemistry" Vol.V,1949;M.Goodman and G.W.Kenner,ibid.,Vol.XII,1957;K.Hofmann and P.G.Katsoyannis,in"The Proteins"Vol.1,ed.H.Neurath,Academic Press,New York,1963;R.Boissonnas,in"Advanes in Organic Chemistry", Vol.3,ed.R.A.Raphael,Interscience,New York,1963;H.D.Law,in"Progress in Medicinal Chemistry",ed.G.P.Ellis and G.B.West,Butterworths, London,1965;E.Schröder and K.Lübke,"The Peptides",Academic Press, New York,1966;M.Bodanszky and M.Ondetti,"Peptide Synthesis", Interscience,New York, 1966.

2)M.Bergmann and L.Zervas,Ber.,65,1192(1932);German Patent 556,798 (1932).

3)(a)Bergmann and L.Zervas,c.f.ref.2;Ber.65,1201(1932);M.Bergmann, L.Zervas,H.Schleich,and F.Leinert,Z.physiol.Chem.Hoppe-Seyler's 212 72(1932);M.Bergmann,L.Zervas,and J.P.Greenstein,Ber.,65,1692(1932); M.Bergmann,L.Zervas, and H.Schleich,Ber.,65,1747(1932);M.Bergmann, L.Zervas,andL.Salzmann,Ber.66,1288(1933);M.Bergmann and L.Zervas, Z.physiol.Chem.Hoppe-Seyler's,224,11(1934);M.Bergmann,L.Zervas,L. Salzmann,and H.Schleich,ibid.,224,17(1934);M.Bergmann,L.Zervas,H. Rinke,and H.Schleich,ibid.,224,26,33(1934);M.Bergmann,L.Zervas,and H.Rinke,ibid.,224,40(1934);M.Bergmann,L.Zervas,and H.Schleich,ibid.

224,45(1934);M.Bergmann,L.Zervas,J.S.Fruton,F.Schneider,and H. Schleich,J.Biol.Chem.,109,325(1935);M.Bergmann,L.Zervas,and J.S. Fruton,ibid.,111,225(1935);M.Bergmann,L.Zervas,and W.F.Ross,ibid., 111,245(1935);M.Bergmann,L.Zervas,and J.S.Fruton,ibid.,115,593 (1936);M.Bergmann and L.Zervas,ibid.,113,341(1936);(b)L.Zervas,M. Winitz,and J.P.Greenstein,Arch.Bioch.Biophys.65,573(1956);J.Org. Chem.,22,1515(1957);L.Zervas,T.Otani,M.Winitz,and J.P.Greenstein, Arch.Biochim.Biophys.,75,290(1958);J.Am.Chem.Soc.,81,2878(1959);L. Zervas,M.Winitz,and J.P.Greenstein,ibid.,83,3300(1961);(c)L.Zervas, L.Benoiton,E.Weiss,M.Winitz,and J.P.Greenstein,ibid.,81,1729(1959); B.Bezas and L.Zervas,ibid.,83,719(1961).

4)R.H.Sifferd and V.du Vigneaud,J.Biol.Chem.,108,753(1935).

5)D.Ben-Ishai and A.Berger,J.Org.Chem.,17,1564(1952);D.Ben-Ishai, ibid.,19,62(1954);G.W.Anderson,J.Blodinger,and A.D.Welcher,J.A. Chem.Soc.,74,5309(1952);R.A.Boissonnas and G.Preitner,Helv.Chim. Acta,36,875(1953).

6)D.M.Theodoropoulos,Thesis,Faculty Natural Sci.Math.,Univ.of Athens,Greece(1953);L.Zervas and D.Theodoropoulos,J.Am.Chem.Soc., 78,1359(1956);G.C.Stelakatos,D.Theodoropoulos,and L.Zervas,J.Am. Chem.Soc.,81,2884(1959).

7)G.Amiard,R.Heymès and L.Velluz,Bull.Soc.Chim.France,191(1955).

8)A.Cosmatos,I.Photaki,and L.Zervas,Chem.Ber.,94,2644(1961).

9)(a)L.Zervas,D.Borovas,and E.Gazis,J.Am.Chem.Soc.,85,3660(1963); (b)L.Zervas and Ch.Hamalidis,ibid.,87,99(1965).

10)Unpublished data of this laboratory.

11)W.Kessler and B.Iselin,Helv.Chim.Acta,49,1330(1966).

12)K.Poduška,H.Maasen van den Brink-Zimmermanová,J.Rudinger,and F. Šorm,"Peptides:Proceedings of the Eighth European Symposium"Noordwijk, The Netherlands,1966,ed.H.C.Beyerman,North-Holland,Publish. Co.Amsterdam,1967,p.38.

13)A.Fontana,F.Marchiori,L.Moroder,and E.Scoffone,Tetrahedron Letters,2985(1966).

14)D.Bradenburg,Tetrahedron Letters,6201(1966).

15)E.Wünsch,A.Fontana,and F.Drees,Z.Naturforsch.,22b,607(1967).

16)I.Phocas,C.Yovanidis,I.Photaki and L.Zervas,J.Chem.Soc.,(C), in press.

17)R.Schwyzer and P.Sieber,Helv.Chim.Acta,40,624(1957).

18)M.Kinoshita and H.Zahn,Ann.,696,235(1966).

19)M.Bergmann,L.Zervas,and L.Salzmann,Ber.,66,1288(1933).

20)(a)E.Gazis,B.Bezas,G.C.Stelakatos,and L.Zervas,"Peptides: Proceedings of the Fifth European Symposium"Oxford,1962,ed.G.T. Young,Pergamon Press,Oxford,1963,p.17;E.Gazis,D.Borovas,Ch.Hamalidis,G.C.Stelakatos,and L.Zervas,"Peptides:Proceedings of the Sixth European Symposium",Athens,1963,ed.L.Zervas,Pergamon,Oxford,1966, p.107;(b)G.C.Stelakatos,A.Paganou,and L.Zervas,J.Chem.Soc.,(C), 1191(1966).

21)A.A.Aboderin,G.R.Delpierre,and J.S.Fruton,J.Am.Chem.Soc.,87, 5469(1965).

22)R.G.Hiskey and J.B.Adams,Jr.,J.Am.Chem.Soc.,87,3969(1965).

23)J.C.Sheehan and G.D.Daves,Jr.,J.Org.Chem.,29,2006(1964).
24)J.Taylor-Papadimitriou,C.Yovanidis,A.Paganou,and L.Zervas,J. Chem.Soc.,(C),in press.
25)St.Guttmann and J.Pless,Chimia(Aarau)18,185(1964).
26)Unpublished data of this laboratory.
27)V.du Vigneaud,L.F.Audrieth,and H.S.Loring,J.Biol.Chem.,87,XX (1930);J.Am.Chem.Soc.,52,4500(1930).
28)(a)L.Zervas and I.Photaki,Chimia(Aarau),14,375(1960);L.Zervas, Collection Czech.Chem.Commun.,27,2229(1962);(b)L.Zervas,I.Photaki, A.Cosmatos,and N.Ghelis,"Peptides:Proceedings of the Fifth European Symposium,Oxford,1962,ed.G.T.Young,Pergamon Press,Oxford,1963,p.27; (c)L.Zervas and I.Photaki,J.Am.Chem.Soc.,84,3887(1962);L.Zervas,I. Photaki,and N.Ghelis,ibid.,85,1337(1963);(d)unpublished data of this laboratory.
29)I.Photaki and V.Bardakos,Experientia,21,371(1965);J.Am.Chem.Soc. 87,3489(1965);Chem.Commun.,818(1966).
30)C.Zioudrou,M.Wilchek,and A.Patchornik,Biochemistry,4,1811(1965) M.Wilchek,C.Zioudrou,and A.Patchornik,J.Org.Chem.,31,2865(1966).
31)I.Photaki,J.Am.Chem.Soc.,88,2292(1966).
32)Unpublished data of this laboratory.
33)Unpublished data of this laboratory.
34)L.Zervas,I.Photaki,A.Cosmatos,and D.Borovas,J.Am.Chem.Soc.,87, 4922(1965).
35)R.G.Hiskey,T.Mizoguchi,and H.Igeta,J.Org.Chem.,31,1188(1966).
36)C.Yovanidis,Thesis,Faculty Natural Sci.Math.,Univ.of Athens, Greece(1964);L.Zervas,I.Photaki,C.Yovanidis,J.Taylor-Papadimitriou, I.Phocas,and V.Bardakos,"Peptides:Proceedings of the Eighth European Symposium"Noordwijk,The Netherlands,1966,ed.H.C.Beyerman,North-Holland,Publishing Co.,Amsterdam,1967,p.28.

# NOVEL TECHNIQUES IN PEPTIDE SYNTHESIS

A. Patchornik

Department of Biophysics

The Weizmann Institute of Science, Rehovoth, Israel

The following topics studied in our laboratory will be discussed briefly: Synthesis of bradykinin; synthesis of cyclic peptides and the potential use of photosensitive blocking groups in peptide synthesis.

## A. SYNTHESIS OF BRADYKININ

Synthesis of linear peptides with the aid of polymeric active esters was recently described by Fridkin et al. (1965, 1966). The synthesis describing the use of polymeric active esters for peptide synthesis is shown in the scheme

$$\text{Y-NH-}\underset{\text{R}_1}{\text{CH}}\text{-CO-}Ⓟ + \text{NH}_2\text{-}\underset{\text{R}_2}{\text{CH}}\text{-COOX} \rightarrow \text{Y-NH-}\underset{\text{R}_1}{\text{CH}}\text{-CO-NH-}\underset{\text{R}_2}{\text{CH}}\text{-COOX} + \text{H}Ⓟ$$

$$\xrightarrow{-\text{Y}} \text{NH}_2\text{-}\underset{\text{R}_2}{\text{CH}}\text{-CO-NH-}\underset{\text{R}_2}{\text{CH}}\text{-COOX}$$

Ⓟ stands for insoluble poly-4-hydroxy-3-nitrostyrene
Y stands for any amino blocking group.

Active esters of N-blocked amino acids bound to poly-4-hydroxy-3-nitrostyrene when stirred for several hours with amino peptide esters in dimethylformamide (D.M.F.) at room temperature give high yields of blocked peptides. Preferential removal of the $\alpha$-amino blocking group followed by coupling with polymeric active esters of N-blocked amino acids allows the stepwise synthesis of peptides with predetermined sequence. Bradykinin was synthesized by this technique (Fridkin et al., and Patchornik et al., 1966) as shown in the scheme.

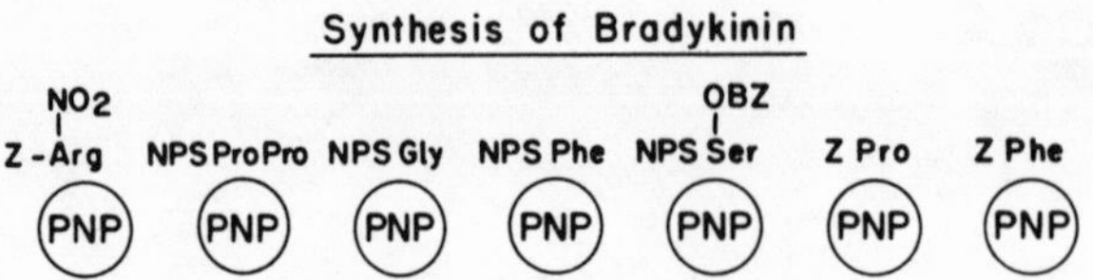


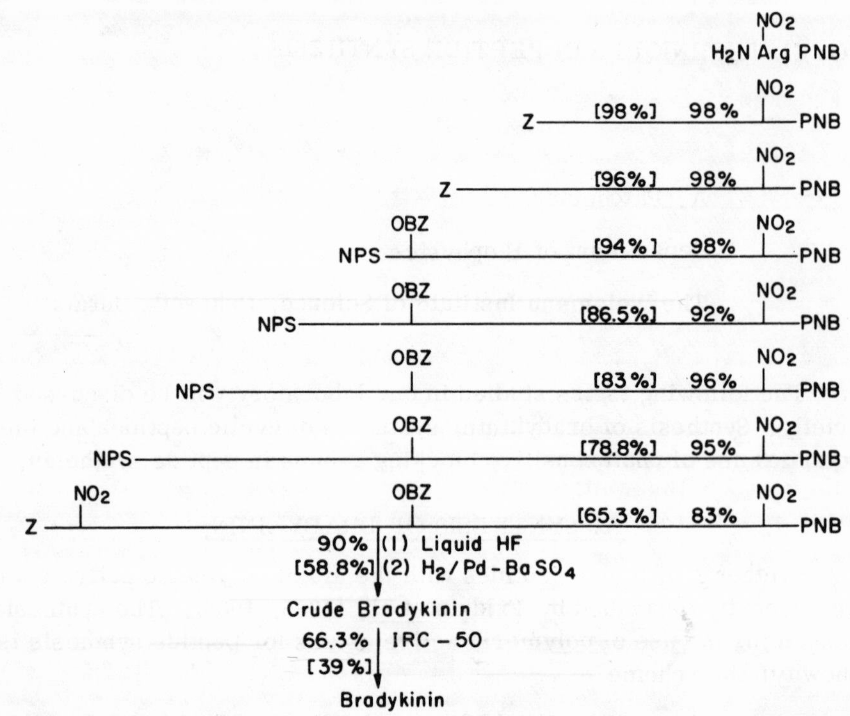


(PNP) stands for the 4-hydroxy-3 nitro styrene polymeric carrier

The N-blocked polymeric active esters shown in the scheme were used in the synthesis of bradykinin. The active esters were prepared by coupling the corresponding N-blocked amino acids with the polymer 4-hydroxy-3-nitrostyrene. The reaction was carried out in D.M.F. and the condensing reagent was dicyclohexylcarbodiimide. The amount of active ester bound to the polymer was between 1-2 mmoles per gram of product.

In the synthesis of bradykinin, 4 equivalents of the polymeric active ester were allowed to react with the free amino peptide ester at room temperature for 8 hours. The crude blocked peptides were isolated in almost quantitative yields after removal of the excess polymeric reagent and evaporation of the solvent. Purification of the crude material was performed by acid and bicarbonate extraction of an ethyl acetate solution of the blocked peptide. The hydrochloride or hydrobromide peptides obtained after removal of the N-blocking groups were purified by precipitation with ether from alcoholic solution. All intermediates were found to be practically pure by chromatography in various solvent systems.

The yields obtained in each step were almost quantitative and some losses were due to procedures of isolation and purification and also probably to some absorption of the blocked peptide to the polymer.

Several methods of deblocking the nonapeptide, in order to obtain bradykinin were tested. Best results were obtained by treatment with HF (S. Sakakibara, 1966, 1967). Under these conditions the carbobenzoxy, benzyloxy, and nitroguanidyl groups were removed. The nitrobenzyl group was removed by exhaustive hydrogenation to yield a crude bradykinin having approximately 80% biological activity when tested on the isolated ileum of the guinea pig (Lewis, 1961) and also by the more sensitive and specific method of Edery (1965). The crude bradykinin was purified by chromatography on IRC 50 (Merrifield, 1964) and was identical in all its properties to samples obtained from Sandoz (Basel) and Dr. Sakakibara. The overall yield of the synthesis of blocked bradykinin was 65%. Big losses were suffered in the steps of deblocking and purification on columns. The overall yield of pure fully active bradykinin was 39%.

It is worthwhile to note that the classical test for bradykinin gives considerable activity with protected bradykinins such as the p-nitrobenzyl ester of bradykinin whereas Edery's sensitization test is specific for the unprotected bradykinin.

## B. SYNTHESIS OF CYCLIC PEPTIDES

Cyclic compounds are usually prepared by allowing a bifunctional linear compound to react in a very dilute solution. Thus, intermolecular reactions are minimized and an intramolecular reaction is favoured. Intramolecular reactions of a bifunctional linear compound will also be favoured when the freedom of motion of the molecule is restricted, for example by binding it to an insoluble carrier. As shown in the scheme, there are two general ways of binding a molecule to a polymeric insoluble carrier.

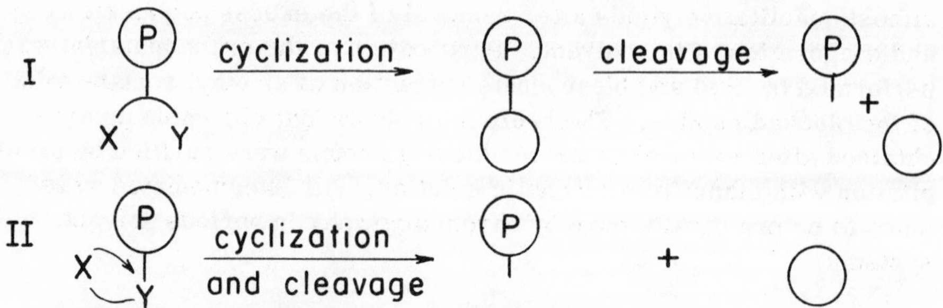


Ⓟ stands for a polymeric carrier
X and Y stand for reactive functional groups.

(I) The molecule is bound in its middle so that after cyclization a cleavage reaction is needed to release the cyclic compound. (II) The molecule is attached to the carrier by Y or X so that an automatic release from the carrier is effected by the reaction of cyclization.

An example of the first type is the synthesis of cystine peptides. Tri-Cys-Cys-Ala-O-$CH_2$-Ⓜ was synthesized on a Merrifield polymer. (The side-chain S of each Cys carries a Tri group: S-Tri, S-Tri.)

After removal of the S-Trityl groups by silver nitrate (Zervas, 1962), the free thiol groups were oxidized by diiodoethane (Weygand, 1962). By releasing the peptides obtained from the polymer by treatment with HBr in trifluoroacetic acid followed by purification on Dowex 50, E. Bondi of our department obtained an 80% overall yield of $H_2N$ Cys-Cys-AlaOH (with an S—S bridge between the two Cys residues).

When the oxidation was performed in dilute solution, however, a large variety of peptides was obtained.

Cystine cyclic compounds might be prepared by reactions of type (II) as follows:

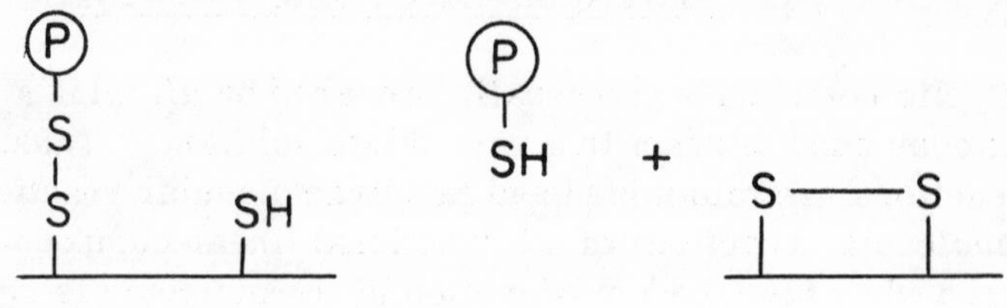


Ⓟ stands for polymer carrier.

Work along these lines is in progress.

A reaction of type (II) has indeed been demonstrated by Fridkin et al. (1965) who synthesized various diketopiperazines and cyclic tetrapeptides by cyclization of polymer bound active esters of appropriate peptides.

## C. POTENTIAL USES OF PHOTOSENSITIVE BLOCKING GROUPS

Photosensitive blocking groups may afford mild deblocking reactions if the groups are sensitive at wavelengths at which proteins are stable. R.B. Woodward has suggested such a photosensitive blocking group, the 6-nitroveratryloxy (NV-O-) group, which can be removed by irradiation with wavelengths longer than 3200 A. In collaboration with Woodward we have succeeded in showing that NV derivatives of amino acids release their amino peptides and $CO_2$ to an extent of 40% to 80% upon irradiation at 3200 A.

$$\text{NV-O-CO-NH} \xrightarrow{h\nu} CO_2 + NH_2\text{-Pep}$$

B. Amit of our laboratory was able to demonstrate quantitative release of a free peptide when the reaction was performed by irradiating a dilute solution of an NV derivative in the presence of hydrazine or hydroxylamine hydrochloride.

An example of a photosensitive ester bond was demonstrated when $H_2N$ Ala-COONV upon irradiation yielded 95% of free alanine. Lower yields of removal of an NV group from a sulfide was shown when S-NV-glutathione was irradiated to yield 50% glutathione.

The NV group is converted on irradiation mainly to the azordicarboxylic acid:

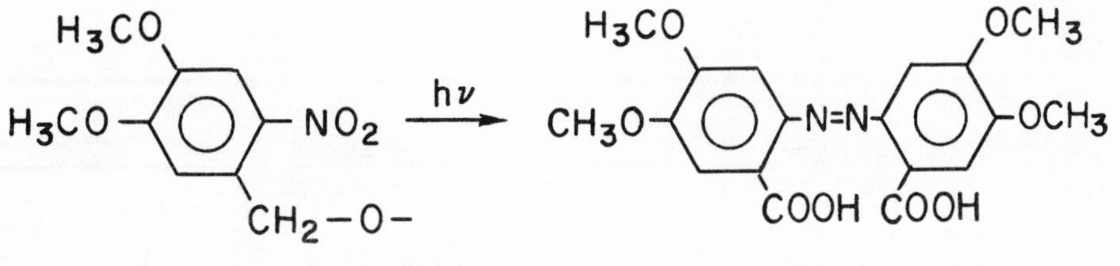

H₃CO / H₃CO–C₆H₂(NO₂)–CH₂–O– $\xrightarrow{h\nu}$ (H₃CO)(CH₃O)C₆H₂(COOH)–N=N–C₆H₂(COOH)(OCH₃)(OCH₃)

## REFERENCES

M. Fridkin, A. Patchornik and E. Katchalski, Israel J. Chem. 3, 69P (1965).

M. Fridkin, A. Patchornik and E. Katchalski, J. Am. Chem. Soc. 88, 3164 (1966).

M. Fridkin, A. Patchornik and E. Katchalski, Israel J. Chem. 4, 59P (1966).

M. Fridkin, A. Patchornik and E. Katchalski, J. Am. Chem. Soc. 87, 4646 (1965).

G.P. Lewis, Nature 192, 596 (1961).

R.B. Merrifield, Biochemistry 3, 1385 (1964).

A. Patchornik and M. Fridkin and E. Katchalski, 8th European Peptide Symposium, Nordwijk, Holland, 1966.

S. Sakakibara, 8th European Peptide Symposium, Nordwijk, Holland, 1966.

S. Sakakibara, Bull. Chem. Soc., Japan, in press, 1967.

F. Weygand and G. Zumach, Z. Naturforsch 17B, 807 (1962).

L. Zervas, I. Photaki, J. Am. Chem. Soc. 84, 3887 (1962).

# Synthetic polypeptides with enhanced biological activities

R.A. Boissonnas

SANDOZ Ltd., Basle, Switzerland

The number of analogues of natural hormones which have been hitherto prepared by synthesis is still small in comparison with the enormous number of analogues which could be theoretically arrived at by modifying even small portions of the original sequences. Nevertheless some interesting results have already emerged from the study of the analogues presently available.

When taking into account the fact that the admitted mechanism of protein synthesis allows for the direct incorporation of only twenty amino acids, it becomes manifest that the natural hormones possess almost optimal active structures arrived at by natural mutation and selection.

Conversly, by including in the structure of synthetic analogues amino acids not normally found in proteins, it has already been possible in some cases to obtain synthetic hormones exhibiting a higher activity than their natural congeners.

# SYNTHESIS OF GASTROINTESTINAL HORMONES

Miguel A. Ondetti, John T. Sheehan & M. Bodanszky[1]

The Squibb Institute for Medical Research

New Brunswick, New Jersey U. S. A.

Even though the gastrointestinal hormones have the rare privilege of being among the very first entries in the fascinating field of chemical messengers, they have had to sustain a hard and lengthy struggle to gain recognition as well defined chemical entities. In some cases, like gastrin, because the fact of its very existence was challenged again and again, and in others, like secretin, because its isolation and purification has taxed heavily the resourcefulness of a long series of investigators. Fortunately the battle has been won and the efforts of sixty years of research have been crowned by the wonderful achievements of Gregory and Tracy in gastrin[2] and of Jorpes and Mutt with secretin[3] and cholecystokinin[4].

After all these sometimes seemingly insurmountable difficulties were finally overcome and structures for these hormones were proposed, the challenge was issued to the synthetic chemist, in this particular case the synthetic peptide chemist, to reproduce in the laboratory, though in a rather clumsy way, the intricate products of nature. The remarkable work of Kenner, Sheppard and collaborators in the synthesis of gastrin[5] marked the beginning of this new period in the study of gastrointestinal hormones. Here I would like to describe our efforts in contributing to the second chapter of this story: the synthesis of secretin[6], and also to show our still very preliminary notes for the third chapter: the synthesis of cholecystokinin .

HIS-SER-ASP-GLY-THR-PHE-THR-SER-GLU-LEU-SER-ARG-LEU-ARG
1 2 3 4 5 6 7 8 9 10 11 12 13 14

ASP-SER-ALA-ARG-LEU-GLN-ARG-LEU-LEU-GLN-GLY-LEU-VAL-$NH_2$
15 16 17 18 19 20 21 22 23 24 25 26 27

SECRETIN

Figure 1.

A long and fruitful relationship between Squibb and Professor Erik Jorpes and Dozent Viktor Mutt paved the way for our entrance into the gastrointestinal field. In a series of personal communications from these investigators we learned the results of those many years of painstaking research: the proposed structure of secretin which is depicted in Figure 1[7].

A close examination of this structure reveals a number of very interesting features. The remarkable resemblance with the pancreatic hormone glucagon has already been pointed out by Mutt and Jorpes[3]. The presence of four arginine residues imparts a strong basic character to the molecule, which, in a way, contrasts with the marked acidic nature of gastrin. The C-terminal

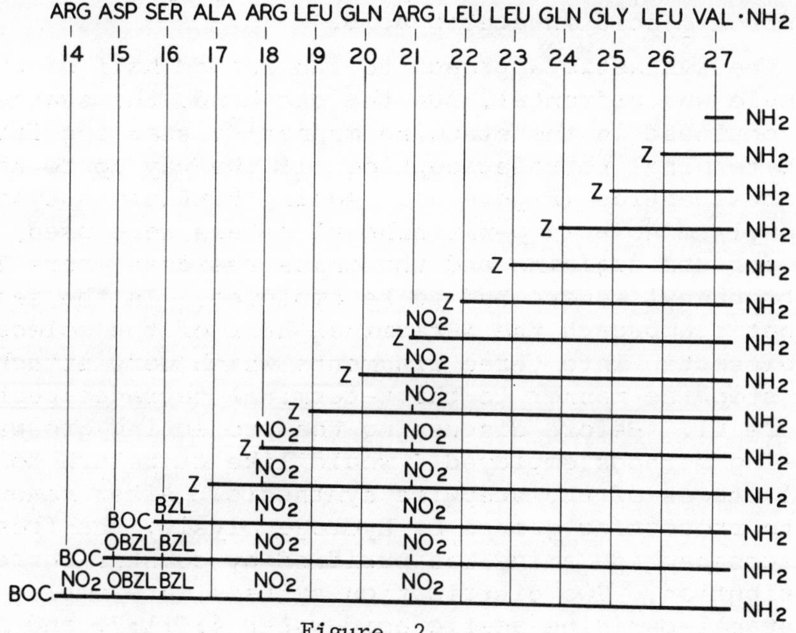


Figure 2.

amino acid valine is present in the amide form, like in several other hormones; oxytocin, vasopressin, gastrin. Finally, there is a significant concentration of hydrophobic side chains at the C-terminal half of the molecule (9 out of 14 residues) and of hydrophilic side chains at the opposite end (9 out of 13 residues).

Our synthetic attempts started with the hydrophobic half, namely, the C-terminal part. The pentapeptide amide corresponding to position 23-27 was synthesized without difficulties using the stepwise strategy[8] starting with the C-terminal amino acid:valine amide. After an unsuccessful attempt in preparing a hydrazide of sequence 14-22 to couple to the aforementioned pentapeptide by the azide procedure, it was finally decided to continue the synthesis in the stepwise manner up to position 14. A schematic description of this synthesis is given in Figure 2[9]. Benzyloxycarbonyl amino acid-p-nitrophenyl esters[10] were used in all steps, except for the introduction of arginine where the 2,4-dinitrophenyl ester was used. From position 16 on, tert.-butyloxycarbonyl[11] amino acid p-nitrophenyl or 2,4-dinitrophenyl esters were used in order to avoid the undesirable acylation of the hydroxyl of serine that can occur during the removal of the benzyloxycarbonyl protection with hydrobromic acid in acetic acid.

The synthetic approach to the second half of the molecule was bifrontal. On the one hand, the synthesis was continued in the stepwise manner[12], starting from the C-terminal tetradecapeptide all the way up to the heptacosapeptide (Figure 3). Again, tertiary butyloxycarbonyl amino acid p-nitrophenyl esters were used, except for the arginine and threonine residues where 2,4-dinitrophenyl esters had to be employed. In the second synthetic approach the N-terminal half of the molecule was dissected into three fragments which were attached in a stepwise manner to the C-terminal tetradecapeptide (Figure 4). Before discussing the protecting groups and coupling methods employed I would like to return to the final stages of the stepwise synthesis. After removal of the protecting groups by hydrogenolysis, the free heptacosapeptide amide was purified by counter-current distribution. Two distribution systems were employed (n-butanol-pyridine-acetic acid-water 4:2:1:7, and n-

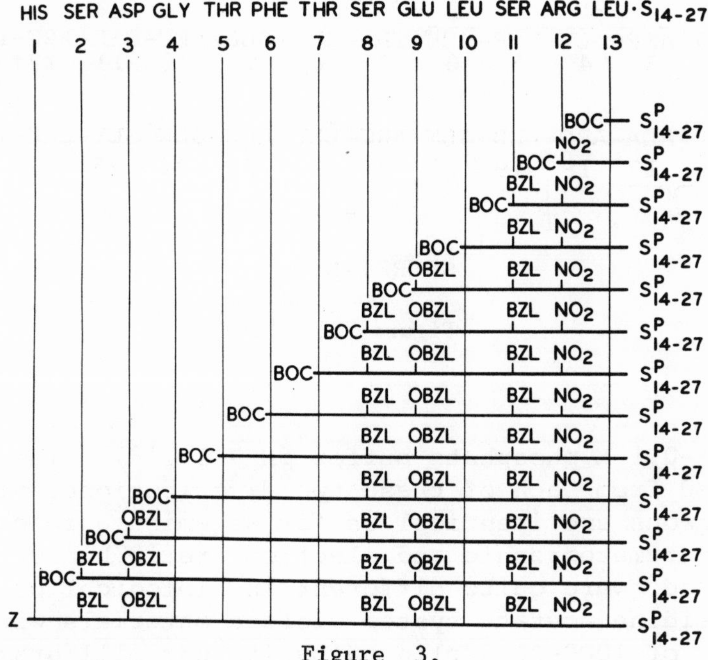

Figure 3.

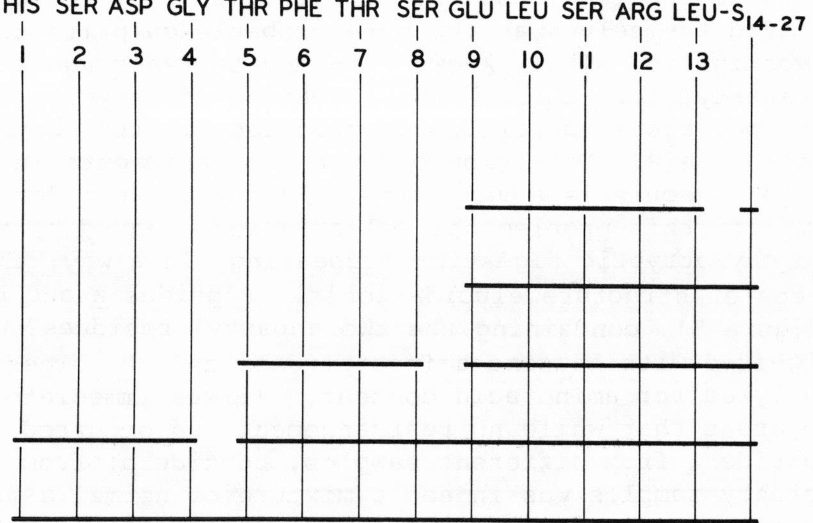

Figure 4.

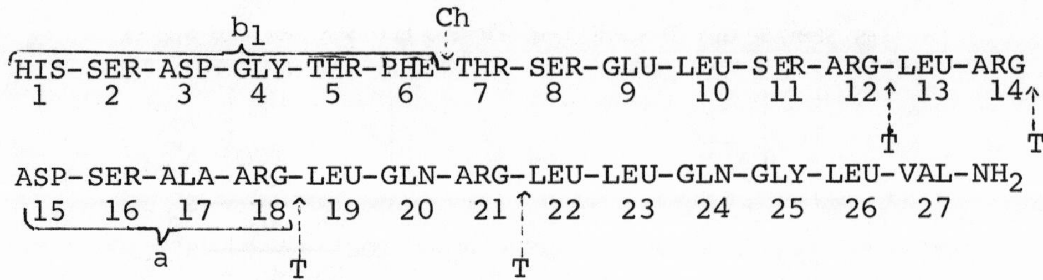


SECRETIN

Figure 5.

butanol-0.1 M phosphate buffer pH 7 1:1[13]). The products isolated from each of these two distributions, although homogeneous and identical as far as amino acid composition, chromatographic and electrophoretic behavior was concerned, were quite different in biological potency. The pyridine acetate system yielded materials with a potency of 1000-2000 clinical units per milligram while the potency of the products purified with the phosphate system was around 4000 clinical units/mg. An intramolecular rearrangement involving a functional group, undoubtedly very critical for the biological activity of the molecule, was considered the most likely explanation of these puzzling results. The two aspartyl residues offered themselves as the more probable culprits to be investigated. Their known tendency to rearrange to the β-aspartyl form through the formation of a cyclic imide intermediate is a well documented fact in the peptide literature[14]. Detection of such rearrangements in a large molecule is a very difficult task. In order to simplify this problem the molecule was cleaved by tryptic and chymotryptic digestion, repeating, in a way, the steps of structure elucidation[15]. Peptides a and $b_1$ (Figure 5), containing the two aspartyl residues, were digested with leucine aminopeptidase and the digests analyzed for amino acid content. It was immediately apparent that while no rearrangement had occurred in peptide a from different samples, peptide $b_1$ from low potency samples was indeed a mixture of normal α-aspartyl and either β or cyclic aspartyl peptides. Further

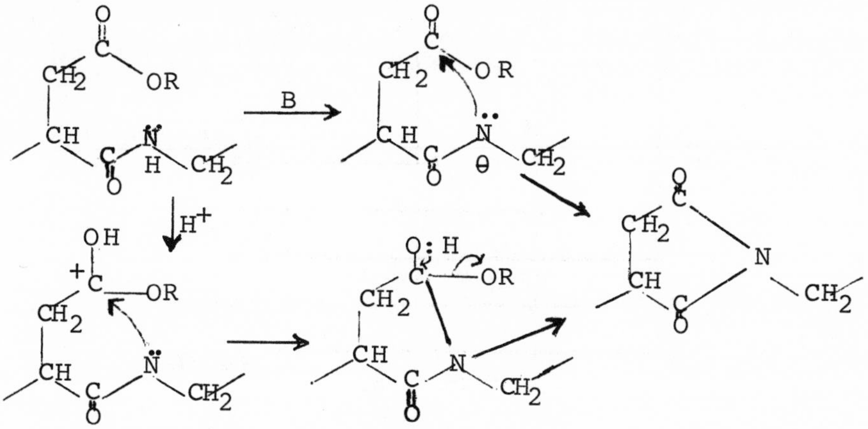

Figure 6.

experiments with model compounds have shown that the sequence aspartyl-glycine is indeed very prone to rearrange to the cyclic imide, particularly when the β-carboxyl group is esterified (Figure 6). This cyclization can occur under anhydrous conditions, either in the presence of base or in strongly acidic medium, e.g., hydrobromic acid-trifluoroacetic acid[16].

These interesting results helped us considerably in clarifying similar problems encountered in the synthesis by fragment condensation to which I now return.

Sequence 9-13 was synthesized in a stepwise fashion starting with the methyl ester of leucine (Figure 7). tert.-Butyloxycarbonyl amino acids p-nitrophenyl esters

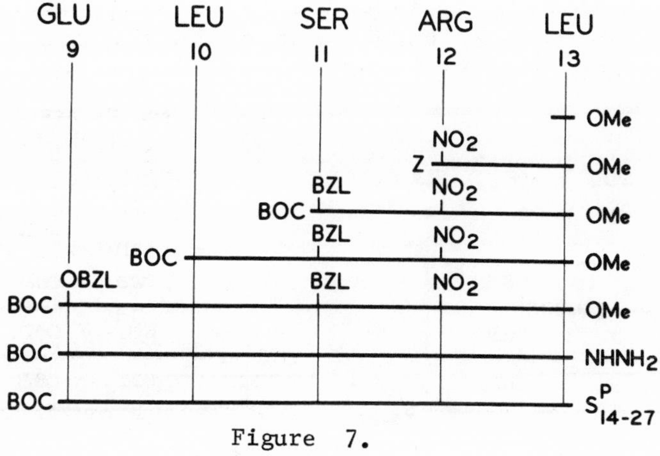


Figure 7.

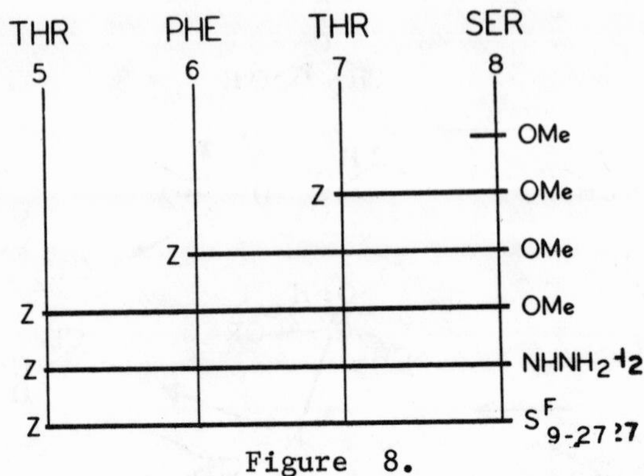


Figure 8.

were used except for arginine where the benzyloxycarbonyl 2,4-dinitrophenyl ester was employed. Hydrogenolysis followed by a treatment with hydrazine gave the desired partially protected pentapeptide hydrazide which was coupled to the partially protected tetradecapeptide amide 14-27 by the azide procedure.

Sequence 5-8 was also synthesized stepwise starting with serine methyl ester (Figure 8). Benzyloxycarbonyl p-nitrophenyl (phenylalanine) and 2,4-dinitrophenyl esters (threonine)[17] were used. Hydrazinolysis of the protected tetrapeptide ester gave the corresponding hydrazide which was coupled to the fully free nonadecapeptide amide 9-27 by the azide procedure. After removal of the N-terminal benzyloxycarbonyl group the free tricosapeptide amide was purified by countercurrent distribution.

The joining of sequence 1-4 and the free tricosapeptide amide 5-27 turned out to be the most crucial step of the fragment condensation approach. In the

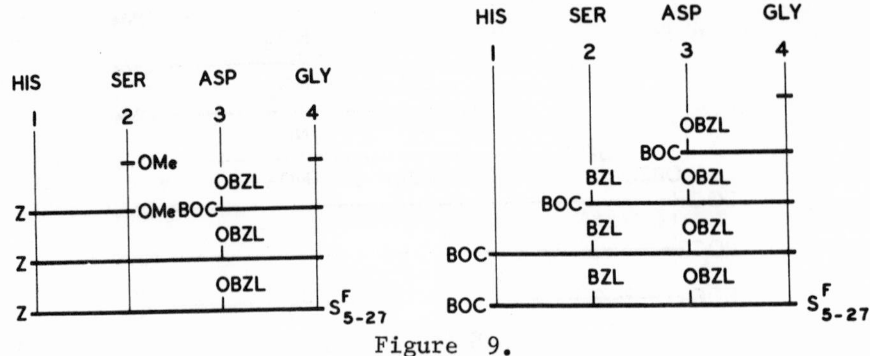


Figure 9.

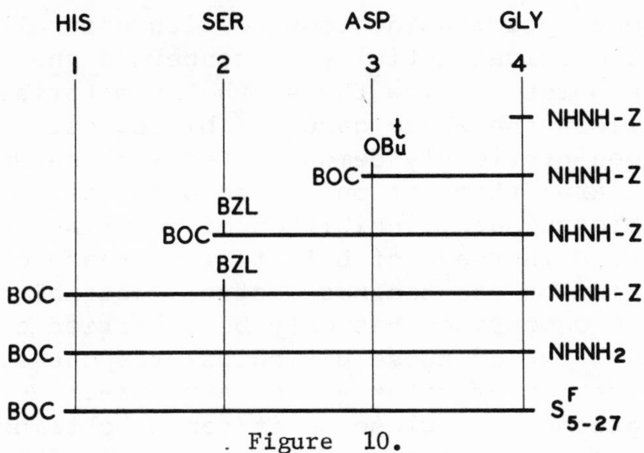


Figure 10.

first attempts two differently protected tetrapeptide acids were coupled to the free tricosapeptide amide by the carbonydiimidazole[18] procedure (Figure 9). Very low yields were obtained and the products, after separation from the unreacted starting materials, still contained significant amounts of rearranged heptacosapeptide requiring an extended purification. Probably the imidazol liberated during the coupling step catalyzed the cyclization of the β-aspartyl benzyl ester to the cyclic imide intermediate.

After these results it seemed desirable to have at hand a tetrapeptide derivative of sequence 1-4 which would allow coupling of the glycine carboxyl to the sequence 5-27, with the β-carboxyl of aspartic acid unprotected, so as to minimize cyclization to the imide under the alkaline conditions of the coupling step. The hydrazide derivative depicted in Figure 10 fulfilled these requirements.

The free heptacosapeptide amide obtained after removal of protecting groups and countercurrent distribution in the n-butanol-phosphate buffer system, was completely degraded by a combined treatment with trypsin and leucine aminopeptidase. These results indicate that no rearranged material was present. The biological potency was again in the range of 4000 to 5000 clinical units per milligram.

We have now to turn to the most important question behind this synthetic enterprise: how does this synthetic heptacosapeptide amide compare with natural secretin? As far as biological activity is concerned and from the qualitative point of view the synthetic material was shown to elicit the whole gamut of biological responses that had been previously demonstrated with natural secretin. These are: stimulation of the pancreas to excrete water and bicarbonate, inhibition of gastrin-induced gastric secretion, increase of bile flow, release of insulin and inhibition of spontaneous gastrointestinal motility[19]. Quantitative comparison has only been carried out in the case of the first of these biological responses, pancreatic secretion of water and bicarbonate. Here, as before, an agreement as close as it can be obtained in a biological assay was observed; the potency of both natural secretin and the synthetic heptacosapeptide amide oscillates between 4000 and 5000 clinical units per milligram.

Independent of this biological scrutiny a physicochemical comparison of natural and synthetic materials was also undertaken. The distribution coefficient in the butanol-phosphate buffer system and the chromotographic mobility in the butanol-pyridine acetate system were found to be identical for both materials. At different stages in the synthesis several synthetic fragments were compared with the corresponding counterparts obtained by tryptic, chymotryptic or thrombic cleavage of natural secretin. No differences were found in any of these fragments which covered practically the whole molecule. Enzymatic degradations of the synthetic heptacosapeptide amide, run in parallel with natural secretin, have failed so far to show any difference.

In summation we can say that the structure proposed by Mutt and Jorpes for pig secretin has been confirmed by synthesis.

## Biological Activity of Partial Sequences of the Secretin Molecule

After the remarkable discovery that a very small part of the gastrin molecule could elicit all the biological effects typical of this hormone, it was natural to raise the same question with respect to secretin. Unfortunately, it seems that here nature does not repeat

| | | Pancreatic Stimulation Potency |
|---|---|---|
| C-TERMINAL SEQUENCES | | |
| Hexapeptide | 22-27 | 0 |
| Tridecapeptide | 15-27 | 0 |
| Tricosapeptide | 5-27 | 0 |
| Hexacosapeptide | 2-27 | ca. 1 c.u./mg. |
| N-TERMINAL SEQUENCES | | |
| Hexapeptide | 1-6 | 0 |
| Tetradecapeptide | 1-14 | ca. 1 c.u./mg. |
| ABBREVIATED SEQUENCE | | |
| Octadecapeptide | 1,4-14,27 | 0 |

Figure 11.

itself. None of the C-terminal sequences assayed, even up to the tricosapeptide stage, showed any significant degree of pancreatic stimulation (Figure 11). Only the hexacosapeptide amide, that is, secretin minus the N-terminal histidine residue, shows a very small, but still measurable, amount of biological activity. Starting from the N-terminal amino acid the situation is very similar. In this case a hexapeptide amide (sequence 1-6) and a tetradecapeptide amide (sequence 1-14) have been tested. Only the tetradecapeptide amide produced pancreatic stimulation at a very low but still measurable level. These results could be taken to mean that both the C-terminal hydrophobic part and the N-terminal hydrophilic portion are simultaneously needed to achieve a significant level of activity. Pursuing this line of thought we succumbed to the temptation of trying to deceive the receptor site by synthesizing an "abbreviated" molecule with an N-terminal tetrapeptide coupled to a C-terminal tetradecapeptide. Apparently receptor sites are not easy to deceive, because this compound was inactive. In brief, the question of which sequence is the "active core" of the secretin molecule, if there is any, remains to be answered.

I TYR-ASP-MET-GLY-TRP-MET-ASP-PHE-$NH_2$

II ASP-TYR-MET-GLY-TRP-MET-ASP-PHE-$NH_2$

GASTRIN:···-GLU-ALA-TYR-GLY-TRP-MET-ASP-PHE-$NH_2$

Figure 12.

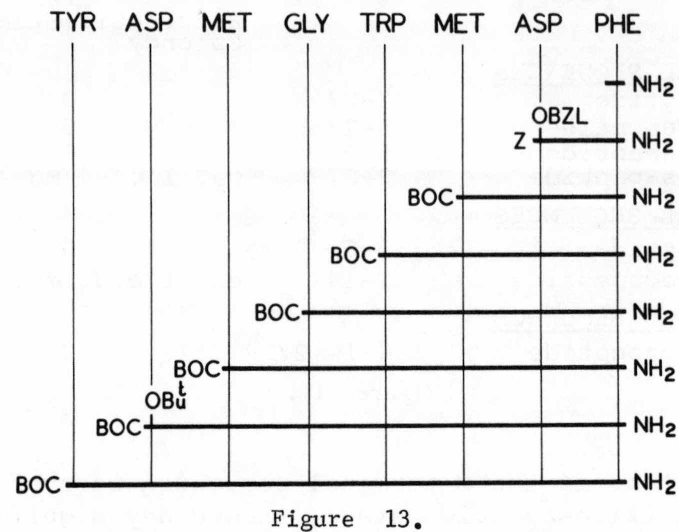


Figure 13.

## Synthetic Work on Cholecystokinin

This chapter is beginning to be written. Cholecystokinin is a single chain polypeptide with 33 amino acid residues[20]. The elucidation of its sequence is in its early stages. However, a very interesting fact has already emerged: the C-terminal pentapeptide of cholecystokinin is identical with that of gastrin[21].

Our efforts in this area have been so far directed towards the synthesis of alternative sequences of the C-terminal octapeptide to aid the clarification of the structure. The two octapeptide amides in point are described in Figure 12.

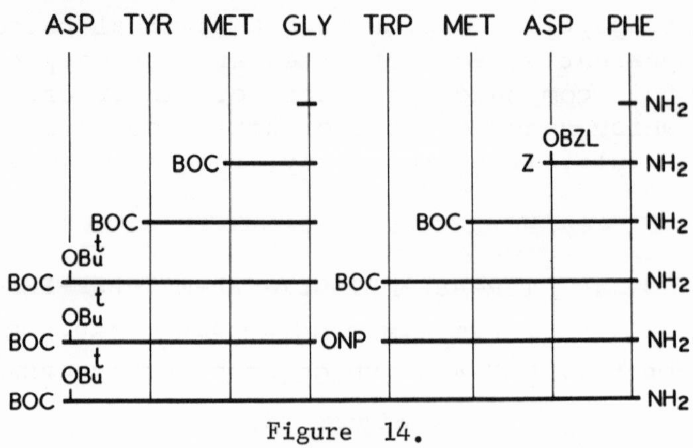


Figure 14.

Sequence I was synthesized stepwise, starting with phenylalanine amide using tert.-butyloxycarbonyl amino acid p-nitrophenyl esters or 2,4,5-trichlorophenyl esters (Figure 13). In the first stages of this synthesis we were, of course, treading along the path followed by Kenner, Sheppard and coworkers in their synthesis of gastrin[22].

Sequence II was synthesized by coupling two tetrapeptides (Figure 14). The stepwise strategy was used for the synthesis of both portions and the coupling of the fragments was achieved by means of the nitrophenyl ester procedure.

Both octapeptides showed cholecystokinin activity. In a preliminary quantitative evaluation[23] octapeptide II showed a potency of approximately 3 Ivy-dog units per milligram and I was only half as active. Pure cholecystokinin has a potency of 3000 Ivy-dog units per milligram.

## References

1. Present address: Department of Chemistry, Case-Western Reserve University, Cleveland, Ohio.

2. R. A. Gregory and H. J. Tracy, Gut, 5, 103 (1964). Structure: H. Gregory, P. N. Hardy, D. S. Jones, G. W. Kenner and R. C. Sheppard, Nature, 204, 931 (1964).

3. a) J. E. Jorpes and V. Mutt, Acta Chem. Scand., 15, 1790 (1961); b) V. Mutt, S. Magnuson, J. E. Jorpes and E. Dahl, Biochemistry, 4, 2358 (1965).

4. J. E. Jorpes, V. Mutt and K. Toczko, Acta Chem. Scand., 18, 2408 (1964).

5. J. C. Anderson, M. A. Barton, R. A. Gregory, P. M. Hardy, G. W. Kenner, J. K. MacLeod, J. Preston and R. C. Sheppard, Nature, 204, 933 (1964).

6. M. Bodanszky, M. A. Ondetti, S. D. Levine. V. L. Narayan, M. von Saltza, J. T. Sheehan, N. J. Williams and E. F. Sabo, Chem. Ind. (London), 1757 (1966).

7. This sequence was presented by V. Mutt and J. E. Jorpes at the fourth International Symposium on the Chemistry of Natural Products, Stockholm, 1966.

8. M. Bodanszky, Ann. N.Y. Acad. Sci., 88, 655 (1960).

9. M. Bodanszky and Nina J. Williams, J. Am. Chem. Soc., 84, 685 (1967).

10. M. Bodanszky, Nature, 175, 685 (1955).

11. F. C. McKay and N. F. Albertson, J. Am. Chem. Soc., 79, 4686 (1957); L. A. Carpino, ibid, 79, 98, 4427 (1957); G. W. Anderson and A. C. McGregor, ibid, 79, 6180 (1957).

12. M. Bodanszky, M. A. Ondetti, S. D. Levine and N. J. Williams, J. Am. Chem. Soc., in press.

13. This system had been originally developed for the purification of natural secretin. Cf. ref. 3a.

14. A. R. Battersby and J. C. Robinson, J. Chem. Soc., 259 (1955); E. Sondheimer and R. W. Holley, J. Am. Chem. Soc., 76, 2467 (1954). S. A. Bernhard, A. Berger, J. H. Carter, E. Katchalski, M. Sela and Y. Shalitin, ibid, 84, 2421 (1962); G. Fölsch, Acta Chem. Scand., 20, 459 (1966).

15. Cf. reference 3b.

16. Unpublished observations from this laboratory.

17. R. Rocchi, F. Marchiori and E. Scoffone, Gazz. Chem. Ital, 93, 823 (1963).

18. H. A. Staab, Ann., 609, 75 (1957); G. W. Anderson and R. Paul, J. Am. Chem. Soc., 80, 4423 (1958).

19. The excretion of water and bicarbonate was assayed by Dr. V. Mutt (Karolinska Institutet) and Dr. S. Engel (Squibb). Inhibition of gastric secretion and enhancement of bile flow were reported by Dr. M. I. Grossman (Veterans Administration, Los Angeles, California). The insulin releasing activity was tested by Dr. H. Elrick (San Diego, California). The inhibition of gastrointestinal motility was studied by Dr. W. Y. Chey (Temple University, Philadelphia, Pennsylvania). We wish to thank all these investigators for their contributions.

20. J. E. Jorpes and V. Mutt. Fourth International Symposium on the Chemistry of Natural Products, Stockholm, 1966.

21. V. Mutt and J. E. Jorpes, Biochem., Biophys. Research Comm., 26, 392 (1967).

22. J. C. Anderson, G. W. Kenner, J. K. MacLeod and R. C. Sheppard, Tetrahedron, 22, Suppl. 8, 39 (1966). Cf. also J. M. Davey, A. H. Laird and J. S. Morley. J. Chem. Soc. (c), 555, (1966).

23. We thank Dr. V. Mutt for these determinations.

THE SYNTHESIS OF CAERULEIN

L. Bernardi, G. Bosisio, R. de Castiglione and O. Goffredo

Istituto Ricerche Farmitalia - Milano - Italy

In the preceding communication by Anastasi, Erspamer and Endean the degradation pattern which led to the complete elucidation of the formula of caerulein (Fig. 1) was presented. As a final proof, a peptide having such a structure was synthesized and found to have the same biological properties as natural caerulein.

The presence of such amino acids as tryptophan, methionine, threonine, tyrosine sulfate and aspartic acid in the molecule of caerulein made the synthesis of this peptide not an easy task: in fact, for instance, the presence of methionine precluded the use of the carbobenzyloxy protecting group since in the presence of sulfur-containing amino acids this group cannot be split by hydrogenolysis. For the same reason the carboxyl group of aspartic acid residues could not be protected as the benzyl ester. On the other hand, these carboxyl groups could not be protected as ter-butyl esters because it was found by preliminary experiments that the conditions for their cleavage, although mild, would split also the tyrosyl-sulfate bond. It was therefore necessary to leave the carboxyl group unprotected; this in turn forbade the use of dicyclohexylcarbodiimi-

$$\text{Pyr-Gln-Asp-}\underset{}{\overset{SO_3H}{\overset{|}{\text{Tyr}}}}\text{-Thr-Gly-Trp-Met-Asp-Phe-}NH_2$$

FIG. 1. STRUCTURE OF CAERULEIN

de. Therefore, less convenient coupling procedures such as the mixed anhydride or the azide methods had to be used.

Finally the sulfate group, being very unstable in acid solution and imparting moreover, some very undesirable solubility characteristics to the molecule, had to be introduced in the last stages of the synthesis. This in turn required suitable protection of the hydroxyl group of threonine to avoid its sulfation. The acetyl group was found to be appropriate since it was proved, by preliminary experiments on model peptides, to be easily removed by aqueous alkali without appreciable β-elimination to α-aminocrotonic acid derivatives [1]

The synthetic procedure is summarized in Figs. 2 and 3. The C-terminal pentapeptide which was synthesized by slight modifications of known procedures [2], was condensed with N-carbo-ter-butyloxy-O-acetyl threonine trichlorophenyl ester to give the protected hexapeptide from which the hydrochloride of O-acetyl-threonyl-glycyl-tryptophanyl-methionyl-aspartyl-phenylalaninamide I was obtained by treatment with hydrochloric acid. The N-terminal tetrapeptide was separately prepared by a stepwise procedure (Fig. 2): the protected hydrazide technique developed by Hofmann [3] was employed and the mixed anhydride method was used. In this way we obtained the fully protected tetrapeptide II which by hydrogenolysis lost three protecting

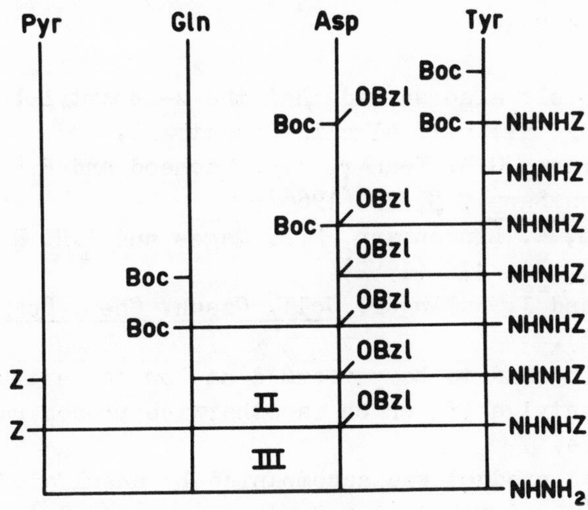


FIG. 2. SYNTHESIS OF N-TERMINAL TETRAPEPTIDE OF CAERULEIN

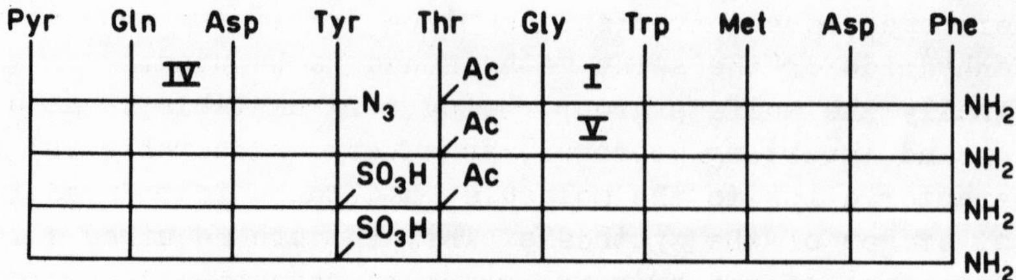


FIG. 3. SYNTHESIS OF CAERULEIN

groups and gave the hydrazide III from which the azide IV (Fig. 3) was obtained by Rudinger's procedure [4]. The fragments I and IV were coupled, in the presence of TEA, at -12° for 4 days to give the decapeptide V [5]. This peptide was next treated with an excess of $SO_3$-pyridine complex and the crude sulfated product was hydrolyzed with sodium hydroxide and extensively purified on DEAE-Sephadex, followed by counter-current distribution [6]. The peptide showed the same electrophoretic and chromatographic properties, the same behaviour towards chymotrypsin, subtilisin and the same degradation pattern and biological properties as natural caerulein, thus confirming the formula deduced from degradation experiments.

References and Notes

1) Work on models also showed that the α-aspartyl bond was not affected by this mild alkaline treatment.
2) J.C. Anderson, G.W. Kenner, J.K. Macheod and R.C. Sheppard, Tetrahedron Supp., 8, 39 (1966).
3) K. Hofmann, A. Lindenmann, M.Z. Magee and N.H. Khan, J. Am. Chem. Soc., 74, 470 (1952).
4) J. Honzl and J. Rudinger, Coll. Czech. Chem. Comm., 26, 2333 (1961).
5) The coupling had to be performed at low temperature to avoid the O-N acetyl shift which is otherwise predominant at room temperature.
6) The desired product was accompanied by unsulfated and polysulfated (mixed anhydrides β-aspartic acid-$SO_3$) peptides and repeated purifications were necessary to eliminate these impurities.

# TECHNIQUES FOR THE SYNTHESIS OF ACTH AND MSH PEPTIDES AND ANALOGUES

W. Rittel

Pharmaceutical Research Laboratories

CIBA Limited, Basle, Switzerland

The manifold biological activities displayed by ACTH peptides - activities that include hormonal as well as extra-adrenal effects - offer the peptide chemist a fruitful field in which to study the relationship between structure and function. Earlier work undertaken in the laboratories of Li, Hofmann, Boissonnas, and Schwyzer was directed towards the synthesis of active ACTH fragments. These efforts culminated in the successful synthesis of hog-type corticotropin by Schwyzer and Sieber (1) and, more recently, of human corticotropin by a Hungarian group (2). In recent years, however, attention has shifted towards the preparation of modified ACTH peptides, and it has been demonstrated that the natural amino acid sequence of corticotropin peptides can be altered considerably with retention of (3,4,5) or even an increase in biological activity (6,7).

After completion in our laboratories of the synthesis of β-corticotropin-(1-24)-tetracosapeptide[1)]

1) The nomenclature and abbreviations used are based mainly on the proposals of the IUPAC-IUB Commission on Biochemical Nomenclature, Tentative Rules, Biochemistry 6, 362 (1967); cf. also: Peptides, Proc. 8th Europ. Peptide Symposium, Amsterdam 1967, p.xi.

(8), which is now manufactured commercially under the trade name "Synacthen®", we attempted to synthesize analogues of this sequence. One approach, which was aimed at the preparation of peptides resistant against enzymatic attack, involved substitution, at or near the amino end, of L-amino acids by their D-isomers (7). Biological testing of this group of analogues led to the surprising finding that the introduction of D-serine (or D-alanine) in position 1 of the chain produces a marked increase in corticotropic activity. Depending upon the test system or route of administration used, levels of corticotropic activity equivalent to up to ten times that of the all-L peptides were observed. Similar results have been reported by Boissonnas and coworkers (6). Recently, in our laboratory another analogue of Synacthen has been found to possess not only potentiated corticotropic activity but also prolonged duration of action (9).

Not only the quantity, but also the quality of the biological action displayed by ACTH peptides is greatly influenced by their chain length. This dependence on chain length has been studied by Li and coworkers (10), and more recently by Bajusz and Medzihradszky (11). Although the results reported are somewhat conflicting, it can nevertheless be concluded that, if *in vivo* corticotropic activity is measured, the octadecapeptide sequence 1-18 in the C-terminal amide form is the shortest ACTH fragment possessing high activity. If the chain is shortened, the activity rapidly diminishes.

These observations prompted us to investigate how the introduction of D-serine in position 1 would affect the corticotropic activity of ACTH peptides with chains differing in length from Synacthen. It appeared to be of special interest to discover whether the low activity of short-chain fragments could be raised by this modification. For this purpose, the following group of ACTH analogues was accordingly prepared:

[D-Ser$^1$-] β-Corticotropin-(1-13)-tridecapeptide amide (I)
[ " -] " " -(1-16)-hexadecapeptide " (II)
[ " -] " " -(1-19)-nonadecapeptide (III)
[ " -] β-Corticotropin (IV)

The corticotropic activities of the corresponding all-L peptides, as measured in the Sayers test, have been reported to be as follows: tridecapeptide amide 1-13: 0,1 I.U./mg (12); hexadecapeptide amide 1-16: 1,4 I.U./mg (11); nonadecapeptide 1-19: 74,2 I.U./mg (13); synthetic β-corticotropin: 140 I.U./mg (14).

The method of synthesis used in the preparation of the protected and free ACTH sequences I-IV is outlined in Figs. 1-4. The procedures employed were those developed and described earlier for the synthesis of β-corticotropin (1) and its active fragments (8). In all cases the protected D-serine[1] decapeptide 1-10 (7) with the free carboxyl group at glycine[10] served as a key intermediate. To this derivative were coupled, using the carbodiimide method, the respective C-terminal fragments 11-13, 11-16, 11-19, and 11-39.

<u>[D-Ser[1]-] β-corticotropin-(1-13)-tridecapeptide amide (I).</u> The synthesis of I, shown in Fig. 1, was carried out by Dr. M. Brugger. It followed closely the one reported from our laboratory (15) for the preparation of α-MSH. Use of the carbodiimide method to couple the protected 1-D-serine decapeptide 1-10 with the hydrochloride of the tripeptide amide 11-13 (15) yielded the protected sequence 1-13, which was purified by column chromatography on silica gel. Removal of the protecting groups by treatment with trifluoro-acetic acid and subsequent ion exchange led to the acetate salt of the free peptide I in pure form.

<u>[D-Ser[1]-] β-corticotropin-(1-16)-hexadecapeptide amide (II).</u> For the synthesis of II (cf. Fig. 2), the decapeptide derivative 1-10 was coupled with the hydrochloride of the hexapeptide amide 11-16. This latter fragment was obtained from Z·Lys(Boc)-Pro-Val-Gly-Lys(Boc)-Lys(Boc)·OMe (16) by ammonolysis and catalytic hydrogenation. The protected peptide sequence 1-16 was purified by countercurrent distribution and the protecting groups removed from the purified product by treatment with trifluoro-acetic acid. After conversion to the acetate salt by ion exchange, the free peptide II was chromatographed on CM-Sephadex C-25. This treatment removed minor amounts of impurities.

<u>[D-Ser[1]-] β-corticotropin-(1-19)-nonadecapeptide (III).</u> In the synthesis of III (cf. Fig. 3), the partially protected fragments 1-10 and 11-19 were employed. In order to obtain the latter fragment[1)] the hexapeptide derivative Z·Lys(Boc)-Pro-Val-Gly-Lys(Boc)-Lys(Boc)·OMe was converted to the crystalline hydrazide. Use of the azide method to couple this derivative with

1) This preparation was carried out by Mr. P. Sieber.

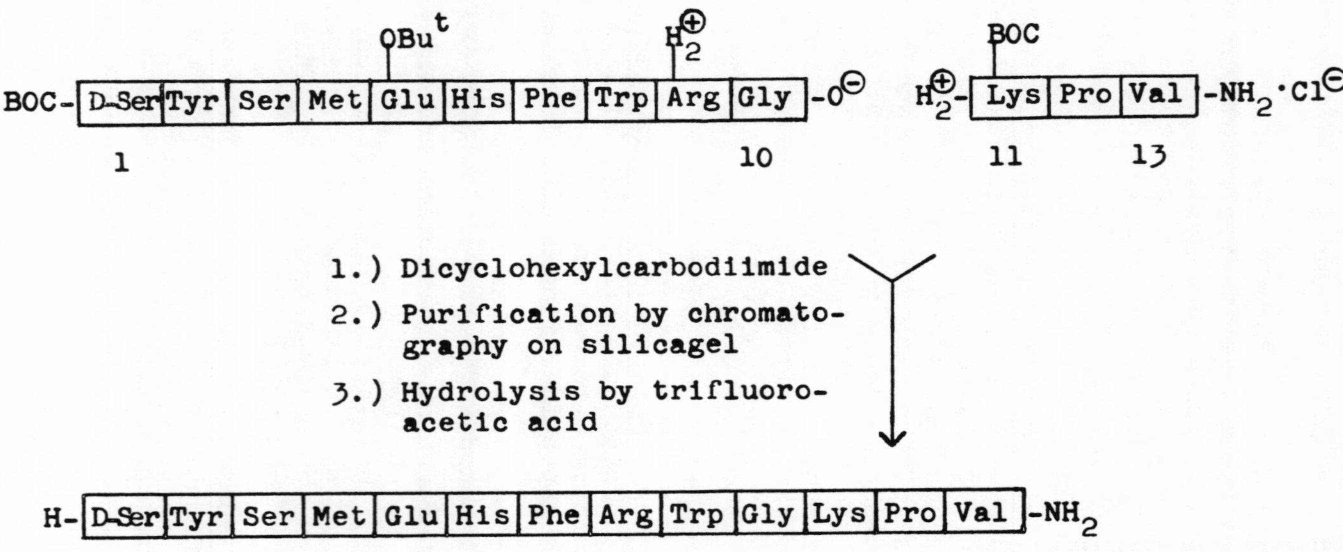


Fig. 1. Synthesis of [D-Ser$^1$] β-Corticotropin-(1-13)-tridecapeptide amide (Desacetyl[D-Ser$^1$]α-MSH).

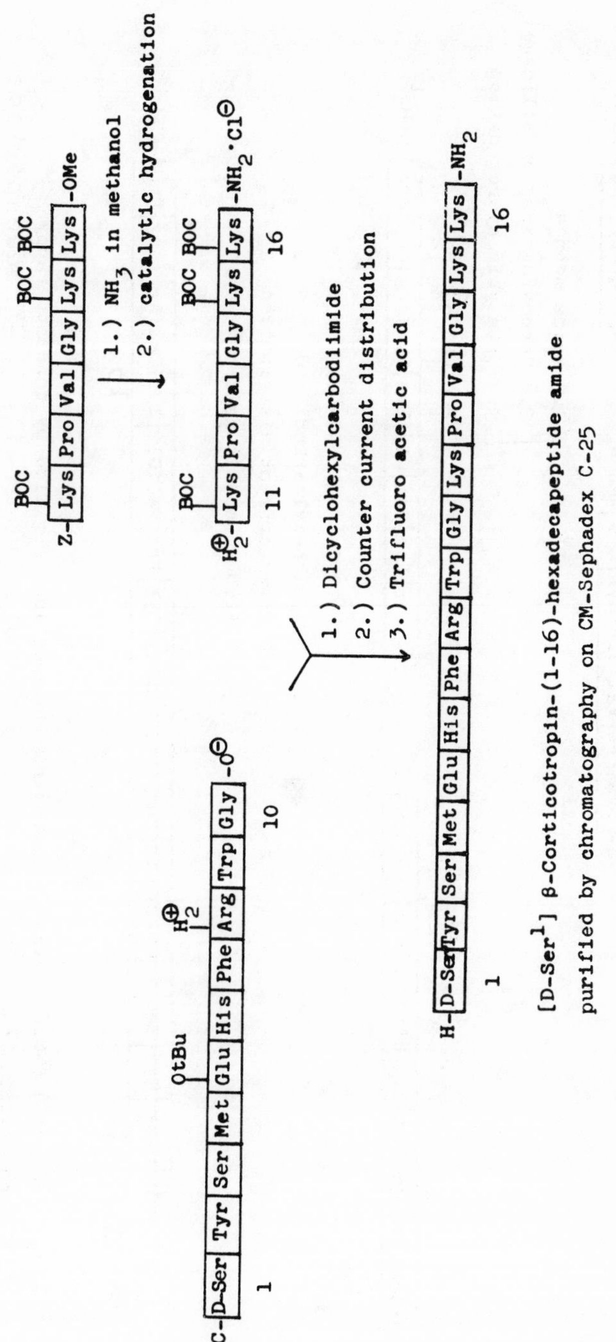


Fig. 2. Synthesis of [D-Ser[1]] β-Corticotropin-(1-16)-hexadecapeptide amide.

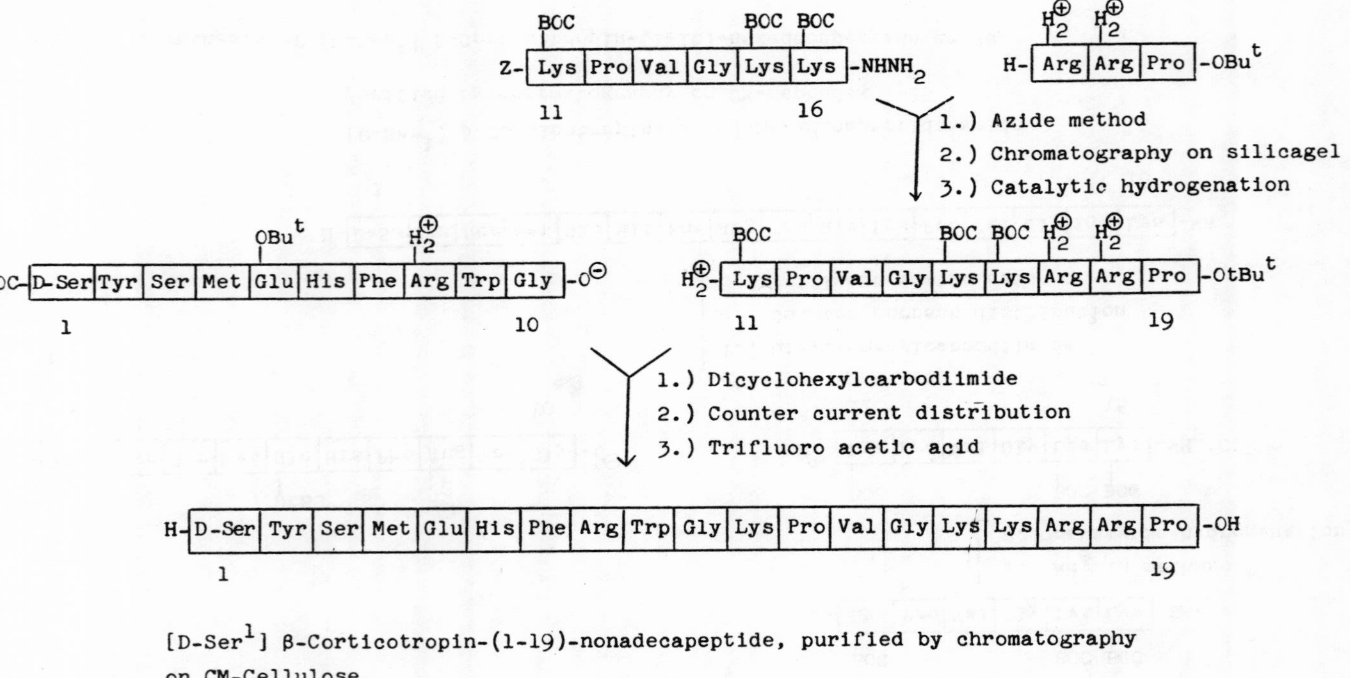


Fig. 3. Synthesis of [D-Ser[1]] β-Corticotropin-(1-19)-nonadecapeptide.

the tripeptide fragment 17-19 led to the protected sequence 11-19, which was obtained in pure form after chromatography on silica gel. The synthesis then proceeded as shown in Fig. 3. The guanidino groups of arginine residues 17 and 18 were protected throughout this synthesis by protonation only. Purification of the free peptide III was achieved by chromatography on CM-cellulose.

<u>[D-Ser$^1$-] β-corticotropin (IV).</u> The synthesis (cf. Fig. 4) of the nonatriacontapeptide IV, an analogue possessing the full amino acid chain length of porcine β-corticotropin, was carried out by Mr. P. Sieber. While the synthesis (1,17) of β-corticotropin itself involved a major chemical effort in which 54 separate chemical steps had to be carried out, preparation of IV was greatly facilitated by the fact that fragment 11-39 was still available from the earlier work. Thus, only two chemical steps - coupling of fragments 1-10 and 11-39 and removal of the protecting groups - led to the free peptide IV. In these steps, and also in the final purification of [D-Ser$^1$-] β-corticotropin, the experimental procedures described in the synthesis of β-corticotropin could be successfully employed.

Taking the syntheses of analogues I-IV as examples, it is evident that the method of fragment condensation makes it possible, with only a limited effort, to build up complex peptides, provided a store of suitable fragments is available. One particular advantage of this method lies in the fact that the experience gained in a given synthesis can be successfully applied to other, analogous cases. Thus the tedious investigation of conditions effecting peptide-bond formation can be avoided. Of equal importance is the fact that the analytical procedures employed in the work up steps, such as thin layer chromatography or countercurrent distribution, are greatly facilitated since the behaviour of the starting materials, by-products to be expected and the end-products is either known or can be anticipated. On the other hand, the use of large peptide fragments as building blocks, especially if these are employed as a carboxyl component in the coupling, may result in low yields and in loss of valuable starting material. From the final coupling step in the synthesis of IV the protected nonatriacontapeptide could be obtained only in a 30-40% yield as calculated on the basis of the amount

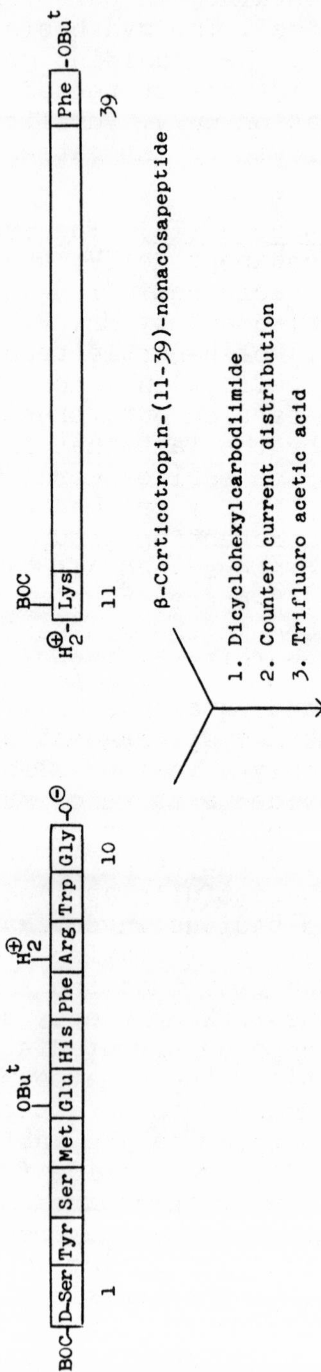


[D-Ser[1]] β-Corticotropin, purified by chromatography on CM-Sephadex C25

Fig. 4. Synthesis of [D-Ser[1]]-β-Corticotropin.

of sequence 11-39 employed. This yield corresponds to that obtained in the synthesis of porcine β-corticotropin (1).

Recently we have observed that in similar cases the use of carbodiimide in the presence of N-hydroxysuccinimide, a method introduced by Weygand and Wünsch (18), is capable of considerably improving the yields.

The results of biological testing of the peptides I-IV cannot be discussed here in detail. When assayed[1] in vivo by reference to adrenal steroidogenesis or adrenal ascorbic acid depletion, all four peptides were found to possess levels of activity which are 5-10 times higher than those of the corresponding all-L peptides.

Synthesis of human β-melanocyte stimulating hormone ($\beta_h$-MSH). As a further example of the fragment-coupling approach, the synthesis of $\beta_h$-MSH, carried out recently in collaboration with Dr. B. Riniker, will be briefly discussed.

$\beta_h$-MSH isolated by Dixon (19) was shown by Harris (20) to possess the docosapeptide sequence shown in Fig. 5. Whereas β-MSH preparations from other species all comprise a sequence of 18 amino acids, in the human hormone this octadecapeptide chain is elongated by four additional amino acid residues.

Isolation procedures have yielded $\beta_h$-MSH in amounts just sufficient for structural work. It therefore appeared of interest to make preparative quantities of this unique hormone available for extensive biological testing.

The final steps of the synthesis of the $\beta_h$-MSH docosapeptide sequence are outlined in Fig. 5. A detailed account of this work, including preparation of the protected fragments 1-6 and 7-11, which have not yet been described, will be given elsewhere.

Our approach was facilitated by the chemical relationship existing between the various β-MSH preparations. In 1963 we described a synthesis of bovine β-MSH (21). From this earlier work we were able to take over the C-terminal undecapeptide sequence 12-22 in protected form, since identical sequences are found in posi-

1) These determinations were kindly performed by Dr. P.A. Desaulles and Dr. M. Staehelin of our Biological Research Department.

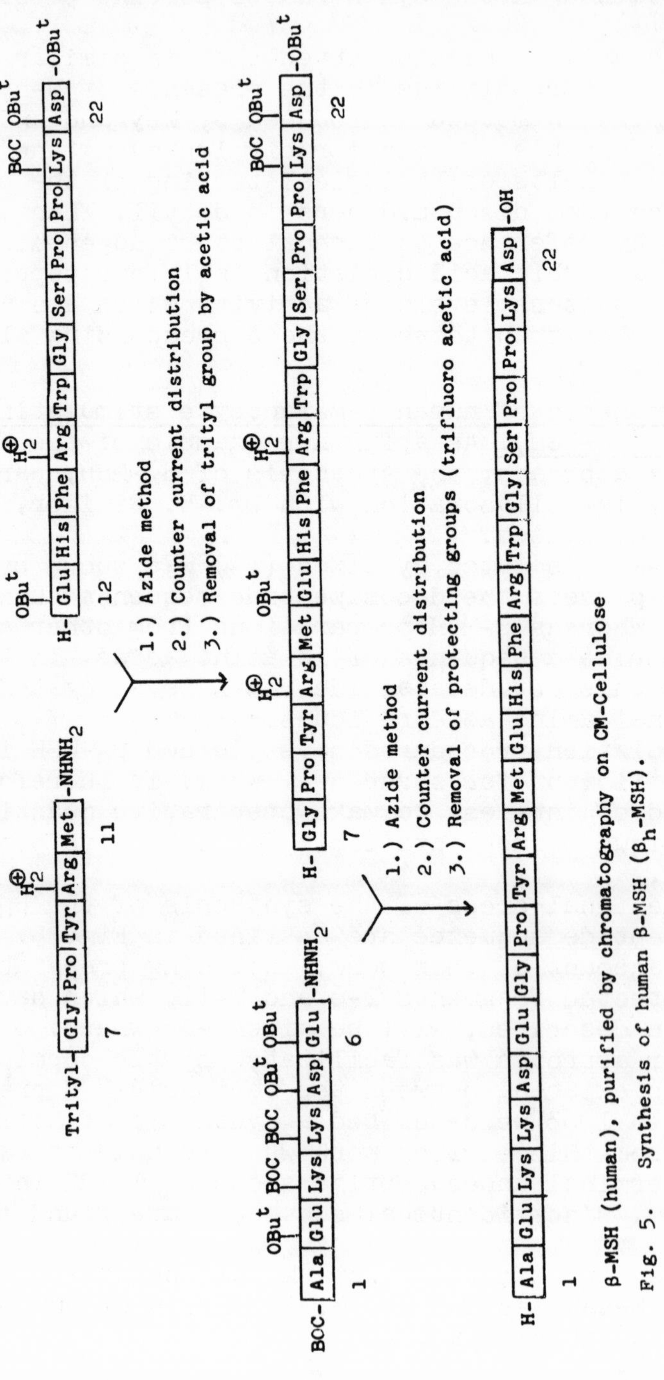


Fig. 5. Synthesis of human β-MSH ($\beta_h$-MSH).

tions 8-18 of bovine and 12-22 of human β-MSH. Using the azide method, the fragment 7-11, which bears a trityl group at the amino end, was coupled to the sequence 12-22. The fragment 7-22 obtained in this way was purified by countercurrent distribution and then detritylated with acetic acid. Prolongation of the chain with the protected fragment 1-6 was again effected by the azide method. It should be noted that a glycine residue occupies position 7. A fragment 1-7 with a free glycine carboxyl group would permit the use of different coupling methods without involving the danger of racemisation. The proline residue in position 8, however, was expected to interfere with coupling at this point owing to steric hindrance. Preference was therefore given to the build-up of fragments 1-6 and 7-11.

From the final coupling step, i.e. between the peptide derivatives 1-6 and 7-22, the protected docosapeptide sequence was obtained in a yield of over 80%; to remove traces of by-products it was subjected to countercurrent distribution. The protecting groups were removed from the purified product with the aid of trifluoroacetic acid containing 10% water. Thin layer chromatography of the free docosapeptide indicated the presence of one compound only. Paper electrophoresis at pH 6.9, however, revealed that a second component was present in a quantity of approximatively 20%. Removal of this by-product was achieved by chromatography on CM-cellulose. The nature of the by-product has not yet been elucidated.

Since $\beta_h$-MSH from natural sources was not at our disposal, a direct comparison of the main product of this synthesis with isolated material could not be made. As shown in Table 1, the properties of the synthetic preparation agree well with the values reported by Pickering and Li (22) for the natural hormone.

| | $[\alpha]_D$ (0.1M acetic acid) | MSH activity in vitro, U./g. |
|---|---|---|
| $\beta_h$-MSH, isolated (22) | $-104^o$ | $3,3 \cdot 10^9$ |
| " , synthetic | $-99.8^o$ | $3-6 \cdot 10^9$ 1) |

Table 1. Properties of natural and synthetic $\beta_h$-MSH.

1) This result was kindly supplied by Dr. P.A. Desaulles.

In order to provide further evidence of its identity with natural $\beta_h$-MSH, the synthetic preparation was subjected to degradation by chymotrypsin according to the method of Pickering and Li (22), who have examined in detail the hydrolysis of natural $\beta_h$-MSH by this enzyme as a means of characterization. Chymotryptic digestion of the synthetic preparation yielded a pattern of fragments similar to the one described by these authors. The peptides were separated by paper electrophoresis and eluted from the electropherograms. Amino acid analyses of these fragments are just being carried out but the results are not available yet.

The outcome of this experiment should yield further evidence concerning the identity of the synthetic docosapeptide sequence with natural $\beta_h$-MSH.

References.

(1) R. Schwyzer & P. Sieber, Nature 199, 172 (1963).
(2) S. Bajusz, K. Medzihradszky, Z. Paulay & Z. Láng, Acta Chimica Acad.Sci.Hung. 52, 335 (1967).
(3) R. Geiger, K. Sturm, G. Vogel & W. Siedel, Z. Naturforsch. 19b, 858 (1964).
(4) S. Guttmann, J. Pless & R.A. Boissonnas, Acta Chimica Acad.Sci.Hung. 44, 23 (1965).
(5) G.I. Tesser & R. Schwyzer, Helv.Chim.Acta 49, 1013 (1966).
(6) R.A. Boissonnas, S. Guttmann & J. Pless, Experientia 22, 526 (1966).
(7) H. Kappeler, B. Riniker, W. Rittel, P.A. Desaulles, R. Maier, B. Schär & M. Staehelin, Peptides, Proc. 8th Europ. Peptide Symposium, North-Holland Publ. Co., Amsterdam 1967, p. 214.
(8) H. Kappeler & R. Schwyzer, Helv.Chim.Acta 44, 1136 (1961).
(9) P.A. Desaulles & W. Rittel, Proc. Royal Soc.Med. (1967), in press; Memoirs Soc.Endocrin. (1967), in press.
(10) J. Ramachandran, D. Chung & C.H. Li, J.Am.Chem. Soc. 87, 2696 (1965).
(11) S. Bajusz & K. Medzihradszky, Peptides, Proc. 8th Europ.Peptide Symposium, North-Holland Publ.Co., Amsterdam 1967, p. 210.
(12) K. Hofmann, Brookhaven Symp.Biol. 13, 184 (1960).
(13) C.H. Li, Rec.Progr.Hormone Res. 18, 1 (1962).
(14) R. Schwyzer, Ann.Rev.Biochem. 33, 276 (1964).
(15) R. Schwyzer, A. Costopanagiotis & P. Sieber, Helv.Chim.Acta 46, 870 (1963).

(16) R. Schwyzer, W. Rittel & A. Costopanagiotis, Helv. Chim.Acta 45, 2473 (1962).
(17) R. Schwyzer & P. Sieber, Helv.Chim.Acta 49, 134 (1966).
(18) F. Weygand, D. Hoffmann & E. Wünsch, Z. Naturforsch. 21b, 426 (1966).
(19) H.B.F. Dixon, Biochim.Biophys.Acta 37, 38 (1960).
(20) J.I. Harris, Nature 184, 167 (1959).
(21) R. Schwyzer, B. Iselin, H. Kappeler, B. Riniker, W. Rittel & H. Zuber, Helv.Chim.Acta 46, 1975 (1963).
(22) B.T. Pickering & C.H. Li, Biochim.Biophys.Acta 74, 156 (1963).

# STUDIES WITH TRYPTOPHAN IN SOLID-PHASE SYNTHESIS*

Garland R. Marshall

Department of Physiology & Biophysics
Washington Univ. Sch. of Medicine
St. Louis, Missouri, U. S. A.

The basic concept of solid-phase synthesis (Merrifield, 1963) is engagingly simple. The synthetic reactions required for the lengthening of a polymer chain are performed while the growing chain is bound through a covalent link to an insoluble support which allows convenient purification by washing. By its very simplicity, however, the chosen methodology imposes stringent restrictions on the use of the synthetic arsenal which peptide chemists have developed. The use of a modified benzyl ester link between the polymer support and the C-terminus of the growing peptide chain determines the type of protecting groups which may be used on the $\alpha$-amino group, i. e., a group which is more labile than the ester link and can be removed without cleaving the chain from the support such as the t-butyloxycarbonyl (BOC) (Merrifield, 1964) and the o-nitrophenylsulfenyl (NPS) (Kessler and Iselin, 1966; Najjar and Merrifield, 1966). The choice of amino-protecting group and its method of removal restrict still further the choice of sidechain-protecting groups.

One becomes acutely aware of the imposed restrictions when one is attempting to synthesize a chain with a group as labile as tryptophan. Tryptophan, whose sensitivity to oxidation under acid conditions has long plagued the protein chemist, is an integral part of the active regions of several hormones (ACTH, gastrin, MSH), and may be equally important to several other peptides and proteins of biological interest (glucagon, growth hormone, parathyroid hormone). The problem of developing the solid-phase method for use in the synthesis of tryptophan-containing

*Supported in part by grants from the American Cancer Society and the United States Public Health Service (AM10836).

peptides can be considered in two parts; first, the selection of a covalent link to the support and a method of cleavage, and second, the selection of the $\alpha$-amino-protecting groups and its method of removal.

While other covalent links such as a modified t-butyl ester to the support were considered, the decision to use the normal benzyl ester linkage was based on the wide experience and successful use of all other commonly-occurring amino acids in solid-phase with this bond. Similarly, the beaded styrene-divinylbenzene copolymer support was chosen over a soluble system which offers the possibility of catalytic or enzymatic cleavage (Shemyakin et al., 1965). Benzyl esters are normally removed by catalytic hydrogenolysis when working with tryptophanyl peptides. This has not proven successful, so far, with peptides attached to the support, probably because of steric hindrance between the support and the catalyst. Cleavage of the chain from the support to give the free carboxyl group is normally accomplished by hydrogen bromide in trifluoroacetic acid, and, recently, by anhydrous hydrogen fluoride (Lenard and Robinson, 1967). With both methods, appropriate conditions have been found which allow tryptophanyl peptides to be cleaved from the support without a large amount of degradation of tryptophan.

In order to determine the amount of degradation during the cleavage reaction, it was necessary to determine conditions for the successful incorporation of tryptophan and measure the extent of that incorporation. Initially, there was considerable difficulty in ascertaining the amount of tryptophan incorporated into the growing peptide chain on the polymer support. Classical alkaline hydrolysis methods as well as acid hydrolysis on model peptides gave variable results, even where recovery of added tryptophan was quite high. Our most consistent method was the removal of the peptide from the support by ammonolysis followed by enzymatic digestion. After it was determined that no special problems were encountered in the incorporation of tryptophan both by dicyclohexylcarbodiimide (BOC and NPS derivatives) and nitrophenyl ester (BOC derivative), the losses of tryptophan due to cleavage and amino-protecting group removal could be ascertained.

Model experiments using BOC-tryptophan were performed initially to check conditions for the cleavage reaction. The loss of tryptophan was determined by following the ratio of tryptophan to histidine on the amino acid analyzer after a standard mixture of BOC-tryptophan and histidine was added to the reaction. By multiple prewashings of the HBr gas with a solution of indole in trifluoroacetic acid, recovery of tryptophan was increased to 85% for a thirty minute exposure to HBr in trifluoroacetic acid. Since this value is almost identical to that obtained during similar experiments with trifluoroacetic acid alone, it was felt that these conditions were nearly optimal for the cleavage with HBr/TFA. Attention then focused on the use of HF and its effect on tryptophan. The stability of tryptophan to anhydrous HF (Sakakibara et al. ,1967) was confirmed when two tryptophanyl peptides, Try-Phe and Try-Leu-Met-Asn-Thr, were synthesized by the solid-phase method and cleaved both by HF and HBr/TFA. In both cases the material resulting from the HF cleavage appeared to have less tryptophan degradation as evidenced spectrophotometrically and by TLC.

Simultaneously, experiments to measure the degradation of tryptophan during the elongation of the growing peptide chain in solid-phase synthesis were underway. Initial observations suggested a loss of as high as 5% for each cycle which extends the peptide by one residue, but more refined measurements indicated 1 to 2% degradation using HCl/HOAc to remove the BOC-protecting group. This is a sufficient rate to seriously hinder the synthesis by the solid-phase technique of peptides which must be extended many residues beyond a tryptophan. The selection of an $\alpha$-amino-protecting group and a method for its removal was limited to a consideration of the BOC and NPS groups. Dilute acid cleavage of the NPS group from tryptophan gives an unwanted rearrangement (Anderson et al., 1965). However, a variety of nucleophilic reagents (Kessler and Iselin, 1966; Fontana et al., 1966; Brandenburg, 1966) have been suggested which appear to overcome this difficulty and make the NPS group quite attractive. One such reagent investigated in collaboration with Prof. G. T. Young is dinitrothiophenol which has some advantages in its speed of reaction. In solution, the removal of the NPS group from $N^{\alpha}$-NPS-S-benzylthiomethyl-cysteine was complete in less than 5 minutes at room temperature with a 10-fold excess of dinitrothiophenol. This rate is considerably retarded when working with the solid-phase method and appears to be very solvent dependent. Further work is necessary to determine optimal conditions for this reagent, but its usefulness has been demonstrated in the removal of the NPS group in the synthesis of the C-terminal hexapeptide of glucagon by solid-phase.

Several alternative methods of BOC removal are available, including HCl/HOAc and trifluoroacetic acid. More recently, formic acid has been suggested (Halpern and Nitecki, 1967) as a superior reagent. The results of model experiments similar to those described for cleavage studies are shown in Fig. I. The recoveries indicate a clear superiority for the HCl/HOAc cleavages over those with TFA. They also indicate a preference for the addition of antioxidants such as thiodiglycol and $\beta$-mercaptoethanol. One per cent $\beta$-mercaptoethanol has been used in the 1 N HCl/HOAc and the HOAc washes after the incorporation of tryptophan. This appears to minimize the extraneous color-development on the resin as well as during the cleavage with HF when a deep purple color was noticed when the mercaptan was omitted during synthesis.

The results of the model experiments were confirmed with the

| | Try Recovery |
|---|---|
| Anhydrous TFA | 87% |
| 99% TFA | 84% |
| 80% TFA | 92% |
| 1 N HCl/HOAc | 95% |
| " +1% thiodiglycol | 97% |
| " +1% mercaptoethanol | 100% |
| Average Recoveries | |
| TFA Group | 86±7% |
| HCl/HOAc Group | 97±3% |

Fig. 1. Recoveries of tryptophan from BOC-tryptophan after a 30 minute exposure to the above reagents. Values are determined by comparison with an internal standard of histidine by amino acid analysis.

Try-Phe*

Try-Leu-Met-Asn-Thr*
Glucagon 25-29

Phe-Val-Gln-Try-Leu-Met-Asn-Thr
Glucagon 22-29

Ser-Arg-Arg-Ala-Gln-Asp-Phe-Val-Gln-Try-Leu-Met-Asn-Thr
Glucagon 16-29

Fig. 2. Tryptophanyl peptides synthesized by the solid-phase method using antioxidants with acidic reagents and cleavage from the support by anhydrous hydrogen fluoride. *Also synthesized using HBr/TFA for cleavage.

synthesis of the peptides shown in Fig. 2. β -mercaptoethanol was used during the acidic steps after the incorporation of tryptophan and the peptides were cleaved from the polymer support by anhydrous hydrogen fluoride. The spectra of the peptides after removal of the anisole by gel filtration showed a characteristic tryptophan spectrum; and leucine aminopeptidase digestion of the peptide gave a quantitative recovery of tryptophan.

Several alternative schemes have been investigated for the synthesis of tryptophan-containing peptides by the solid-phase method. Of these, there appear to be two choices open to the peptide chemist depending on his synthetic goals for amino-protecting group and cleavage method. Either the NPS or BOC groups can be used with the solid-phase method for the synthesis of tryptophanyl peptides which can be removed from the polymer support by either HF or HBr/TFA. Hopefully, these studies with tryptophan in solid-phase synthesis will open the door to the successful synthesis of large tryptophan-containing hormones.

## References

1. Anderson, J. C., Barton, A. M., Hardy, P. M., Kenner, G. W., MacLeod, J. K., Preston, J., Sheppard, R. C., and Morley, J. S. Acta Chim. Acad. Sci. Hung. 44:187 (1965).
2. Brandenburg, D. Tetrahedron Letters 49:6201 (1966).
3. Fontana, A., Marchiori, F., Moroder, L., and Scoffone, E. Tetrahedron Letters 26:2985 (1966).
4. Halpern, B., and Nitecki, D. E. Tetrahedron Letters 31:3031 (1967).
5. Kessler, W., and Iselin, B. Helv. Chim. Acta 49:1330 (1966).
6. Lenard, J., and Robinson, A. B. J. Am. Chem. Soc. 89:181 (1967).
7. Merrifield, R. B. J. Am. Chem. Soc. 85:2149 (1963).
8. Merrifield, R. B. Biochemistry 3:1385 (1964).
9. Najjar, V. A., and Merrifield, R. B. Biochemistry 5:3765 (1966).
10. Sakakibara, S., Shimonishi, Y., Okada, M., and Kishida, Y. in Peptides Ed. Beyerman, H. C., Van de Linde, A., and Van den Brink, W. M. North-Holland Publishing Co., Amsterdam, 1967, p. 44.
11. Shemyakin, M. M., Ovchinnikov, Yu. A., Kiryuskin, A. A., and Kozhevnikova, I. V., Tetrahedron Letters 27:2323 (1965).

# OXYTOCIN - MOLECULAR ASPECTS

B. BERDE

Biological and Medical Research Division

Sandoz Ltd., Basle, Switzerland

The subject "Oxytocin - molecular aspects" can be treated from various points of view. One of the possible approaches is the following: To-day we have quantitative data on a number of well-defined pharmacological properties for more than 200 synthetic peptides of the neurohypophysial type, which can all be considered to be derivatives of oxytocin. Therefore, it is tempting to try to discover, on the basis of the pharmacological properties of these derivatives, which parts of the oxytocin molecule are essential for the characteristic biological effects of this hormone.

However, when trying to do this, one encounters a fundamental difficulty: There would appear to be considerable differences in the sensitivity, i.e. the structure, of the oxytocin receptors. It is widely recognised that important species differences exist in the structure of the receptors of a given organ (e.g. the uterus or the vessels responsible for blood pressure). But there seem to be similar important differences between the receptors of different organs (e.g. uterus, mammary gland and vessels) in the same species. This is probably one of the major reasons why it is so difficult to establish generally applicable structure/activity relationships.

Another difficulty is that synthetic analogues of oxytocin have been produced and tested in different laboratories and the methods used are not all identical; consequently they do not yield strictly comparable results. Hence only those activities can be considered which have been established with reasonable accuracy for a considerable number of compounds under at least roughly comparable experimental conditions.

For these reasons only three oxytocin-like activities will be discussed here, and then only where each of them was measured by a particular experimental procedure. For the present purposes uterotonic activity is the potency measured on the isolated non-pregnant (preferably oestrous) rat uterus suspended in a magnesium-free physiological solution at a working temperature around 30°C (for possible sources of errors in this widely used method see BERDE & SAAMELI, 1966); vasodilatator potency is avian blood pressure lowering activity, as determined on the systemic arterial pressure of roosters (COON, 1939); and milk-ejection activity is the potency determined on the lactating mammary gland of anaesthetised rabbits after intravenous administration of the peptides (for the method see e.g. van DYKE et al., 1955; BERDE & CERLETTI, 1960). In the tables these three activities are given under the headings "rat uterus", "chicken blood pressure" and "rabbit mammary gland". The values are in International Units (≃ U.S.P. Units) per micromole peptide. "Anti" means that the compound antagonises oxytocin in the test and in the given quantitative relationships.

The following system of abbreviations for amino acids will be used: Abu = L-α-aminobutyric acid; Ala = L-alanine; β-Ala = β-alanine; Arg = L-arginine; Asn = L-asparagine; Asp = L-aspartic acid; Cys = L-cysteine (or L-half cystine); Dbu = L-α,γ-diamino-butyric acid; Gln = L-glutamine; Glu = L-glutamic acid; Gly = glycine; His = L-histidine; Ile = L-isoleucine; aIle = L-allo-isoleucine; Leu = L-leucine; Lys = L-lysine; Met = L-methionine; Nle = L-norleucine; Nva = L-norvaline; Orn = L-ornithine; Pro = L-proline; Sar = sarcosine; Ser = L-serine; Thr = L-threonine; Trp = L-tryptophan; Tyr = L-tyrosine; Val = L-valine.

## The General Structural Pattern

If a rough characterisation of the shape of oxytocin is to be given, this molecule may be said to consist of a large cyclic moiety (6 amino acid residues) and a main peptide side-chain (3 amino acid residues). This general structure has a number of

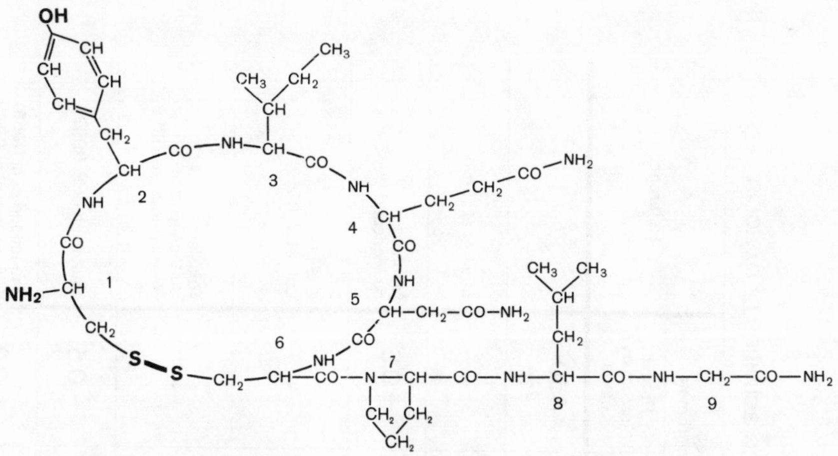

Fig. 1: The structure of oxytocin. (The chemically reactive groups are printed in bold type.)

smaller side-chains. These structural elements of the oxytocin molecule are illustrated in Figure 1.

## The Twenty-membered Ring

Studies of the importance of the size of the cyclic moiety for the oxytocin-like activities have revealed that any enlargement of this twenty-membered disulphide ring leads to loss of the specific activities. Hemihomocystine$^{1}$-oxytocin (JARVIS et al., 1961), $(\beta Ala)^{4}$-oxytocin (MANNING & Du VIGNEAUD, 1965a), Isoasparagine$^{5}$-oxytocin (LUTZ et al., 1959), Isoglutamine$^{4}$-oxytocin (RESSLER & Du VIGNEAUD, 1957), Homo-Tyr$^{2/3}$-oxytocin (GUTTMANN et al., 1957), Homo-Ile$^{3/4}$-oxytocin (JAQUENOUD et al., 1958; KONZETT & BERDE, 1958) are compounds with twenty-one, twenty-two or twenty-three atoms in the disulphide ring. All are biologically inactive as is evident from the data in Table 1.

There are fewer synthetic analogues to illustrate the influence of reducing the size of the disulphide ring on the biological activities of oxytocin. It seems nevertheless that decreasing the size of the ring from twenty to nineteen members sharply reduces all acti-

## Table 1

The influence of the enlargement of the disulphide ring on the activities of oxytocin

| Name and abbreviated chemical formula | Number of atoms in the ring | Oxytocin-like activities in International Units per micromole | | | References |
|---|---|---|---|---|---|
| | | Rat uterus | Chicken blood pressure | Rabbit mammary gland | |
| Cys-Tyr-Ile-Gln-Asn-Cys-Pro-Leu-Gly-$NH_2$<br>Oxytocin | 20 | 450 ±30 | 450 ±30 | 450 ±30 | |
| Homocys-Tyr-Ile-Gln-Asn-Cys-Pro-Leu-Gly-$NH_2$<br>Hemihomocystine$^1$-oxytocin | 21 | $\sim$ 0.75 | $<$ 0.01 | – | Jarvis et al. (1961) |
| Cys-Tyr-Ile-βAla-Asn-Cys-Pro-Leu-Gly-$NH_2$<br>(βAla)$^4$-oxytocin | 21 | $\sim$ 0.12 | $\sim$ 0.4 | $\sim$ 0.5 | Manning, du Vigneaud (1965) |
| Cys-Pro-Leu-Gly-$NH_2$<br>Cys-Tyr-Ile-Gln-Asp-$NH_2$<br>Isoasparagine$^5$-oxytocin | 21 | $<$ 0.01 | $<$ 0.05 | – | Lutz et al. (1959) |
| Asn-Cys-Pro-Leu-Gly-$NH_2$<br>Cys-Tyr-Ile-Gln-$NH_2$<br>Isoglutamine$^4$-oxytocin | 22 | $<$ 0.01 | $<$ 0.05 | – | Ressler, du Vigneaud (1957)<br>Ressler, Rachele (1958) |
| Cys-Tyr-Tyr-Ile-Gln-Asn-Cys-Pro-Leu-Gly-$NH_2$<br>Homo-Tyr$^{2/3}$-oxytocin | 23 | Anti (500:1) | Anti (500:1) | $\sim$ 0.5 | Guttmann et al. (1957) |
| Cys-Tyr-Ile-Ile-Gln-Asn-Cys-Pro-Leu-Gly-$NH_2$<br>Homo-Ile$^{3/4}$-oxytocin | 23 | $\sim$ 0.1 | $\sim$ 0.05 | $\sim$ 0.2 | Jaquenoud et al. (1958)<br>Boissonnas et al. (1961)<br>Konzett, Berde (1958) |

## Table 2

The influence of the shortening of the lateral peptide chain on the activities of oxytocin

| Name and abbreviated chemical formula | Oxytocin-like activities in International Units per micromole: Rat uterus | Chicken blood pressure | Rabbit mammary gland | References |
|---|---|---|---|---|
| Cys-Tyr-Ile-Gln-Asn-Cys-Pro-Leu-Gly-$NH_2$<br>Oxytocin | **450**<br>± 30 | **450**<br>± 30 | **450**<br>± 30 | |
| Cys-Tyr-Ile-Gln-Asn-Cys-Pro-Leu-$NH_2$<br>De-Gly⁹-oxytocin | **4.0**<br>± 0.7 | **Anti (10000:1)** | **~ 3** | Jaquenoud, Boissonnas (1962)<br>Berde, Stürmer (1962) |
| Cys-Tyr-Ile-Gln-Asn-Cys-Pro-Gly-$NH_2$<br>De-Leu⁸-oxytocin | **4.0**<br>± 1 | **Anti (1000:1)** | **~ 3** | Jaquenoud, Boissonnas (1962)<br>Berde, Stürmer (1962) |
| Cys-Tyr-Ile-Gln-Asn-Cys-Leu-Gly-$NH_2$<br>De-Pro⁷-oxytocin | **3.5**<br>± 0.8 | **Anti (1000:1)** | **1.5**<br>± 0.2 | Jaquenoud, Boissonnas (1962)<br>Berde, Stürmer (1962) |
| Cys-Tyr-Ile-Gln-Asn-Cys-$NH_2$<br>De-(Pro-Leu-Gly)⁹-oxytocin | **~ 2.4** | **< 0.02** | **~ 0.8** | Ressler (1956) |

vities: Mercapto-acetic acid$^{1}$-oxytocin (JARVIS & Du VIGNEAUD, 1967), an analogue of De-amino$^{1}$-oxytocin with a nineteen-membered ring displayed the following, greatly attenuated activities per milligram: 25 I.U. uterotonic, 4 I.U. avian depressor and 50 I.U. milk-ejecting activity. The corresponding potencies of De-amino$^{1}$-oxytocin are 10 to 250 times higher (see Tab.5)

Thus the data available at the present time seem to indicate that the size of the cyclic moiety of oxytocin, the twenty-membered ring, may well be the "optimal" size. Deviation from this optimum is deleterious to biological activity. An increase in size seems to be less compatible with the retention of some activities than a moderate decrease in size.

## The Tripeptide Side-chain

Synthetic analogues with a lengthened peptide side-chain have not so far been investigated. The activities of synthetic analogues with a shortened peptide side-chain would appear to indicate that the length of this peptide chain attached to the disulphide ring in position 6, i.e. the distance between the half cystine residue and the terminal amide group is "optimal" or at least "critical": Any shortening practically eliminated all oxytocin-like biological activities: De-Gly$^{9}$-oxytocin, De-Leu$^{8}$-oxytocin, De-Pro$^{7}$-oxytocin (JAQUENOUD & BOISSONNAS, 1962; BERDE & STÜRMER, 1962) as well as De-(Pro-Leu-Gly)$^{9}$-oxytocin, i.e. the amidified hexapeptide ring of oxytocin (RESSLER, 1956) are all more than 100 times less active than oxytocin, as shown by the data contained in Table 2. Indeed several of these compounds have some oxytocin antagonistic effect on the chicken blood pressure.

## The Smaller Side-chains

The disulphide ring as well as the tripeptide side-chain of oxytocin have a number of smaller side-chains, e.g. in positions 1, 2, 3, 4, 5 and 8 (see Fig. 1). These side-chains have been much manipulated in several laboratories in order to study the influence of limited chemical variations on biological properties. Neither time nor space allow us to discuss here the huge body of detailed information collected. Those interested are referred to recent review articles which contain tabulated data (e.g. BERDE & BOISSONNAS, 1966, 1968). Generally speaking it may be said that numerous modifications of the natural structure are compatible with the conservation of considerable biological activity.

As an example the influence of "stepwise" shortening of the side-chain in position 8 is given in Table 3, which also contains the relevant references. Roughly speaking it may be said that on reducing the side-chain in this position the oxytocin-like activities are attenuated but by no means eliminated. Even after replacement of the original isobutyl structure by a methyl group ($Ala^8$-oxytocin) activities in excess of 100 I.U. per micromole remain and after the removal of the whole side-chain ($Gly^8$-oxytocin) a milk-ejecting activity 10% that of oxytocin is still present.

Many other examples, which cannot be enumerated here, show that limited chemical modifications of the smaller side-chains are often compatible with relatively high biological activities. Moderate reduction of the size of a naturally occurring side-chain appears to be less deleterious to the biological effects in many cases than increasing the size of (lengthening) these side-chains. The effect of lengthening a side-chain is particularly well documented for position 1 of oxytocin: A number of analogues ranging from (N-Me)$Cys^1$-oxytocin (JOŠT et al., 1963), which represents a modest increase in length to (Leu-Gly-Gly-$Cys)^1$-oxytocin (JOŠT et al., 1963) which has a very long side-chain, has been studied. They show very low oxytocin-like activities. (The possibility that these compounds may act as "hormonogens" in the body does not concern us here.)

## D-amino Acids

All amino acid residues in oxytocin are in the L-configuration. The replacement of an L-amino acid residue by the same residue in D-configuration is in a sense manipulation of a side-chain: The modification affecting not its length but its position in space. As is evident from the figures in Table 4, which also lists the relevant literature, oxytocin analogues containing D-amino acid residues in lieu of the same L-amino acid residues show little or hardly any biological activities. It seems therefore that the

Table 3

The influence of the shortening of the side-chain in position 8 on the activities of oxytocin

| Name and abbreviated chemical formula | Structure of amino acid in position 8 | Oxytocin-like activities in International Units per micromole | | | References |
|---|---|---|---|---|---|
| | | Rat uterus | Chicken blood pressure | Rabbit mammary gland | |
| Cys-Tyr-Ile-Gln-Asn-Cys-Pro-Leu-Gly-$NH_2$<br>Oxytocin | $CH_3$ $CH_3$<br>CH<br>$CH_2$<br>-HN-CH-CO- | **450**<br>± 30 | **450**<br>± 30 | **450**<br>±30 | |
| Cys-Tyr-Ile-Gln-Asn-Cys-Pro-Ile-Gly-$NH_2$<br>Ile$^8$-oxytocin = mesotocin | $CH_3$<br>$CH_3$ $CH_2$<br>CH<br>-HN-CH-CO- | **291**<br>±21 | **502**<br>±37 | **330**<br>± 21 | Jaquenoud, Boissonnas (1961)<br>Berde, Konzett (1960) |
| Cys-Tyr-Ile-Gln-Asn-Cys-Pro-Val-Gly-$NH_2$<br>Val$^8$-oxytocin | $CH_3$ $CH_3$<br>CH<br>-HN-CH-CO- | **199**<br>±15 | **278**<br>±17 | **308**<br>±20 | Jaquenoud, Boissonnas (1961)<br>Berde et al. (1960) |
| Cys-Tyr-Ile-Gln-Asn-Cys-Pro-Abu-Gly-$NH_2$<br>Abu$^8$-oxytocin | $CH_3$<br>$CH_2$<br>-HN-CH-CO- | – | **430**<br>± 31 | – | Jaquenoud (1965)<br>Berde, Stürmer (1965) |
| Cys-Tyr-Ile-Gln-Asn-Cys-Pro-Ala-Gly-$NH_2$<br>Ala$^8$-oxytocin | $CH_3$<br>-HN-CH-CO- | **136**<br>±17 | **130**<br>±13 | **201**<br>±8 | Jaquenoud (1965)<br>Berde, Stürmer (1964) |
| Cys-Tyr-Ile-Gln-Asn-Cys-Pro-Gly-Gly-$NH_2$<br>Gly$^8$-oxytocin | H<br>-HN-CH-CO- | **15**<br>±1.4 | **7**<br>±1.4 | **44**<br>±7 | Jaquenoud (1965)<br>Berde, Stürmer (1965) |

Table 4

The influence of the replacement of an L-amino acid residue by a D-amino acid residue on the activities of oxytocin

| Name and abbreviated chemical formula | Oxytocin-like activities in International Units per micromole | | | References |
|---|---|---|---|---|
| | Rat uterus | Chicken blood pressure | Rabbit mammary gland | |
| Cys-Tyr-Ile-Gln-Asn-Cys-Pro-Leu-Gly-$NH_2$<br>Oxytocin | 450 ± 30 | 450 ± 30 | 450 ± 30 | |
| D-Cys-Tyr-Ile-Gln-Asn-Cys-Pro-Leu-Gly-$NH_2$<br>D-Cys[1]-oxytocin | 1.9 ± 0.1 | ~ 0.2 | 6.2 ± 0.2 | Jošt et al. (1963)<br>Hope et al. (1963)<br>Yamashiro et al. (1966) |
| Cys-D-Tyr-Ile-Gln-Asn-Cys-Pro-Leu-Gly-$NH_2$<br>D-Tyr[2]-oxytocin | 6.6 ± 1 | 34 ± 3 | 34 ± 9 | Drabarek, du Vigneaud (1965) |
| Cys-Tyr-Ile-D-Gln-Asn-Cys-Pro-Leu-Gly-$NH_2$<br>D-Gln[4]-oxytocin | ~ 0.1 | ~ 1.0 | – | Dutta, Anand (1965)<br>Dutta, Anand (1966) |
| Cys-Tyr-Ile-Gln-D-Asn-Cys-Pro-Leu-Gly-$NH_2$<br>D-Asn[5]-oxytocin | ~ 0.2 | ~ 0.36 | – | Dutta, Anand (1966) |
| Cys-Tyr-Ile-Gln-Asn-D-Cys-Pro-Leu-Gly-$NH_2$<br>D-Cys[6]-oxytocin | ~ 0.62 | <0.01 | < 0.002 | Manning, du Vigneaud (1965) |
| Cys-Tyr-Ile-Gln-Asn-Cys-Pro-D-Leu-Gly-$NH_2$<br>D-Leu[8]-oxytocin | ~ 20 | ~ 20 | ~ 50 | Schneider,<br>du Vigneaud (1962) |

L-configuration of the amino acid residues is of considerable importance for the biological effects of oxytocin.

All-D-oxytocin, i.e. a molecule corresponding to oxytocin but built up from D-amino acids has also been synthetised (FLOURET & Du VIGNEAUD, 1965). This compound is biologically inactive.

## The Chemically Reactive Groups

The oxytocin molecule contains three chemically reactive groups: a free amino group in position 1, a phenolic hydroxy group in position 2 and a disulphide group in position 1/6 (see Figure 1). None of these groups is essential for the biological activities of the hormone. In Table 5 synthetic analogues of oxytocin are listed, from which one of these chemically reactive groups is absent. The table also gives their pharmacological activities and references to the relevant publications. The de-amino-derivative is more active than oxytocin itself. The de-oxy-derivative ($Phe^2$-oxytocin) is less active than the natural hormone but its potencies are still considerable. One of the sulphur atoms of the disulphide bond can be replaced by a methylene group without complete loss of the characteristic biological activities.

Even after the disulphide bond has been reduced to - SH HS - ,this opening the twenty-membered ring, some biological activity still remains: this compound, oxytoceine, is reported to have 80 I.U./mg avian depressor activity (YAMASHIRO et al., 1966). This figure is higher than the comparable figures for analogues with an enlarged disulphide ring (see Table 1). There is, of course, reason to believe that oxytoceine may have a tertiary structure which is similar to the structure of oxytocin in shape and size.

## CONCLUSIONS

From the three dominant oxytocin-like pharmacological qualities of a considerable number of synthetic analogues and homologues of oxytocin known to-day, the following tentative conclusions may be drawn as to the relative importance of certain parts of the oxytocin

Table 5

The influence of the elimination of one of the three chemically reactive groups on the activities of oxytocin

| Name and abbreviated chemical formula | Oxytocin-like activities in International Units per micromole | | | References |
|---|---|---|---|---|
| | Rat uterus | Chicken blood pressure | Rabbit mammary gland | |
| Cys-Tyr-Ile-Gln-Asn-Cys-Pro-Ile-Gly-$NH_2$ <br> Oxytocin | **450** <br> ± 30 | **450** <br> ± 30 | **450** <br> ± 30 | |
| S-$(CH_2)_2$-CO-Tyr-Ile-Gln-Asn-Cys-Pro-Leu-Gly-$NH_2$ <br> De-Amino[1]-oxytocin | **795** <br> ± 36 | **965** <br> ± 24 | **536** <br> ± 13 | du Vigneaud et al. (1960) <br> Hope et al. (1962) <br> Chan, du Vigneaud (1962) <br> Ferrier et al. (1965) |
| Cys-Phe-Ile-Gln-Asn-Cys-Pro-Leu-Gly-$NH_2$ <br> Phe[2]-oxytocin | **32** <br> ± 2 | **63** <br> ± 9 | **140** <br> ± 21 | Jaquenoud, Boissonnas (1959) <br> Konzett, Berde (1959) |
| $(CH_2)_3$-CO-Tyr-Ile-Gln-Asn-Cys-Pro-Leu-Gly-$NH_2$ <br> Methylene[1]-De-Thio[1]-oxytocin | **~ 60** | **~ 25** | **–** | Rudinger, Jošt (1964) |

molecule for the biological activities of this hormone:
The size of the twenty-membered disulphide ring seems to be "optimal".
The length of the tripeptide side-chain is "critical" in the sense that any shortening is deleterious.
Many limited modifications of the smaller side-chains are compatible with the conservation of considerable biological activities.
The L-configuration of the amino acid residues is of great importance.
None of the three chemically reactive groups of the molecule (free amino group, phenolic hydroxy group, disulphide bond) are essential for the biological activities.

## LITERATURE

BERDE B., BOISSONNAS R.A.: Synthetic analogues and homologues of the posterior pituitary hormones. In: The Pituitary Gland (G.W.HARRIS, B.T.DONOVAN, ed.). Butterworths, London 1966, p.624.

BERDE B., BOISSONNAS R.A.: Basic pharmacological properties of synthetic analogues and homologues of the neurohypophysial hormones. In: Neurohypophysial Hormones and Similar Polypeptides (B.BERDE, ed.), Vol.XXIII of "Handbuch der Experimentellen Pharmakologie". Springer, Berlin 1968 (in press).

BERDE B., CERLETTI A.: Acta endocr. 34, 543 (1960).

BERDE B., KONZETT H.: Med.Exp. 2, 317 (1960).

BERDE B., SAAMELI K.: Evaluation of substances acting on the uterus. In: Methods in Drug Evaluation (P.MANTEGAZZA, F.PICCININI, ed.). North-Holland Publ.Co., Amsterdam 1966, p.481.

BERDE B., STÜRMER E.: Personal communication (1962).

BERDE B., STÜRMER E.: Personal communication (1964).

BERDE B., STÜRMER E.: Personal communication (1965).

BERDE B., KONZETT H., STÜRMER E.: Personal communication (1960).

BOISSONNAS R.A., GUTTMANN St., BERDE B., KONZETT H.: Experientia 17, 377 (1961).

CHAN W.Y., Du VIGNEAUD V.: Endocrinology 71, 977 (1962).

COON J.M.: Arch.int.pharmacodyn. 62, 79 (1939).

DRABAREK S., Du VIGNEAUD V.: J.Amer.chem.Soc. 87, 3974 (1965).

DUTTA A.S., ANAND N.: Ind.J.Chem. 3, 232 (1965).

DUTTA A.S., ANAND N., KAR K.: J.Med.Chem. 9, 497 (1966).

Van DYKE H.B., ADAMSONS K.Jr., ENGEL S.L.: Recent Progr.Hormone Res. 11, 1 (1955).

FERRIER B.M., JARVIS D., Du VIGNEAUD V.: J.Biol.Chem. 240, 4264 (1965).

FLOURET G., Du VIGNEAUD V.: J.Amer.chem.Soc. 87, 3775 (1965).

GUTTMANN St., JAQUENOUD P.-A., BOISSONNAS R.A., KONZETT H., BERDE B.: Naturwiss. 44, 632 (1957).

HOPE D.B., MURTI V.V.S., Du VIGNEAUD V.: J.Biol.Chem. 237, 1563 (1962).

HOPE D.B., MURTI V.V.S., Du VIGNEAUD V.: J.Amer.chem. Soc. 85, 3686 (1963).

JAQUENOUD P.-A.: Helv.chim.Acta 48, 1899 (1965).

JAQUENOUD P.-A., BOISSONNAS R.A.: Helv.chim.Acta 42, 788 (1959).

JAQUENOUD P.-A., BOISSONNAS R.A.: Helv.chim.Acta 44, 113 (1961).

JAQUENOUD P.-A., BOISSONNAS R.A.: Helv.chim.Acta 45, 1462 (1962).

JAQUENOUD P.-A., GUTTMANN St., BOISSONNAS R.A.: Personal communication (1958).

JARVIS D., Du VIGNEAUD V.: J.Biol.Chem. 242, 1768 (1967).

JARVIS D., BODANSZKY M., Du VIGNEAUD V.: J.Amer.chem. Soc. 83, 4780 (1961).

JOŠT K., RUDINGER J., ŠORM F.: Coll.Czechoslov.Chem. Commun. 28, 2021 (1963).

KONZETT H., BERDE B.: Brit.J.Pharmacol. 14, 133 (1959).

KONZETT H., BERDE B.: Personal communication (1958).

LUTZ W.B., RESSLER C., NETTLETON D.E.Jr., Du VIGNEAUD V.: J.Amer.chem.Soc. 81, 167 (1959).

MANNING M., Du VIGNEAUD V.: Biochemistry 4, 1884 (1965a).

MANNING M., Du VIGNEAUD V.: J.Amer.chem.Soc. 87, 3978 (1965b).

RESSLER Ch.: Proc.Soc.Exper.Biol.Med. 92, 725 (1956).

RESSLER Ch., RACHELE J.R.: Proc.Soc.Exper.Biol.Med. 98, 170 (1958).

RESSLER Ch., Du VIGNEAUD V.: J.Amer.chem.Soc. 79, 4511 (1957).

RUDINGER J., JOŠT K.: Experientia 20, 570 (1964).

SCHNEIDER C.H., Du VIGNEAUD V.: J.Amer.chem.Soc. 84, 3005 (1962).

Du VIGNEAUD V., WINESTOCK G., MURTI V.V.S., HOPE D.B., KIMBROUGH R.D.Jr.: J.Biol.Chem. 235, PC 64 (1960).

YAMASHIRO D., GILLESSEN D., Du VIGNEAUD V.: Biochemistry 5, 3711 (1966a).

YAMASHIRO D., GILLESSEN D., Du VIGNEAUD V.: J.Amer. chem.Soc. 88, 1310 (1966b).

# OXYTOCIN AND VASOPRESSIN: BIOCHEMICAL CONSIDERATIONS

J. Rudinger, V. Pliška, I. Rychlík and F. Šorm

Institute of Organic Chemistry and Biochemistry
Czechoslovak Academy of Science, Prague

Hormone action under physiological conditions is a complex process of many stages, including biosynthesis of the hormone, release, transport and distribution, excretion, and metabolic process (inactivation and sometimes activation) as well as the interaction of the hormone with its tissue receptor and the processes triggered by this interaction. The relations between structure and biological activity of the peptide hormones and their synthetic analogues have been analysed mainly in terms of the interactions at the receptor. This paper will be concerned with some approaches to the metabolic phase of hormonal regulation involving the design of synthetic analogues of oxytocin and vasopressin directed specifically toward this end.

Because of the metabolic lability of peptide bonds, metabolic inactivation is likely to play a particularly important role in the overall action of the peptide hormones. Structural alterations in the hormone molecule which increase metabolic stability would be expected to increase the duration of hormone action. Oxytocin is well known to be inactivated by an enzyme present in the serum of pregnant primates ("serum oxytocinase"; see (1) but also more generally by homogenates of various tissues including the kidneys and liver. The major pathway of oxytocin inactivation by liver cell-sap has been shown to involve reductive fission of the disulphide bond followed by aminopeptidase degradation of the peptide chain (2). In an effort to obtain peptides resistant to aminopeptidase action we introduced D-hemicystine and N-methylhemicystine into the amino terminal position of oxytocin in place of L-hemicystine (3) (see also (4)). The potent analogue, deamino-oxytocin (Ia), obtained by du Vigneaud and his coworkers (5) is also resistant to aminopeptidases (6). Nevertheless, the biological responses to all these analogues show the same time course as the responses to oxytocin (7, 8).

These results appeared to exclude aminopeptidase degradation as the process responsible for the decay of the physiological response; there yet remained the possibility that it was reduction of the disulphide bond which was the critical process. In the course of investigations on the molecular mechanism of neurohypophysial hormone action we have synthesized analogues isosteric with deamino-oxytocin (Ia) in which one or other of the sulphur atoms in the disulphide bridge, or both these sulphur atoms, are replaced by methylene groups (9, 10, 11). The first two of these analogues which for convenience may be called 1-deamino-1-carba-oxytocin (Ib) and 1-deamino-6-carba-oxytocin (Ic) show oxytocin-like activity of the same order as the parent hormone whereas the sulphur-free 1-deamino-1, 6-dicarba-oxytocin (Id) shows lower but qualitatively normal activity. What is, however, important in the present context is the finding that in the assay for antidiuretic activity, which is particularly suitable for following the time course of the effect, the responses to the "carba" analogues show exactly the same course as the responses to oxytocin (12). This can be established both by direct inspection of the experimental records and by analysis of the dose-response curves. We are therefore forced to the conclusion that it is not reduction of the disulphide bond, either, which governs the rate at which the response fades under physiological conditions.

Two possible explanations of this situation come to mind. Either, there is some as yet unrecognised enzyme or enzyme system to which oxytocin, its deamino-analogues, and the carba-analogues are all equally susceptible (e. g., an endopeptidase or amidase) and which plays a decisive role in the overall inactivation; or, the invariance of the time course of the response is the result of some particular pharmacokinetic situation which makes this time course independent of the enzymic inactivation of the hormone.

The second possibility has been examined by computation (13). A three-compartment model was considered consisting of a vascular space in communication on the one hand with a receptor compartment and on the other hand with an extravascular compartment (excluding the receptor compartment). Elimination

$$\begin{array}{l} CH_2 \text{———————} X \text{———————} CH_2 \\ | \qquad\qquad\qquad\qquad\qquad\qquad\qquad\quad | \\ CH_2.\,CO\text{-}Tyr\text{-}Ile\text{-}Gln\text{-}Asn\text{-}Nh.\,CH.\,CO\text{-}Pro\text{-}Leu\text{-}Gly\text{-}NH_2 \end{array}$$

Ia: X = S-S Ic: X = S-$CH_2$

Ib: X = C-$CH_2$ Id: X = $CH_2$-$CH_2$

from each compartment )due to excretion, or inactivation, or both) was characterised by rate constants kappa, transfer between the compartments constants k. The drug was assumed to be introduced into the vascular compartment and the time course of the response, expressed as an appropriate function of the drug concentration in the receptor compartment, was plotted by an analogue computer for various combinations of kappas, k's and parameters of the concentration-response relation. By a suitable choice of constants (some of them based on experimental data) the time course of an antidiuretic response to oxytocin could be imitated very accurately. It now transpored that although any change in the assumed rate of elimination of the hormone from the receptor compartment profoundly affected the time-response curve, changes in the rate of elimination from the extravascular compartment had much less effect and variations in kappa for the vascular compartment affected the maximal intensity of the computed response but not the general shape of the curve. It thus appears that under certain circumstances even a considerable decrease in the rate of inactivation of a drug may not lead to a protracted response but may instead result in an apparent increase in potency. This model, therefore, could account not only for the failure of enzyme-resistant analogues of oxytocin to show protracted activity but also for the remarkably high potencies of deamino analogues of the hormone (5, 7) at least *in vivo*. It should be stressed that these computations, based as they are on a wealth of hypothesis, merely demonstrate that such an explanation is possible in principle but do not themselves provide evidence that it is correct.

On the other hand, it has also been found (14) that 1-deamino-1-carba-oxytocin (Ib) is inactivated by liver homogenates, though very much more slowly than oxytocin. This finding demonstrates the existence of mechanisms other than disulphide reduction and aminopeptidase action capable of inactivating oxytocin-like peptides. The biochemical features and physiological significance of these mechanisms remain to be examined.

Metabolic activation may be a feature in the biosynthesis of the neurohypophysial hormones (15) but there is no evidence for an activation step after release of the hormones from the neurohypophysis. However, a phase of metabolic activation may be artificially introduced by constructing analogues with "hormonogen" characteristics - derivatives which themselves have little or no hormone-like activity but are capable of releasing active hormone under physiological conditions. We have prepared a whole series of such derivatives of oxytocin (II) with additional amino-acid residues or short peptide chains attached to the terminal amino group (3,16,17). All these compounds showed the expected protracted action when examined in vivo for uterotonic, milk-ejecting, antidiuretic, and vasodepressor activity (8). Evidence that this effect is, indeed, due to enzymic conversion of the analogue to the hormone was obtained in a variety of ways, including structure-activity relations within the series of hormonogens (8) and enzymic activation experiments in vitro (18,19). An unexpected property shown by the hormonogens was their ability to inhibit the response to oxytocin in the milk-ejection and avian depressor assays but this property, too, can be explained by the gradual release of oxytocin - it can be closely mimicked by infusing oxytocin at suitable rates (8).

A similar series of derivatives of lysine vasopressin (III) (17,20) also has protracted antidiuretic and pressor activity in the rat (21). On the other hand, an analogue (IV) in which a leucyl residue was attached to the epsilon amino group of lysine in lysine vasopressin (17) did not appear to have protracted action.

Generally the response to the hormonogens not only decays more slowly but also develops more slowly than the response to oxytocin. Occasionally, however, a two-phase effect is seen: The response develops and begins to fade with the same time course as the response to oxytocin but before returning to baseline it turns into a typical protracted response. We believe that this type of effect is due to an initial response to the analogue itself followed, when the hormone concentration has been built up to a sufficient level, by the effect of the released hormone (8). Such a response would be favoured when the inherent activity of the analogue is high, or when its conversion to the hormone (particularly in the receptor compartment) is slow, or when both conditions are met. By an extension of the three-compartment model discussed above to include distribution and elimination of the precursor as well as the drug, and also rate constants for conversion of the precursor to the drug, the time course of the response to the hormonogens can be successfully modelled, including the two-phase type of response (13). Again, such an

X-Cys-Tyr-Ile-Gln-Asn-Cys-Pro-Leu-Gly-$NH_2$

II

X = Gly, Leu, Phe, Lys, Pro, Gly-Gly, Leu-Leu, Gly-Gly-Gly, Leu-Gly-Gly

X-Cys-Tyr-Phe-Gln-Asn-Cys-Pro-Lys-Gly-$NH_2$

III

X = Gly, Ala, Leu, Phe, Tyr, Trp, Pro, Gly-Gly, Gly-Pro, Leu-Leu, Sar-Gly, Gly-Gly-Gly

Leu

Cys-Tyr-Phe-Gln-Asn-Cys-Pro-Lys-Gly-$NH_2$

IV

agreement between the computed and experimentally found time course of the response does no more than demonstrate the plausibility of the explanation offered; further work will show whether features predicted by the model can be experimentally confirmed.

A more thorough study of the vascular responses to hormonogen analogues of oxytocin and vasopressin, more particularly derivatives in which glycine was one of the "additional" amino acids, suggested that these compounds might have favourable effects on the circulation in haemorrhagic shock and this prediction has been confirmed experimentally (22). Further work, using chiefly $N^{\alpha}$-glycyl-glycyl-clycyl-8-lysine-vasopressin ("glypressin"), has shown that single injections of this peptide increase the survival rates of rats and monkeys in haemorrhagic shock (23) and of rats in traumatic shock (24). The favourable effect of glypressin on the circulatory and metabolic parameters of dogs in irreversible haemorrhagic shock has been confirmed in detailed studies (25) and it has been found that the peptide protects against the effect of lethal doses of histamine (26). Most of these actions cannot be adequately duplicated by lysine vasopressin.

It therefore appears that deliberate modification of the metabolic phases of hormone action, together perhaps with modification of the distribution behaviour, may lead to qualitatively altered properties of both theoretical and practical interest.

## REFERENCES

1. H. Tuppy and E. Wintersberger, in Oxytocin, Vasopressin and their Structural Analogues (ed. J. Rudinger), p. 143. Proc. 2nd Intern. Pharmacol. Meeting, Prague 1963, Vol. 10. Czechoslovak Medical Press and Pergamon Press, Prague and Oxford, 1964.

2. I. Rychlík, In Oxytocin, Vasopressin and their Structural Analogues (ed. J. Rudinger), p. 153. Proc. 2nd Intern. Pharmacol. Meeting, Prague 1963, Vol. 10. Czechoslovak Medical Press and Pergamon Press, Prague and Oxford 1964.

3. K. Jošt, J. Rudinger and F. Šorm, Collection Czechoslov. Chem. Commun. 26, 2496 (1961); 28, 1706 (1963).

4. D. B. Hope, V. V. S. Murti and V. du Vigneaud, J. Am. Chem. Soc. 85, 3886 (1963).

5. D. B. Hope, V. V. S. Murti and V. du Vigneaud, J. Biol. Chem. 237, 1563 (1962).

6. J. Golubow and V. du Vigneaud, Proc. Soc. Exptl. Biol. Med. 112, 218 (1963); J. Golubow, W. Y. Chan and V. du Vigneaud, Proc. Soc. Exptl. Biol. Med. 113, 113 (1963).

7. W. Y. Chan and V. du Vigneaud, Endocrinology 71, 977 (1962).

8. Z. Beránková-Ksandrová, G. W. Bisset, K. Jošt, I. Krejčí, V. Pliška, J. Rudinger, I. Rychlík and F. Šorm, Brit. J. Pharmacol. 26, 615 (1966).

9. J. Rudinger and K. Jošt, Experientia 20, 570 (1964).

10. K. Jošt and J. Rudinger, Collection Czechoslov. Chem. Commun. 32, 1229 (1967).

11. K. Jošt and P. Moritz, unpublished results.

12. V. Pliška, J. Rudinger, T. Dousa and J. H. Cort, Am. J. Physiol., in press.

13. V. Pliška, unpublished results.

14. V. Pliška and I. Gašparović, unpublished results.

15. H. Sachs, Am. J. Med. 42, 687 (1967).

16. E. Kasafírek, K. Jošt, J. Rudinger and F. Šorm, Collection Czechoslov. Chem. Commun. 30, 2600 (1965).

17. E. Kasafírek, V. Rábek and J. Rudinger, unpublished results.

18. Z. Beránková-Ksandrová, I. Rychlík and F. Šorm, in Oxytocin, Vasopressin and their Structural Analogues (ed. J. Rudinger), p. 181, Proc. 2nd Intern. Pharmacol. Meeting, Prague 1963, Vol. 10. Czechoslovak Medical Press and Pergamon Press, Prague and Oxford 1964.

19. V. Pliška, J.H. Cort, T. Douša, I. Rychlík and F. Šorm, unpublished results.

20. E. Kasafírek, V. Rábek, J. Rudinger and F. Šorm, Collection Czechoslov. Chem. Commun. 31, 4581 (1966).

21. J. Kynčl, V. Pliška and V. Jelínek, Českoslov. Fysiol. 15, 398 (1966); K. Řežábek and J. Kynčl, Českoslov. Fysiol. 15, 399 (1966); J. Kynčl and K. Řežábek, unpublished results.

22. J.H. Cort, J. Hammer, M. Ulrych, Z. Píša, T. Dousa and J. Rudinger, Lancet 1964, II, 840.

23. V. Trčka and J. Hladovec, Abstr. 3rd Intern. Pharmacol. Congress, Sao Paulo, 1966.

24. V. Lichardus, M. Vigaš, Š. Németh and J. Rudinger, Českoslov. Fysiol. 15, 521 (1966).

25. J.H. Cort, M.R. Jeanjean, A.E. Thomson and M. Nickerson, Am. J. Physiol., in press.

26. K. Řežábek, La Thérapei, in press.

# NEUROPHYSIN, OXYTOCIN AND VASOPRESSIN IN NEUROSECRETORY GRANULES AND IN CRYSTALLINE COMPLEXES

D. B. Hope

Department of Pharmacology

University of Oxford

Sometime ago it was noticed in this laboratory that neurophysin was composed of several proteins separable by electrophoresis in starch gels.[1] Previously bovine neurophysin had been considered to be a homogeneous protein[2] which could bind both of the pituitary polypeptide hormones, oxytocin and vasopressin. A complex containing all three constituents isolated by van Dyke and his coworkers[3] possessed oxytocic and pressor activities in a ratio of 1:1. Landgrebe, Ketterer and Waring[4] pointed out that the biological activities of the van Dyke protein corresponded to a complex of a Mole of oxytocin and of vasopressin bound per Mole of protein. The fact that neurophysin was a mixture of proteins was of interest because of evidence for the independent release of oxytocin and vasopressin [5,6]. If this interpretation were to prove correct then the all-or-none nature of neurone activation requires that oxytocin cannot be stored in the same granule as vasopressin. No histological methods for distinguishing oxytocin from vasopressin in neurones exists. A biochemical approach would be the isolation of two kinds of neurosecretory granules, the one containing only oxytocin and the other containing only vasopressin. Although there is some evidence that vasopressin containing granules sediment more readily than those containing oxytocin when centrifuged in a sucrose density gradient[7,8], nothing like a complete separation had been achieved. If neurophysin were a homogeneous protein, the "carrier" would be found in all the neurosecretory granules, as a complex with oxytocin in one location and with vasopressin in another. The fact that on electrophoresis in a starch gel neurophysin gave rise to several protein bands suggested to us that a second biochemical approach was possible. One constituent of neurophysin might be specific for oxytocin and another for vasopressin as suggested in 1963 [9]. However, there was at the time we began this work no good evidence for the localization of neurophysin within the same granules as the hormones. I will, therefore, discuss the results obtained by my two collaborators Mr. C.R. Dean and Mr. M.D. Hollenberg by pursuing two approaches to the problem of the biological significance of neurophysin.

## FRACTIONATION OF NEUROPHYSIN

Hormone-protein complex was isolated from acetone-desiccated bovine pituitary posterior lobes [2, 10]. Gel-filtration on Sephadex G-25 under mild acid conditions separated the hormones from the protein "crude neurophysin". Starch gel electrophoresis was carried out during the early stage of the work by the method of Smithies [11] with the buffer devised by Poulik [12]. This system revealed the presence of four components [1, 13]. Amino acid analysis of an acid hydrolysate of "crude neurophysin" revealed the presence not only of cystine but also of methionine and in addition tryptophan was found in an alkali hydrolysate. These results were surprising because the protein isolated by van Dyke and his coworkers [3] was reported to be entirely free from methionine and tryptophan [14, 15]. "Crude neurophysin" was further fractionated by molecular-sieve chromatography and then the fraction possessing affinity for the hormones was submitted to ion-exchange chromatography [16, 17].

Two fractions were obtained by chromatography on a column of Sephadex G-75. The first emerged with the void volume and contained protein with a molecular weight greater than 50,000; this protein appeared to lack affinity for either oxytocin or vasopressin. The second fraction, accounting for over 80 per cent of the protein recovered had a molecular weight of less than 50,000 and in dialysis experiments bound both hormones. Amino acid analysis of hydrolysates of the two fractions of protein showed that all of the tryptophan originally present in "crude neurophysin" could be recovered in the protein with the higher molecular weight. On the other hand methionine was present in both fractions. Further, the amino acid analyses showed that the protein of higher molecular weight was relatively poor in cystine, was clearly not a neurophysin and has not been studied further. Ion-exchange chromatography on a column of CM Sephadex C-50 resolved the main fraction into five subfractions. A major subfraction accounted for more than 50 per cent of the protein recovered. Amino acid analysis showed that in composition all the subfractions were extremely similar: more than 10 per cent of the weight of each protein was accounted for by cystine, and all appeared to possess a strong affinity for the polypeptide hormones. The major fraction was studied in greater detail partly because of the quantity available and also it contained 93 per cent of the methionine found in the lower molecular weight fraction from Sephadex G-75. In this and other respects e.g. histidine was absent, it was different from the so-called van Dyke protein. We decided to call the major sub-fraction neurophysin-M until more was known about it and a better name could be found. Neurophysin-M behaved as a homogeneous protein in the ultracentrifuge and assuming a partial specific volume of 0.749 ml. per g. a molecular weight of 23,000 was calculated. During the course of the work we began to study the electrophoretic behaviour of the proteins in the buffer devised by Ferguson and Wallace [18]. In this system neurophysin-M was resolved into two closely adjacent bands. The complexity of "crude neurophysin" as revealed by ion-exchange chromatography and starch gel electrophoresis was puzzling.

## ISOLATION OF PURIFIED NEUROSECRETORY GRANULES FROM BOVINE PITUITARY POSTERIOR LOBES

While the work described above was in progress the fractionation by differential and density gradient centrifugation of homogenates of bovine pituitary posterior lobes led to the isolation of purified neurosecretory granules [19, 20]. The posterior lobes taken from fresh bovine pituitary glands were homogenized in 0.3M-sucrose. The nuclei, blood cells and cell debris were removed by centrifugation at 15,000 g-min. The supernatant was centrifuged and two particulate fractions were collected: Fraction II sedimented between 15,000 g-min and 87,000 g-min, and Fraction III sedimented between 87,000 g-min. and 543,000 g-min. Most of the hormonal activities present in the tissue was recovered in these particulate fractions. Fraction III was selected for further fractionation because the proportion of mitochondria and lysosomes to neurosecretory granules was low compared with Fraction II and with whole tissue. Fraction III was resuspended in 0.3M-sucrose solution, layered over a linear gradient ranging in density from 0.9 to 2M-sucrose and centrifuged for 1 hr. at $1.45 \times 10^5$g. Small fractions were collected from the bottom of the tube through a hypodermic needle inserted into the plastic centrifuge tube. Although the distributions of pressor and succinate dehydrogenase activities were similar we found that about half of the hormonal activities in regions of density greater than 1.9 g/ml with only a small proportion of mitochondria. A discontinuous density gradient was used to collect the neurosecretory granules in a small volume. This gradient behaved better than expected and a Sub-fraction (C) consisting of neurosecretory granules free from contamination with mitochondria or lysosomes was obtained. The advantage of using the discontinuous gradient is almost certainly due to the exposure of neurosecretory granules to a higher concentration of sucrose (1.3M at the top) than was used in the continuous gradient (0.9M at the top). Thus, although only approximately 6 per cent of the total hormonal activities of the gland was recovered in

Sub-fraction C highly purified neurosecretory granules containing $11.61 \pm 1.30$ units of oxytocic and $10.73 \pm 1.74$ units of pressor activity per mg of protein were obtained. The soluble granule proteins were extracted by lysing Sub-fraction C in a hypotonic solution at pH 8.0. The insoluble material was removed by centrifugation and the soluble proteins were separated by electrophoresis in a starch gel with the buffer devised by Ferguson & Wallace [18]. The pattern of bands showed that the granule protein consisted of two main components with traces of some others.

When the electrophoretogram of granule protein was compared with that of neurophysin it was seen that only one intense band was common to both. The most surprising finding was that neurophysin contained four constituents which were not present in neurosecretory granules. Further analysis of the other sub-cellular fractions isolated from tissue homogenates by differential and density gradient centrifugation failed to reveal the presence of the four missing constituents. However, when Fractions II and III were subjected to the conditions used by Acher _et al._ [2] and van Dyke _et al._ [3], where the pH of the extract was in both cases close to 4, the missing constituents appeared. The nature of the reaction responsible has not been elucidated but pH 4 is close to the optimum for the activity of cathepsin a group of proteolytic enzymes present in lysosomes. Cathepsin activity is present in the bovine pituitary posterior lobe [7, 20, 21, 22] and the activity was found to persist in acetone desiccated tissue. The activity of cathepsin was irreversibly destroyed at pH 1.6 when an acetone powder was extracted in 0.1 N-HCl. Electrophoresis of extracts of bovine posterior lobe tissue prepared in this way revealed the presence of only two proteins identical in mobility with those present in neurosecretory granules [24]. It is an interesting possibility that the origin of the large amount of peptide material present in extracts of the posterior pituitary is the proteolytic degradation of neurophysin [22, 25].

As already discussed neurophysin-M consisted of two protein components separable by electrophoresis. One of the two constituents occurred in neurosecretory granules and the other appeared to be a degradation product. When attempts were made to separate them by fractional precipitation in the presence of vasopressin by the addition of increasing amounts of salt the proportion of the two remained constant. However, after a few days the amorphous precipitate kept at 4° crystallized. This was interesting because it was the first time that vasopressin had been obtained in crystalline form [26]. The need to separate the two proteins in neurophysin-M was removed when it was found that only one of them was present in extracts of an acetone powder of pituitary tissue.

## ISOLATION OF NEUROPHYSIN-I & II

A hormone-protein complex was precipitated from an acidic extract (pH 1.6) of acetone powder by addition of salt at pH 3.9. The methods previously used for the fractionation of "crude neurophysin" were then applied and two protein fractions were obtained. Electrophoresis in a starch gel showed that they corresponded to the two granule proteins. The one with greater electrophoretic mobility emerged from the column first and contained a trace of another protein while the second appeared to be homogeneous. In accordance with the recommendation of the Standing Committee on Enzymes of the International Union of Biochemistry for the naming if isoenzymes [27] the protein with highest mobility at pH 8.1 was designated neurophysin-I and the other neurophysin-II.

The two proteins were studied in the Spinco analytical ultracentrifuge, model E, where they appeared to be homogeneous on sedimentation in a sodium acetate buffer at pH 4.8, I 0.1. Their sedimentation coefficients were $S_{20,w}$, 1.66 (neurophysin-I) and 2.02 (neurophysin-II) in Svedberg units. The molecular weights measured by the equilibrium sedimentation method of Yphantis [28] were 19,000 (neurophysin-I) and 21,000 (neurophysin-II).

The amino acid analyses of the two bovine neurophysins have been compared with the figures reported for the so-called van Dyke protein. The latter contains the polypeptide hormones in addition to neurophysins: this could explain the higher content of aromatic amino acids. The amino acid residues present in greatest numbers were glycine, glutamic acid, proline and cystine. Neurophysins appear to contain one disulphide residue per 1,500 g of protein; this is similar to the disulphide content of arginine vasopressin (M.W. 1,100). The proline content is also remarkably high, one residue of proline in 1,000 g. of neurophysin-I and in 1,200 g. of neurophysin-II. The high proline content of the proteins is of interest in relation to their secondary structure. Low & Edsall [29] showed that proline cannot be accommodated in an α-helix except at the N-terminal end. Further, it was predicted by Szent-Györgi and Cohen [30] that proteins containing more than 8 per cent proline would behave as a random coil. In 1962 Perutz [31] reported that in myoglobin and haemoglobin the proline residues all lie in the corners or in non-helical regions of the chain. The high proline content of neurophysins can explain the results of optical rotatory dispersion measurements which have indicated a low content of α-helix (Hope & Hollenberg, unpublished observations).

The amino acid composition of neurophysin -I and -II are similar to that of neurophysin-M. The methionine content of neurophysin-M is no longer a characteristic feature because it is present in both the native neurophysins. Neurophysin-II, a constituent of neurophysin-M contains two residues of methionine per mole whereas neurophysin-M contains only one. Thus the degraded protein in neurophysin-M must be methionine-free. Neurophysin-I possesses one residue of methionine and in addition contains two residues of histidine which distinguishes it from neurophysin-II and neurophysin-M.

## CRYSTALLINE COMPLEXES OF NEUROPHYSIN-I AND -II WITH 8-ARGININE VASOPRESSIN

The conditions used for the crystallization of neurophysin-M with 8-arginine vasopressin proved suitable for neurophysin-I and -II. Neurophysin-II formed long needle shaped crystals which were indistinguishable from those of neurophysin-M. Crystallization of these two proteins with vasopressin was completed within a few days, whereas neurophysin-I crystallized more slowly forming clusters of small needles. Thus all the proteins formed crystalline complexes with 8-arginine vasopressin. Although all three proteins formed complexes with oxytocin which were precipitated on addition of salt none of them have yet crystallized. Protein-hormone complexes containing both oxytocin and 8-arginine vasopressin have been prepared. The double complexes of neurophysin-M and neurophysin-II have been crystallized but as yet the one containing neurophysin-I has not.

The composition of the crystalline complexes is summarised below.

Moles of Polypeptide Bound per Mole of Protein

| | Neurophysin | | |
|---|---|---|---|
| | I | II | M |
| Vasopressin | 3 (3.10) | 2 (2.12) | 3 (3.16) |
| Oxytocin | 3 (2.97) | 2 (2.46) | 2 (2.07) |
| Vasopressin | 1 (0.67) | 1 (1.24) | 2 (2.39) |
| plus oxytocin | 2 (2.52) | 2 (2.10) | 1 (0.94) |

The complexes formed by neurophysin-M contained 3 Moles of vasopressin, or 2 Moles of oxytocin or 2 Moles of vasopressin and 1 Mole of oxytocin per Mole of protein. It appears that oxytocin competes with vasopressin for one of the binding sites. Neurophysin-II, a constituent of neurophysin-M, forms crystalline complexes containing 2 Moles of vasopressin or 2 Moles of oxytocin and 1 Mole of vasopressin per Mole of protein. Thus in neurophysin-II the third binding site is available only to oxytocin; as in the complex of neurophysin-M oxytocin completes for and replaces one Mole of vasopressin. All the complexes of neurophysin-II and -M containing vasopressin have crystallized readily. On the other hand, neurophysin-I crystallized when only vasopressin was present; the crystals contained 3 Moles of polypeptide per Mole of protein. An amorphous complex of neurophysin-I with both hormones contains 2 Mole of oxytocin and 1 Mole of vasopressin. In the absence of vasopressin all of the three binding sites appeared to be available for oxytocin. Neurophysin-I was similar to neurophysin-II in possessing three polypeptide binding sites with a preference for oxytocin at two of them.

The binding of two Moles of oxytocin by neurophysin-II and neurophysin-M is in agreement with the finding of Breslow and Abrash [32] for a fraction (E) of bovine neurophysin. The native neurophysin-I has been found to bind three Moles of oxytocin per Mole of protein. Neurophysin-I would have been absent from the preparations of neurophysin used by Breslow and Abrash [32]. They also reported that neurophysin fraction E bound two Moles of 8-lysine vasopressin per Mole of protein agreeing with the results obtained by us with neurophysin-II but not for neurophysin-I or M which both bind three Moles of the polypeptide. The results of Breslow and Abrash [32] revealed the presence of only two peptide binding sites in their protein fraction whereas those reported here show that there are three in neurophysin-I, -II and -M. The number of moles of vasopressin bound per mole of any one of the three proteins was reduced in the presence of oxytocin. This is almost certainly because of a competition for certain peptide binding sites by oxytocin. It is interesting to note that the non-mammalian hormone 8-arginine vasotocin is bound by mammalian neurophysin [33, 34]. In addition, Breslow and Abrash [32] found that several synthetic analogues of oxytocin were bound by neurophysin. The most important amino acid residue for binding is the half-cystine residue which bears a primary amino group. When this group is replaced by a hydrogen atom either in deamino-oxytocin [35] or in deamino-8-arginine-vasopressin [36] no binding whatever occurs. It seems therefore that the binding of both hormones occurs via the same basic mechanism. However, we have recently shown that a variety of other basic peptides, e.g., bradykinin are not bound by neurophysin [36]. The precise relationship between the structure of the octapeptide hormones and binding sites in neurophysins remains to be explored. A related phenomenon is crystallization of neurophysins with vasopressin but not with oxytocin.

## ISOLATION OF NEUROPHYSIN FROM NEUROSECRETORY GRANULES

Recent work with neurosecretory granules from the bovine pituitary gland showed that centrifugation over a sucrose density gradient for long periods of up to 5 hr. led to a virtually complete separation of granules from mitochondria. This effect of the duration of centrifugation is partly owing to the small size of the granules (150 nm) and to an increase in sedimentation coefficient in hypertonic sucrose. Sufficient granules have been prepared to permit the isolation of the proteins precipitable by low concentrations of salt from lysates. Fractionation on a column of Sephadex G-75 gave one main peak of protein which was shown by amino acid analysis to be a mixture of neurophysins-I and -II in equal proportions. This fraction accounted for 60 per cent of the soluble granule protein and for 50 per cent of the total granule protein. The neurophysins of the granules are associated with 19.1 units of oxytocin and 21.1 units of pressor activity per mg of protein. These figures correspond to a complex of one Mole of oxytocin, one Mole of vasopressin to one Mole of neurophysin. Comparison with the composition of the crystalline and amorphous complexes of neurophysin-I and -II suggests that all the polypeptide binding sites present in the proteins are not utilized in the neurosecretory granule. Before this finding can be evaluated it will be necessary to learn more about the localization of neurophysin in neurosecretory granules and about the half-lives of the proteins and polypeptides of the pituitary gland.

The presence of two similar hormone binding proteins in bovine neurophysin recalls the occurrence of isoenzymes. Since the neurophysins were isolated from pituitary posterior lobes pooled from many animals it was possible that individual animals of the same species would produce a single neurophysin. Neurath [37] and his co-workers reported that the two isoenzymes of bovine carboxypeptidase Aα occurred separately and also together in the pancreas of different animals. When a study was made of the neurophysins of 15 individual bovine pituitaries both varieties of protein were found in all. Two intriguing problems remain as to whether neurophysins-I and -II are stored in different neurosecretory granules and the biological roles of the two proteins.

We know of several instances where hormones are stored in association with a protein in endocrine glands, e.g. thyroglobulin and chromogranin [38] as well as neurophysin. The strength of binding between hormone and protein differs considerably. In thyroglobulin a covalent bond is present; in neurophysin-hormone complexes ionic and secondary forces are involved. With chromogranin no form of binding withcatechol-amines has been demonstrated yet it is known that the protein is secreted together with the amines following splanchnic stimulation, even so, no function has yet been found for chromogranin. It would be of great interest to know whether neurophysin is secreted together with a polypeptide hormone now that we know that they are stored together. It is known that certain proteins are depleted from the pituitary posterior lobe following dehydration or NaCl administration [39, 40] but it is not known whether these are neurophysins or not.

In conclusion recent work has shown that the polypeptide hormones, oxytocin and 8-arginine vasopressin, are normally closely associated with neurophysin within neurosecretory granules. It has been shown that neurophysin of bovine origin is a mixture of two closely related proteins. There are indications that further studies of this interesting protein will lead to a greater understanding of the synthesis, storage and release of the hormones themselves.

## REFERENCES


1. Hope, D.B., Schacter, B.A. & Frankland, B.T.B. (1964). Biochem. J. 93, 7P.

2. Acher, R., Chauvet, J. & Olivry, G. (1956). Biochim. Biophys. Acta, 22, 421.

3. van Dyke, H.B., Chow, B.F., Greep, R.O. & Rothen, A. (1942). J. Pharmacol., 74, 190.

4. Landgrebe, F.W., Ketterer, B. & Waring, H. (1955). In: The Hormones, Vol. 111 Eds. Pincus, G. & Thimann, K.V. Academic Press, New York pp 389-431.

5. Ginsburg, M. & Smith, M.W. (1959). Brit. J. Pharmacol. 14, 327.

6. Gaitan, E., Cobo, E. & Mizrachi, M. (1964). J. clin. Invest. 43, 2310.

7. La Bella, F.S., Reiffenstein, R.J. & Beaulieu, G. (1963). Arch. Biochem. Biophys. 100, 399.

8. Barer, R., Heller, H. & Lederis, K. (1963). Proc. Roy. Soc. B., 158, 388.

9. Hope, D.B. (1964). In: Oxytocin Vasopressin and their structural analogues. p. 105. Ed. by Rudinger, J. London: Pergamon Press.

10. Chauvet, J., Lenci, M.-T. & Acher, R. (1960). Biochim. Biophys. Acta, 38, 266.

11. Smithies, O. (1955). Biochem. J. 61, 629.

12. Poulik, M.D. (1957). Nature, Lond. 180, 1477.

13. Frankland, B.T.B., Hollenberg, M.D., Hope, D.B. & Schacter, B.A. (1966). Brit. J. Pharmacol. 26, 502.

14. Block, R.J. & van Dyke, H.B. (1950). Nature, Lond. 165, 975.

15. Block, R.J. & van Dyke, H.B. (1952). Arch. Biochem. Biophys. 36, 1.

16. Hope, D.B. & Hollenberg, M.D. (1966). Biochem. J. 99, 5P.

17. Hollenberg, M.D. & Hope, D.B. (1967). Biochem. J. 104, 122.

18. Ferguson, K.A. & Wallace, A.L.C. (1961). Nature, Lond. 190. 629.

19. Dean, C.R. & Hope, D.B. (1966). Biochem. J. 101, 17P.

20. Dean, C.R. & Hope, D.B. (1967). Biochem. J. In Press.

21. Adams, E. & Smith, E.L. (1951). J. biol. Chem. 191, 651.

22. Ramachandran, L.K. & Winnick, T. (1957). Biochim. biophys. Acta, 23, 533.

24. Dean, C.R., Hollenberg, M.D. & Hope, D.B. (1967). Biochem. J. 104, 8C.

25. Winnick, T., Winnick, R.E., Acher, R. & Fromageot, C. (1955). Biochim. biophys. Acta, 18, 488.

26. Hollenberg, M.D. & Hope, D.B. (1966). J. Physiol. 185, 51P.

27. Webb, E.C. (1964). Nature, Lond., 203, 821.

28. Yphantis, D.A. (1960). Ann. N.Y. Acad. Sci., 88, 586.

29. Low, B.R. & Edsall, J.T. (1956). In: Currents in Biochemical Research. p. 378. Ed. Green, D.E. New York: Interscience Publishers Inc.

30. Szent-Györgyi, A. & Cohen, C. (1957). Science, 126, 697.

31. Perutz, M.F. (1962). Nature, Lond., 194, 914.

32. Breslow, E. & Abrash, L. (1966). Proc. Nat. Acad. Sci. Wash. 56, 640.

33. Acher, R., Chauvet, J. & Lenci, M.-T. (1960). Biochim. biophys. Acta, 38, 344.

34. Acher, R., Chauvet, J., Lenci, M.-T., Morel, F. & Maetz, J. (1960). Biochim. biophys. Acta, 42, 379.

35. Stouffer, J.E., Hope, D.B. & du Vigneaud, V. (1962). In: Perspectives in Biology pp 75-80. Ed. by Cori, C.F., Foglia, V.G., Leloir, L.F. & Ochoa, S. The Netherlands: Elsevier Publishing Co.

36. Hollenberg, M.D. & Hope, D.B. (1967). Biochem. J. In Press.

37. Walsh, K.A., Ericsson, L.H. & Neurath, H. (1966). Proc. Nat. Acad. Sci., Wash. 56, 1339.

38. Blaschko, H., Comline, R.S., Schneider, F.H., Silver, M. & Smith, A.D. (1967). Nature, Lond. 215, 58.

39. Rennels, M.L. (1966). Endocrinology, 78, 659.

40. Friesen, H.G. & Astwood, E.B. (1967). Endocrinology, 80, 278.

# MECHANISM OF RELEASE AND MECHANISM OF ACTION OF VASOPRESSIN

Niels A. Thorn
Institute of Medical Physiology C
University of Copenhagen, Denmark

Vasopressin and oxytocin or one or several precursors of them are produced in the pericarya of cells in certain nuclei of the hypothalamus and maybe also within the axons leading from these cell bodies to the neural lobe of the pituitary. The hormones or their precursors are transported in neurosecretory granules from the perikarya of the cells to the neural lobe where they are stored and from where the controlled release takes place. The appropriate (or inappropriate) stimuli for release produce nerve signals, the last pathway of which is the nerve membrane surrounding the terminal swellings of the axon. There is no information on the speed with which the neurosecretory granules move within the axons.

The neural lobe has morphological and functional characteristics common to nerve cells and endocrine cells. Fig. 1 shows an electron microscopic picture of nerve terminals from the neural lobe of a rat, containing, among other structures, some large and many small neurosecretory granules. The relations of the nerve terminals to the blood vessels are shown. An unusually wide perivascular space containing two condensed layers of basement membrane can be seen. This has also been described in the neurohypophysis of the rabbit (Barer & Lederis (1966) and is apparently characteristic of many endocrine organs. Figure 1 shows the neurohypophysis of a normal rat. The relationships between neurosecretory axons (A-1-5), a pituicyte (P) and a blood capillary (C) are seen. The capillary wall consists of a

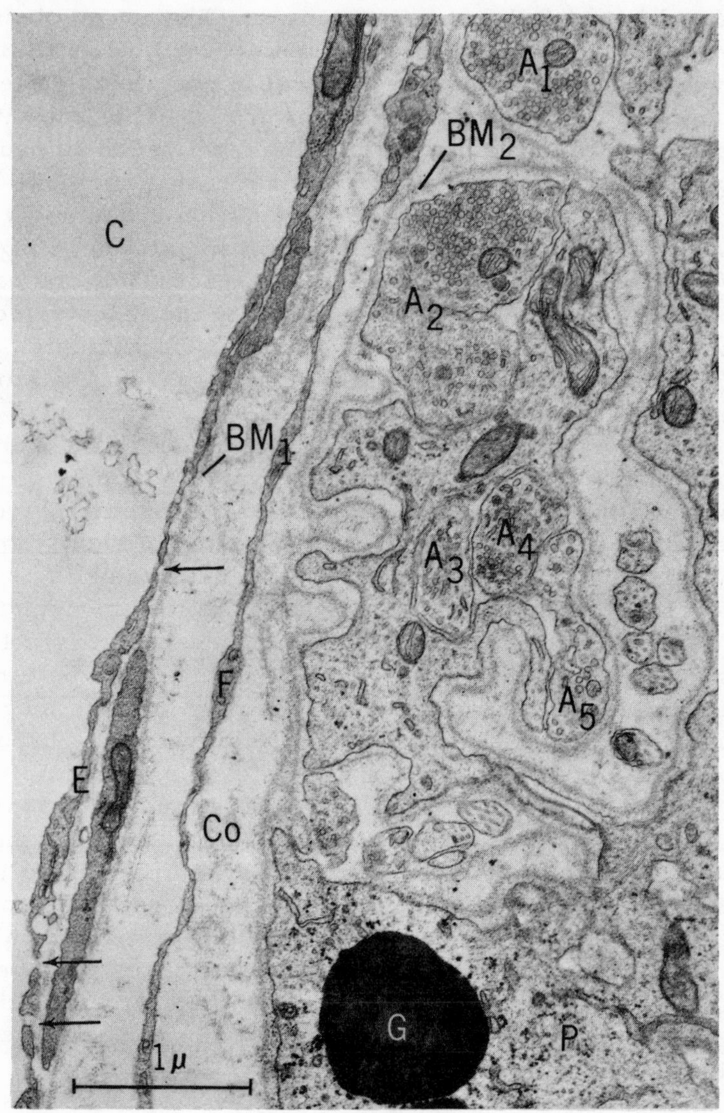


Fig. 1. Neurohypophysis of a normal rat.

very thin endothelium (E) surrounded by a thin basement ($BM_1$). The endothelium displays pinocytotic vesicles and fenestrations (arrows) and are of the same type as found in the kidney, the intestine and in several endocrine organs. The neural tissue is separated from the capillary by a basement membrane ($BM_2$) and an intervening connective tissue space (Co) containing a slender cellular element, probably a fibroblast process (F), but without the interposition of any other cellular element, such as neuroglia or a pituicyte process. The figure indicates that hormone in axon $A_1$ and $A_2$ before reaching the circulation presumably has to pass several barriers: The thin basement membrane ($BM_2$) of the nervous tissue, the connective tissue space (Co), the basement membrane ($BM_1$) of the capillary and finally the fenestrated endothelial cell. G, characteristic granule of a pituicyte. x 30.000 (Electron micrograph by Dr. J. Rostgaard, Anatomy Department C, University of Copenhagen).

The secretion of vasopressin and oxytocin from the neurohypophysis has been considered to be a typical example of neurosecretion, and the cells have been termed neurosecretory cells. Recent evidence (Bargmann et al. 1967) seems to show, however, that they may not differ so much from other nerve cells as has previously been thought. Bargmann et al. have demonstrated synapses involving anatomically typical neurosecretory cells in the pars intermedia of the cat. It may thus be that a better term would be peptidergic neurons, since they may function much like e.g. adrenergic neurons. It is in agreement with this hypothesis that there does not seem to occur any release of hormone from the cell body. Bie & Thorn (1967) tried to stimulate vasopressin release from isolated pieces of rat hypothalamus containing the supraoptic and paraventricular nuclei by stimuli which do release hormone from isolated neural lobes. Although hormone could be extracted from the tissue, no activity could be released on stimulation. It would appear that studies of the mechanism of release of these hormones might function as model experiments for the release of other peptidergic neurone substances, such as e.g. hypothalamic releasing factors.

The ultrastructure of the neurohypophysis has been described in such detail to provide a background for a discussion of the mechanism of release of the hormone. Several anatomists seem to agree that it is little likely that secretion occurs by a reverse pinocytosis. If the granules are not released to the blood, one would have to explain what becomes of the granule membrane after hormone has left it. The rate of turnover of granules in the neural lobe is unknown. Several possibilities seem to exist for a release of hormones from the granules within the cells. In this connection it might be mentioned that in pinocytotic vesicles the membranes become permeable to substances to which they were previously impermeable, when they increase in size (Holter 1966).

In the neurohypophysis we have a unique situation concerning our knowledge of the physicochemical state of the hormone. There has for long been known a carrier or precursor protein, neurophysin, which binds the hormone, and which may have an important function in restricting the hormone to the small pool within the neurohypophysis, from which secretion seems to take place (Thorn 1966).

Table I

POSSIBILITIES
RELEASE MECHANISM FOR VASOPRESSIN

1. Release of intact granules (Reverse pinocytosis)
2. Disintegration of granules within nerve endings (with ensueing 3) or 4)
3. Release of neurophysin with octapeptides
   - via axoplasm from granules
   - directly from axoplasm
4. Release of octapeptides
   - from a free state (via axoplasm from granules)
   - from a bound state (directly from axoplasm)

One present hypothesis for the mechanism of release is the following one: Neurosecretory granules in the neural lobe "mature" and release hormone to the axoplasm. Consequently, there exists at the site of release two "pools" of hormone, a large one in the granules, and a small one outside the granules. In both cases the hormone is bound to neurophysin. Nerve stimuli arrive at the terminals in the neural lobe. In the depolarization procedure calcium enters the axoplasm or is released from subcellular sites. The slight increase in the intracellular concentration of calcium ions causes a release of octapeptide from binding to the carrier protein.

This is assumed to occur from the small pool, since calcium can presumably not enter the granules. The octapeptide follows a step concentration gradient, diffuses to the blood and is quickly washed away by the high blood flow in that region. This latter process requires a reasonable permeability to hormone for the neural lobe-to-blood barriers, at least during stimulation. Repletion of the readily releasable pool could take place from granular hormone. Some of the evidence which can support this hypothesis is outlined in Table II.

Table II

EVIDENCE FOR THE ROLE OF CALCIUM IN THE EXCITATION-SECRETION COUPLING FOR VASOPRESSIN

1. In vivo experiments (intracarotid infusion of calcium releases vasopressin)

2. In vitro experiments (calcium is needed for release of vasopressin on depolarization)

3. Biochemical experiments concerning the influence of calcium on the binding of vasopressin to neurophysin
   a) Gelfiltration experiments ($Ca^{++}$ inhibits the binding of arg-vasopressin to neurophysin)
   b) Film-dialysisexperiments ($Ca^{++}$ inhibits the binding of lys-vasopressin to neurophysin) ($Ca^{++}$ inhibits the binding of oxytocin to neurophysin). No effect of $Mg^{++}$, $Na^{+}$, acetylcholine, noradrenaline or serotonin).

There is no agreement on the question whether neurophysin is released with the octapeptide or not. Ginsburg & Ireland (1966) and Thorn (1966) did not find evidence for this, but on the other hand Sachs (1967) relates results of some preliminary experiments which seemed to indicate that after labelling of the hormone and neurophysin in vivo, labelled neurophysin was released on stimulation in vitro.

In making studies of the possible participation of neurophysin in the mechanism of release, it should be recalled, that recent studies have indicated that the chemical properties of neurophysin may vary with the preparation technique (Dean et al. 1967). It seems that special precautions should be taken when isolating neurophysin, to obtain a product similar to the natural one. It may be, however, that studies of binding properties of neurophysin obtained by previous techniques have still yielded reasonable information, since nearly all of the fractions, produced by different preparation procedures, do bind hormones. There has been some disagreement about the quantitative binding properties for various preparations of neurophysin. This might be explained by the effects of different preparation procedures.

It seems to be an essential question whether the release mechanism is dependent on the availability of energy (whether it is an "active process" or not). Douglas et al. (1965) stated that the release was dependent on energy, since it was practically abolished when isolated neural lobes, incubated with metabolic inhibitors in a calcium free, high K Locke medium, were exposed to calcium. It would seem little likely that this inhibition is due to an inhibition of an active process which brings the calcium ions into the interior of the axon. Douglas et al. interpreted their experiments to show that an energy-dependent link is essential for the release of the hormone, or, alternatively, that the inhibitor might act by blocking production of some intermediate required to prime the hormone extrusion mechanism.

We have repeated the experiments of Douglas et al. We have found that depolarization of the neural lobes by a high potassium concentration in the medium does cause a release of hormone, and that a release takes place when calcium is introduced into the medium of glands, incubated in a high K, Ca-free medium. After preincubation with iodoacetic acid and DNP, no total inhibition was found, but a considerable release of hormone did take place. Since cold is a stimulant for release, it seems that release can take place independently of the availability of energy (provided the cold has not damaged the organization of the tissue).

The concept that there is a readily releasable pool comprising only some 10% of the total hormone content of the gland has found considerable support. Both in experiments in vivo (Sachs 1967) and in experiments in vitro (Thorn 1966) it has been shown that violent stimuli to the release can only mobilize this fraction of extractable hormone. The decrease in the rate of release with continued stimulation does not seem to be due to a decrease in the total contents of hormone in the gland. (Sachs 1967). The cellular or subcellular localization of readily releasable hormone is unknown, as are the processes presumably responsible for transforming the non-readily releasable part to the easily mobilizable part. It might be possible that this transformation involves an energy dependent step, and that the inhibition of release first reported by Douglas et al. after incubation in a medium containing metabolic inhibitors might be due to inhibition of the transformation from the non-releasable to the releasable pool, possibly in combination with a depletion of the readily releasable pool prior to stimulation.

In close connection with the problem of the readily releasable pool stands the peculiar fact that various stimuli cause a disappearance of the neurosecretory material, both by light microscopic and electron microscopic criteria, whereas very little hormone has disappeared (see e.g. Barer & Lederis 1966). Transformation of carrier to a different substance within the granules is one possibility, but a release of hormone to the axoplasm or even the perivascular spaces surrounding the nerve endings would seem possible.

It has been known for a great number of years that hypercalcemia inhibits the action of vasopressin on the kidneys (for references see Thorn 1960). This made Thorn try the simple reasoning: If calcium can inhibit the action of vasopressin it might be possible that vasopressin acts by removing membrane-bound calcium from the cells. In experiments in rats and dogs (Thorn 1960, 1961) it turned out that vasopressin (or oxytocin) injected in doses which gave antidiuresis produced a slight, but definite increase in the rate of excretion of calcium in the urine.

This effect was well correlated with the antidiuretic effect, both as concerned the magnitude as the duration.

Due to the complexity of the kidney functions and the difficulties of locating so small changes in the rate of excretion to any of the possible partial functions, it was tested in a simplier system whether hormone affected cellular calcium (Thorn & Schwartz 1965).

In these experiments toad blatter tissue from Bufo Marinus was"loaded up" with $Ca^{45}$ by incubation for 60 min. in a toad Ringer fluid, after which washout-curves for the radioactive calcium were obtained. When hormone was added to the medium during the washout procedure, there was a sudden increase in the release of $Ca^{45}$ to the medium. This increased release was influenced by pH changes and changes in the concentration of calcium in the medium in the same way as was affected the permeability to water by the hormone. There thus seemed to be a good correlation between effect of hormone on the permeability to water and its effect on release of $Ca^{45}$ to the medium during the washout.

Other experiments on amphibian membranes have demonstrated that calcium can interfere with the action of vasopressin.

Bentley(1959) showed that calcium inhibits the effect of Pitressin on the isolated toad bladder. Similar effect in the isolated frog skin have been noted by Gill & Nedergaard (see Gill & Bartter 1961). A corresponding effect in the intact frog skin was found by Pigeon & Epstein (1963). Findings of a similar nature have been reported for kidney slices of Necturi by Whittembury et al. (1960).

The hypothesis that calcium is crucially involved in the mechanism of action of vasopressin seems to be attractive for different reasons: It has been known for long that calcium is very important for the permeability to water of a number of animal membranes (e.g. Fukuda 1935, Bozler 1959). It has been shown by Lassiter et al. (1965) that calcium inhibits the permeability to water in the distal part of the nephron, but not in the more proximal parts where the hormone presumably does not act on the water permeability. It is known that calcium is crucial for the action of oxytocin on the uterus. The concentration of calcium is crucial for the inhibitory action of certain analogues of oxytocin (Rudinger 1964).

The facts supporting the hypothesis that calcium is involved in the mechanism of action of vasopressin are summarized in Table III.

Table III

EVIDENCE THAT CALCIUM IS INVOLVED IN THE MECHANISM OF ACTION OF VASOPRESSIN

1. Calcium can inhibit the action of vasopressin in man.
2. Calcium can inhibit the action of vasopressin on $H_20$ flow in the toad bladder.
3. Calcium inhibits the action of vasopressin on the equivalent pore radius of kidney slices from Necturus.
4. Calcium inhibits the water uptake through intact frog skin.
5. Calcium decreases the permeability to water in sections of the kidney tubules which are sensitive to vasopressin (here it presumably acts on the membrane on which the hormone acts and which is different from the one through which it enters the cells).
6. Vasopressin-induced antidiuresis is correlated well with increased renal excretion of calcium.
7. Vasopressin causes release of $Ca^{45}$ from isolated toad bladder tissue.

Several hypotheses could be advanced for the biochemical mechanism of action of vasopressin - involving calcium. A displacement of calcium from crucial membrane sites might open up channels or pores and allow a more free passage of water. Displacement of calcium from pore regions might be brought about e.g. by interaction of the disulphide groups or the peptide links of the hormones with lipid constituents of the membrane (Robinson 1966, Sanyal & Snart 1967). Calcium induced activation of cyclic AMP might also be involved in a pore reaction.

## REFERENCES

Barer, R. & K. Lederis: Zeitschr. f. Zellf. 75:201, 1966
Bargmann, W. et al.: Zeitschr. f. Zellf. 77:282, 1967
Bentley, P.: J. Endocrinol. 18:327, 1959.
Bie, P. & N.A. Thorn: Acta endocr. 56:139, 1967
Bozler, E.: Amer. J. Physiol. 197:505, 1959
Dean, C.R. et al.: Biochem. J. 104:8c, 1967
Douglas, W.W. et al.: J. Physiol. 181:753, 1965
Fukuda, T.R.: J. cell. comp. Physiol. 7:301, 1936
Gill, J.R., Jr & F.C. Bartter: J. clin. Invest. 40:716, 1961
Ginsburg, M. & M. Ireland: J. Endocrinol. 35:289, 1966
Holter, H.: Physiologie der Pinocytose bei Amoben in: Wohlfarth - Bottermann, K.E. (ed.): Sekretion und Exkretion. Springer Verlag 1965
Lassiter, W.E. et al.: Pflugers Arch. 285:90, 1965
Pigeon, G. & F.H. Epstein: Amer. J. Physiol. 204:217, 1963
Robinson, J.D.: Nature 212:199, 1966
Rudinger, J.: Excerpta Medica Congr. Ser. 83:1202, 1964
Sachs, H.: Amer. J. Med. 42:687, 1967
Sanyal, N.N. & R.S. Snart: Nature 1:798, 1966
Thorn, N.A.: Danish Med. Bull. 7:109, 1960
Thorn, N.A.: Acta endocr. 38:563, 1961
Thorn, N.A. & I.L. Schwartz: Gen. Comp. Endocr. 5:710, 1965
Thorn, N.A.: Acta Endocr. 53:644, 1966
Whittembury, G. et al.: Nature 187:699, 1960

# RADIOIMMUNOASSAY OF OXYTOCIN

Seymour M. Glick, M.D., Mary Wheeler, M.D., Avir Kagan, M.D. and Perianna Kumaresan, Ph.D.

Department of Medicine of the Coney Island Hospital, affiliated with the Maimonides Medical Center, Brooklyn, New York

The development of immunoassays for the various peptide hormones has greatly enhanced our understanding of endocrine physiology. The present paper reports progress in the development of an immunoassay for the relatively small peptide hormone, oxytocin.

Antibodies have been produced in rabbits by multiple subcutaneous injections of synthetic oxytocin in Freund's adjuvant and by repeated, closely spaced, injections of oxytocin without adjuvant. Of eight rabbits and three guinea pigs originally injected, two rabbits developed antibody detectable at a 1:100 dilution in an equilibrium dialysis system, and of these, one antiserum, with considerably the better affinity for oxytocin, was used. Subsequently six more rabbits have been injected, and two of these developed minimal titers after three injections. The sequence of inoculations in the animal whose antiserum was used in the present studies is shown in figure 1. The feasibility of producing antibodies by the injection of oxytocin without conjugation to a larger molecule has been reported previously by Gilliland and Prout(1), and by Klein and Roth(2). Antibodies similarly produced to vasopressin have been the basis of a radioimmunoassay for vasopressin reported by Roth, Klein & Peterson(3).

Radioiodination of oxytocin to a specific activity of 400 mc/mg was accomplished by the chloramine T method, originally applied to the iodination of human growth hormone by Hunter and Greenwood(4). The addition of sodium metabisulfite to terminate the iodination altered the migration of the iodinated oxytocin on

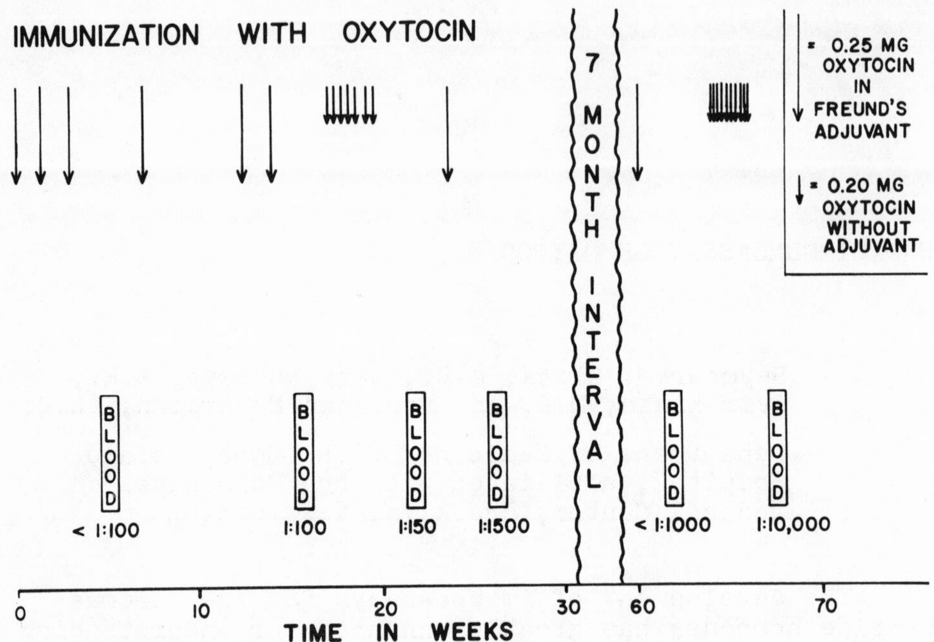


Figure 1: Immunization schedule in rabbit whose serum was used for radioimmunoassay.

thin layer ascending chromatography in a butanol/acetic acid mixture, and on chromatoelectrophoresis on Whatman 3MC paper (figure 2). The metabisulfite was, therefore, omitted.

After iodination the hormone was purified by passage through a Sephadex G-.25 column (figure 3). The high molecular weight material, consisting for the most part of damaged, aggregated and protein-bound oxytocin passes through first, followed by the intact oxytocin. The oxytocin peak is not homogeneous, with significant retardation of more highly iodinated hormone of higher specific activity. This selective binding to the dextran of hormone with high specific activity may be used to advantage in selecting iodinated hormone with high specific activity. On the other hand the selection of the retarded fraction from a preparation with an already high overall specific activity may result in an overiodinated preparation with significant impairment of binding to antibody.

Iodinated oxytocin is not adsorbed well to the site of application to Whatman 3MC paper. Thus chromatoelectrophoresis on Whatman 3MC paper which may be used for the radioimmunoassay of insulin and of growth

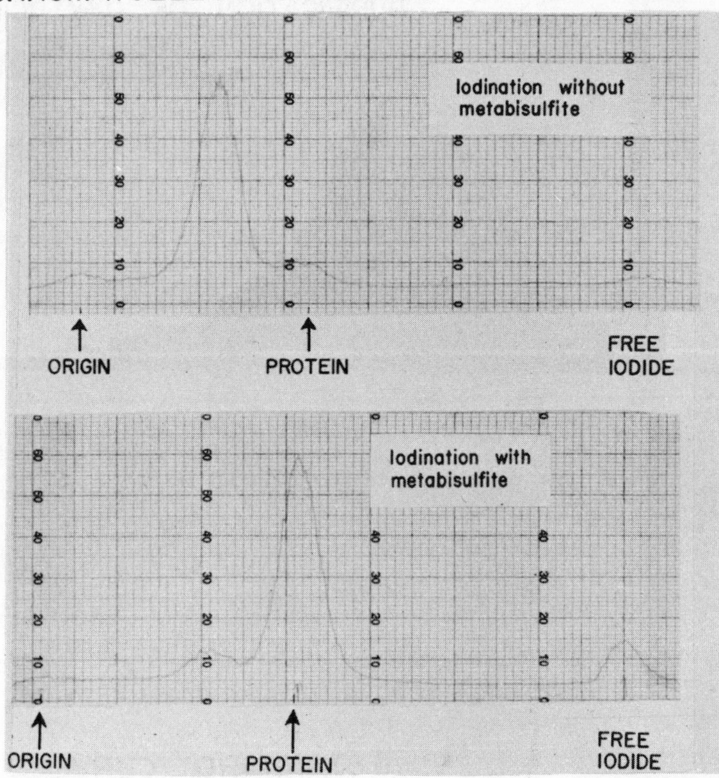


Figure 2: Migration of oxytocin - $^{131}I$ on chromatoelectrophoresis on Whatman 3MC paper. Oxytocin - $^{131}I$ untreated with metabisulfite (top). Oxytocin - $^{131}I$ treated with metabisulfite (bottom).

hormone will not satisfactorily separate bound oxytocin from free oxytocin. Iodinated oxytocin, after reduction by metabisulfite does bind moderately well to ecteola paper at pH 8.6, but the separation attained by the use of ecteola paper is unsatisfactory with respect to completeness and reproducibility.

In early experiments equilibrium dialysis was used to separate bound from free hormone $I^{131}$, but more recently we have been using the double antibody method with anti-rabbit gamma globulin antiserum for precipitation of the antibody-bound hormone. The assay can presently detect .05 mug/ml of oxytocin (figure 4). Since the injection of 10 mU (20 mug) in a human can produce a detectable physiologic effect and since oxytocin has a large volume of distribution and is degrad-

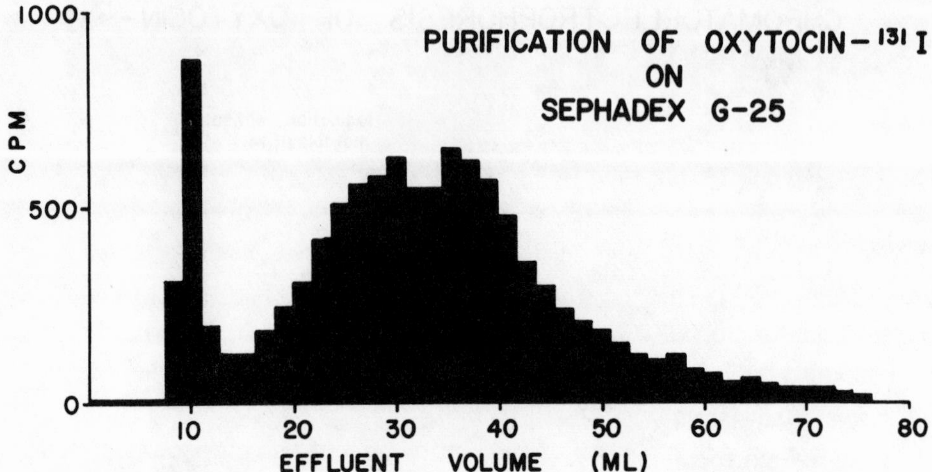


Figure 3: Elution pattern of oxytocin - $^{131}$I on Sephadex G-25.

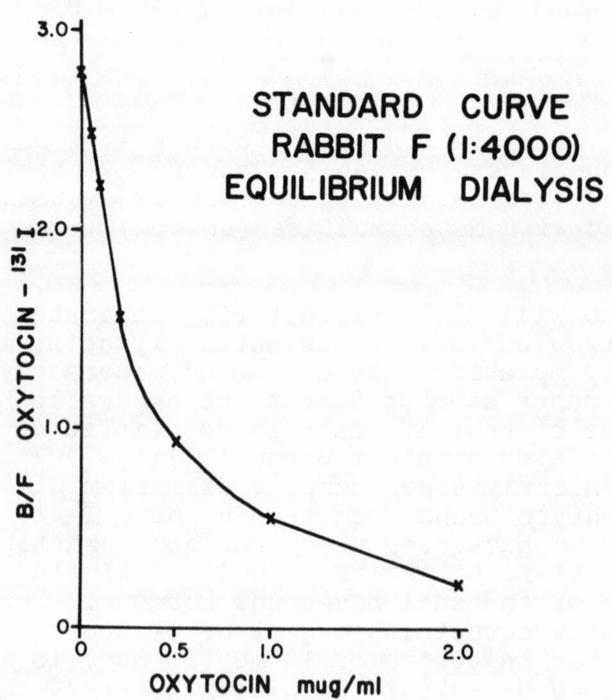


Figure 4: Standard curve for immunoassay of oxytocin.

Table 1

**COMPARISON OF BIOLOGIC AND IMMUNOLOGIC ACTIVITY OF OXYTOCIN ANALOGUES**

| SUBSTANCE | FOWL DEPRESSOR ACTIVITY | RAT OXYTOCIC ACTIVITY | IMMUNOLOGIC ACTIVITY |
|---|---|---|---|
| OXYTOCIN | 507 U/MG | 546 U/MG | 100 % |
| 1-DEAMINO-OXYTOCIN | 975 | 803 | 110 % |
| 4-VALINE-OXYTOCIN | 240 | 140 | 80 % |
| 1-DEAMINO-4-VALINE-OXYTOCIN | 800 | 350 | 90 % |
| LYSINE-VASOPRESSIN | 5 | 40 | <1 % |
| 5-VALINE-OXYTOCIN | 0.1 | 0.3 | <1 % |
| 1-DEAMINO-5-VALINE-OXYTOCIN | 0.3 | 0.7 | <1 % |
| 8-ALANINE-OXYTOCIN | 240 | 166 | <1 % |
| 8-ALANINE-OXYPRESSIN | 38 | 15 | <1 % |
| 1-DEAMINO-8-ALANINE-OXYPRESSIN | 47 | 25 | <1 % |

IMMUNOLOGIC REACTIVITY WITH RABBIT ANTI-OXYTOCIN ANTIBODY =

5-VALINE OXYTOCIN > 8-ALANINE OXYPRESSIN > 1-DEAMINO 5-VALINE OXYTOCIN > LYSINE VASOPRESSIN > 8-ALANINE OXYTOCIN > 1-DEAMINO 8-ALANINE OXYPRESSIN

ed very rapidly it is apparent that the present sensitivity is not nearly adequate for the detection and quantitation of physiologic concentrations of oxytocin in unextracted plasma.

Present efforts are concentrated on the production of an antiserum with greater affinity for oxytocin and on the development of a satisfactory extraction procedure. Preliminary results using a single acetone extraction of plasma have been promising with extractions ranging between 50-90%. Although Yamashiro, Aaning & DuVigneaud(5) have reported significant chemical alterations of oxytocin by acetone, the immunoreactivity of acetone-treated oxytocin seems grossly unimpaired.

The specificity of the antigen-antibody reaction is rather high, and lysine vasopressin, for example, cross-reacted very poorly in the assay system. A variety of structural analogues of oxytocin varied in their cross-reactivity from almost zero to a reactivity actually exceeding that of oxytocin itself (table 1). 1-Deamino-oxytocin, whose biologic potency exceeds that of oxytocin, exhibited also increased immunoreactivity.

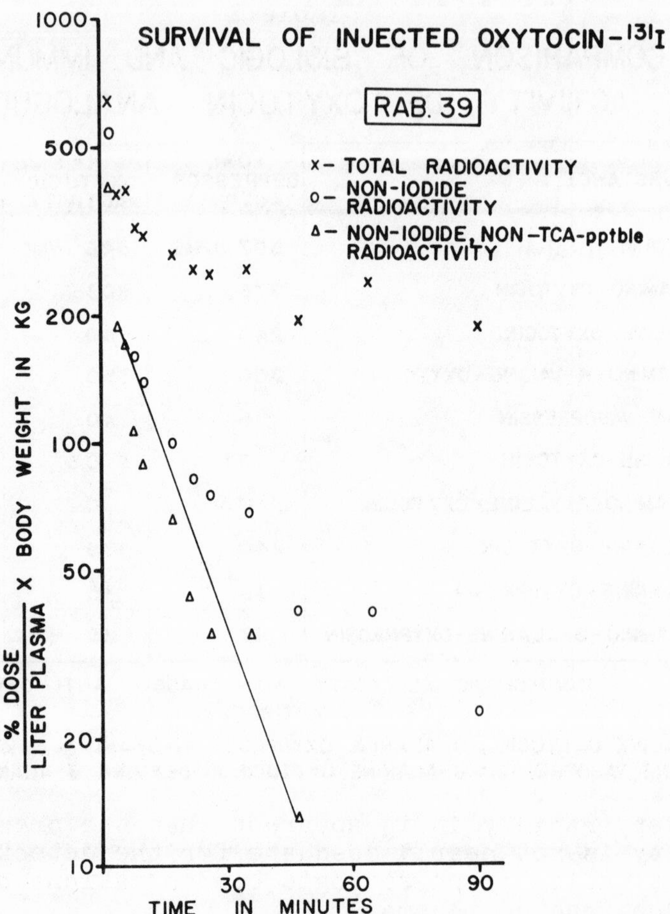


Figure 5: Decline of radioactivity in plasma of rabbit injected with oxytocin - $^{131}$I.

The 4-valine substituted oxytocin and its deamino derivative retain both immunologic and biologic activity whereas the 5-valine substituted oxytocin and its deamino derivative have markedly reduced immunologic and biologic activity. Most of the other analogues tested cross-reacted very poorly. While for some of the analogues there was reasonable correlation between biologic and immunologic activities, there was on the whole no parallelism between the two activities, and identity of immunologic and biologic sites is most unlikely.

There have been numerous reports regarding the half life in the plasma of injected oxytocin using chiefly biologic assays. The reported 1/2 times have

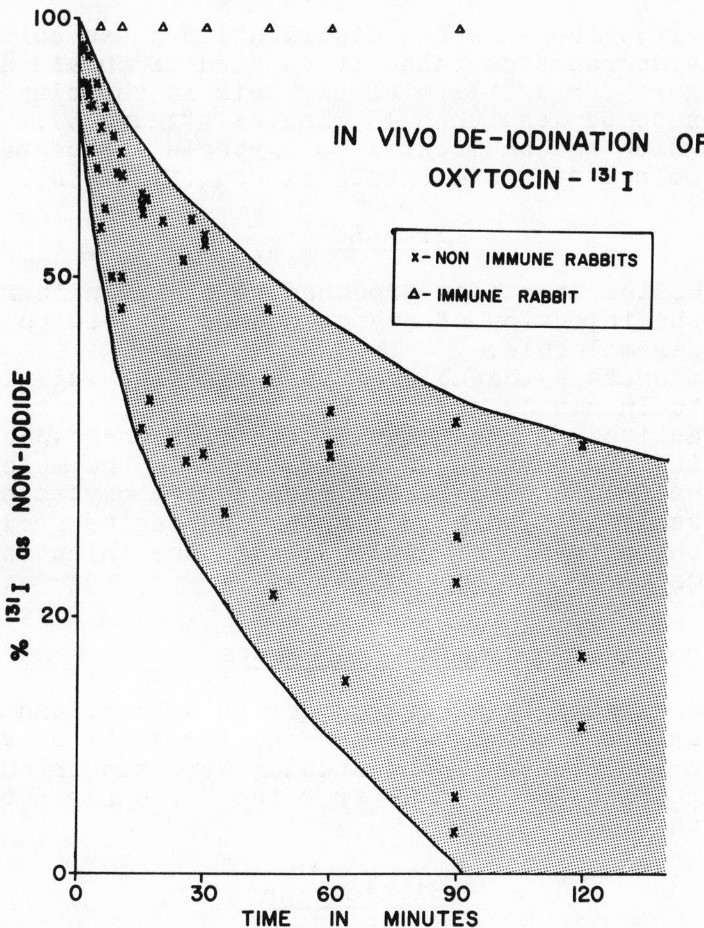


Figure 6: Comparison of deiodination of oxytocin $^{131}I$ in immune rabbit and in non-immune rabbits.

almost uniformly been in the range of a few minutes. We have injected radioiodinated oxytocin into several rabbits and studied the decrease of radioactivity. Since oxytocin is not precipitable with trichloracetic acid (TCA) the usual technique of determining TCA precipitable radioactivity is not useful in distinguishing iodide attached to hormone from that which is free. Instead chromatoelectrophoresis was used for quantitation of the percentage of free iodide. The TCA precipitable radioactivity in the present system was regarded as altered oxytocin, which is attached non-specifically to plasma protein. The content of oxytocin was then considered to be the noniodide, non-TCA-precipitable radioactivity. Because the distribution phase coincides with the degradation phase it is hard to obtain a precise figure for 1/2 life of oxytocin in the plasma, but it seems to be less than 15 minutes (figure 5). In an animal which had antibodies to oxytocin the degradation of oxytocin $I^{131}$ was markedly slowed (figure 6).


## SUMMARY

1. Antibodies have been produced to oxytocin in rabbits by the injection of oxytocin unconjugated to a larger molecule.
2. An immunoassay capable of measuring .05 mug/ml of oxytocin has been developed.
3. The antibody formed to oxytocin is rather specific in its recognition of alterations in the molecule.
4. The half life of injected radioactive oxytocin is probably less than 15 minutes and can be prolonged by the presence of antibodies in the injected animal.



## Acknowledgements

The authors thank Drs. Roderich Walter, and Irving L. Schwartz and the Sandoz Company for gifts of various oxytocin analogues. These studies were supported by Training Grant AM 05551-01 from the National Institutes of Health.


## Bibliography


1. Gilliland, P.F. and T.E. Prout, Metabolism 14, 918, 1965
2. Klein, L.A. and Roth, J., (in preparation), cited in Roth, J., S.M. Glick, L.A. Klein and M.J. Peterson, J. Clin. Endocr. and Metab. 26, 671, 1966
3. Roth, J., L.A. Klein and M.J. Peterson, J. Clin. Invest. 45, 1064, 1966 (abstract)
4. Hunter, W.M. and F.C. Greenwood, Nature (London) 194, 495, 1962
5. Yamashiro, D., H.L. Aaning and V. DuVigneaud, Proc. Nat. Acad. of Sci. 54, 166, 1965

# COMPARISON OF THE PHARMACOLOGICAL PROPERTIES OF TWO HIGHLY POTENT, CRYSTALLINE NEUROHYPOPHYSEAL HORMONE ANALOGS: DEAMINO-1-SELENO-OXYTOCIN AND DEAMINO-OXYTOCIN

Roderich Walter and Irving L. Schwartz

Physiology Department, Mt. Sinai Medical and Graduate Schools, New York, and Medical Department, Brookhaven National Laboratory, Upton

Investigations over many years have repeatedly drawn attention to the question of what --- if anything --- is the biological significance of the disulfide grouping in neurohypophyseal hormones. Two groups of analogs appear to be of special interest in this connection: [1] derivatives lacking both the disulfide grouping and the cyclic component of the molecule (1-8) and [2] derivatives lacking the disulfide grouping but retaining the cyclic component of the molecule (9-11). To gain deeper insight into the function of the disulfide bridge as a unit and to explore the function of each sulfur center individually, one or both of the sulfur atoms was replaced in oxytocin and deamino-oxytocin by selenium atoms (12-14) and the resultant selenium-containing analogs are being studied by chemical, physical and pharmacological methods.

In the course of these studies we have observed differences in the biological activities of amorphous preparations of deamino-, deamino-1-seleno-, deamino-6-seleno- and deamino-diseleno-oxytocin (12-14). Particularly striking was the potentiation of an oxytocin reference standard and the autopotentiation in an <u>in vitro</u> rat uterotonic assay system which amorphous deamino-1-seleno-oxytocin --- unlike deamino-6-seleno-oxytocin --- exhibited (14). It was of special interest to determine if the observed potentiating effect was caused by impurity in the amorphous deamino-1-seleno-oxytocin or, alternatively, if the effect was an inherent property of the molecule. We felt that it was therefore essential to apply the most rigorous criteria in our continued study of the selenium-containing analogs; for this reason we addressed ourselves to the crystallization of these analogs in order to avoid the ever-present ambiguities which plague all interpretations of pharmacological data obtained with amorphous preparations.

Table I. Biological Potencies[a] of Crystalline Deamino-oxytocin and Crystalline Deamino-1-seleno-oxytocin

| Compound | Depressor[b] (fowl) | Oxytocic[c] (rat) | Pressor[d] (rat) | Antidiuretic[e] (rat) |
|---|---|---|---|---|
| Deamino-oxytocin[f] | 975±24 | 803±36 | 1.44±0.06 | 19.0±1 |
| Deamino-1-seleno-oxytocin | 985 | 1100 | 4.8 | 29 |

[a]Expressed in USP units/mg.

[b]Were performed on conscious chickens according to the procedure employed by R.A. Munsick, W.H. Sawyer, and H.B. van Dyke, *Endocrinology*, **66**, 860 (1960).

[c]Performed on isolated uteri from rats in natural estrus according to the method of P. Holton, *Brit. J. Pharmacol.*, 3, 328 (1948) as modified by R.A. Munsick, *Endocrinology*, **66**, 451 (1960) with the use of magnesium-free van Dyke-Hastings solution as the bathing fluid.

[d]Carried out on atropinized, urethane-anesthetized male rats as described in "The Pharmacopeia of the United States of America," 17th Revision, Mack Printing Co., Easton, Pa., 1965, p. 749.

[e]Performed on anesthetized, hydrated male rats according to the method of W.A. Jeffers, M.M. Livezey, and J.H. Austin, *Proc. Soc. Exptl. Biol. Med.*, **50**, 184 (1942) as modified by W.H. Sawyer, *Endocrinology*, **63**, 694 (1958).

[f]Values reported by B.M. Ferrier, D. Jarvis, and V. du Vigneaud, *J. Biol. Chem.*, **240**, 4264 (1965).

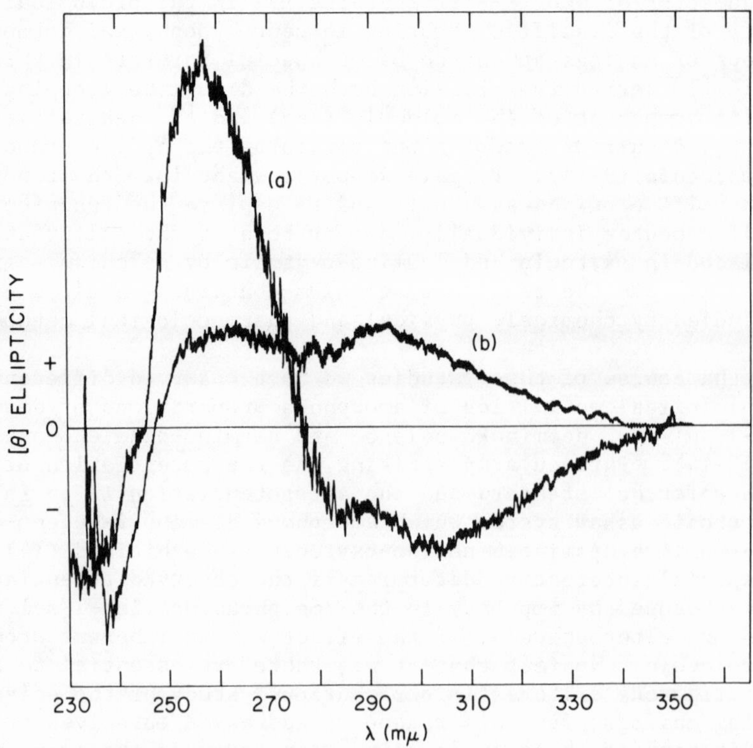


Figure I. - Circular dichroism curves of deamino-1-seleno-oxytocin (a) [1.0 mg/ml in water, pH 5.5] and deamino-6-seleno-oxytocin (b) [1.03 mg/ml in water, pH 5.5] showing the effect of replacing the sulfur atom by a selenium atom in deamino-oxytocin in position 1 or position 6.

As a first step towards this objective we repeated the synthesis of deamino-1-seleno-oxytocin (14). The amorphous product (50 mg) was dissolved at 80° in 1.5 ml of 0.1% acetic acid. A small amount of insoluble material was filtered off after the solution had reached room temperature and the filtrate was kept for 1 day at 4°. During this time crystals separated which were isolated by filtration and kept *in vacuo* over $P_2O_5$ until the weight remained constant; yield 22 mg; mp 248° dec. (15); $[\alpha]_D^{23}$ -83.7° (*c* 0.5, *N* acetic acid). The material gave the expected elementary analysis and amino acid analysis. Upon paper chromatography deamino-1-seleno-oxytocin gave only one spot at concentrations as high as 100γ with an $R_f$ of 0.75 in the solvent system butanol-acetic acid-water (5:1:5, v/v/v, upper phase).

Upon evaluation for biological activity in several assay systems crystalline deamino-1-seleno-oxytocin was found to be a highly active compound; a comparison of its potency values with those for crystalline deamino-oxytocin, evaluated in the same assay systems, reveals that the biological activity spectra of these two analogs differ [Table I]. These differences in activity may be associated with the larger diameter and an increased polarizability of the selenium moiety as compared with the sulfur moiety.

Furthermore, the potentiating effect in the oxytocic assay system previously found for amorphous deamino-1-seleno-oxytocin was retained by the crystalline analog, thus indicating a non-equivalence of the selenium moieties in position 1 and 6. In this context it should be mentioned that we demonstrated the non-equivalence of both selenium moieties also by another approach, namely, by comparing the circular dichroism spectra of deamino-1-seleno-oxytocin and deamino-6-seleno-oxytocin (16). While the spectrum of deamino-1-seleno-oxytocin exhibits negative peaks at 300, 280 and 235 mμ, and a positive peak at 258 mμ, that of deamino-6-seleno-oxytocin exhibits positive peaks at 290 and 260 mμ and negative peaks at 280 and 237 mμ (Fig. I).

Intrigued by the question of the functional equivalence or non-equivalence of the individual sulfur atoms in disulfide-containing peptides and proteins in general, we were tempted to interpret the non-equivalence of both selenium centers in terms of a non-equivalence of both sulfur centers in the hormone analogs. The fact that each sulfur atom of the dissymmetric disulfide bridge in neurohypophyseal hormones is influenced to a different degree by its neighboring asymmetric center does not necessarily mean that there will be a difference in the contribution of the individual sulfur *loci* to the expression of biological activity. The differences in the biological activities of deamino-1-seleno-oxytocin and deamino-6-seleno-oxytocin may be due to a disparity between the two sulfur atoms in the bridge, but it also may result from an unequal interaction with the asymmetric environment.

If there is a difference in the biological function of the two sulfur loci in deamino-oxytocin, we may have detected this disparity through the use of a selenium "marker". However, at present we have not excluded the possibility that the disparity in the biological activities of deamino-1-seleno-oxytocin and deamino-6-seleno-oxytocin result from an additional factor introduced by the "marker", namely its unequal interaction with the asymmetric environment.

In summary, deamino-1-seleno-oxytocin has been crystallized, and its biological properties have been shown to differ from those of crystalline deamino-oxytocin.

## References


(1) R. R. Sealock and V. du Vigneaud, J. Pharmacol. Exptl. Therap., 54, 433 (1935).
(2) S. Gordon and V. du Vigneaud, Proc. Soc. Exp. Biol. Med., 84, 723 (1953).
(3) K. Jošt, V. G. Debabov, H. Nesvadba and J. Rudinger, Coll. Czech. Chem. Commun., 29, 419 (1964).
(4) R. A. Turner, J. G. Pierce and V. du Vigneaud, J. Biol. Chem., 193, 359 (1951).
(5) Z. Beránková and F. Šorm, Coll. Czech. Chem. Commun., 26, 2557 (1961).
(6) R. L. Huguenin and St. Guttmann, Helv. Chim. Acta, 48, 1885 (1965).
(7) J. M. Mueller, J. G. Pierce, H. Davoll and V. du Vigneaud, J. Biol. Chem., 191, 309 (1951).
(8) D. Yamashiro, D. Gillessen and V. du Vigneaud, Biochemistry, 5, 3711 (1966).
(9) K. Jošt and J. Rudinger, Coll. Czech. Chem. Commun., 32, 1229 (1967).
(10) J. Rudinger and K. Jošt, Experientia, 20, 570 (1964).
(11) I. L. Schwartz, H. Rasmussen and J. Rudinger, Proc. Natl. Acad. Sci. U.S., 52, 1044 (1964).
(12) R. Walter and V. du Vigneaud, J. Am. Chem. Soc., 87, 4192 (1965).
(13) R. Walter and V. du Vigneaud, ibid., 88, 1331 (1966).
(14) R. Walter and W. Y. Chan, ibid., 89, 3892 (1967).
(15) The melting point was determined in an open capillary. The crystals changed contours at 178° and turned brown at 240°.
(16) R. Walter, F. Quadrifoglio and D. Urry, presented before the 154th National Meeting of the American Chemical Society, Chicago, Sept. 10-15, 1967.


Acknowledgment - This work was supported by Grant AM-10080 of the National Institute of Arthritis and Metabolic Diseases, U. S. Public Health Service and by the U. S. Atomic Energy Commission. The authors wish to thank Dr. W. Y. Chan for independently evaluating the biological activities of crystalline deamino-1-seleno-oxytocin.

# STRUCTURE AND FUNCTION OF PITUITARY GRAFTS, RELATED TO THE PROBLEM OF HYPOTHALAMIC CONTROL OF HYPOPHYSEAL ACTIVITY

L. Desclin and J. Flament-Durand

Laboratoire d'Anatomie-pathologique-Université Libre de Bruxelles, Fondation Médicale Reine Elisabeth, Bruxelles

The neurohumoral control of hypophyseal activity is at present perfectly established. Several mediators reaching the anterior lobe through the portal vessels have been extracted from hypothalamic tissue acting selectively on the elaboration and liberation of the various pituitary hormones.

In the development of this present conception, the study of transplanted pituitaries played a decisive role. The first indication of the importance of the hypothalamic contact for a normal pituitary activity came probably from the early morphologic study of Hohlweg and Junkmann on pituitary transplants into the kidney. After castration, the grafted pituitary did not show the well known castration cells whereas regular castration changes did appear in the anterior lobe in its normal location. From these results, Hohlweg and Junkmann proposed the existence of an hypothalamic sexual center regulating the gonadotrophic activity of the pituitary (Hohlweg and Junkmann, 1932).

A few years later, we could confirm these results of the German workers, but, in addition we observed that the administration of oestrogen to normal rats bearing a transplanted pituitary, induced definite morphological changes in the transplanted pituitary as well as in the pituitary of the host (Desclin and Gregoire, 1936). These changes were the same as those observed in the anterior lobe of normal females injected with oestrogen when the corpora lutea persist and become hypertrophic. They were regularly coincident with the so-called luteotrophic activity of oestrogens (Desclin, 1935).

Later, we studied pituitary grafts under the kidney capsule in hypophysectomized rats (Desclin, 1950). In those animals, the genital tract remained atrophic. Injected with oestrogen however, they did not show the oestrus reaction of the vagina, but a vaginal mucification; at the same time, the corpora lutea became hypertrophic as they did in a normal female injected with the same dosage of oestrogenic hormone.

These early esperiments already pointed to the dual nature of the hypothalamic control of the gonadotrophic hypophyseal activity. At the same time numerous results stressed the importance of the hypothalamic contact for the elaboration of LH and the other pituitary hormones. Harris and Jacobsohn studying pituitary transplants showed that the graft could restore regular cycles only when it was located under the median eminence and received its vascularization from the portal vessels (1952). Everett confirmed our results (1954) and in addition made the very important observation we could also confirm, (Desclin, 1956), that the hypophyseal graft without any oestrogenic treatment, liberates luteotrophin and activates progesterone secretion.

In very ingenious experiments, Nikitovitch-Winer and Everett showed also that in hypophysectomized female rats, pituitary transplanted under the kidney capsule, apparently secretes luteotrophin only, but, retransplanted under the median eminence restores oestrus cycles and contains gonadotrophic basophils.

Everett and Harris from these and many other evidences, concluded that the hypothalamus has two opposite actions in its gonadotrophic control. One concerns the elaboration and liberation of FSH and LH and is an excitatory mechanism. The other one is inhibitory and concerns the elaboration and liberation of prolactin.

Many results have confirmed this conception. A LH- and a FSH-releasing factor have been isolated. A prolactin-inhibiting factor has been convincingly demonstrated. A corticotrophic releasing factor and a TSH releasing-factor have also been isolated.

It is not my purpose to insist on these aspects of the problem. One of the main points which remains obscure is surely the site of elaboration of these various mediators in the hypothalamus.

Some very important results have emerged from the study of hypothalamic lesions and numerous researches have been devoted to this aspect of the problem. Results coming from the study of hypophyseal grafts have been fewer and, for this reason, I should like to report the results of some of our experiments on this line, concerning hypophyseal grafts implanted into the hypothalamus. As I hope to show, this technique may give interesting indications on the site of elaboration of the releasing factors.

But even some points concerning the pituitary graft under the kidney capsule are worth mentionning before we consider the study of the hypothalamic pituitary transplants.

As we have already said, there is ample evidence that the pituitary transplanted under the kidney capsule produces great amounts of prolactin and it is generally stated that this is the only hormone produced under these conditions.

Some discording results however derive from the study of pituitary grafts in hypophysectomized males which, after a rather long time, are irregularly able to restore normal testicles and spermatogenesis (Cutuly, Courrier and Colonge, Ahren, etc...).

The histologic study of pituitary transplants has shown that the tissue contains one single type of cells. These are poorly granulated eosinophilic elements. Electron microscopy has defined the caracteristics of these cells. They are undoubtedly prolactin cells. However, in these long standing hypophyseal transplants in the male, Herlant, Courrier and Colonge were able to detect some gonadotrophs and it is fair to say that these grafts are able to elaborate some FSH and also some LH as well.

But even in females, these grafts seem also to elaborate some FSH and some LH.

We have with G. Carpent studied pituitary transplants under the kidney capsule in pregnant rats. These animals were hypophysectomized the first day of pregnancy. It has been usually stated that, under these conditions, the blastocyst does not implant unless some oestrogen is administered. In our experiments, without any treatment, implantation could take place, though regularly delayed and sometimes asynchronous.

The actual evidence is that blastocyst implantation is only possible if some oestrogen has been produced under the influence of the pituitary graft. Electron microscopic studies of pituitary grafts under the kidney capsule show that FSH and LH cells are still present in this tissue (Rennels, Potvliege). The overwhelming dominant type of cells is however the typical prolactin cell.

In conclusion, we might say that the pituitary graft without contact with the hypothalamus has a very uniform structure and does not, with the usual techniques show basophilic cells, either gonadotrophs or thyrotrophs. When present, these elements are very rare. If some FSH and LH and also TSH is secreted, this is only a residual secretion. This structure is entirely different if the graft is in contact with the median eminence as has been shown in the experiments of Nikitovitch-Winer and Everett. Basophils are numerous under these conditions.

So we now come to the study of pituitary grafts implanted, not under the median eminence, in contact with the portal vessels, but within the hypothalamus itself. These experiments have been done in my laboratory by J. Flament-Durand. Small pituitaries from 10 days old rats have been implanted with a Krieg-Johnson stereotaxic apparatus into the hypothalamus of adult male and female rats. Some of these rats were hypophysectomized. Various types of experiments have been done to study the types of cells, gonadotrophs, thyrotrophs and prolactin cells, and also to study the physiologic activity of the transplanted gland.

If we consider first the morphologic results, we can summarize them by saying that the transplant outside a definite zone of the hypothalamus shows the same structure as the pituitary grafted under the kidney capsule and contains only finely granulated eosinophilic prolactin cells. However, if the graft has been implanted in a territory including the ventral and median parts of the hypothalamus and extending from behind the optic chiasma to the median eminence, its morphological appearance is entirely different. Numerous basophilic cells are clearly evident in the grafted tissue. Gonadotrophs are numerous and located mainly in contact with the arcuate nuclei. They change to castration cells after spaying or castration in the male. They also become castration cells in hypophysectomized animals. Thyrotrophs are more anteriorly and more dorsally located in a distinctly different zone. They are particularly impressive after thiouracyl administration and transform to thyroidectomy cells.

There is thus, in the hypothalamus, an hypophyseotrophic area whose localisation agrees with the observations of Halász and col. However, as far as the thyrotrophic area is concerned, our results differ from those of the Hungarian workers. Thyrotrophs may surely be present even in grafts without any contact with the portal vessels as is also the case for the gonadotrophs. Their localisation is different and can be easily demonstrated in castrated animals given thiouracyl and treated with thyroxin. In these animals,the vacuolated gonadotrophs remain unchanged around the arcuate nuclei whereas the thyroidectomy cells anteriorly located have regressed.

We have also studied the gonadotrophic activity of the grafts as well as their thyrotrophic function and also their luteotrophic capacity. As for the first of these functions, among 93 grafted hypophysectomized females, 8 only showed the return of oestrus cycles after hypophysectomy. They were put together with males. Six accepted the male and spermatozoa were found in the vagina several times. They never became pregnant, no pseudo-pregnancy resulted from these infertile coïtus. The ovaries of these animals were definitely heavier than those of the other hypophysectomized grafted animals and significantly heavier than those of hypophysectomized non grafted rats. The hypophyseal grafts were well within the hypophyseotrophic area and contained many vacuolated basophils with the appearance of castration cells. More than 40 animals whose graft was equally within the hypophyseotrophic area and contained also many vacuolated basophils, remained in dioestrus and the ovaries were small without significant difference from those of hypophysectomized controls.

Among 24 hypophysectomized grafted males, 4 had normal testicles with full spermatogenesis. Three of them had pituitary grafts located in the gonadotrophic area and showed very numerous castration cells even with fully developed testicles. In the other case, the graft was suprachiasmatic and no gonadotrophic cells could be observed in spite of the development of the gonads.

Several remarks seem to be pertinent concerning these results. 1°) The rather poor functional activity of these hypothalamic

grafts which in spite of their very active morphologic appearance are unable to restore fully functional activity of the ovaries in the female. A relation between the size of the graft and its functional activity seems to be unimportant.
2°) Lesions produced by placing the graft and disrupting hypothalamic connexions afford no explanation for this puzzling situation. Previous to hypophysectomy the animals have shown regular cycles and the graft "per se" did not disturb the cyclic recurrence of oestrus. In normal animals bearing an hypothalamic graft whatever the site of the graft could be, no pseudogestation was observed. Apparently the most plausible explanation would be the difficulty for these grafts to liberate their hormones into the blood due to a disturbance in their venous drainage. This situation applies also to the males. In addition however, in one case, restoration of normal testicles was induced when the graft was outside the hypophyseotrophic area. This particular observation can be compared with similar results reported in the male for pituitary grafted under the kidney capsule.

There must be also a definite alteration in the feed-back mechanisms of the sexual hormones. Many castration cells are present in the grafts located in the gonadotrophic area. They are absent in normal animals but quickly developed after castration. These gonadotrophs are apparently very sensitive to oestrogen deprivation. However, they are also present in hypophysectomized animals whose gonads have not been removed. Even when the gonads were fully restored in the male, as judged from their weight and from their structure, very numerous castration cells are still to be found.

We should like to mention also briefly our results concerning the luteotrophic activity of the pituitary grafts where we are facing a similar situation. It is well established and we have confirmed that the transplantation of a pituitary under the kidney capsule in a normal female induces a succession of pseudogestations resulting from the continuous production of prolactin released from the transplant.

In very numerous females bearing hypophyseal grafts in the hypothalamus, outside the so called hypophyseotrophic area, we never observed pseudopregnancies. We also mentioned that in hypophysectomized rats whose anterior lobe had been transplanted under the kidney capsule, administration of oestrogen does not induce vaginal oestrus but results in mucification of the vagina. We tried to see if oestrogen treatment would equally be luteotrophic in female rats with hypothalamic grafts. These animals were first implanted and then treated with 15 gammas a day of oestradiol benzoate. These animals showed very rapidly a dioestrus. Their ovaries contained large corpora lutea and the vagina was mucified. If during this treatment, the hypophysis is removed, oestrus reappears within a few days.

The intrahypothalamic hypophyseal graft seems not to be able to release into the blood the prolactin necessary to maintain active corpora lutea. The structure of the graft however has been modified and is very impressive. Prolactin cells, very active, are the dominant element in the pituitary tissue. This change takes place in every graft and is completely independant of its location.

These last results seems to be in favour of a direct action of oestrogen on the pituitary itself, rather than an indirect hypothalamic action, as you know a very controversial question. Here again, the pituitary tissue seems to have difficulties in the liberation of its products into the blood. The situation is different as far as the thyrotrophic activity is concerned. Providing that the graft has been located in the thyrotrophic area, thyroid weight and height of the epithelial cells is restored to normal.

Administration of propylthiouracyl in implanted hypophysectomized animals is definitely goitrogenic. Hyperplasia of the thyroid takes place even when the grafts have no contact with the portal vessels. These results are thus different from those of Halåsz and col. Our functional results as well as our morphological observations give thus a more precise aspect of the location of the thyrotrophic area.

## Bibliography


Ahren, K., Arvill, A. and Hjarlmarson.
Endocrinology, 71, 176-178, 1962.

Carpent, G. and Desclin, L.
C.R. Acad. Sci.(Paris) 260, 4618, 1965.

Courrier, R. and Colonge, A.
C.R. Acad. Sci. (Paris) 245, 388, 1957.

Cutuly, E.
Anat. Rec. 80, 83-97, 1941.

Desclin, L.
C.R. Soc. Biol. (Paris) 120, 526-528, 1935.

Desclin, L.
Ann. Endocrin. (Paris) 11, 656-659, 1950.

Desclin, L.
Ann. Endocrin. (Paris) 17, 586-595, 1956.

Desclin, L. and Grégoire, Ch.
C.R. Soc. Biol. (Paris) 121, 1366-1368, 1936.

Everett, J.W.
Endocrinology 54, 685-690, 1954.

Flament-Durand, J.
Ann. Soc. roy. Sci. med. et natur. (Bruxelles) 19,1-119,1966.

Halasz, B., Pupp, L. and Uhlarik, S.
J. Endocrin. 25, 147-154, 1962.

Halasz, B., Pupp, L., Uhlarik, S. and Tima, L.
Endocrinology 77, 343-355, 1965.

Harris, G.W. and Jacobsohn, D.
Proc. roy. Soc. B. 139, 263-276, 1952.

Herlant, M., Courrier, R. and Collonge, A.
C.R. Acad. Sci. (Paris) 250, 1770-1774, 1960.

Hohlweg, W. and Junkmann, K.
Klin. Wschr. 11, 321-323, 1932.

Nikitovitch-Winer, M.B. and Everett, J.W.
Endocrinology 65, 357-368, 1959.

Rennels, E.G.
Endocrinology 71, 713, 1962.

Potvliege, P.
Journal de Microscopie 4, 485, 1965.

# THE PHYSIOLOGY AND BIOCHEMISTRY OF LUTEINIZING HORMONE-RELEASING FACTOR AND FOLLICLE STIMULATING HORMONE-RELEASING FACTOR

S. M. McCann,[1] S. Watanabe,[2] D. B. Crighton, D. Beddow and A. P. S. Dhariwal
University of Texas Southwestern Medical School at Dallas, Texas

Hypothalamic control over the secretion of the anterior pituitary gland has become almost universally accepted and attention has shifted to attempts to clarify the mechanism by which the hypothalamus influences the gland. A great deal of indirect evidence has supported the concept that this control is exerted by a neurohumoral pathway via the hypophysial portal vessels. The following discussion will summarize the evidence obtained in our laboratory for the existence, physiological significance, chemical nature, and mode of action of two neurohumoral agents which appear to regulate the output of gonadotrophins. The first of these to be discovered was a luteinizing hormone-releasing factor (LRF), the second a follicle stimulating hormone-releasing factor (FRF). Each of these substances appears to regulate the output of the respective pituitary trophic hormone.

Luteinizing hormone-releasing factor. In 1959 we began an intensive search for a possible LH-releasing action of hypothalamic extracts and employed a new sensitive and specific assay for LH, the ovarian ascorbic acid depletion test of Parlow (1). In immature female rats which have been pretreated with gonadotrophins, minute doses of LH evoke a decline in ovarian ascorbic acid concentration. Other anterior lobe hormones are without effect.

Crude acidic extracts of rat stalk-median eminence (SME) tissue were capable of depleting ovarian ascorbic acid and a dose-response relationship was obtained (2). Similar activity was found in ovine or bovine hypothalamic extracts (2,3). Since the extract exhibited little or no activity upon injection into hypophysectomized rats, it appeared that contamination with LH could not explain these results and that a release of LH from the pituitaries of the test rats had occurred. Furthermore, heating for 10 min in a boiling water bath inactivated rat pituitary LH while leaving unchanged the activity of the hypothalamic extract. The results also could not be explained by the content of known physiologically active substances which were present in the extract, such as vasopressin, oxytocin, substance P, histamine, serotonin, or epinephrine. It should be pointed out, however, that sufficient doses of vasopressin can deplete ovarian ascorbic acid, apparently by a direct action on the ovary (4) and can even release LH in one sensitive test system (5).

Similarly prepared extracts of cerebral cortex were ineffective which indicated that the action of hypothalamic extracts was a specific one. Extraction of various hypothalamic areas revealed that the major LH-releasing activity was concentrated in the SME region. There was minimal activity in the overlying ventral hypothalamus (2). At this point the unknown substance responsible for the LH-releasing effect was designated LRF.

Since the immature test rats used in evaluating LH-releasing activity were treated with large doses of gonadotrophins and had relatively low stores of hypophysial LH (2,6), it was advisable to determine if LH-releasing activity could be demonstrated in more physiological circumstances. The LH-releasing action of hypothalamic extracts has now been evaluated in a variety of test animals. In these experiments, the extract was injected intravenously into the test rats which were bled 10 min later from the jugular vein. The effect of the extract on plasma LH activity of the test animals was estimated by the ovarian ascorbic acid depletion test. The factor has been found to be effective in a variety of situations. It is active in normal female rats and in ovariectomized rats in which the release of LH has been inhibited either by administration of gonadal steroids or by lesions in the median eminence (ME) (2). This latter observation is important because it indicates that the LRF acts directly on the anterior pituitary to release LH. An indirect action via the nervous system would have been blocked by these lesions which interrupted neural control over LH secretion. Further evidence that the LRF acts directly on the gland to release LH has been provided by the experiments of Campbell *et al*. (7) in rabbits and Nikitovitch-Winer (8) in rats. They showed that infusion of hypothalamic extract directly into the anterior lobe of the pituitary could evoke ovulation. Systemic administration of the same dose of extract was without effect. Schally & Bowers (9) have demonstrated an LH-releasing action of hypothalamic extracts on pituitaries incubated *in vitro*, and we have recently confirmed this observation (Watanabe & Ratner, unpublished data, 1965). This provides further evidence that LRF acts directly on the hypophysis.

Rather surprisingly, LRF failed to elevate plasma LH in untreated ovariectomized rats. Plasma LH is already elevated in the ovariectomized rat because of the elimination of ovarian steroid negative feedback, so we have postulated that this animal is already responding maximally to release of endogenous LRF and is incapable of responding further to exogenous releasing factor (2).

Physiological significance of LRF. It is important to determine if changes in the rate of secretion of LRF are responsible for bringing about the alterations in secretion of LH which occur during the estrous cycle and after administration of gonadal steroids.

LH secretion fluctuates during the estrous cycle and reaches a peak at proestrus just prior to ovulation. Pituitary LH begins to decline at this time and reaches a minimum after ovulation. Accompanying the rise in LH release at proestrus is a fall in content of hypothalamic LRF (10,11). This observation is consistent with the view that LRF triggers the preovulatory discharge of LH.

One could also explain the ovulatory discharge of LH by postulating increased responsiveness of the pituitary at proestrus to a constant release of LRF. Consequently, we have evaluated the responsivity of the pituitary to purified LRF at various stages of the estrous cycle. Doses of 0.3 or 0.1 ml of LRF were clearly effective in elevating plasma LH within 10 min of their intravenous injection at all stages of the cycle (12). A dose of 0.02 ml was ineffective at all stages of the cycle although there was a suggestion of a positive response during proestrus.

The data to date are consistent with the thesis that an increased release of LRF mediates the discharge of LH that triggers ovulation. Conclusive proof will require demonstration of increased levels of LRF in hypophysial portal blood at this stage of the cycle.

A hypothalamic site for the inhibitory action of gonadal steroids on gonadotrophin secretion has been suggested from the results of implanting minute amounts of estrogen or testosterone into the hypothalamus (13,14). Estrogen implants can inhibit LH secretion in gonadectomized rats and can augment prolactin release in intact females (15,16); however, implants of estrogen in the anterior pituitary are also effective (17,15). This latter observation clearly establishes that these steroids can alter gonadotrophin and prolactin secretion by an action directly on the pituitary gland. Bogdanove (17) has stressed the possibility that the results obtained with hypothalamic implants may be due to uptake of steroid in portal vessels with distribution to the pituitary where their effect is mediated. Although this possibility seems unlikely, it made it necessary to employ other approaches to establish a hypothalamic site for the inhibitory effects of gonadal steroids.

If the inhibitory effect of gonadal steroids were exerted primarily on the pituitary, one would predict that large doses of these steroids would block the action of administered LRF. Such is not the case. Even after pretreatment of ovariectomized rats for 3 days with massive doses of estrogen and progesterone, LRF was effective in elevating plasma LH. If anything, it appears that these steroid-blocked rats are supersensitive to the transmitter (5). These results speak for an inhibitory effect of the steroids on the hypothalamus rather than the pituitary gland itself. In further recent studies of this question, it has been shown that addition of increasing doses of soluble equine estrogens to male rat pituitaries incubated for 6 hrs *in vitro* failed to block the action of added LRF., although they produced a slight reduction in the "basal" release of LH in the absence of added releasing factor (Crighton & McCann, unpublished data).

If changes in gonadal steroid titer were accompanied by alterations in the level of LRF stored in the hypothalamus, evidence for an effect of these steroids on secretion would be provided. We have been unable to detect an effect of castration on the hypothalamic content of LRF (10). Administration of estrogen also failed to alter the stored LRF, but a large dose of testosterone propionate produced a decline in the level of stored factor. Implants of either estradiol or testosterone in the median eminence were capable of lowering the content of LRF. Implants of testosterone in the pituitary produced no alteration in the level but estradiol implants in the pituitary rather surprisingly produced a significant increase in stored LRF. A speculative explanation for this latter observation involving a possible negative feedback of LH itself on the hypothalamus has been advanced (10). Taken altogether, the observations to date are consistent with an effect of altered levels of gonadal steroids on the content of stored LRF.

Direct evidence for an effect of these steroids on LRF secretion would be provided if alterations in the level of circulating LRF could be produced by administration of gonadal steroids. No detectable LRF can be found in plasma of normal rats; however, several months after hypophysectomy the plasma contains a factor which will deplete ovarian ascorbic acid (18,19). The factor vanishes within 24 hrs of coagulation of the ME. Since this procedure would eliminate the site of storage and release of LRF, we concluded that the factor circulating in blood of chronically hypophysectomized rats was LRF. Unfortunately the circulating LRF levels are rather low, so that they are frequently undetectable with present methods. This has hampered our investigation of the factors which alter the level of the circulating LRF; however, preliminary experiments suggest that the level of circulating LRF may be lowered by pretreatment of the hypophysectomized rats with either LH itself (Nallar, unpublished data, 1964) or a combination of estrogen and progesterone (Antunes-Rodrigues, unpublished data, 1965). Recently, Frankel *et al*. (20) have reported an ovarian ascorbic acid depleting activity in hypophysectomized chicken plasma which may also represent circulating LRF.

Ramirez & Sawyer (21) have recently found a decrease in the content of stored LRF at puberty which would suggest that an increased discharge of this factor may stimulate the release of LH which occurs at this time in the life cycle.

Although more experimentation is needed, it appears highly likely that most changes in LH secretion are mediated by variations in the release of LRF. Some changes may be exerted via direct inhibitory effects of gonadal steroids on the pituitary gland.

Follicle stimulating hormone releasing factor. It has now been possible to demonstrate an FSH-releasing action of hypothalamic extracts(22,23). For most of these experiments a new, sensitive method for assay of FSH was employed, the mouse uterine weight augmentation assay (24), but we have also obtained similar results with the rat ovarian weight augmentation method of Steelman and Pohley (25,26). Hypothalamic extracts elevated plasma FSH when injected iv into ovariectomized rats in which the release of FSH had been inhibited, either by hypothalamic lesions, or by treatment with both estrogen and progesterone. Extracts from cerebral cortex were inactive and the results were not caused by contamination of the extract with FSH, itself, or with vasopressin or oxytocin.

Kuroshima et al. (27) have confirmed these results and have also shown as have Mittler and Meites (28) that the FRF will increase the release of FSH into the media of pituitaries incubated *in vitro*. Watanabe & McCann (29) have been able to confirm the *in vitro* FRF activity of rat and sheep hypothalamic extracts. In these latter studies the method of Steelman & Pohley was used for assay of FSH.

Our procedure has been essentially that of Mittler & Meites (30). The period of incubation has been 6 hrs. Crude or purified murine or ovine hypothalamic extracts can produce a several fold augmentation in the release of FSH from the pituitaries which is not accompanied by any significant reduction in pituitary FSH content. Cerebral cortical extracts and muscle extracts in equivalent doses have been without effect; however, a significant stimulation of FSH release above basal levels is obtained with a minimal effective dose of 1/16 - 1/8 of a rat's SME/pituitary and a dose-response relationship can be demonstrated. The *in vitro* assay appears to be the most sensitive yet devised for FRF, and in our hands has been shown to be quite specific. Martini's group (31) and Ishida *et al*. (32) have reported that hypothalamic extracts can deplete pituitary FSH in normal males and ovariectomized steroid-blocked females, respectively.

The *in vitro* method has also been employed to determine if the content of FRF in the SME is affected by castration (Watanabe & McCann, unpublished data, 1966). A small but significant ($P < .01$) decline in the content of stored FRF has been observed 15-29 days after spaying in Sherman strain rats Mittler & Meites (30) reported a rise in content of the factor in spayed Sprague-Dawley rats examined 14 days after the operation and also observed a decline in activity following testosterone therapy. The reason for the different response to castration observed by the 2 groups remains to be eludicated; however, the fact that altered levels of gonadal steroids change the level of stored FRF suggests that the factor plays a physiological role in altering FSH secretion by the adenohypophysis.

Localization of the gonadotrophin releasing factors. We have have employed the *in vitro* assay for FRF (Watanabe & McCann, unpublished data) and LRF (Crighton & McCann, unpublished data) to determine the localization of these factors in the hypothalamus. Frozen sections were cut at 100$\mu$ along 3 planes, i.e., sagittal, horizontal and frontal. The sections were pooled to give tissue 0.7-1.5 mm in thickness, and were assayed at a dose of 0.075-0.2 hypothalamic equivalents/pituitary. In the case of FSH release, horizontal sections were biologically active only when the basal zone was assayed. Activity was found only with the medial zone obtained with sagittal sections. Two sections were active with the frontal cut. These corresponded to the sections 1.5-2.8 and 2.8-3.6 mm caudal to the rostral edge of the optic chiasm. Thus, biological activity was localized to the medial, basal, anterior, tuberal region, an area corresponding to the median eminence of the tuber cinereum. The results clearly indicate that the FRF is stored in the median eminence region. The absence of activity in other areas of the hypothalamus suggests that the cell bodies of the neurons which secrete the FRF are probably located in this region as well, perhaps in the vicinity of the arcuate mucleus; however, it is possible that neurons at a greater distance may snythesize the factor but store it in too small amounts to be detectable at the dose of extract used so far.

When the localization of the LRF was determined by this method, activity was also localized to a basal zone of the hypothalamus but extended over a wider area. Activity was found as far rostrally as the optic chiasm and persisted as far caudally as the point of separation of the pituitary stalk from the hypothalamus. Thus neurons which synthesize LRF are still found rostral to the median eminence and arcuate nucleus in the hypothalamic region which would contain the suprachiasmatic and anterior hypothalamic nuclei.

One could postulate that the cell bodies of the neurosecretory neurons which secrete this transmitter are localized to this rostral region and that their axons extend to the ME where the factor is released into portal vessels; however, it has been shown that lesions which destroy this rostral region fail to block the castration-induced rise in plasma and pituitary LH secretion (33). This must mean that at least some of the neurosecretory neurons are located caudal to this region. We would postulate that these neurons are rather widely distributed in the medial, basal,tuberal region and that all of their axons project to the median eminence, there to release LRF into the portal vessels.

Our localization for the LRF and FRF is consistent with the findings that pituitaries grafted to this region retain gonadotrophs (34,35) and that gonadotrophin secretion can continue after isolation of the medial basal tuberal region from the rest of the CNS (36).

Residual function of the pituitary in the absence of hypothalamic influence. When the pituitary gland is removed from its normal position and grafted to a distant site, gonadotropin secretion is markedly impaired in agreement with the concept of neurohumoral control of the gland via the hypophysial portal vessels; however, if multiple grafts are so located, considerable residual function, which is roughly proportional to the mass of transplanted tissue, can be demonstrated (37). It was important to determine if this represented a residual autonomous functional capacity of the pituitary, or if it was caused by circulating releasing factors which reached the glands in their new location via the systemic circulation. Recent studies (Beddow & McCann, unpublished data) have confirmed the finding that multiple pituitary grafts in hypophysectomized immature rats can result in considerable testicular and accessory organ maintenance, although the degree of maintenance was quite variable. When these hypophysectomized grafted rats were unilaterally castrated, the opposite testis increased slightly in size on removal 3 weeks later; however, when ME lesions were placed at the time of unilateral castration, a marked reduction in the weight of the remaining testis and of the accessory organs followed. Sham lesions in several animals and a unilateral hypothalamic lesion in one rat failed to produce the effect. These results provide strong support for the view that the residual gonadotrophin secretion of the grafted pituitary is the result of stimulation by peripheral circulating releasing factors. Additional evidence for a circulating LRF in the hypophysectomized rat was cited earlier.

Mechanism of action of releasing factors. Little attention has been paid to the possible mechanism of action of the various releasing factors at the level of the pituitary cell. Most studies have dealt with release of hormone from the gland, but a few *in vitro* and *in vivo* studies point up the fact that these agents probably influence synthesis as well as release of pituitary hormones. This suggested the possibility that agents which inhibit protein synthesis might influence the action of the releasing factors. Two such agents are the antibiotics puromycin and actinomycin. Puromycin is thought to exert its blocking effect by acting as a structural analogue of esterified sRNA, whereas actinomycin is thought to block DNA-directed RNA synthesis. Both of these antibiotics have been found capable of blocking a variety of hormonal effects.

The addition of either actinomycin or puromycin to the incubation mediam has been found to inhibit the FSH release in response to either crude rat hypothalamic extract or partially purified ovine FRF (38). The additions produced a somewhat more reproducible effect if a 30 min preincubation period in the presence of the antibiotic preceded the 6 hrs of incubation in the presence of both hypothalamic extract and antibiotic; however, significant results were obtained by either procedure. Approximately equal effects were obtained by doses of 10-40$\mu$g/ml of actinomycin and 10-100$\mu$g/ml of puromycin. These dosages are in the range found to be effective in blocking the response of adrenal cortex to ACTH *in vitro* (39,40) and were effective in blocking protein synthesis in the pituitary in the case of puromycin and RNA synthesis in the case of actinomycin. They did not interfere with the assay of the

FSH released into the incubation medium. Interestingly enough, the effective doses of actinomycin and puromycin did not influence the small basal release of FSH from the incubated pituitaries, but only blocked the enhanced release obtained with hypothalamic extract.

When pituitaries were incubated in the presence of both hypothalamic extract and antibiotic, pituitary FSH either was unchanged or increased somewhat on comparison with levels found with hypothalamic extract alone. This suggests that the effects of the antibiotics were primarily to block the augmented release of FSH as a result of the action of hypothalamic extract. It is tempting to suggest that the action of the FRF in promoting FSH release from the pituitary cell requires new protein synthesis.

We have observed (Crighton, Watanabe, Dhariwal & McCann, unpublished data, 1967) that neither of the antibiotics effects the release of LH under identical conditions to those which interfere with FSH release. Apparently the mechanism of action of LRF in evoking LH release differs from that of FRF in promoting FSH discharge.

While these experiments were in progress we became aware of similar experiments carried out by Dr. M. Jutisz of the College of France (41). He also reported that both puromycin and actinomycin blocked the FSH discharge provoked by FRF, but he noted similar results with LRF. Our findings are not in agreement with his on this latter point.

Studies on the chemistry of the factors which alter gonadotrophin secretion. Considerable information is now available on the chemical nature of the various hypothalamic chemotransmitters which affect anterior pituitary secretion. FRF, LRF and PIF all retain their activity after heating for 10 min in a boiling water bath. By contrast, rat LH is completely inactivated, and bovine and ovine LH are partially inactivated after similar treatment (2, 42,23). The LRF and FRF are partially or completely inactivated by the proteolytic enzymes, pepsin or trypsin (2,42, Watanabe, Dhariwal & McCann, unpublished data, 1966). This observation indicates that intact peptide bonds are required for their activity. Carboxypeptidase failed to inactivate the FRF in one experiment. Thioglycollate splits the disulfide bridge in oxytocin and vasopressin, thus inactivating the molecules, but this treatment is without influence on LRF or hypothalamic CRF (16). Consequently, it appears likely that the chemical structures of LRF and CRF, at least with respect to the disulfide bridge, are dissimilar from that of the known neurohypophysial polypeptides.

Recently, several attempts have been made to purify the LRF, using ovine or bovine extracts as the starting material. Gel filtration on Sephadex G-25 was first used by Porath & Schally (43) for separation of posterior lobe hormones and has proved to be a good method for purification of the various hypothalamic releasing factors. In our hands, using ovine SME extract and ammonium acetate as the eluting buffer, LRF has been eluted from the Sephadex column just prior to the emergence of vasopressin (3). With a sufficiently long column it has been possible to separate completely LRF from vasopressin by gel filtration alone. Similar results were obtained with bovine or ovine extracts and acetic acid or pyridine acetate as the eluent (44,45,46). Further purification of LRF has been achieved by chromatography on carboxymethylcellulose (CMC) (3). After careful desalting of the extract with glacial acetic acid and by several passages through the CMC column, LH-releasing activity was retained on the ion exchanger and was eluted by application of ammonium acetate buffers of increasing pH and ionic strength. The LRF was eluted from the column just prior to elution of residual vasopressin which still contaminated the fractions obtained from the Sephadex column. The retention of the LRF on the CMC column suggests that it is a basic polypeptide. By this means LRF, essentially free of vasopressin (<5mU/dose) was obtained, which was active at a dose of $<3\mu g$ of peptide. The highly purified material migrated on paper electrophoresis as a major component with a trailing minor component. A similar purification of bovine LRF has recently been announced by Schally _et al_. (45).

Gel filtration on Sephadex is sufficient to separate the FRF from the LRF (46). In these experiments we employed a tall column of Sephadex G-25 and used ammonium acetate as the eluting buffer. Earlier experiments with a shorter column and employing acetic acid as the eluent had failed to separate the two activities (26). Further chromatography of the active fractions on CMC has resulted in the preparation of highly purified FRF (47), which was active in the _in vitro_ assay system at a dose of $<1\mu g$ of peptide. Residual LRF which had contaminated the FRF was separated from it by the ion exchange chromatography.

Prolactin-inhibiting activity has also been localized in the effluent from the Sephadex column (48); however, to date it has not been possible to separate consistently this activity from the LRF. The 2 activities have resided in the same tubes although there has been a tendency in 3 of 5 runs for the PIF to begin emerging from the column just prior to the LRF. Even CMC-purified LRF, active at a dose of only a few $\mu g$ of peptide, possessed prolactin-inhibiting activity as indicated by blockade of nursing-induced depletion of pituitary prolactin. Clearly, a manifold purification of PIF was obtained without achieving a separation from LRF. Since there is often a reciprocal relationship between the secretion of LH and prolactin, i.e., after ME lesions, in lactation or in pseudopregnancy, it is possible that LRF and PIF represent 2 activities of a single molecule. On the other hand further purification may clearly separate the 2 activities.

Little difficulty has been encountered in separating the gonadotrophin-releasing factors and PIF from the other hypothalamic pituitary stimulating hormones. Gel filtration through Sephadex is sufficient to achieve the separation. The order of elution of the various factors has been as follows: growth hormone-releasing factor (GRF), CRF, melanocyte stimulating hormone-releasing factor (MRF), FRF, PIF and LRF. LRF is followed by vasopressin. Since molecular size is the principal factor affecting mobility on Sephadex, the fact that LRF is eluted from these columns just prior to vasopressin suggests that this factor is a small polypeptide slightly larger than vasopressin and oxytocin. The other releasing factors would by this reasoning appear to be somewhat larger molecules than LRF. Highly purified CRF and GRF also differ from the LRF in amino acid composition (48).

By contrast with vasopressin and oxytocin, these highly purified factors (CRF, GRF, and LRF) had an absence of sulfur-containing amino acids in their molecules. We have already alluded to the resistance of CRF and LRF reduction by thioglycollate which splits the disulfide bridge in the molecules of vasopressin and oxytocin. Consequently, it would appear that the hypothalamic factors which affect pituitary secretion constitute a new family of relatively small, basic polypeptides.

## REFERENCES

1. Parlow, A. F., In Human Pituitary Gonadotrophins, Albert, A., ed., C. C. Thomas, Springfield, 1961, p. 300-10.
2. McCann, S. M., and V. D. Ramirez, Rec Prog Hor Res 20: 131, 1964.
3. Dhariwal, A. P. S., J. Antunes-Rodrigues, and S. M. McCann, Proc Soc Exp Biol and Med 118: 999, 1965.
4. McCann, S. M., and S. Taleisnik, Am J Physiol 199: 847, 1960.
5. Ramirez, V. D., and S. M. McCann, Endocrinology 73: 193, 1963.
6. Novella, M., J. J. Alloiteau, and P. Ascheim, Compt Rend 259: 1953, 1964.
7. Campbell, H. J., G. Feuer, and G. W. Harris, J Physiol 170: 474, 1964.
8. Nikitovitch-Winer, M. B., Endocrinology 70: 350, 1962.
9. Schally, A. V., and C. Y. Bowers, Endocrinology 75: 312, 1964a.
10. Chowers, I., and S. M. McCann, Endocrinology 76: 700, 1965.
11. Ramirez, V. D., and C. H. Sawyer, Endocrinology 76: 1158, 1965.
12. Antunes-Rodrigues, J., A. P. S. Dhariwal, and S. M. McCann, Proc Soc Exp Bio and Med 122: 1001, 1966.
13. Lisk, R. D., Canad J Biochem 38: 1381, 1960.
14. Davidson, J. M., and C. H. Sawyer, Proc Soc Exp Biol and Med 107: 4, 1961.
15. Ramirez, V. D., R. Abrams, and S. M. McCann, Endocrinology 75: 243, 1964.
16. Ramirez, V. D., and S. M. McCann, Am J Physiol 207: 441, 1964a.
17. Bogdanove, E. M., Endocrinology 72: 638, 1963.
18. Nallar, R., and S. M. McCann, Endocrinology 76: 272, 1965.
19. Callantine, M. R., R. R. Humphrey, S. L. Lee, B. L. Windsor, N. H. Schotten, and O. P. O'Brien, Endocrinology 79: 153, 1966.
20. Frankel, A. I., W. R. Gibson, J. W. Graber, D. M. Nelson, L. E. Reichert, Jr., and A. V. Nalbandov, Endocrinology 77: 651 1965.

## References Continued

21. Ramirez, V. D., and C. H. Sayer, Endocrinology 78: 958, 1966.
22. Igarashi, M., and S. M. McCann, The Endocrine Society, Program of the 45th Meeting, p. 29, 1963 (Abstract).
23. Igarashi, M., and S. M. McCann, Endocrinology 74: 446, 1964a.
24. Igarashi, M., and S. M. McCann, Endocrinology 74: 440, 1964.
25. Steelman, S. L., and F. M. Pohley, Endocrinology 53: 604, 1953.
26. Igarashi, M., R. Nallar, and S. M. McCann, Endocrinology 75: 901, 1964.
27. Kuroshima, A., Y. Ishida, C. Y. Bowers, and A. V. Schally, Endocrinology 76: 614, 1965.
28. Mittler, J. C., and J. Meites, Proc Soc Exp Biol and Med 117: 309, 1964.
29. Watanabe, S., and S. M. McCann, Fed Proc 26: 365, 1967 (Abstract).
30. Mittler, J. C., and J. Meites, Endocrinology 78: 500, 1966.
31. David, M. A., F. Fraschini, and L. Martini, Compt Rend 261: 2249, 1965.
32. Ishida, Y., A. Kuroshima, C. Y. Bowers, and A. V. Schally, Proc XXIII Int Congr Physiol Sci (Tokyo), p. 276, 1965.
33. Antunes-Rodrigues, J., and S. M. McCann, Endocrinology(In press
34. Halasz, B., W.H. Florsheim, N.L. Corcorran, and R.A. Gorski, Endocrinology 80: 1075, 1967.
35. Flament-Durand, J., and L. Desclin, Endocrinology 75: 22, 1964.
36. Halasz, B., and R. A. Gorski, Endocrinology 80: 608, 1967.
37. Gittes, R. F., and A. J. Kastin, Endocrinology 78: 1023, 1966.
38. Watanabe, S., A. P. S. Dhariwal, and S. M. McCann, Endocrinology (submitted for publication) 1967.
39. Ferguson, J. J., J Biol Chem 238: 2754, 1963.
40. Farese, R. V., Endocrinology 78: 929, 1966.
41. Jutisz, M., and P. de la Llosa, Compt Rend 264: 118, 1967.
42. Guillemin, R., Rec Progr Hor Res 20: 89, 1964a.
43. Porath, J., and A. V. Schally, Endocrinology 70: 738, 1962.
44. Ramirez, V. D., R. Nallar, and S. M. McCann, Proc Soc Exp Biol and Med 115: 1072, 1964a.
45. Schally, A. V., and C. Y. Bowers, Endocrinology 75: 608, 1964b.
46. Guillemin, R., M. Jutisz, and E. Sakiz, Compt Rend 256: 504, 1963.
47. Dhariwal, A. P. S., S. Watanabe, A. Ratner, and S. M. McCann, Neuroendocrinology (In press).
48. Dhariwal, A. P. S., C. Grosvenor, J. Antunes-Rodrigues, and S. M. McCann, Program Endocrine Society Meeting (Abstract) 1965.

## FOOTNOTES

[1]This study was supported by the Ford Foundation and USPHS grant AM 10073-02

[2]Population Council Fellow. Present Address: Cancer Center,Niigata Hospital, Dept. of Ob-Gyn, Kawagishicho, Niigata City, Japan

# STUDIES ON THE ISOLATION OF LUTEINIZING HORMONE RELEASING FACTOR (LRF)

H. GREGORY, A. L. WALPOLE,

H. M. CHARLTON, G. W. HARRIS and MAY REED

I.C.I. Ltd.(Pharmaceuticals Division), Macclesfield

M.R.C. Neuroendocrinology Research Unit, Oxford

Our work has been concerned mainly with attempts to obtain highly purified ovine LRF and to establish the chemical nature of the active material.

## Methods of Assay

Two methods have been employed in assessing LRF activity. In the first of these, samples are infused directly into the anterior pituitary of isolated rabbits in oestrus and the release of LH is indicated by ovulation. This method has been used to assay regions of activity in the various stages of purification.

In addition to this test, however, we required a simple and rapid method of screening large numbers of fractions for following chromatographic separations. For this purpose we have used androgen-sterilized "constant oestrus" rats. Test samples were given i.v. to the rats and again LH releasing activity was shown by ovulation. In our experiments the two methods have behaved in a parallel manner.

## Purification Procedure

Frozen ovine hypothalamic fragments were used as the source of the releasing factor and the procedure we have adopted is shown in Table 1.

TABLE 1

| STAGE | PROCEDURE | YIELD(%) | ACTIVITY |
|---|---|---|---|
| 1 Preparation of crude extract | a) Acetone powder<br>b) 2N Acetic acid, $100^{o}$.<br>c) Acetic acid, acetone, ether. | 14.5<br><br>0.95 | <br><br>10ME/ml 5/5 |
| 2 Solvent extraction | 0.5N Acetic acid, isopropanol, petroleum ether. | 0.35 | 15ME/ml 3/3 |
| 3 Gel filtration | Sephadex G-25; 0.3N Acetic acid. | 0.20 | 15ME/ml 2/2+2*<br>30ME/ml 4/4 |
| 4 Partition chromatography | Sephadex G-25; n-Butanol, pyridine, acetic acid, water. | 0.002 | 20ME/ml 4/4<br>60ME/ml 5/5 |
| 5 Ion exchange chromatography | DEAE Sephadex A-25; Ammonium acetate pH 8.0. | 0.0001 | 20ME/ml 2/2+2*<br>60ME/ml 5/5 |
| 6 Electro-phoresis | Sephadex G-50; Pyridine, acetic acid, water, pH 3.4. | 0.00004 (60ng/ME) | 20ME/ml 2/2+2$^{+}$<br>60ME/ml 5/5 |
| 7 Gel filtration | Sephadex G-25; Water. | - | 20ME/ml 0/6<br>60ME/ml 1/6 |

The yield at each stage is expressed as a percentage by weight of the starting material. Under 'activity' is shown the concentration of the material infused - expressed as the number of median eminence fragments (ME) from which it was derived, per ml. of infusate, and the number of rabbits ovulating. The volume infused, 0.13 ml, was constant throughout. The concentrations shown assume that no losses occur between stages. Losses certainly do occur, and the figures shown indicate the recovery at each stage.

The initial extraction was carried out along the lines described by Guillemin, Schally and co-workers.[1]

In earlier work[2] the crude extract was applied directly to G-25 Sephadex but this gave rather a wide band of activity. A second stage was introduced, therefore, whereby a solution in aqueous acetic acid was diluted with isopropanol to give an inactive precipitate. Dilution of the supernatent solution with petroleum ether gave an aqueous solution which was used for gel filtration.

Gel filtration was carried out in aqueous acetic acid and the effluent was scanned at 280 mμ. This served as a control of the column performance and allowed the active region to be identified.

This gave a product amounting to 0.2% of the starting material. Unfortunately two of the rabbits* in this test were found later to have been unable to respond.

Some groups have followed gel filtration by using ion exchange materials but with mixed success, apparently because the active material is contaminated with salts. We continued our purification by resorting to partition chromatography. This was carried out on G-25 Sephadex using a system derived from n-butanol, pyridine, acetic acid and water and the effluent fractions were scanned using the Folin-Lowry reaction. This process gave a substantial increase in the purity of the releasing factor.

In stage 5 the material was applied to a short column of DEAE Sephadex at pH 8. The releasing factor was not retarded. After this stage the amount of material present was too small to be weighed and the figure given is the peptide content as determined by amino-acid analysis of an acid hydrolysate.

Further purification was effected by electrophoresis on G-50 Sephadex. The activity was found in a narrow band which had moved a considerable distance to the cathode. Again the yield was based upon the amino-acid content. At the lower dose tested (ca 0.2 μg per rabbit) two of the rabbits+ did not ovulate.

Up to this stage no substantial loss of activity had occurred but attempts to carry out further gel filtration caused considerable losses. Only small amounts of active material (rat test) were found in the expected region. Further work indicated that this is probably due to adsorption on the Sephadex. Experiments with Bio-Gel were not more successful.

It was found earlier[2] that desalting of a less pure sample on Sephadex gave rise to two regions of activity. This may well be explained by adsorption upon the gel from the faster moving active material and then reversal of this process during passage of the ionic "salt" region thus giving rise to the two regions of activity.

## Nature of LRF

We had assumed so far that we were dealing with a biologically active peptide (previous workers have shown that the activity is destroyed by pepsin and trypsin[3]) but this merited further investigation. We know that the releasing factor is eluted from G-25 Sephadex in the region associated with molecular weights of 1-2000, it is found in regions giving a positive Folin-Lowry reaction, it behaves as a strongly basic entity on electrophoresis, and amino acids are found in an acid hydrolysate of active material from stage 6.

TABLE 2 REACTIONS OF LRF

| | |
|---|---|
| ACTIVITY RETAINED | a) pH 3.5, 25°, 3 hr. |
| | b) pH 7.7, 25°, 3 hr. |
| | c) Formic Acid, -10°, 2.5 hr. |
| ACTIVITY DESTROYED | a) pH 7.7, 25°, 3 hr, Trypsin. |
| | b) pH 7.7, 25°, 3 hr, Chymotrypsin. |
| | c) pH 7.7, 25°, 3 hr, 2:4-Dinitro-1-fluoro-benzene. |
| | d) Performic Acid, -10°, 2.5 hr. |

Using material from stage 6, the experiments shown in table 2 were carried out. Activity is retained at different pH values but is rapidly lost, within the limits of our test system using rats, under the influence of the proteolytic enzymes trypsin and chymotrypsin. The latter enzyme is somewhat less specific than trypsin but the finding of tyrosine and phenylalanine, together with the basic amino acids, in the hydrolysate is consistent with inactivation by chymotrypsin.

Inactivation by DNFB is consistent with the releasing factor being of a peptide nature.

The action of performic acid may be taken as an indication that sulphur containing amino-acids are present (it has previously been shown that thioglycollate does not affect the activity therefore cystine is not directly involved[3]). Examination of an acid hydrolysate of oxidised LRF showed that methionine sulphone was present but cysteic acid was only present in trace amounts. Previous figures[4] for the amino acid content of LRF preparations have not included methionine, and the aromatic amino-acids were reported in trace amounts.

These experiments provide further evidence of the association between luteinizing hormone releasing activity and peptide structure.

## Acknowledgement

We are greateful to Mrs. B. E. Valcaccia for her continued and skilful cooperation in carrying out many tests using "constant oestrus" rats.

## References


1. Guillemin, R., Schally, A. V., Lipscomb, H. S., Andersen, R. N., and Long, J. M. Endocrinology, 1962, 70, 471.

2. Fawcett, C. P., Reed, May, Charlton, H. M., and Harris, G. W. Biochem. J. in press.

3. McCann, S. M., and Ramirez, V. D., Rec. Prog. Hormone Research, 1964, 20, 131.
   Guillemin, R., ibid., 1964, 20, 89.

4. Schally, A. V. and Bowers, C. Y. Endocrinology, 1964, 75, 608.
   Guillemin, R. Proc. XXIII int. Congr. physiol. Sci., Tokyo, 1965, 4, 284.

# STUDIES ON LUTEINIZING HORMONE RELEASING FACTOR (LRF) IN HYPOPHYSIAL PORTAL BLOOD OF RATS

G. FINK

DEPARTMENT OF HUMAN ANATOMY, OXFORD, ENGLAND

In 1966 Worthington described a method for the collection of hypophysial portal blood from the cut pituitary stalk of rats. This method was used (Fink, Nallar & Worthington, 1967) to demonstrate the presence of LRF in portal blood obtained from pro-oestrous and hypophysectomized rats. The possibility that the activity of portal plasma was due to vasopressin or a non specific factor was largely excluded, but it was pointed out that a significant proportion of the activity of portal plasma from pro-oestrus rats could have been due to LH contained in back-flow blood from the pituitary sinusoids.

In order to determine the level of LRF in hypophysial portal blood from animals which had not been hypophysectomized, it was important to inactivate LH. Since this could not be achieved by surgery, the following method, devised by Dr. C.P. Fawcett, was employed. The plasma was mixed with 5 volumes of acid alcohol (HCL, $H_2O$, $C_2H_5OH$; V/V; 0.5, 25, 75) shaken for 30 min and centrifuged at 10,000 RPM for 5-10 min. The precipitate was resuspended in acid alcohol, shaken for 30 min and centrifuged at 10,000 RPM for 5-10 min. The supernatants were combined, evaporated in a rotary evaporator and the residue was dissolved in 5-10 ml. deionized water and applied to a column of 'Sephadex G-25', (2.5 x 35-41 cm.). The sample was eluted at 30 ml/hr with 0.33N acetic acid and the eluate was collected in 5 ml. fractions. The optical density of the eluate was determined at 280mμ. The elution volume of the electrolytes was determined by the flame test for sodium. The extraction procedure was based on that used by Grodsky and Forsham (1960) to extract insulin from plasma.

On the basis of studies on extracts of hypothalamus, it was thought that any LH which remained after the extraction would be

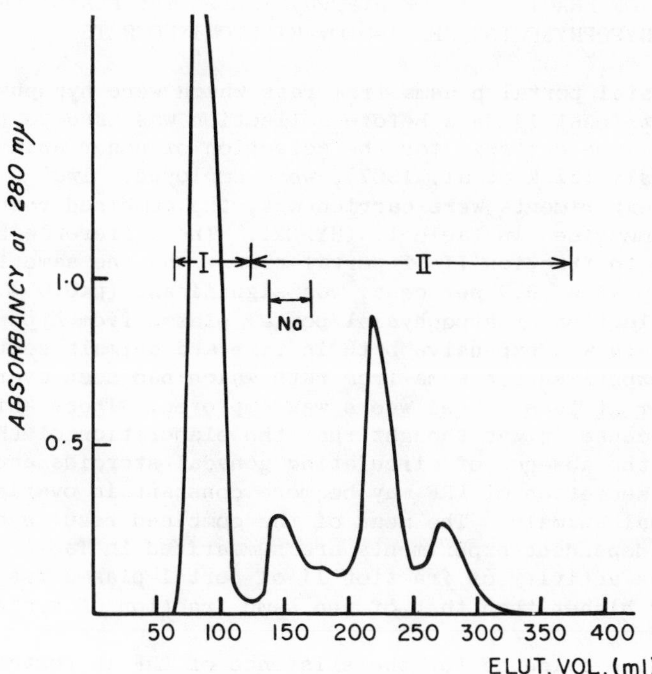


Fig. 1. Elution profile on 'Sephadex G-25'. 9 ml. portal plasma from nine ovariectomized rats extracted with acid alcohol. Supernatant evaporated, redissolved in 4.5 ml. of distilled water and loaded on column of 'Sephadex G-25' (fine), 35.5 x 2.5 cm. Eluted with 0.3 molar acetic acid; 5.0 ml.fractions; rate 0.5 ml./min. Fraction I contains substances of molecular weight >5,000; Fraction II contains substances of molecular weight <5,000.

eluted with substances excluded from the gel (fraction I, Fig. 1) while LRF would be retarded by the gel to appear in fraction II (Fig.1). In three independent experiments it was found that extraction with acid ethanol inactivated relatively large amounts of both ovine and rat LH (saline extract of rat anterior pituitary glands) to levels which could not be detected by the OAAD assay. However, the fractionation step was retained because it was found that fraction I of both portal and systemic plasma from hypophysectomized and ovariectomized rats caused a significant depletion of ovarian ascorbic acid (OAA). The activity of fraction I was reminiscent of the OAAD evoked by proteins (Pelletier, 1964, 1965).

## LRF ACTIVITY OF FRACTION II OF HYPOPHYSIAL PORTAL PLASMA FROM HYPOPHYSECTOMIZED AND OVARIECTOMISED RATS

Hypophysial portal plasma from rats which were hypophysectomized at least 14 days before collection was used to test the method. The same criteria for the selection of donor animals as used previously (Fink et al. 1967), were employed. Two+ independent experiments were carried out, the combined results of which are summarized in Table 1. (HYPOX). The difference between the response to fraction II of portal plasma and the same fraction of systemic plasma, 9.7 per cent, was significant ($p<0.05$).

The collection of hypophysial portal plasma from hypophysectomized rats was expensive both in time and animals so that in subsequent experiments,plasma from rats which had been ovariectomized for at least three weeks was employed. These animals were used because it was thought that the elaboration of LRF may increase in the absence of circulating gonadal steroids and also because the secretion of LRF may be more constant in ovariectomized than in normal animals. The mean of the combined results of a number of independent experiments are summarized in Table 1. (OVARX). The activity of fraction II of portal plasma was consistently higher than that of the same fraction of systemic plasma.

So far the evidence for the existence of LRF in portal blood rested on the ability to cause the depletion of ovarian ascorbic acid. It was important to assay the plasma in another system with a more physiological end point.

## THE ASSAY OF FRACTION II OF PORTAL PLASMA BY THE RABBIT OVULATION TEST

The rabbit ovulation test of Campbell, Feuer, Garcia and

Table I

Assay of Fraction II of Hypophysial Portal and Systemic Plasma from Hypophysectomized and Ovariectomized Rats

| Source | Conc X1 | | Conc X2 | |
|---|---|---|---|---|
| | Portal | Systemic | Portal | Systemic |
| HYPOX | 13.0 ± 1.7(9)* <br> $p<0.001$ | 3.3 ± 3.2(4) <br> n.s. | -- | - |
| OVARX | 10.8 ± 1.6(13) <br> $p<0.001$ | 4.4 ± 3.9(9) <br> n.s. | 13.4 ± 1.4(7) <br> $p<0.001$ | 8.3 ± 0.8(7) <br> $p<0.001$ |

* Mean percent OAAD ± SEM (Number of animals)

+ The first of these was carried out in collaboration with Drs. C.P. Fawcett and R. Nallar.

and Harris (1961) was employed.

In the first experiment, a 10-fold concentrate of fraction II of portal plasma was infused into the anterior pituitary glands of 4 rabbits of which 2 ovulated. Each rabbit received the equivalent of 2.5 ml. of portal plasma in 0.13 ml in 2 hrs. In the second experiment, which was carried out in collaboration with Dr. H.M. Charlton, rabbits were infused with a 20-fold concentrate of fraction II of either portal or systemic plasma. Each rabbit received the equivalent of 5.0 ml. of plasma in 0.13 ml in 2 hrs. Four of five animals infused with fraction II of portal plasma ovulated compared to one of five of those infused with the same fraction of systemic plasma.

The fact that the infusion of fraction II of portal plasma into the anterior pituitary gland caused ovulation in rabbits suggested that this fraction contained a substance which was capable of releasing LH. This conclusion received support from the observation that fraction II was also capable of raising the concentration of LH in the systemic plasma of ovariectomized, oestrogen, progesterone treated rats (see below).

## THE LRF ACTIVITY OF HYPOPHYSIAL PORTAL PLASMA AT VARIOUS PHASES OF THE OESTROUS CYCLE OF THE RAT

Three independent experiments were carried out to determine the LRF activity of hypophysial portal plasma at various phases of the oestrous cycle of the rat. Adult female rats of the Wistar strain which had exhibited at least 2 successive 4-day vaginal cycles were used. Portal blood was collected over a 2 hr. period in the afternoon of dioestrus, oestrus, and metoestrus and during the critical period of prooestrus. The plasma from each phase (7-10 donors/day/experiment) of the cycle was pooled, extracted and fractionated. Fraction II was assayed by the OAAD method. In the first experiment each test animal received the equivalent of 1.6 ml. of plasma while in the second and third experiments the animals received the equivalent of 1.8 hr. of collection irrespective of the volume. The results are presented in Table 2; experiments 2 and 3 are combined.

In all the experiments the activity of fraction II of portal plasma from animals in oestrus was lower than that of the same fraction of portal plasma from animals in other phases. The difference from dioestrus and prooestrus was significant ($p<0.05$) in the first experiment.

The collection of hypophysial portal blood for the first experiment was carried out with particular care. There was no significant difference between the means of the volumes obtained for each phase.

These results suggest that both the concentration of LRF in hypophysial portal blood and the amount of LRF secreted into the portal vessels per unit time is less during the afternoon of

Table 2

| Experi-ment | Dioestrous | Prooestrous | Oestrous | Metoestrous |
|---|---|---|---|---|
| 1 | 11.3 ± 2.6(4)* | 11.6 ± 2.2(4) | 1.5 ± 2.6(4) | 8.4 ± 3.7(4) |
| | $p<0.05$ | $p<0.02$ | n.s. | $p<0.1$ |
| 2 + 3 | 14.7 ± 3.6(6) | 12.2 ± 2.5(8) | 8.1 ± 2.1(8) | 11.5 ± 1.4(8) |
| | $p<0.02$ | $p<0.01$ | $p<0.01$ | $p<0.001$ |

* Mean per cent ȮAAD ± SEM (Number of animals)

oestrus than during the afternoon of other phases of the oestrus cycle of 4-day cyclic rats.

The finding by Ramirez and Sawyer (1965) that there was a marked drop in the LRF activity of the median eminence of rats between prooestrus and oestrus together with the results reported here, suggest that the secretion of LRF is reduced late on the day of prooestrus. However, Chowers and McCann (1965) found that the only significant change in the LRF activity of the median eminence, was a rise at dioestrus. If this were confirmed, a more complicated explanation would have to be invoked.

The other feature which invites consideration is the absence of an expected peak of LRF activity of portal plasma during the critical period of prooestrous. The most likely explanation is that the anaesthetic and trauma of the operation permitted a basal secretion of LRF but inhibited a surge. There are, however, at least two alternative explanations.

1. It is possible, as Rothchild (1965) suggested, that the critical period represents the terminal phase of an exponentially rising rate of secretion of the hormone. If this were the case, little, if any, increase in the secretion of LRF would be required during this period.

2. Another possibility, proposed by Brown-Grant (1967), is that the release of LH during the critical period may depend on gonadal steroids which may alter the sensitivity of the pituitary to LRF. The blockade of ovulation by drugs would still be due to an inhibition of the secretion of LRF but the critical period could be considered to be of a hormonal rather than a neural nature.

It is of interest that neither Ramirez and Sawyer (1965) nor Chowers and McCann (1965) reported a significant change in the LRF activity of the median eminence during the critical period.

## THE NATURE OF LRF IN HYPOPHYSIAL PORTAL BLOOD

The aim of the following experiments was to examine the behaviour of LRF from hypophysial portal blood when subjected to gel filtration in order to compare it to LRF from extracts of hypothalamic tissue and to obtain an estimate of the molecular weight. In all experiments, portal plasma was obtained from rats which had been ovariectomized 3-8 weeks before collection.

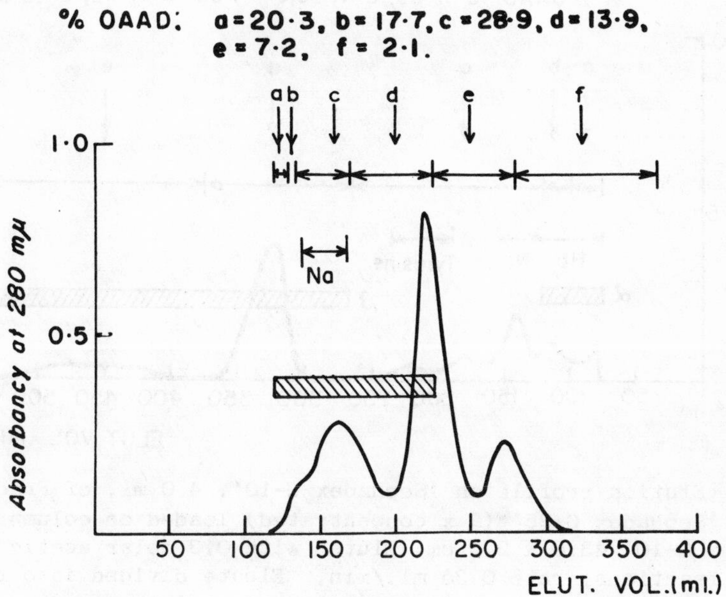


Fig. 2. Elution profile on 'Sephadex G-25'. 9 ml. portal plasma from nine animals extracted with acid alcohol. Supernatant evaporated and redissolved in 4.5 ml. distilled water. Loaded on to a column of 'Sephadex G-25' (fine) 35.5 x 2.5 cm; eluted with 0.3 molar acetic acid; 5.0 ml. fractions; rate 0.5 ml./min. Fraction II lyophilized, redissolved in 4.5 ml. distilled water and refractionated on same column. Eluate divided into six fractions and assayed by OAAD method. LRF activity found in shaded region. Note activity in d suggesting adsorption (Kd >1).

## Behaviour of LRF on 'Sephadex G-25'

Figure 2 summarizes the results obtained when subfractions of fraction II (from 'Sephadex G-25') of portal plasma were assayed by the OAAD method. The region of LRF activity (Ve $\simeq$ 2 x Vo) was similar to that reported for LRF activity of extracts of hypothalamic tissue (for example, see Schally & Bowers 1964; Fawcett, Harris & Reed, 1965; Guillemin, 1965). The fact that adsorption probably played a part in the retardation of LRF (activity in d; Kd>1) precluded the use of calibrated columns of 'Sephadex G-25' to estimate the molecular weight of this factor. It was therefore decided to use 'Sephadex G-10' which has an exclusion limit of 700-900 to determine whether the molecular weight of LRF was above or below this range. Evidence from studies which employed extracts of hypothalamus suggested that LRF was a peptide with a molecular weight of 1200-1600 (see Harris, Reed & Fawcett, 1966;

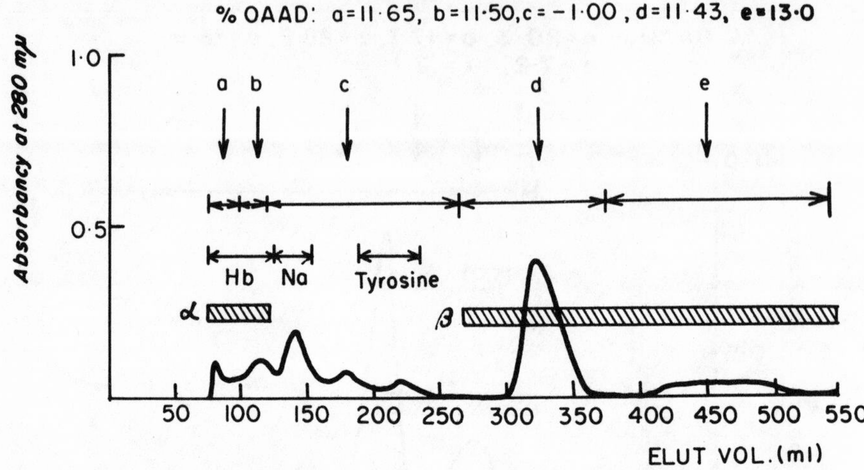


Fig. 3. Elution profile on 'Sephadex G-10'. 4.0 ml. of Fraction II from 'Sephadex G-25' (2 x concentrated) loaded on column of 'Sephadex G-10' 43.0 x 2.5 cm. Eluted with 0.3 molar acetic acid; 5.0 ml. fractions; rate 0.35 ml./min. Eluate divided into five fractions and assayed by OAAD method. LRF activity in shaded region. Note retention of β beyond tyrosine.

McCann, & Dhariwal, 1966). Consequently, it was predicted that LRF would be excluded from 'Sephadex G-10'. The results, however, were quite unexpected.

## Behaviour of LRF on 'Sephadex G-10'

Hypophysial portal plasma was extracted with acid ethanol and fractionated on 'Sephadex G-25'. Fraction II was lyophilized to dryness, redissolved in distilled water and fractionated on 'Sephadex G-10' (Fig.3). LRF activity (OAAD) was found in two regions, α and β, separated by an interval which included the electrolytes. Whereas α was largely excluded by the gel, β was markedly adsorbed (beyond tyrosine). This experiment was repeated twice using longer columns (to facilitate the fractionation of larger quantities of portal plasma) of 'Sephadex G-10' (Fig.4). Similar results were obtained.

It is noteworthy that the OAAD caused by active fractions from the longer columns of 'Sephadex G-10' (Fig.4) and also from the column of 'Sephadex G-25' was much higher than that evoked by similar concentrations of the whole of fraction II.

In order to exclude the possibility that α and β had caused depletion of ovarian ascorbic acid by direct action on the ovaries

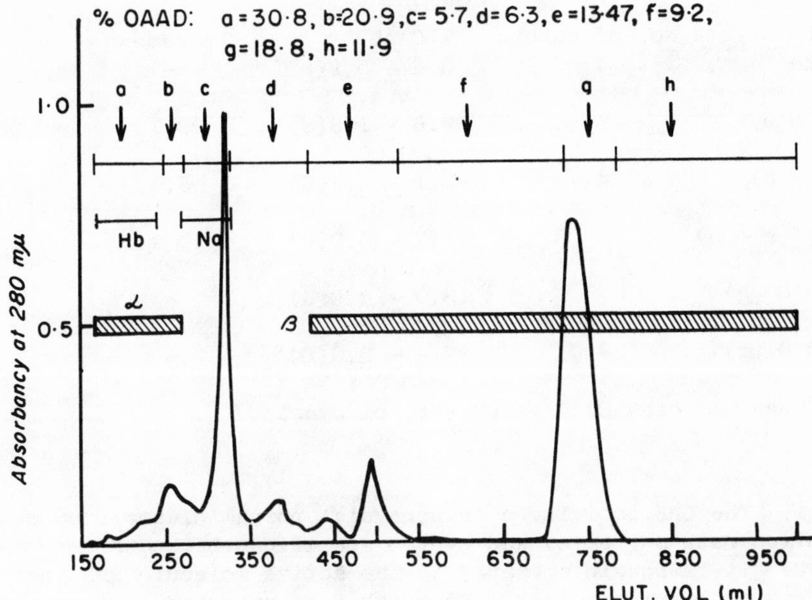


Fig. 4. Elution profile on 'Sephadex G-10'. 4.0 ml. of Fraction II (concentrated from 24 ml. of portal plasma) from 'Sephadex G-25' loaded on column of 'Sephadex G-10', 94.0 x 2.5 cm. Eluted with 0.3 molar acetic acid; 5.0 ml.fractions; rate 0.5 ml./min. Eluate divided into eight fractions, aliquots of which were assayed by the OAAD method and by determining release of LH from the anterior pituitary glands of ovariectomized, oestrogen, progesterone blocked rats. LRF activity found in shaded region.

rather than by the release of LH from the anterior pituitary glands of the test animals, aliquots of the fractions from the experiment presented in Fig.4. were bulked to make 3 major fractions, α (a plus b), S (c plus d) and β (e,f,g and h), which were lyophilized to dryness, redissolved in enough deionized water to make a three-fold concentrate and (after neutralization) injected intravenously (1 ml./animal) into ovariectomized,oestrogen progesterone treated rats (Ramirez & McCann, 1963). The systemic plasma of the intermediate animals was assayed by the 2 ovary, 4 hr, modification of the OAAD method (Schmidt-Elmendorff & Loraine 1962). The results are presented in Table 3.

The activity of the plasma from the intermediate animals treated with α and β was significantly different from the activity of the plasma of animals treated with saline only (control). Two doses of LH-NIH-S10 were included in the assay. Although the slope was not significant the responses are presented for comparison.

The results are open to a number of interpretations;

Table 3

| Treatment | No. of donors | % OAAD | Increment | P |
|---|---|---|---|---|
| Saline | 4 | 3.4 ± 5.2(8)* | - | - |
| | | n.s. | | |
| α (a + b) | 4 | 32.6 ± 1.8(8) | 29.2 | <0.001 |
| | | p<0.001 | | |
| S (c + d) | 4 | -2.1 ± 4.4(8) | -5.5 | n.s. |
| | | n.s. | | |
| β (e,f,g + h) | 4 | 21.3 ± 5.0(6) | 17.9 | <0.05 |
| | | p<0.01 | | |
| LH (0.6 μg) | - | 19.0 ± 4.9(6) | - | - |
| | | p<0.02 | | |
| LH (3.0 μg) | - | 27.4 ± 3.0(6) | - | - |
| | | p<0.001 | | |

*Mean per cent OAAD ± SEM (No. of ovaries)

1. The LRF activity of hypophysial portal plasma from ovariectomized rats may be caused by two specific substances.

2. It is possible that β is the active molecule and that α consists of carrier molecule to which β is attached. Dissociation of a proportion of α would explain the finding of two active regions.

3. Substance α may be a polymer of β.

The molecular weight of α probably lies in the range of 900-5,000; however, because of the strong adsorption of β by 'Sephadex G-10', a valid estimate of the molecular weight of this substance may not be made.

## SUMMARY

1. A method for the inactivation of LH in the plasma of hypophysial portal blood collected from the cut pituitary stalk of rats has been described.

2. The LH-free fraction of portal plasma from hypophysectomized rats exhibited LRF activity in the OAAD assay.

3. LRF activity was demonstrated in the LH-free fraction of portal plasma from ovariectomized rats by the OAAD assay, the rabbit ovulation test and by the use of ovariectomized, oestrogen, progesterone treated rats.

4. The LH-free fraction of portal plasma from 4-day cyclic rats was assayed by the OAAD method. There was a drop in LRF activity at oestrus but an expected peak at prooestrus did not occur.

5. The behaviour of LRF from portal plasma on 'Sephadex G-25' was similar to that of LRF from hypothalamic tissue.

6. The LH-free fraction of portal plasma from ovariectomized rats was fractionated on 'Sephadex G-10'. Two regions of LRF activity were found. The first (α) was largely excluded by the gel while the second (β) was markedly retarded.

## ACKNOWLEDGEMENTS

It is a preasure to record my indebtedness to Professor G.W. Harris, C.B.E., F.R.S., for his interest, helpful suggestions and valuable advice throughout this project. The gift of LH from the Endocrinological study section of the NIH, Bethesda, is gratefully acknowledged. This study was carried out under the tenure of a Nuffield Dominions Demonstratorship and the technical expenses were defrayed in part by a grant from the Medical Research Council and a grant from the U.S. Air Force to Professor Harris.

## REFERENCES

Brown-Grant, K. (1967) J. Physiol. 190, 101.

Campbell, H.J., Feuer, G., Garcia, J., & Harris, G.W., (1961) J. Physiol. 157, 30p.

Chowers, I., & McCann, S.M. (1965) Endocrinology 76, 700.

Fawcett, C.P., Harris, G.W., & Reed, M., (1965) Proc. 23rd Intern. Cong. Physiol. Sci., Tokyo, 275.

Fink, G., Nallar, R., & Worthington, W.C.Jr. (1967) J. Physiol. 191, 407.

Grodsky, G.M., & Forsham, P.M., (1960) J. Clin. Invest. 39, 1071.

Guillemin, R., (1965) Proc. 23rd Intern. Cong. Physiol. Sci. Tokyo. 284.

Harris, G.W., Reed, M., & Fawcett, C.P., (1966) Brit. Med. Bull. 22, 266.

McCann, S.M. & Dhariwal, A.P.S. (1966). In Martini, L. & Ganong, W.F., Neuroendocrinology. New York; Academic Press.

Pelletier, J., (1964) C.r.hebd. Seanc.Acad.Sci. Paris. 258, 5979.

Pelletier, J., (1965) C.r.hebd. Seanc.Acad.Sci. Paris. 260, 5624.

Ramirez, V.D., & McCann, S.M., (1963) Endocrinology 73, 193.

Ramirez, V.D., & Sawyer, C.H., (1965) Endocrinology 76, 282.

Rothchild, I. (1965) Vitams. Horm. 23, 209.

Schally, A.V., & Bowers, C.Y. (1964) Endocrinology 75, 608.

Schmidt-Elmendorff, H., & Loraine, J.A., (1962) J. Endocr. 23, 413.

Worthington, W.C., Jr. (1966) Nature, Lond. 210, 710.

# STUDIES ON THE MECHANISM OF ACTION OF HYPOTHALAMIC FSH- AND LH-RELEASING FACTORS.

M. JUTISZ, A. BERAULT and M.P. de la LLOSA

Laboratoire de Morphologie Expérimentale et Endocrinologie, Collège de France, PARIS V°

Although hypothalamic LH- and FSH- releasing factors (LRF and FRF) have been discovered since 1960 (1, 2, 3, 4, 5) and 1964 (6, 7, 8) respectively, only a little work was done until now on their mechanism of action. Three possibilities at least can be envisaged (9) : 1. a stimulation of excretion (or discharge) of hormone stored in the pituitary, 2. a direct or indirect action on synthesis of hormone in the appropriate cells, 3. an effedt on activation of hormonal precursors.

In the work I report here, we have approached the problem of the mode of action of RF's using an <u>in vitro</u> method.

## EXPERIMENTAL

LRF and FRF fractions used in this study were from ovine origin. They have been purified according to a method published elsewhere (10, 11). For the work on LRF, we used only a highly purified product. In the case of FRF, a less purified fraction was used, in some experiments.

The technique of incubation was similar to that used by Schally and Bowers (12, 9, 11) : female rats were ovariectomized and 4 to 6 weeks later they were injected with estradiol benzoate and progesterone (13). Three days after this treatment, the anterior pituitaries were removed, hemisected and incubated in a Krebs-Ringer-bicarbonate-glucose buffer. After 15 min of preincubation and changing the medium, incubation proceeded generally for

2 hr. When puromycin and actinomycin D were employed they were already introduced before the preincubation. At the end of the incubation, gonadotropins were assayed in the supernatants and occasionally in the tissues : LH by the Parlow OAAD method (14) and FSH by the Steelman-Pohley augmentation test (15). As standards for these hormones we used NIH-LH-S3 and NIH-FSH-S3 (supplied by the Endocrinology Study Section, NIH, Bethesda Md., U.S.A.), and the results are expressed in terms of these standards.

## 1. Effect of Graded Doses of LRF and FRF on the Release of LH and FSH.

There is a log dose-response relation between the amounts of LH and FSH released and the doses of LRF and FRF (11, 16). Figure 1 shows the release of FSH, in terms of ovarian weights in the augmentation test, as function of graded dose of FRF. The curve is linear for doses between about 5 to 80 nanograms of a highly purified fraction of FRF per mg of pituitary tissue. For higher doses of FRF a plateau was observed or even a slight decrease of FSH release. For a highly purified LRF fraction the log dose response curve is linear between approximately 60 (minimal effective dose) to 800 nanograms. The *in vitro* technique can be used as a specific and sensitive method for assay of these RF's.

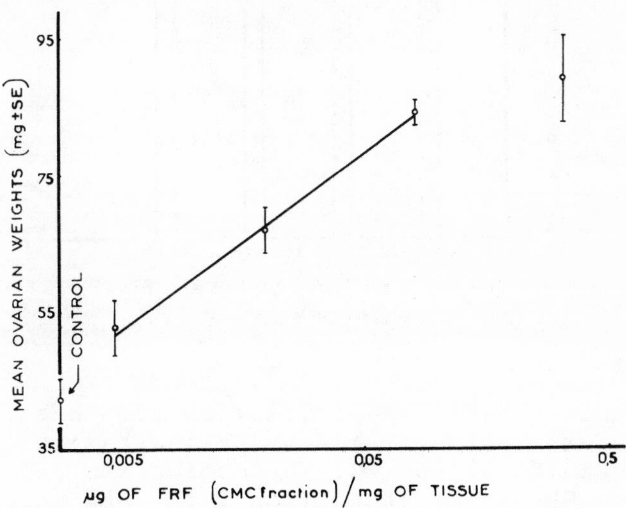


Fig.1. Effects of graded doses of a highly purified FRF on FSH release by rat pituitaries *in vitro*.

## 2. Evidence of Biosynthesis of Gonadotropins in Pituitary Tissue Incubated with LRF and FRF.

To bring a direct evidence of the synthesis of FSH in FRF treated tissue (16), 4 sets of pituitary halves were incubated at the same time for 2 hr periods. Two groups were made and in each one of these groups one set of pituitaries served as control and the second was treated with FRF. In the first group the medium was replaced after 30, 60 and 90 min of incubation ; the treated set receiving each time 0.015 µg of FRF/mg of tissue. In the second group the incubation was performed for 2 hr without interruption, the treated set receiving 0.05 µg of FRF/mg of tissue. At the end of incubation, tissues were ground in supernatants, then extracted in the same media for FSH assay. The results are given in Fig. 2.

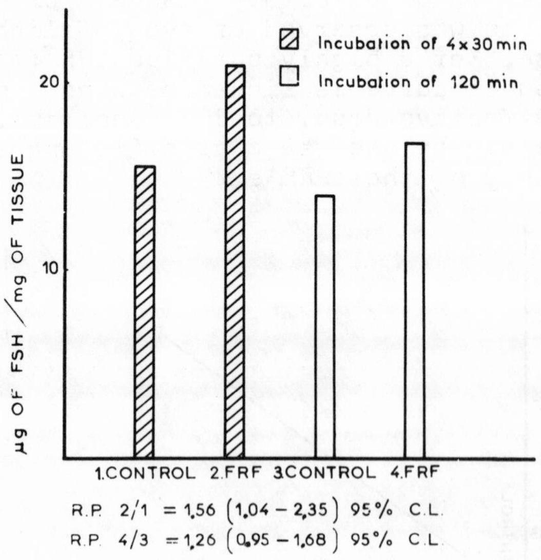


Fig. 2. FSH content of pituitaries ground and extracted in the supernatants after 2 hr of incubation. In the first group (no. 1 and 2), the medium was replaced after 30, 60 and 90 min of incubation ; one set receiving each time 0.015 µg of FRF/mg of tissue, another set serving as control. The second group (no. 3 and 4) was incubated 2 hr without interruption, the treated set receiving 0.05 µg of FRF/mg of tissue.

As can be seen from Fig. 2, the relative potency of the two first sets of pituitaries, (media changed during incubation) is 1.56 and the difference between these two sets is significant (cf. Confidence limits). The relative potency of the second group, (incubated without interruption) is 1.26 and the difference between the two sets is not significant. It is evident from these experiments that synthesis occured in FRF treated tissue and that by changing the medium the synthesis was accelerated.

A question arises : why replacing the medium during incubation stimulates the synthesis of FSH ? Two possibilities were visualized : either an internal feed-back by the FSH released into the medium on its synthesis or release, or the destruction of FRF during the incubation process. It seems from preliminary experiments that a rat pituitary extract added to the medium before incubation is capable of inhibiting the release of FSH under the action of FRF. The second hypothesis has not been tested until now and therefore can not be eliminated.

LRF gave similar results.

### 3. Failure of LRF and FRF, when Incubated with Pituitary Homogenate, to Increase the Amount of LH and FSH.

One of the possibilities envisaged for the mechanism of the action of RF's was an effect on activation of hormonal precursors. This hypothesis was taken into account for two major reasons : 1. LH molecule consisting, as we have shown in our laboratory, of two biologically inactive sub-units (17), one could imagine that LRF acted on the activation of an enzyme which catalyzes the binding of the sub-units, 2. carbohydrates being bound to polypeptide chains of glycoproteins by enzymes catalyzed reactions, RF's could be expected to promote that process. To check these two hypothesis, pituitary homogenates were incubated with LRF or FRF, and gonadotropins were assayed after incubation. The results of these experiments were given in Table I.

As shown in Table I, neither LRF nor FRF are able to increase significantly the amount of gonadotropins, when incubated with pituitary homogenates.

### 4. Failure of Puromycin and Actinomycin D to Abolish the Action of LRF and FRF on LH and FSH Release *in Vitro*.

It was considered of interest to test the possibility that puromycin and actinomycin D might interfere with the stimulating action of RF's. If so, this would signify

TABLE I

Failure of LRF and FRF, when Incubated (2 hr) with Pituitary Homogenate, to Increase the Amount of LH and FSH

| N° | Treatment | μg of LH or FSH/mg of pituitary tissue (a) |
|---|---|---|
| | LH : | |
| 1. | Control homogenate | 5.5 ( 3.25 - 9.42 ) |
| 2. | Homogenate + LRF (b) | 5.9 ( 2.92 -11.93 ) |
| | FSH : | |
| 3. | Control homogenate | 19.8 (15.32 -25.77 ) |
| 4. | Homogenate + FRF (c) | 21.2 (16.20 -27.00 ) |

(a) In terms of NIH-LH-S3 and NIH-FSH-S3 with 95 % C.L.
(b) 0.63 μg/mg of pituitary tissue.
(c) 0.08 μg/mg of pituitary tissue.

that LRF and FRF might exert a direct effect on the synthesis of the LH and FSH molecules. As a first approach to this problem, we have verified that $^{14}$C-aminoacids and $^{3}$H-guanosine were incorporated respectively into the protein and RNA synthesized _de novo_ in pituitary tissue. Puromycin at a concentration of $2 \times 10^{-4}$M inhibited the incorporation of $^{14}$C-aminoacids to an extent of 95 % and actinomycin D at 10 μg/ml inhibited the incorporation of $^{3}$H-guanosine to an extent of 84 % (9). On the other hand we have verified that these antibiotics did not interfere with assay methods (16).

To control a possible effect of puromycin and actinomycin D on the release of FSH under action of FRF, pituitary halves were incubated for 2 hr with either FRF or antibiotics and FRF (18, 16). The whole experiment was done in 4 flasks and the pituitary halves were randomized in order to compare the results statistically.

Fig. 3 presents graphically the results obtained. The statistical analysis of the results is recorded in the legend of Fig. 3. Puromycin diminished significantly the releasing action of FRF (compar. 3/2) but did not abolish this action (compar. 3/1). Actinomycin gave similar results. There was no significant difference between the control tissues and tissues treated with FRF but the FSH content of the tissues treated with puromycin and actino-

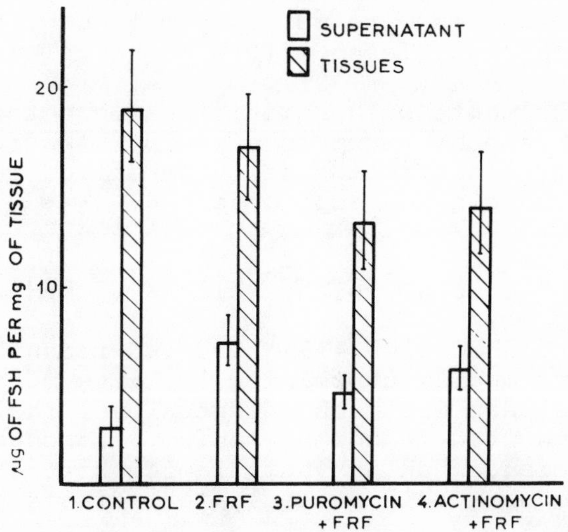


Fig. 3. FSH content of incubation media and tissues: 1. Control, 2. Incubation with 9.0 µg of FRF per mg of tissue, 3. Incubation with puromycin ($2 \times 10^{-4}$M) and with 8.1 µg of FRF per mg of tissue, 4. Incubation with actinomycin D (10 µg/ml) and with 8.7 µg of FRF per mg of tissue. Bars represent 95 % confidence limits.

Relative potencies with 95 % confidence limits for 4 incubation groups:

| Compared Groups | Supernatants | Tissues |
|---|---|---|
| 2/1 (FRF/Control) | 3.14(1.84-5.36) | 0.89(0.76-1.04) |
| 3/1 (Puromycin+FRF/Contr.) | 1.63(1.13-2.35) | 0.68(0.55-0.83) |
| 4/1 (Actinom.+FRF/Contr.) | 2.59(1.49-4.50) | 0.72(0.59-0.88) |
| 3/2 (Puromycin+FRF/FRF) | 0.61(0.49-0.78) | 0.77(0.62-0.94) |
| 4/2 (Actinom.+FRF/FRF) | 0.79(0.64-0.99) | 0.82(0.67-1.00) |

(From : Jutisz et de la Llosa, (16)).

mycin D was significantly lower as compared with the control and with the tissue incubated with FRF. Thus, the results obtained with puromycin support the conclusion that FSH is synthesized in pituitary tissue incubated with FRF. Since synthesis of proteins was blocked in puromycin treated tissues (9), the total amount of FSH found in the medium and in the tissues incubated with this antibiotic, corresponded to the amount of FSH present originally in the pituitary. This suggests that the excess of FSH in the FRF treated medium and tissues was produced during the incubation process.

Similar results were obtained with LRF, but actinomycin D had no significant effect on LH releasing action of LRF (9).

The experiments with antibiotics lead to the conclusion that FRF and LRF seem to stimulate the release of corresponding gonadotropins as their action is not abolished by puromycin and actinomycin D.

5. Effect of Potassium on the Release of FSH *in Vitro*.

SAMLI and GESCHWIND (19) reported recently that in a Krebs-Ringer medium containing 10 times more $K^+$ than normal, LH release was much enhanced over that from pituitaries incubated in normal medium, although synthesis was not affected. When a hypothalamic extract and $K^+$ enriched medium were both present, the effects were additive.

In a preliminary experiment we checked the effect of potassium on the release of FSH. The results are recorded in Table II in terms of mean ovarian weights in the Steelman-Pohley test.

TABLE II

Effect of Potassium on the Release of FSH *in Vitro*
(Results of FSH Assay by the Steelman-Pohley Method (a))

| N° | Treatment | Ovarian Wt.: mg $\pm$ SE | p vs N°4 |
|---|---|---|---|
| 1. | Control HCG | 44.1 ± 2.96 | - |
| 2. | NIH-FSH-S3 75 µg | 73.3 ± 9.62 | - |
| 3. | " 150 µg | 122.8 ± 12.21 | - |
| 4. | Incub.Contr.(5.12mM in $K^+$) | 64.9 ± 6.62 | - |
| 5. | Medium 4 + FRF (b) | 126.4 ± 6.80 | 0.001 |
| 6. | Incub.Contr.(without $K^+$) (c) | 66.0 ± 4.12 | NS |
| 7. | Medium 6 + FRF (b) | 69.6 ± 3.88 | NS |
| 8. | Incub.Contr.(51.2mM in $K^+$) | 90.9 ± 8.39 | 0.05-0.1 |
| 9. | Medium 8 + FRF (b) | 73.7 ± 9.23 | NS |

(a) Equivalents of 15 mg of pituitary tissue per rat were injected.
(b) FRF : 0.065 µg/mg of pituitary tissue.
(c) $NaH_2PO_4$ was used instead of $KH_2PO_4$.

As can be seen from Table II, the FRF caused, in a normal Krebs-Ringer medium (gr. 5), a highly significant release of FSH over that of incubation control (gr. 4). In the absence of $K^+$, and in an isotonic Krebs-Ringer medium containing all other ions and glucose, the same amount of FRF, as previously, did not release FSH. When the amount of $K^+$ was increased 10 times over that of normal medium and in isotonic conditions, a small but still significant amount of FSH was released. By adding FRF into the last medium, instead of additive effect as for LH (19), a small inhibition was observed.

It can be concluded from these results : 1. that the potassium is necessary for the FSH releasing action of FRF, 2. that an excess of potassium is able to release a small but significant amount of FSH, 3. that potassium enriched medium and FRF have no additive effect for FSH release.

## D I S C U S S I O N

It seems possible now to draw some conclusions as to the mode of action of LRF and FRF, considering experiments discussed in the present lecture and other datas found in the litterature. Most information on this subject was obtained using <u>in vitro</u> methods, although <u>in vivo</u> techniques allowed also some important observations.

When highly purified preparations of RF's are incubated with pituitary halves, they stimulate in certain conditions the release from the tissue of corresponding gonadotropins. The amount of a released gonadotropin is a linear function (within certain limits) of the log dose of hypothalamic factor.

Some evidence was found indicating that an acceleration of the synthesis of FSH occured in the pituitary incubated with FRF. The synthesis seems to be enhanced when the medium is frequently changed during the incubation process, with addition of FRF. The reason for this last phenomenum may be either an internal feed-back by the FSH released into the medium on its synthesis or release, or some destruction of FRF during the incubation process.

As shown by experiments with pituitary homogenates, the RF's have no effect on a possible enzymatic system capable of acting on the activation or completion of the FSH molecule.

Experiments with inhibitors of proteins or RNA biosynthesis indicate that the stimulating action of RF's on the release of gonadotropins, even if it is slightly diminished during the incubation with puromycin and actinomycin D, is not abolished. This leads to the conclusion that the principal action of RF's is on the release of the corresponding gonadotropins.

It is interesting to point out that potassium is necessary for the releasing action of FRF ; this ion may play an active physiological role together with RF's in the secretion of gonadotropins.

If LRF and FRF acts only on the release of LH and FSH, how is the fact explained that synthesis of these hormones is enhanced in the pituitary incubated with RF's ? Several authors reported recently (20, 21, 22, 23) that intracarotid or intrajugular injection of a hypothalamic extract or a purified FRF preparation in the rat induced, 15 to 45 min after treatment, a depletion of pituitary FSH. This initial effect was followed by the resynthesis of this hormone in the pituitary (22). It is possible consequently that the phenomena observed *in vitro* in our work are composite results : RF's inducing the release of gonadotropins, which in the second stage, are resynthesized in the tissue. Both phenomena of synthesis and release of gonadotropins are probably closely related; the resynthesis being consequential of the release. If so, the precise mechanism that controls the synthesis of gonadotropins still remains to be discovered.

On the basis of this study a double action of RF's on release and synthesis of corresponding pituitary hormones can not be excluded.

## ACKNOWLEDGEMENTS

It is a pleasure to acknowledge the help of Miss Lynn Rapley in the correction of the english text. We express our thanks to Mr. G. Vassent for the statistical analysis of our results and to Mr. B. Kerdelhué and Miss Wisnewsky for their valuable technical assistance.

## REFERENCES

1. HARRIS, G.W., in Control of ovulation, C.A. VILLEE ed., Pergamon Press, London 1961, p. 56.
2. McCANN, S.M., S. TALEISNIK and H.M. FRIEDMAN, Proc. Soc. Exptl. Biol. Med., 104 : 432, 1960.
3. CAMPBELL, H.J., C. FEUER, G. GARCIA and G.W. HARRIS, J. Physiol., London, 157 : 30P, 1961.
4. McCANN, S.M. and S. TALEISNIK, Endocrinology, 68 : 1071, 1961.
5. COURRIER, R., R. GUILLEMIN, M. JUTISZ, E. SAKIZ and P. ASCHHEIM, C.R. Acad. Sc., Paris, 253 : 922, 1961.

6. IGARASHI, M. and S.M. McCANN, Endocrinology, 74 : 446, 1964.
7. MITTLER, J.C. and J. MEITES, Proc. Soc. Exptl. Biol. Med., 117 : 309, 1964.
8. KUROSHIMA, A., Y. ISHIDA, C.Y. BOWERS and A.V. SCHALLY, Endocrinology, 76 : 614, 1965.
9. JUTISZ, M., A. BERAULT, M.-A. NOVELLA and F. CHAPEVILLE, C.R. Acad. Sc., Paris, 263 : Ser. D, 664, 1966.
10. GUILLEMIN, R., M. JUTISZ and E. SAKIZ, C.R. Acad. Sc., Paris, 256 : 504, 1963.
11. JUTISZ, M., A. BERAULT, M.-A. NOVELLA and G. RIBOT, Acta Endocrinol., 55 : 481, 1967.
12. SCHALLY, A.V. and C.Y. BOWERS, Endocrinology, 75 : 312, 1964.
13. RAMIREZ, V.D. and S.M. McCANN, Endocrinology, 73 : 193, 1963.
14. PARLOW, A.F., in Human pituitary gonadotropins, A. ALBERT ed., C.C. Thomas, Springfield, Ill., 1961, p. 300.
15. STEELMAN, S.L. and F.M. POHLEY, Endocrinology, 53 : 604, 1953.
16. JUTISZ, M. and M.P. de la LLOSA, Endocrinology, accepted for publication.
17. de la LLOSA, P., C. COURTE and M. JUTISZ, Biochem. Biophys. Res. Commun., 26 : 411, 1967.
18. JUTISZ, M. and M.P. de la LLOSA, C.R. Acad. Sc., Paris, 264 : Ser. D, 118, 1967.
19. SAMLI, M.H. and I.I. GESCHWIND, Program of the 49th Meeting of the Endocrine Society, 1967, no.59.
20. DAVID, M.A., F. FRASCHINI and L. MARTINI, Experientia, 21 : 483, 1965.
21. CORBIN, A. and J. STORY, Experientia, 22 : 694, 1966.
22. KUROSHIMA, A., A. ARIMURA, T. SAITO, Y. ISHIDA, C.Y. BOWERS and A.V. SCHALLY, Endocrinology, 78 : 1105, 1966.
23. SAITO, T., A. ARIMURA, E.E. MULLER, C.Y. BOWERS and A.V. SCHALLY, Endocrinology, 80 : 313, 1967.

# THYROTROPIN-RELEASING FACTOR (TRF) OF HYPOTHALAMIC ORIGIN. A REVIEW OF PHYSIOLOGICAL AND BIOCHEMICAL STUDIES. [1]

Roger Guillemin

Department of Physiology, Baylor University College of Medicine, Texas Medical Center, Houston, Texas USA

As requested by the Organizing Committee of this meeting, I will restrict myself in this lecture entirely to a discussion of our studies on the hypothalamic control of TSH-secretion. At this point may I say that in fact, for the last couple of years we have been working almost exclusively on this problem and have practically abandoned our earlier studies of the mechanisms involved in the secretion of ACTH, LH, growth hormone because of the difficulties and what I consider the unreliability of all the assay methods currently available for the study of these problems (1,2). I want to emphasize here the remarkable reliability, reproducibility, and above all specificity of all the assay methods both *in vivo* and *in vitro*, that we have been using and that are available to study TSH secretion and the nature of its hypothalamic control (3,4,5). The results which form the basis of this short review have been obtained over the last five years with several collaborators to whom of course all the credit must go: Dr. Sakiz, Dr. Yamazaki, Dr. Gard, and for the early purification studies, Dr. Jutisz while I was in Paris; Dr. Sakiz again, Dr. Burgus, Dr. Ducommun and Mr. Vale in Houston, Texas.

When we decided in 1962 to approach the problem of the mechanisms involved in the hypothalamic control of TSH secretion there was good evidence from excellent physiological studies coming over the preceding years from many laboratories (reviews in 6,7,8,9) that such a control did exist. An area of the anterior hypothalamus appeared to be more particularly involved in the

---

[1] Research currently supported by USPHS grants No. AM 08290-04 and HD 02577-01.

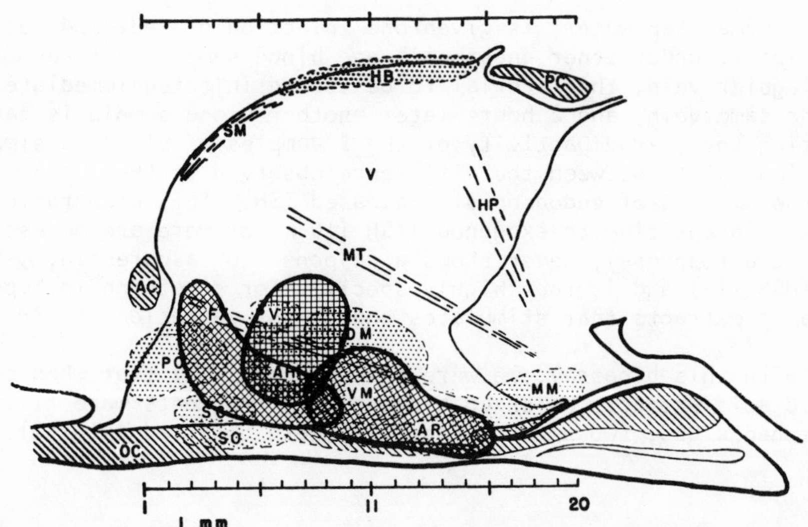


Legend to Figure 1: This simplified diagram shows 3 areas circled in black lines on a sagittal section of the hypothalamus (brain of the rat). These areas are related respectively from left to right with secretion of TSH, gonadotropins and ACTH. The diagram attempts to show that these areas overlap considerably; further, that these "hypophysiotropic" areas do not coincide with the classical nuclei of the hypothalamus as described by neuroanatomists. (OC) optic chiasm, (AC) anterior commissure, (PC) posterior commissure—other abbreviations as in de Groot atlas of the rat brain.

acute secretion of TSH (Fig. 1) and the existence of a specific hypothalamic neurohumor, TRF, for TSH releasing factor had been postulated for some time by several investigators. Indeed Shibusawa in Japan, Schreiber in Czechoslovakia, Saito and Tani here, had concluded to having demonstrated the existence of such a substance and its "isolation" had even been reported as a small polypeptide related to oxytocin (10,11). In several reviews (1,7,8) I have clearly stated my own critical appraisal of these many elegant conclusions relating to TRF and which I think belong to what we call in my laboratory, "<u>the prophetic literature</u>"; I will not reiterate these comments here. Thus, in 1962, we started searching for a then hypothetical TSH-releasing factor or TRF. First of all we devised a bioassay that logically should be adequate and specific for this substance. With Dr. Sakiz and Dr. Yamazaki we reported on a simple <u>in vivo</u> method (3) based on the concepts involved in the assay for TSH described earlier by Purves and Griesbach and McKenzie (12). This assay for TRF which over the years has been considerably simplified from the original description is simply as follows: a normal male rat kept on normal diet, normal tap water, is given one injection of radioiodine; 3 days later, under ether anesthesia one blood sample is taken from the jugular vein, the material to be tested injected immediately in the same vein, and 2 hours later another blood sample is taken. Counting the $\gamma$-radioactivity of the 2 samples of blood, a simple relation exists between the difference observed in the 2 samples and the amounts of endogenously released TSH. This preparation is grossly insensitive to exogenous TSH (10 mu or more are necessary to give a response), never shows a response to vasopressin, $\alpha$-MSH or $\beta$-MSH (13) and is thus highly specific for something in hypothalamic extracts that stimulates endogenous secretion of TSH.

With this bioassay, we were able to show (14) that when crude acetic acid extracts of sheep hypothalamic fragments were filtered on Sephadex G25, two zones of activity were observed (Fig. 2).

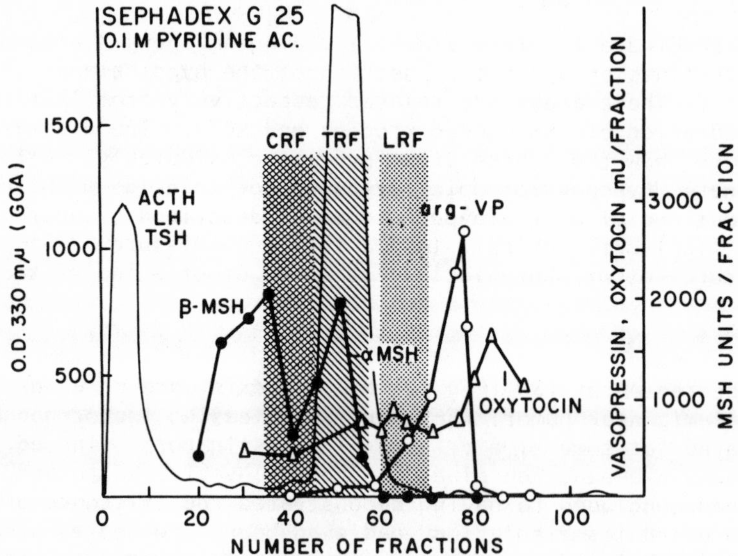


Legend to Figure 2: Filtration chromatography on Sephadex G25 of an acetic acid extract of sheep hypothalamus. The biological activities due to arginine-vasopressin (arg-VP), oxytocin, $\alpha$- and $\beta$-melanophoretic hormones (MSH) are shown on the graph with their elution volumes. Three discrete zones of the effluent have biological activities corresponding to CRF, TRF, LRF. At that early stage of purification, FRF is not well separated from LRF. The first peak of substances appearing in the effluent contains activities corresponding to ACTH, LH, TSH or like substances. These activities are undoubtedly present in the hypothalamic tissues (no pituitary tissue contamination); their true origin (hypothalamic or hypophysial) is unknown at this time.

One, in the peak of materials not retarded on the gel, another, in a zone of the effluent corresponding to strongly retarded substances. To our surprise the first zone remained active in hypophysectomized animals or in mice given large doses of thyroxine as in McKenzie assays. This activity we think is due to TSH or TSH-like substances present in these hypothalamic extracts. The significance of these substances is still unclear but we have good evidence that they are truly in hypothalamic tissues and do not come from some contamination with pituitary tissues. The other zone of the effluent was inactive in hypophysectomized animals or in animals given large doses of thyroxine. Similarly, materials from this zone of the effluent would liberate the secretion of TSH from anterior pituitary fragments incubated in vitro (4) and this TSH-releasing activity could be inhibited by adding $T_4$ to the incubation medium (4). Thus we concluded that these results were best explained by the presence in the hypothalamic extracts of a substance that stimulated the secretion of TSH and which we called TRF for TSH-releasing factor.

Over the last few years we have spent considerable efforts towards the isolation of this substance, the elucidation of its chemical structure and more recently about its mode of action at the cellular or subcellular level.

It became rapidly obvious that we were dealing with a potent material of which only minute quantities were to be found in one hypothalamic fragment. Hence, we started organizing the collection of hypothalamic fragments by the hundreds of thousands and we devised a methodology to cope with these large quantities of materials. The stage of filtration on G25 was enlarged to semi-industrial size. On a column 150 cm high and 15 cm diameter we can in one pass, filter the extract of 25,000 to 50,000 sheep hypothalamus. The material obtained here has by definition, 1 unit of TRF activity/mg dry weight (see 15). Over the last 3 years we have collected more than 1 million fragments of sheep hypothalamus. From this large quantity of starting material we have now only a few hundred micrograms of a highly purified substance active to release TSH at picogram level in vitro or nanogram level in vivo. Table 1 summarizes the stages of purification involved in starting with a total of 1/2 million hypothalamic fragments.

Recently we have replaced the early stages by a method more practical even though no more efficient, based on the differential solubility of the releasing factors particularly TRF in organic solvents using sequentially several mixtures of methanol, chloroform and ether (16).

The material obtained at the last stage reported in Table 1 by adsorption on activated charcoal and elution with ethanol was

Table 1

Sequence of purification for TRF

| | | Step | Amount |
|---|---|---|---|
| | | Sheep hypothalamus<br>500,000 fragments | 5000 Kg |
| | | ↓ acetone powder | 52 Kg |
| | | ↓ 2 N HOAc extract | 12 Kg |
| | | ↓ gel filtration G25<br>0.5 M pyridine acetate | |
| ↓ LRF<br>AVP | ↓ CRF<br>MSH | ↓ TRF Concentrate | 180 g<br>1 U/mg |
| | | ↓ CM-Sephadex C50 | 30 g |
| | | ↓ G25 in 1 N HOAc | 3 g |
| | | ↓ liquid partition<br>HOAc, pyridine, butanol | 30 mg |
| | | ↓ Norit A<br>$H^2O$ → ethanol | 2.0 mg<br>15-25,000 U/mg |

rechromatographed on the same system; it gave a symmetrical peak which after lyophilization yielded a material with a specific activity of 20,000-50,000 units/mg, i.e., showing biological activity <u>in vivo</u> at doses smaller than 1.0 $\mu$g. This material behaved as a single spot in 5 different solvent systems on thin-layer chromatography or on thin-layer electrophoresis at pH 2.0, 7.0 and 8.5, was ninhydrin negative and strongly positive

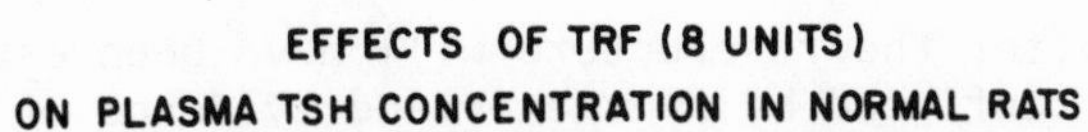


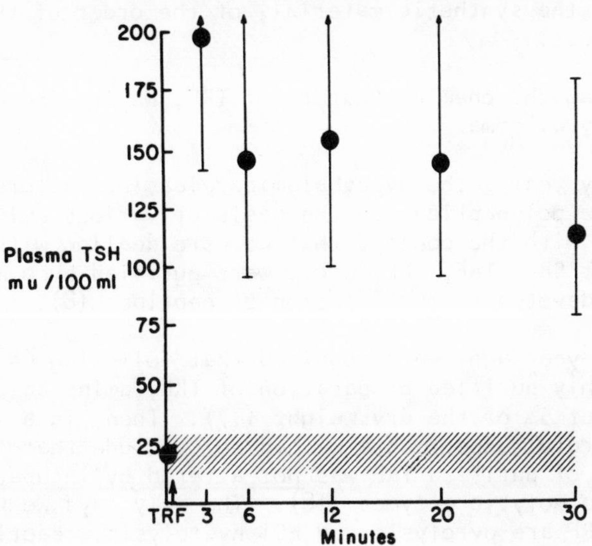


Legend to Figure 3: Liquid partition chromatography in the system butanol-acetic acid-water (4-1-5) on a column of Sephadex G25 (0.5 x 25 cm) of the TRF fraction obtained previously from adsorption on charcoal. Volume of each fraction (tube) of the effluent: 0.17 ml.

with Pauly's reagent (diazotized sulfanilic acid) or iodine in chloroform.

When this material was further chromatographed by liquid partition chromatography on a microcolumn using the system butanol-acetic acid-water (4-1-5), it was however further resolved in 3 distinct peaks, with all the biological activity located in the smallest one (Fig. 3). From the results obtained after careful weighing _in vacuo_ under infrared drying with an electrobalance, we are now the somewhat dismayed owners of less than 300 μg of a substance with a specific activity of between 50,000 and 75,000 TRF units/mg (with by the way complete overlapping of confidence limits of all the bioassays of the 2 preceding stages).

At this point let me say that we are still not ready to claim _isolation_ of TRF. When dealing with such minute quantities of substances, proving homogeneity is an extremely difficult task. These restrictions should apply similarly for other laboratories working on this problem, even though they may have taken a somewhat more liberal view of the problem. I think that it will be safe to report the isolation of any one of these hypothalamic substances only after their structure will have been established and better still, confirmed by synthesis and evidence of specific activity for the synthetic material, of the order of that of the natural product.

Regarding the chemical nature of TRF, we are indeed in an extraordinary dilemma.

For many years, the hypothalamic releasing factors were considered to be polypeptides on the basis of various evidence. And indeed it is with the concept that we were dealing with basic polypeptides that CRF, TRF, LRF, etc., were purified by a methodology classically devoted to purification of peptides (8).

About a year ago, we recognized that following 6N HCl hydrolysis of a highly purified preparation of TRF, amino acids accounted for only about 5% of the dry weight (17). Then, in a series of extremely careful experiments we had to conclude that the biological activity of purified TRF was not altered by incubation with a series of proteolytic enzymes (18). The only ways we know of to inactivate TRF are pyrolysis, 6N HCl hydrolysis, reacting it with Pauly reagent or, of all things, incubating it in plasma, some times as shortly as 60 seconds. With the preparations of purified TRF which we have studied so far, the material has appeared to be non-volatile at atmospheric pressure which precludes the use of gas chromatography, or in high vacuum of the order of $10^{-7}$ thor even at 130°C, which precludes the use of mass spectrography to study it. The molecule of TRF may contain 3 amino acids, as we still find after 6N HCl hydrolysis of a highly purified preparation of TRF, proline, histidine and glutamic acid. The quantities and by the way, molar ratios do not quite agree with what was reported by Schally *et al*. (19,20) but I certainly would not want to make a statement at this moment as to who is right and who is wrong. Nuclear magnetic resonance spectra of highly purified TRF at 50, 100 or 220 megacycles, with time averaging have not yielded any meaningful information except that we may be dealing with a highly saturated alicyclic or heterocyclic structure with peripheral $CH_3$ groups without completely ruling out a polyamide structure. About all we can say with some certainty about TRF is that the molecular weight is below 2,000.

More rewarding than this difficult probing of the chemical nature of TRF have been a series of experiments dealing with the physiological and cellular mode of action of TRF. A rapid, intravenous injection of TRF produces a striking increase in the plasma levels of TSH within a few seconds (Fig. 4). This can be prevented by pretreatment with thyroxine and there is some sort of a competition between thyroxine and TRF, which we have observed both *in vivo* and *in vitro* (4, 21).

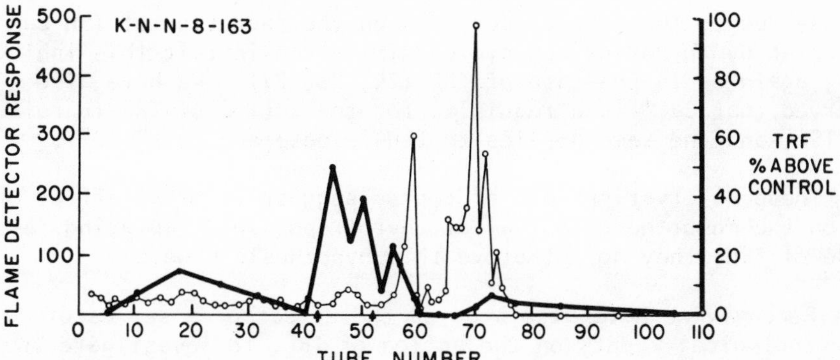


Legend to Figure 4: Effects of a rapid single intravenous injection of purified TRF on plasma TSH concentration—study in normal rats. In less than 3 minutes following the administration of TRF, plasma TSH concentration has increased almost ten-fold. A similarly rapid response of the pituitary has been shown to occur following injection of CRF or LRF, also following exposure to acute stress (secretion of ACTH) or to cold (secretion of TSH).

In contradistinction to the conclusions published (*) earlier by Jutisz _et al_. regarding LRF and FRF (22, 23) neither cycloheximide nor actinomycin D modify or inhibit the release of TSH as stimulated by TRF (24, 25, 26). On the other hand, we have good evidence that cycloheximide can prevent and or reverse the thyroxine-induced inhibition of TSH-release stimulated by TRF, whereas actinomycin D can prevent it but not reverse it (24, 25, 26). Thus, the competition between thyroxine and TRF at the pituitary level for the release of TSH appears to be not a direct one, but involves some intermediate step probably induced by thyroxine.

The release of TSH stimulated by TRF thus does not require the _de novo_ synthesis of TSH or any other substance. It is a rapid and immediate effect on TSH pre-existing in the adenohypophysial cells. In fact, a similar effect can be reproduced by elevating the $K^+$ concentration of the incubation fluid in which pituitary fragments are incubated (26, 27). As in the case of the TSH-

---

(*) We refer here to the conclusions of Jutisz _et al_. as in the _published_ text (in French) of their two notes. The mimeographed abstracts in English included with reprints sent by the authors present conclusions opposite to those of the French text. At this meeting, on the basis of recent data, Jutisz reported to be in agreement with our earlier conclusions studying TRF, which thus would appear to apply equally to LRF, FRF and TRF.

release due to TRF, the effect of K on the secretion of TSH can be prevented by thyroxine and cycloheximide can inhibit this inhibition, again as in the case of TRF (24, 26, 27). We have also observed that $Ca^{++}$ is a requisite for the action of TRF in releasing TSH, and the same applies to TSH-release due to $K^{+}$ (28).

These observations are of course suggestive of an effect of TRF on the membrane potential of thyrotrophs in stimulating release of TSH; they do not prove this hypothesis however.

Recently we have also studied the effect of a series of pharmacological agents on the action of TRF, to investigate the source of energy involved in the release of TSH. DNP does not inhibit TRF, which is however inhibited by iodoacetamide. Along the same line, theophylline potentiates considerably the effects of TRF, and in several preliminary experiments, soluble dibutyryl-AMP appears to stimulate release of TSH. Neither ouabain nor tetrodotoxin have been able to modify the activity of TRF in our hands and in the conditions studied so far (Vale, Burgus, Guillemin, to be published).

As a summary, we can make the following statements. The existence of a TSH-releasing factor (TRF) of hypothalamic origin is now unquestionably established. The material has been obtained in a state of high purity; its chemical characterization however remains to be elucidated. The preparations of highly purified TRF are active to release TSH at nanogram levels in various in vivo assay systems. The well known feedback control between circulating thyroxine levels and pituitary secretion of TSH has been shown to be exerted at the level of the adenohypophysial cells in some sort of a variable set point competition involving thyroxine, TRF, and a proteinic or polypeptidic intermediate induced by thyroxine. So far, on the basis of bioassay studies we have no evidence that TRF might be involved in the biosynthesis of TSH; this point requires further investigation with a methodology more refined than that using bioassays alone. Thus, in spite of considerable progress over the last few years, we still have much more to clarify before closing the chapter of the hypothalamic control of the secretion of the thyrotropic hormone.

## References


1. Guillemin, R., Ann. Rev. Physiol., 29, 313, 1967.
2. Rodger, N. W., J. C. Beck, R. Burgus and R. Guillemin, Prog. Am. Endoc. Soc., 1967, p. 88.
3. Yamazaki, E., E. Sakiz and R. Guillemin, Experientia, 19, 480, 1963.

4. Guillemin, R., E. Yamazaki, D. A. Gard, M. Jutisz and E. Sakiz, Endocrinology, 73, 564, 1963.
5. Guillemin, R., and W. Vale, Endocrinologia Experimentalis, 1, 137, 1967.
6. D'Angelo, S. A., Advances in Neuroendocrinology, A. V. Nalbandov, Ed., Univ. of Illinois Press Pub., 1963, pp. 158-194.
7. Guillemin, R., J. Physiol., (Paris), 55, 7, 1963.
8. Guillemin, R., Recent Progress in Hormone Research, Vol. 20, Academic Press, New York, 1964, pp. 89-130.
9. Brown-Grant, K., The Pituitary Gland, Harris and Donovan, Eds, Univ. of California Press, Vol. 2, p. 235, 1966.
10. Saito, S., and F. Tani, Endocrinol. Jap., 7, 13, 1960.
11. Schreiber, V., M. Rybak, A. Eckertova, V. Jirgl, J. Koci, Z. Franck and V. Kmentova, Experientia, 18, 338, 1962.
12. McKenzie, J. M., Endocrinology, 63, 372, 1958.
13. Yamazaki, E., E. Sakiz and R. Guillemin, Ann. Endocr. (Paris), 24, 795, 1963.
14. Guillemin, R., E. Yamazaki, M. Jutisz and E. Sakiz, C. R. Acad. Sci. (Paris), 255, 1018, 1962.
15. Guillemin, R., and E. Sakiz, Nature, 207, 297, 1965.
16. Burgus, R., M. S. Amoss and R. Guillemin, Experientia, 23, 417, 1967.
17. Guillemin, R., R. Burgus, E. Sakiz and D. N. Ward, C. R. Acad. Sci. (Paris), 262, 2278, 1966.
18. Burgus, R., D. N. Ward, E. Sakiz and R. Guillemin, C. R. Acad. Sci. (Paris), 262, 2643, 1966.
19. Schally, A. V., T. W. Redding, J. F. Barrett and C. Y. Bowers, Fed. Proc., 25, 348, 1966.
20. Schally, A. V., C. Y. Bowers, T. W. Redding and J. F. Barrett, Biochem. Biophys. Res. Commun., 25, 165, 1966.
21. Vale, W., R. Burgus and R. Guillemin, Proc. Soc. Exp. Biol. Med., 125, 210, 1967.
22. Jutisz, M., A. Berault, M-A. Novella and F. Chapeville, C. R. Acad. Sci. (Paris), 263, 664, 1966.
23. Jutisz, M., and P. de la Llosa, C. R. Acad. Sci. (Paris), 264, 118, 1967.
24. Vale, W., R. Burgus and R. Guillemin, Prog. Am. Endocr. Soc., 1967, p. 89.
25. Vale, W., R. Burgus and R. Guillemin, Neuroendocrinology, accepted for publication, 1967.
26. Vale, W., M. Amoss, R. Burgus and R. Guillemin, Int. Symposium on the Pharmacology of Hormonal Polypeptides, Sept., 1967, Milan, Italy, p. 122.
27. Vale, W., and R. Guillemin, Experientia, in press, 1967.
28. Vale, W., R. Burgus and R. Guillemin, Experientia, in press, 1967.

# PHYSIOLOGICAL AND BIOCHEMICAL STUDIES ON SOME HIGHLY PURIFIED HYPOTHALAMIC RELEASING FACTORS

A.V. Schally, A. Arimura, C.Y. Bowers, S. Sawano, A.J. Kastin, T.W. Redding and T. Saito

V.A. Hospital and Dept. of Med., Tulane University School of Medicine, New Orleans, La., USA

and

W.F. White and A.I. Cohen

Abbott Laboratories, North Chicago, Ill., USA

Since our group has been scheduled to deliver the concluding presentation, we would like to mention topics not discussed by previous speakers. For this reason, our presentation had to be improvised in part during the preceding talks, so that it will be composed of many topics.

First, we would like to propose a new nomenclature for hypothalamic releasing factors. Two weeks ago, we made these proposals at the Laurentian Hormone Conference in Canada. We will repeat these proposals for our European colleagure and those who did not attend this Conference.

The term "releasing factor," proposed by Saffran and Schally in 1955 (1, 2), is well established and convenient, but creation of another nomenclature for hypothalamic principles in which they are designated as hormones is desirable. Although several criteria should be satisfied before a substance is called a hormone, the properties of releasing factors (3, 4, 5), particularly their action in minute amounts to regulate release and, in some cases, synthesis of tropic hormones, support classification of hypothalamic releasing factors as hormones. This nomenclature is shown in Table 1. The new nomenclature could have as its basis the name or the abbreviation of the respective adenohypophysial principle, followed by "releasing" or "inhibiting" (since either may be the **primary** effect of a given hypothalamic hormone) and then the word **hormone**.

TABLE 1

THE NOMENCLATURE OF SOME HYPOTHALAMIC NEUROHUMORS

| Present Name | | Proposed Name | |
|---|---|---|---|
| Hypothalamic Factor | Abbreviation | Hypothalamic Hormone | Abbreviation |
| Corticotropin-releasing factor | CRF | Corticotropin-releasing hormone | CRH |
| Luteinizing hormone-releasing factor | LRF or LH-RH | Luteinizing hormone-releasing hormone | LH-RH or LRH |
| Follicle-stimulating hormone-releasing factor | FSH-RF | Follicle-stimulating hormone-releasing hormone | FSH-RH or FRH |
| Thyrotropin-releasing factor | TRF | Thyrotropin-releasing hormone | TRH |
| Growth hormone releasing factor or somatotropin-releasing-factor | GRF or SRF | Growth hormone-releasing hormone or somatotropin-releasing hormone | GRH or SRH |
| Prolactin inhibiting factor (mammals) | PIF | Prolactin release inhibiting hormone | PRIH |
| Melanocyte stimulating hormone (MSH) release-inhibiting factor | MIF | MSH-release-inhibiting hormone | MRIH |

From Schally *et al.* (3).

The role of hypothalamic releasing hormones on synthesis of pituitary hormones cannot be fully described at present and it is possible that if it is established that release is secondary to synthesis, some modification of this nomenclature may be desirable. The term "factor" should be retained only for those principles the specificity of action of which is in doubt or not established (MRF, MSH-releasing factor; Growth hormone-inhibiting factor, GIF), but not for those that are obviously artifacts.

We believe that at present, it is best to express activity of releasing hormones in terms of the minimal active dose reported as dry weight which produces a statistically significant response in a specific test. Specific details of these proposals will be incorporated in the published proceedings of the Laurentian Hormone Conference (3).

<u>Corticotropin releasing hormone (CRH).</u> Corticotropin releasing hormone has not yet been discussed at this meeting. Perhaps the first direct evidence supporting the neurohumoral theory of hypothalamic control of anterior pituitary gland postulated by Professor G.W. Harris came from the work of Saffran and Schally in 1955 (1, 2). It was concluded that neurohypophysial and hypothalamic extracts contain a corticotropin-releasing factor (CRF) which stimulates release of adrenocorticotropic hormone (ACTH) from the anterior pituitary gland (1, 2). Guillemin, on the basis of independent work, reached similar conclusions (6). Support for the concept that vasopressin is the principal ACTH releasing hormone has now been abandoned (3-5). In discussing biochemical studies on CRF, it should be mentioned that the partial structure of $\beta$-CRF from pig posterior pituitaries was elucidated by us in collaboration with Dr. Boissonnas (5, 7). Instability of CRF and difficulty of assays, caused mainly by contamination with vasopressin and ACTH-like peptides, and other problems have so far delayed isolation of CRF in a homogenous form. Recent work in our laboratory may give considerable impetus to the elucidation of CRF problem. We have prepared $\alpha$ - and $\beta$ -CRF from beef neurohypophyses (8). We have also prepared CRF from pig hypothalami (3, 4, 9). The last step in the preparation consisted of rechromatography on carboxymethylcellulose (CMC) which separated CRF from LRF and FSH-RF (3, 4).

The CRF activity of this hypothalamic material was carefully characterized and the responses were compared to those obtained with neurohypophysial, $\alpha$ - and $\beta$ -CRF (9). $\beta$ -CRF showed CRF activity in a log-dose response relationship when assayed in rats treated with chlorpromazine, morphine and Nembutal (CPZ-M-N). $\beta$ -CRF was active in doses of 1 ug in CPZ-M-N treated rats, but inactive at twice this dose in hypophysectomized animals. In agreement with previous results, the response to $\beta$ -CRF was inhibited by pretreating the rats with dexamethasone. The effects of 4 ug $\alpha$ -CRF and 4 ug hypothalamic CRF on levels of plasma corticosterone in rats pretreated with CPZ-M-N, dexamethasone, morphine and Nembutal (Dex-M-N) or hypophysectomized were also investigated (9). Although $\alpha$-CRF had some ACTH-like activity, possibly inherent, the response in CPZ-M-N rats was significantly greater than in either Dex-M-N rats or in rats hypophysectomized for 24 hrs. Since adrenocortical stimulation following administration of ACTH is the same in all three animal preparations (9), we must conclude that the $\alpha$-CRF preparation released ACTH. Similarly in the case of hypothalamic CRF, two factor factorial analysis as well as Duncan's multiple-range comparison tests showed that the responses obtained in CPZ-M-N treated rats were significantly larger than in Dex-M-N rats and that the responses in the latter were larger than those obtained in hypophysectomized rats (9). This supports the conclusion that this hypothalamic material acted as CRH and that it may have been somewhat related in its properties to neurohypophysial $\alpha$-CRF (3). Although we do not know whether this material represents true physiological CRH, it appears to be the most potent hypothalamic CRF so far obtained (3).

Thyrotropin releasing hormone (TRH). Dr. Guillemin has presented today an excellent account of his studies on sheep TRH. We would simply like to add that the properties of bovine and porcine TRH are similar to those described by Dr. Guillemin for ovine TRH. We have prepared what we think is homogenous pig TRH (3, 10). Porcine TRH is not a polypeptide, but studies by nuclear magnetic resonance and mass spectra have not yet revealed its full structure.

In vivo Effects of TRH. Porcine TRH is active in vivo at doses as low as 1 nanog in mice treated with codeine and 1 ug thyroxine (3). Porcine and bovine TRH show linear log dose relationship in this system (3, 11). TRH is also active in a linear dose response relationship in mice treated with 0.1 ug Triiodothyronine (3) and will increase plasma TSH in thyroidectomized rats pretreated with 1 ug $T_3$ (3). Intravenous administration of a highly purified preparation of bovine or porcine TRH will also cause a significant depletion of pituitary TSH content in mice (12). The minimum active dose of TRH required for significant pituitary TSH depletion is 4 nanograms.

In vitro Effects of TRH. TRH releases TSH from rat anterior pituitary glands incubated in vitro according to the method of Saffran and Schally (1). The results of this in vitro method correlate perfectly with the in vivo data. In four different experiments, 0.01 nanograms of TRH significantly increased release of TSH in vitro as compared with controls (3, 13). The results of other in vitro experiments show that as the doses of TRH are increased, greater amounts of TSH are found in the media. The released TSH was assayed by the McKenzie method (14). Simple calculations showed that if the potency of pure TSH is taken as 60 U/mg, then TRH released 200-2000 times its weight of TSH. Such a significant "multiplier effect" strongly supports the concept that TSH is a hormone (3, 13).

Addition of Actinomycin-D in doses as high as 10 $\mu$g/ml to preincubation and incubation media did not affect the release of

TSH induced by TRH (3, 13). This suggests that TRH acts mainly by stimulating release rather than synthesis of TSH and that *de novo* synthesis of TSH is not required for TRH to exert its effect. However, the addition of 5 nanog TRH to incubation media significantly increased the incorporation of $C^{14}$ amino acids into pituitary tissue. This may be indicative that TRH can also induce TSH synthesis (3).

*Effect of thyroxine ($T_4$) and triiodothyronine ($T_3$) on the response to TRH.* Relatively large doses of $T_3$ and $T_4$ can inhibit the stimulatory effect of TRH on TSH release *in vivo* and *in vitro*. Inhibition by $T_3$ and $T_4$ is dose related and can be overcome by increasing the doses of TRH (3). Similarly addition of Actinomycin-D to preincubation media *in vitro* impairs the ability of $T_4$ or $T_3$ to inhibit the TRH-induced stimulation of the TSH release (3, 13).

*Growth hormone-releasing hormone (GRH).* The chemical and physiological properties of GRH were discussed in detail at the preceding Symposium on Growth Hormone (15). Perhaps it could be mentioned here that our studies carried out with Dr. Eugenio E. Muller and Dr. Sanford L. Steelman revealed that GRH is an acidic polypeptide with a M.W. of about 2500 (16). In rats, highly purified GRH will deplete pituitary GH content, increase plasma GH levels (3) and release GH *in vitro* (17). GRH can also stimulate the synthesis of GH *in vitro* (17). These results support the concept that the same substance is responsible for both release and synthesis of Growth Hormone (3, 17).

*Follicle-Stimulating hormone-releasing hormone (FSH-RH).* FSH-RH was discussed at this meeting by Dr. McCann and Dr. Jutisz. Our preparations of porcine and bovine FSH-RH seem to be considerably more potent than their preparations of ovine FSH-RH and are also free of LH-RH. A complete separation of FSH-RH from LH-RH was achieved by preparative column electrophoresis (3, 18, 19). Thus fractions which release LH and have no effect on FSH secretion and conversely fractions which release FSH and have no effect on LH secretion. Highly purified FSH-RH at doses of 4-10 nanog depletes pituitary FSH content in castrated male rats pretreated with testosterone (3, 19). FSH-RH will also deplete pituitary FSH in normal male rats (3, 19). Some amines such as putrescine are also capable of depleting pituitary FSH content *in vivo* (20). Mittler and Meites (36) were the first to demonstrate the stimulatory effect of rat hypothalamic extracts on the release of FSH *in vitro*. We have shown by *in vitro* experiments that purified porcine FSH-RH stimulates the release of FSH by a direct action on pituitary tissue (21). Thus, addition of partially purified FSH-RH, but not of histamine, greatly enhanced the release of FSH. In these experiments, partially purified FSH-RH (at least 10 times less potent than our most purified material) was calculated to have caused the release of at least 100-200 x its own weight of FSH expressed as NIH standard S-3. The addition of Actinomycin-D to incubation media at a concentration of 2 µg/ml did not block the stimulatory effect of FSH-RH on FSH release. Similarly pretreatment with Actinomycin-D _in vivo_ does not block the depletion of pituitary FSH after administration of FSH-RH (19). This indicates that FSH-RH acts mainly by stimulating the release and not synthesis of FSH.

_Luteinizing hormone-releasing hormone (LH-RH)._ This was also already discussed by Dr. McCann, Dr. Gregory and Professor Harris, and Dr. Jutisz. Porcine LH-RH prepared by us is free of FSH-RH and other releasing factors (3, 18). Highly purified LH-RH does not appear to be a peptide and is not destroyed by pepsin and trypsin. Our preparations of porcine LH-RH seem to be much more potent than those reported by previous speakers. Partially purified porcine LH-RH releases LH _in vitro_ at doses of 0.02 μg/pituitary (18). This is equivalent to a few nanograms of LH-RH per mg of pituitary tissue. Although this LH-RH was at least 10 times less potent than our most purified preparations, it released 10-50 times its own weight of LH expressed as NIH standard-S-1. This multiplier effect supports the concept of the hormonal nature of LH-RH (3, 18).

_In vivo_ studies indicate that Act-D does not block the response to LH-RH, indicating that the primary effect of this neurohormone must be on release rather than synthesis. Highly purified LH-RH will increase plasma LH levels in ovariectomized female rats pretreated with estrogen and progesterone according to the technique of Ramirez and McCann (21). LH-RH is in this test active at doses of a few nanograms. When the amount of LH was assayed by 4 point assays, it was calculated that partially purified LH-RH had released from 5-10 times its own weight of LH. LH-RH also increases plasma LH levels in ovariectomized female rats or castrated male rats pretreated with testosterone (22). In the latter assay, dose response relationship to LH-RH was obtained as well as a depletion of pituitary LH. As little as 1 μg of partially purified LH-RH will induce ovulation in cyclic rats blocked with Nembutal, before 2 p.m. on the day of pro-estrus (23).

_Prolactin release-inhibiting factor (PRIH)._ PRIH was not discussed so far at this meeting. Preparations of PRIH of ovine, porcine and bovine origin inhibit the release of prolactin from anterior pituitary tissue _in vitro_. We have also shown that porcine preparations of PRIH inhibit the depletion of prolactin from pituitary of estrus rats, following cervical stimulations. These results confirm the conclusions of Meites _et al._ (24) concerning the existence of PRIH in mammalian hypothalamus. Our _in vivo_ and _in vitro_ data indicates that PRIH and LH-RH are distinct substances (25, 26, 27).

_MSH-release-inhibiting hormone (MRIH)._ In the past few years, evidence has been accumulating from our laboratory that the predominant control of MSH release in mammals is exerted by a hypothalamic hormone designated MRIH (28, 29, 30). We have also demonstrated a feedback control of MSH release and presented evidence indicating that the pineal gland is involved in control of MSH release (31, 32). Data from the laboratory of Taleisnik and Orias (33), as well as our own (29), raised the possibility that an MSH-releasing factor (MRF) may also exist.

MRIH has now been found in the hypothalamic extracts of 8 species of animals including man (29, 30). Fragments of beef hypothalamic tissue were extracted and the extracts concentrated (34). The resulting material was separated into multiple fractions by gel filtration on Sephadex G-25 columns. The MRIH was concentrated approximately 11,000 times (34). The molecular weight of this material is probably only of the order of several hundred. Only 0.06 μg of this purified MRIH was able to greatly increase MSH content in the pituitaries of normal rats (34). Injection of MRIH in rats which had been pretreated with tranquilizers (35) resulted in pituitary MSH levels greater than those found in rats treated with tranquilizers only. Recent evidence, moreover, indicates that purified MRIH also inhibits release of MSH from rat posterior lobes (including intermediate lobe) incubated *in vitro* in Krebs-Ringer bicarbonate. Administration of MRIH, made in our laboratory, has been found to decrease plasma levels of MSH in mice.

*Hypothalamus and Disease.* Just as diseases have been described for an excess or deficiency of almost every pituitary hormone, it can be predicted that some cases of these diseases may be due to an excess or deficiency of a hypothalamic hormone. It is not difficult to conceive possible roles of hypothalamic hormones in the diagnosis and treatment of disease.

*Conclusions.* Hypothalamic extracts from a variety of animals contain distinct substances which stimulate release of ACTH, TSH, GH, FSH, LH, and inhibit the release of prolactin and MSH. Various physiological and biochemical studies indicate their important physiological role and support the concept that these substances are hormones.

## Acknowledgments

The studies reported here were supported in part by grants from the U.S. Veterans Administration and NIH-USPHS grants AM-07467, AM-09094 and AM-08743.

## References

1. Saffran, M. and Schally, A.V. (1955). Can. J. Biochem. Physiol. 33, 408.

2. Saffran, M., Schally, A.V. and Benfey, B.G. (1955). Endocrinology 57, 439.

3. Schally, A.V., Arimura, A., Bowers, C.Y., Kastin, A.J., Sawano, S. and Redding, T.W. Recent Progress in Hormone Research 24, 1968.

4. Schally, A.V., Kastin, A.J., Locke, W. and Bowers, C.Y. Hormones in Blood (C.H. Gray ed.) vol 2, 1967 London, Academic Press.

5. Schally, A.V., Bowers, C.Y. and Locke, W. (1964). Am. J. Med. Sci. 248, 79.

6. Guillemin, R., Hearn, W.F., Cheek, W.R. and Housholder, D.E. (1957). Endocrinology 60, 488.

7. Schally, A.V. and Bowers, C.Y. (1964). Metabolism 13, 1190.

8. Schally, A.V., Carter, W.H., Hearn, I.C. and Bowers, C.Y. (1965). Am. J. Physiol. 209, 1169.

9. Arimura, A., Saito, T. and Schally, A.V. Endocrinology (1967) 81, 235.

10. Schally, A.V., Bowers, C.Y., Redding, T.W. and Barrett, J.F. (1966). Biochem. Biophys. Res. commun. 25, 165.

11. Redding, T.W., Bowers, C.Y. and Schally, A.V. (1966). Proc. Soc. Exptl. Biol. Med. 121, 726.

12. Redding, T.W. and Schally, A.V. Endocrinology 81, October, 1967.

13. Schally, A.V. and Redding, T.W. Proc. Soc. Exp. Biol. Med. 1967 or Jan. of 1968.

14. McKenzie, J.M. (1958). Endocrinology 63, 372.

15. Schally, A.V., Sawano, S., Muller, E.E., Arimura, A., Bowers, C.Y., Redding, T.W. and Steelman, S.L. Proc. Int. Symposium on Growth hormone, Milan, Sept., 1967, Excerpta Medica, Amsterdam.

16. Schally, A.V., Muller, E.E., Arimura, A., Saito, T., Sawano, S., Bowers, C.Y. and Steelman, S.L. (1967). Ann. N.Y. Acad. Sci. In print.

17. Schally, A.V., Muller, E.E. and Sawano, S. (1968). Endocrinology in press.

18. Schally, A.V., Bowers, C.Y., White, W.F. and Cohen, A.I. (1967). Endocrinology 81, 77.

19. Schally, A.V., Saito, T., Arimura, A., Sawano, S., Bowers, C.Y., White, W.F. and Cohen, A.I. (1967). Endocrinology 81, October, 1967.

20. White, W.F., Cohen, A.I., Storey, J. and Schally, A.V. Endocrinology in press.

21. Ramirez, V.D. and McCann, S.M. (1963). Endocrinology 73, 193.

22. Schally, A.V., Carter, W.H., Arimura, A. and Bowers, C.Y. (1967). Endocrinology, 81, November, 1967.

23. Arimura, A., Schally, A.V., Saito, T., Muller, E.E., and Bowers, C.Y. (1967). Endocrinology 80, 515.

24. Talwalker, P.K., Ratner, A. and Meites, J. (1963). Am. J. Physiol. 205, 213.

25. Schally, A.V., Meites, J., Bowers, C.Y. and Ratner, A. (1964). Proc. Soc. Exptl. Biol. Med. 117, 252.

26. Arimura, A., Saito, T., Muller, E., Bowers, C.Y., Sawano, S. and Schally, A.V. (1967). Endocrinology 80, 972.

27. Schally, A.V., Kuroshima, A., Ishida, Y., Redding, T.W. and Bowers, C.Y. (1965). Proc. Soc. Exptl. Biol. Med. 118, 350.

28. Kastin, A.J. (1965). Program Endocrine Society Meeting p. 98.

29. Kastin, A.J. and Schally, A.V. (1966). Gen. Comp. Endocrinol. 7, 452.

30. Kastin, A.J. and Schally, A.V. (1967). Gen. Comp. Endocrinol. 8, 344.

31. Kastin, A.J. and Schally, A.V. (1967). Nature 213, 1238.

32. Kastin, A.J., Redding, T.W. and Schally, A.V. (1967). Proc. Soc. Exptl. Biol. Med. 124, 1275.

33. Taleisnik, S. and Orias, R. (1965). Am. J. Physiol. 208, 293.

34. Schally, A.V. and Kastin, A.J. (1966). Endocrinology 79, 768.

35. Kastin, A.J. and Schally, A.V. (1966). Endocrinology 79, 1018.

36. Mittler, J.C. and Meites, J. (1964). Proc. Soc. Exptl. Biol. Med. 117, 309.

# ROLE OF THE CEREBRAL CORTEX IN THE CONTROL OF ACTH SECRETION

G. Lugaro and M.M. Casellato
Department of Organic Chemistry, University of Milan
and
M.Motta, F. Piva and L. Martini
Department of Pharmacology, University of Milan ,Italy

Data are accumulating which indicate that nervous centers located above the hypothalamus may exert an inhibitory effect on the hypothalamic-pituitary-adrenal axis (Mangili,Motta and Martini, 1966). It has been shown, for instance, that the interruption of all nervous pathways going to the hypothalamus from higher brain structures ("hypothalamic deafferentation") results, in the rat, in a significant increase of adrenal weight (Halasz and Pupp, 1965) and of resting levels of plasma and adrenal corticosterone (Halasz, Slusher and Gorski, 1967). Data obtained by Egdhal (1967) suggest that the cerebral cortex (CC) may be one of the structures involved in maintaining the hypothalamic-pituitary-adrenal axis under a constant tonic inhibition. He has shown that, in the dog, the surgical elimination of the CC brings about a significant increase of the secretory activity of the adrenal gland; consequentely, he has postulated that an inhibitory hormone,opposite in effect to the hypothalamic Corticotropin Releasing Factor (CRF), is present in the CC.

An experimental validation of this hypothesis is provided by the studies here reported, which demonstrate that principle(s) exerting an inhibitory action on the hypothalamic-pituitary-adrenal axis may be extracted from cerebral cortex tissue. The attempt to purify from CC principle(s) inhibiting ACTH secretion was started after preliminary experiments had shown that the intravenous administration of crude extracts of rat CC reduces plasma corticosterone levels in normal female rats and enhances the inhibitory activity exerted on the pituitary-adrenal axis by dexamethasone. Due to the obvious difficulty in collecting sufficient quantities of rat brain, bovine CC were routinely used as the starting material in the experiments here to be described.

A summary of the procedures adopted in the early stages of purification is given in Table 1. An acetone powder of calf CC was extracted with 2N acetic acid and lyophilized; the lyophilized material was further extracted with glacial acetic acid and lyophilized again; the material obtained at this stage was dissolved in 0.1 M ammonium acetate buffer and gel-filtrated through columns (4x98 cm) of Sephadex G-25 Fine (New bead form) using 0.1 M ammonium acetate buffer (pH 5) as the eluting medium; each column was charged with 1.5 g of material dissolved in 18 ml of ammonium acetate; the effluent was separated into several fractions of 7.5 ml each; all the fractions obtained were lyophilized; materials eluted from the column were monitored by means of a continuous U.V. analyzer (Uvicord LKB); lyophilized fractions were redissolved in 0.1 M ammonium acetate buffer and neutralized immediately before being bioassayed. The inhibiting activity on the hypothalamic-pituitary-adrenal axis of crude materials, of acetic acid extracts, and of the fractions obtained from Sephadex columns were assayed by evaluating their ability to block the 4.00 p.m. rise of plasma corticosterone levels in female rats following intravenous injections performed at 12.00 noon.

TABLE 1. SUMMARY OF EARLY STEPS IN THE EXTRACTION AND PURIFICATION OF ACTH-INHIBITING PRINCIPLES FROM CALF CC

| Description of starting material | Weight of material obtained at each step |
|---|---|
| Calf cerebral cortex | 5 Kg |
| Acetone powder | 800 g |
| 2N Acetic acid extract | 150 g |
| Glacial acetic acid extract | 90 g |
| 0.1 M Ammonium acetate extract | 75 g |
| Gel-filtration on Sephadex G-25 in 0.1 M ammonium acetate buffer pH 5 | 1,3 g (inhibiting activity) |

In this bioassay advantage was taken of the fact that the secretory activity of the hypothalamic-pituitary-adrenal complex of the rat has a diurnal cyclicity with a peak in the middle of the afternoon (Guillemin, Dear and Liebelt, 1959); female rats were routinely used since they have a higher diurnal excursion of plasma corticosterone values (Critchlow, Liebelt, Bar-Sela, Mountcastle and Lipscomb, 1963); plasma corticosterone levels were evaluated according to the procedure of Guillemin, Clayton, Lipscomb and Smith (1959), as modified by Fraschini, Mangili, Motta and Martini (1964).

Table II shows the results obtained in a pilot experiment in which 4 mg/100 g of body weight of a lyophilized 2N acetic acid extract or 2 mg/100 of body weight of a lyophilized glacial acetic acid extract were injected intravenously at 12.00 noon into normal female rats. It will be noted that both extracts are able to reduce the 4.00 p.m. increase of plasma corticosterone levels, although they do not suppress it completely; in this experiment the glacial acetic acid extract was more effective than the other preparation even if given in smaller amounts.

TABLE II. EFFECT OF ACETIC ACID EXTRACTS OF CALF CC ON THE 4.00 p.m. RISE OF PLASMA CORTICOSTERONE LEVELS IN NORMAL FEMALE RATS

| Groups | | Plasma Corticosterone µg/100 ml |
|---|---|---|
| Controls | a.m. | 25.6 ± 2.7 (59)° |
| | p.m. | 61.7 ± 3.3 (50) |
| 2N Acetic acid extract (4mg/100g b.w.) | | 45.0 ± 7.2 (45) |
| Glacial acetic acid extract (2mg/100 g b.w.) | | 36.4 ± 6.4 (45) |

Values are means ± SE
(°) No. of rats in parentheses

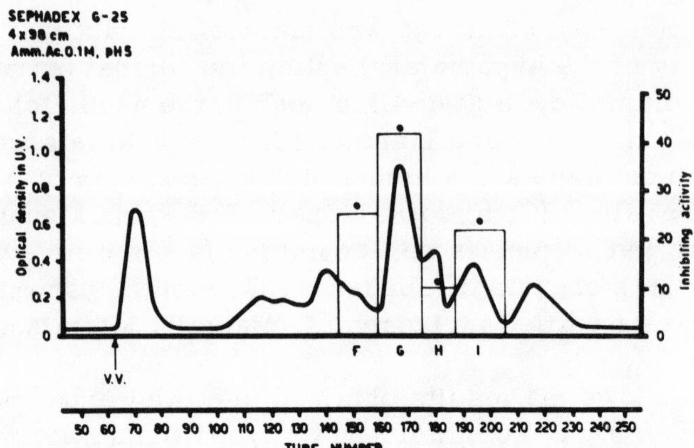


Fig. 1- Gel-filtration of 1.5 g of glacial acetic acid extract of calf cerebral cortex; Sephadex G-25; 0.1 M ammonium acetate buffer (pH5); flow rate 75 ml/hr; column size 4x98 cm; fraction size 7.5 ml; optical density measured with a continuous U.V. analyzer.

The inhibiting activity is expressed as the % decrease of plasma corticosterone levels (at 4.00 p.m.) in treated animals versus saline-injected controls.

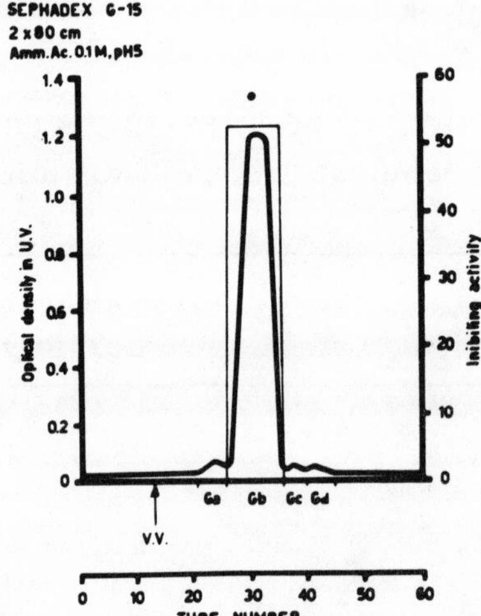


Fig.2- Gel-filtration of 15 mg of fraction G; Sephadex G-15; 0.1 M ammonium acetate buffer (pH5); flow rate 37.5 ml/hr; column size 2x80 cm; fraction size 7.5 ml; optical density measured with a continuous U.V. analyzer.

The inhibiting activity is expressed as the % decrease of plasma corticosterone levels (at 4.00 p.m.) in treated animals versus saline-injected controls.

After completion of gel-filtration seven major functions of U.V. absorbing materials were obtained (Fig. 1); principles reducing the 4.00 p.m. rise of plasma corticosterone levels have been found to be associated with three of these peaks; the corresponding fractions have been provisionally called F, G and I. When these fractions were tested (at a dose level of approximately 50 $\mu$g per 100 g body weight) it was found that the bulk of activity resided in fraction G. It is interesting to note that fraction H which is located between two areas containing inhibitory activity (fractions G and I) is practically devoid of such effect.

The next step in the purification of the active principle(s) has been the gel-filtration through columns of Sephadex G-15 (2x 80 cm); the active principle (15 mg of fraction G) was dissolved in 2 ml of 0.1 M ammonium acetate buffer (pH5); the same buffer was also used as the eluting material.

Fig. 2 shows the results of this further purification. Four fractions of U.V. absorbing materials were obtained; inhibitory activity on ACTH secretion was found associated only with the major fraction, which has been provisionally called fraction Gb.

The data here reported, although preliminary in nature, indicate that it is possible to extract from bovine CC principle(s) which are able to inhibit the hypothalamic-pituitary-adrenal axis of the rat. It is obviously too early to draw conclusions on the chemical nature of the compound(s) involved. Preliminary experiments performed using a crude preparation (acetone powder) seem to indicate that triptic and peptic digestion reduces the inhibitory activity of CC extracts. However, this is not felt to be a clear-cut demonstration that the activity is linked to material of pure peptidic nature; it has actually been reported from the laboratories concerned with the isolation of the hypothalamic Releasing Factors that proteolytic enzymes may destroy the activity of crude or impure preparations, while leaving unmodified that of more purified materials (Guillemin, Burgus, Sakiz and Ward, 1966; Burgus, Ward, Sakiz and Guillemin, 1966). It is also premature to state whether the principles having an inhibitory effect on the hypothalamic-pituitary-adrenal axis which were found in fractions G, F and I are similar in nature or whether different principles are involved. Apparently ACTH-inhibiting principles are present only in the CC, since similarly prepared extracts of other organs (liver, tongue, etc.) do not show this type of activity.

Much more work also is obviously needed in order to establish the exact physiological significance and the mechanism of action of the material(s) involved. It is possible that the ACTH-inhibiting principle(s) of cerebro-cortical origin play a role in the control of the circadian adrenal rhythm as well as in the inhibition of the anterior-pituitary induced by adrenocortical hormones. It has been shown that cortisol, corticosterone and other similar steroids are accumulated and metabolized in the brain (Grosser 1966, Grosser and Bliss, 1966; Peterson, Chaikoff and Jones, 1965; Touchstone, Kasparow, Hughes and Horwitz, 1966); it is tempting to suggest that they might alter the synthesis or the release of CC ACTH-inhibiting principle(s). These principle(s) might participate in the control of ACTH secretion,either entering the general circulation and acting as humoral mediators,or being released at nerve endings and acting as synaptic mediators. Neurons liberating materials of polypeptidic nature at their terminals have already been described in the central nervous system; one example is provided by the hypothalamic neurosecretory system (Bern and Knowles,1966; Krivoy and Kroeger, 1963) which liberates the peptides vasopressin and oxytocin; another one by a sensory pathway in the posterior columns of the spinal cord, in which the polypeptide Substance P acts as the synaptic mediator (Krivoy, Lane and Kroeger, 1963; Stern,1963).

It has not been established so far at which level of the hypothalamic-pituitary adrenal axis CC principle(s) exert their inhibitory effect; it is suspected that they might operate through the suppression of the secretion of the hypothalamic CRF, but no experimental data are available yet to support this assumption. The data showing the existence of ACTH-inhibiting principle(s) in the CC provide a satisfactory explanation for the results of the decortication experiments reported by Egdhal (1967). They also fit in very well with a few preliminary results which have shown that CC humoral factors may play some role in the control of gonadotropins and of Melanocyte Stimulating Hormone (MSH) secretion. Hopkins and Pincus (1965,1967) have found that extracts of rat cerebral cortex inhibit experimentally-induced ovulation in the immature female rat; Endröczi and Hilliard (1965, and personal communication) have shown that the intrapituitary infusion of extracts of rabbit and dog cerebral cortex reduces in the rabbit the secretion of gonadotropins (as measured by ovarian progestin output) below control values; Tixier-Vidal (personal communication) has observed that extracts of pigeon cerebral cortex block the release of prolactin from pigeon pituitary. Brinkley and Bercu (1965), Bercu and Brinkley (1967) and Ralph and Sampath (1966) have reported that cerebral cortex extracts prepared from frog brains inhibit the secretion of MSH from frog pituitaries.

## REFERENCES

Bercu, B.B., and Brinkley, H.J. (1967). Endocrinology, 80, 399.

Bern, H.A., and Knowles, G.W. (1966). In: "Neuroendocrinology" (L. Martini and W.F. Ganong, eds.), vol. 1, p. 139. Academic Press, New York.

Brinkley, H.J., and Bercu, B.B. (1965). Am. Zool., 5, Abstract N. 88.

Burgus, R., Ward, D.N., Sakiz, E., and Guillemin, R. (1966). C.R. Acad. Sci. (Paris), 262, 2643.

Critchlow, V., Liebelt, R.A., Bar-Sela, M., Mountcastle, W., and Lipscomb, H.S. (1963). Am. J. Physiol., 205, 807.

Egdhal, R.H. (1967). In: "Proceedings of the Second International Congress on Hormonal Steroids (L. Martini, F. Fraschini and M. Motta, eds.), p. 990. Excerpta Medica International Congress Series, Amsterdam.

Endröczi, E., and Hilliard, J. (1965). Endocrinology, 77, 667.

Fraschini, F., Mangili, G., Motta, M., and Martini, L. (1964). Endocrinology, 75, 765.

Grosser, B.I. (1966). J. Neurochem., 13, 475.

Grosser, B.I., and Bliss, E.L. (1966). Steroids, 8, 915.

Guillemin, R., Clayton, G.W., Lipscomb, H.S., and Smith, J.D. (1959). J. Lab. Clin. Med., 53, 830.

Guillemin, R., Dear, W.E., and Liebelt, R.A. (1959). Proc. Soc. Exptl. Biol. Med., 101, 394.

Guillemin, R., Burgus, R., Sakiz, E., and Ward, D.N. (1966). C.R. Acad. Sci. (Paris), 262, 2278.

Halasz, B., and Pupp, L. (1965). Endocrinology, 77, 553.

Halasz, B., Slusher, M.A., and Gorski, R.A. (1967). Neuroendocrinology, 2, 43.

Hopkins, T.F., and Pincus, G. (1965). Endocrinology, 76, 1177.

Hopkins, T.F., and Pincus, G. (1967). Fed. Proc., 26, 366.

Krivoy, W.A., and Kroeger, D.C. (1963). Experientia, 19, 366.

Krivoy, W.A., Lane, M., and Kroeger, D.C. (1963). Ann. N.Y. Acad.Sci., 104, 312.

Mangili, G., Motta, M., and Martini, L. (1966). In: "Neuroendocrinology" (L. Martini and W.F. Ganong, eds.), vol. 1, p.297. Academic Press, New York.

Peterson, N.A., Chaikoff, I.L., and Jones, C. (1965). J. Neurochem., 12, 273.

Ralph, C.L., and Sampath, S. (1966). Gen. Comp. Endocrin., 7, 370.

Stern, P. (1963). Ann. N.Y. Acad. Sci., 104, 403.

Touchstone, J.C., Kasparow, M., Hughes, P.A., and Horwitz, M. R. (1966). Steroids, 7, 205.

THE PHARMACOLOGICAL APPROACH TO THE STUDY OF THE MECHANISMS REGULATING ACTH SECRETION

Joan Vernikos-Danellis

Environmental Biology Division, Ames Research Center

NASA, Moffett Field, California 94035

The common association of sympathoadrenal and pituitary adrenocortical activity in stress has suggested that catecholamines play a special role in regulating ACTH secretion. Several theories have been proposed through the years to explain this relationship. These included (a) Long's concept (1952) that emotional stimuli activate the hypothalamus which in turn stimulates the adrenal medulla via the spinal cord and splanchnic nerves, and the subsequently released epinephrine stimulates the release of ACTH; (b) Fortier et al.'s (1957) classification of emotional (epinephrine mediated) versus systemic stresses; (c) Smelik's (1959) suggestion that "neurotropic" stresses result in the release of epinephrine from the adrenal medulla which would in turn activate "hypothalamic nervous pathways leading to the neurohypophysis; and (d) involvement of hypothalamic norepinephrine in the regulation of pituitary function (Vogt, 1954; Carlsson et al., 1962; Vernikos-Danellis, 1965). There is almost as much evidence for, as against any one of these theories and the vast number of reports in the literature using indiscriminately various psychodepressants and psychic energizers to "elucidate" the involvement of catecholamines in pituitary ACTH secretion have merely added to the confusion. This is brought out nicely in the review by De Wied (1967) on the effects of chlorpromazine on endocrine function.

In a recent series of experiments (Vernikos-Danellis, 1966) it was observed that pretreatment of rats with one of the methyl xanthines, caffeine or theophylline, enhanced the stress-induced secretion of ACTH and antagonized the ability of steroids to inhibit hypothalamic-pituitary ACTH secretion. Since these methyl xanthines have been shown in a variety of tissues to inhibit *in vitro* the 3'5' nucleotide phosphodiesterase that breaks down cyclic 3'5' adenosine monophosphate (AMP) and to potentiate the cyclic AMP-mediated effects of various hormones both *in vitro* and *in vivo* (Sutherland and Rall, 1958; Butcher and Sutherland, 1962; Hynie et al., 1966; Hess et al., 1963), the possibility existed that these drugs exerted their effects on the hypothalamic-pituitary unit by a similar mechanism. Studies on the distribution of the phosphodiesterase in various tissues of the rat (Vernikos-Danellis and Harris, 1966, unpublished) indicated high activity in both the median eminence and the anterior pituitary gland as compared to other organs. Furthermore, this enzyme was markedly depressed following incubation with one of the methyl xanthines in relatively large concentrations but also showed a 30% decrease in activity in the pituitary in *in vivo* experiments under the conditions that showed enhancement of the pituitary ACTH stress response. β-adrenergic blocking drugs are generally believed to exert their effects by inhibiting the activation of adenyl cyclase (Murad et al., 1962; Robison et al., 1967). Of these MJ-1999 [(2-Isopropylamino, 1-Hydroxyethyl) methanesulfonanilide)] offers advantages by being devoid of intrinsic adrenergic activity. Figure 1 shows the results of an experiment designed to determine whether MJ-1999 could prevent the increase in ACTH secretion following ether stress and the potentiation of this response by caffeine.

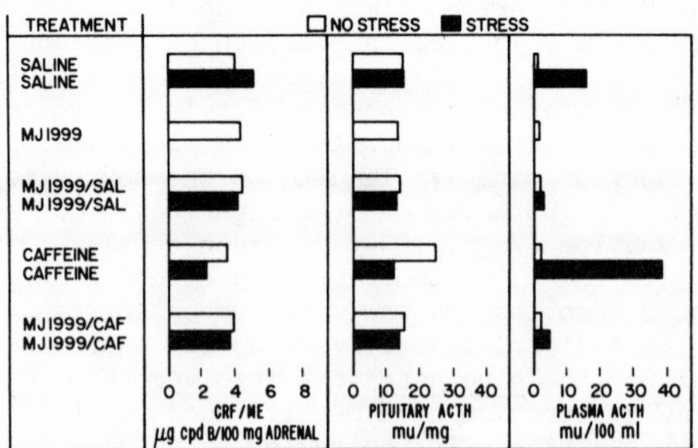


Figure 1. Changes in the concentration of corticotropin releasing activity, pituitary and plasma ACTH before and 2.5 minutes after stress (ether 1 min.) in rats pretreated with saline, caffeine, or MJ-1999 30 minutes previously. CRF content is expressed as μg corticosterone per 100 mg adrenal tissue in rats given a subcutaneous injection of 2.5 mg prednisolone 4 hours earlier, resulting from the intravenous injection of a crude acid extract of 1 rat median eminence. Pituitary and plasma ACTH was expressed as mU/mg tissue or mU/100 ml. plasma.

Throughout this work female Sprague-Dawley rats weighing 100 to 120 g. were used as the donors. Plasma and pituitary ACTH concentrations were determined in male rats of the same weight and strain 4 hours after hypophysectomy or 4 hours after a subcutaneous injection of 2.5 mg prednisolone per 100 g. body weight for CRF determinations. In these methods the increase in adrenal corticosterone concentration produced by one or more dilutions of the sample is compared to two doses of standard ACTH. (Vernikos-Danellis et al., 1966).

A dose of 4 mg/100 g. body weight of MJ-1999 was given subcutaneously 30 minutes prior to caffeine (2 mg/100 g. s.c.) or saline administration. Thirty minutes later half the animals in each group were stressed with ether (one minute) and decapitated 2.5 minutes after the beginning of the ether. The remaining animals served as unstressed controls. This dose of MJ-1999 is 4 times as great as that which is reported to prevent the rise in the isoproterenol-induced activation of myocardial phosphorylase in the rat (Kvam et al., 1965). The results show that it also prevented the stress-induced secretion of ACTH, and markedly depressed the increase in pituitary ACTH and the enhancement of the ether stress usually seen after caffeine.

Since catecholamines, vasopressin, and histamine share the property of activating adenyl cyclase in different tissues or exerting their effects through the mediation of cyclic AMP (Robison et al., 1967) and have also been implicated at one time or another in the mechanism regulating ACTH secretion, it became of interest to determine whether these substances could also potentiate the secretion of ACTH in response to a subsequent stress and whether a straightforward pharmacological study using α- and β-adrenergic blocking agents would yield any information about the mechanism of action of these substances, in the endocrine response to stress.

Female rats kept under controlled environmental conditions were given a single intraperitoneal injection of saline solutions of isoproterenol (80 μg), epinephrine (80 or 160 μg), norepinephrine (80 or 160 μg), dopamine (1 mg), vasopressin (500 mU), histamine (300 μg), or 0.9% normal saline (0.2 ml). Throughout this paper all doses are expressed as weight units of free base. Ten minutes after the injection they were decapitated and the plasma separated from the pooled heparinized blood, was frozen and stored at -12° C until assayed for its ACTH content. The results as shown in Table 1, indicate that norepinephrine is more potent than epinephrine which is in turn more potent than isoproterenol in stimulating ACTH secretion, and that dopamine, histamine, and vasopressin in larger amounts have a similar effect. Since the potency of the catecholamines in this respect appeared to be

Table 1. Changes in plasma ACTH concentrations 10 minutes after the intraperitoneal injection of various substances in saline-, phentolamine-, or MJ-1999-pretreated rats. Doses are expressed as μg free base per rat. B.P. = blood pressure. Pooled samples from 10 rats were assayed in 5 to 20 assay animals. (Fiducial limits at P = 0.95 are given in parentheses.)

| Treatment (i.p. in 0.2 ml) | Receptor type | Change in B.P. | Plasma ACTH mU/100 ml | | |
|---|---|---|---|---|---|
| | | | Saline (0.2 ml/rat) | Phentolamine (1 mg/rat) | MJ-1999 (4 mg/rat) |
| Saline | - | - | 1.7 (0.8-2.1) | 1.5 (1.0-1.9) | 1.8 (1.4-2.0) |
| Isoproterenol 80 μg | β | 0 - ↓ | 1.9 (1.6-2.4) | 2.0 (1.6-2.5) | 1.4 (1.0-1.7) |
| Epinephrine 160 μg | α + β | ↑ | 8.3 (6.5-9.8) | 2.5 (2.0-2.9) | 3.7 (3.1-4.4) |
| 80 μg | α + β | ↑ | 5.5 (4.9-6.4) | - | - |
| Norepinephrine 160 μg | α | ↑ | 12.3 (10.3-14.0) | 2.8 (2.0-3.4) | 14.0 ( - ) |
| 80 μg | α | ↑ | 3.4 (2.6-4.1) | - | - |
| Dopamine 1 mg | Dopaminergic ? | 0 - ↑ | 3.7 (3.1-4.1) | 1.9 (0.9-2.7) | 4.0 (3.2-4.8) |
| Vasopressin 500 mU | ? | ↑ | 5.6 (4.8-6.2) | 5.8 (5.0-6.3) | 3.9 (3.2-4.6) |
| Histamine 300 μg | ? | ↓ | 4.8 (4.1-5.6) | 4.2 (3.8-4.8) | 4.5 (4.0-5.1) |
| Ether 1 min. | | | 4.2 (3.8-5.0) | 5.0 (4.2-5.8) | 1.2 (0.7-1.6) |

related to their pressor activity, a similar series of experiments was performed in rats given i.p. 30 minutes earlier 1 mg of the α-adrenergic blocking agent phentolamine. The results are in agreement with previous observations (Tepperman and Bogardus, 1948; Guillemin, 1955; George and Way, 1957; Van Peenen and Way, 1957) that α-adrenergic blocking agents do not inhibit the response to stresses other than epinephrine and extend this observation to include norepinephrine and dopamine. The increase in circulating ACTH concentration caused by histamine or vasopressin administration or exposure to ether was not affected by pretreatment with phentolamine. In contrast, previous experiments (Figure 1) had shown that pretreatment of animals with the β-adrenergic blocking agent MJ-1999 effectively inhibited the stress-induced ACTH secretion following ether. It therefore appeared possible that this drug might inhibit other types of stresses and thus suggest a role for catecholamines common to all types of stress in the mechanism regulating ACTH secretion. The results in Table 1 show that β-adrenergic blocker injected subcutaneously 30 minutes earlier in a dose of 4 mg/100 g·body weight, markedly inhibits the acute pituitary response to epinephrine and ether and surprisingly reduces that to vasopressin.

Since those substances with greatest β-adrenergic activity were least effective in stimulating ACTH secretion and since the β-adrenergic blocker was previously shown to inhibit the ability of caffeine to enhance ether stress, experiments were designed to compare the ability of the substances studied to enhance a second stress and to determine whether α- or β-adrenergic blockade affected this property. Thirty minutes before exposure to one minute of ether stress rats were given the following substances: saline (0.2 ml), isoproterenol (80 μg), epinephrine (80 μg), norepinephrine (80 μg), dopamine (1 mg), vasopressin (250 or 500 mU), histamine (300 μg). All animals were decapitated 2.5 minutes after the beginning of the ether and their plasma ACTH content determined. Where α-blockade was required phentolamine was administered 30 minutes before the injection of the neurohumor or one hour before the ether stress. The time sequence was therefore: phentolamine followed 30 minutes later by the neurohumor, followed 30 minutes later by one minute ether and decapitation at 2.5 minutes after the beginning of the ether stress. Since MJ-1999 inhibited the ether stress _per se_, the timing was selected in such a way as to allow the response to ether to return yet maintain adequate β-adrenergic blockade at the time of the injection. It is of interest here to note that the inhibition of the ether stress was very transient as compared to the effect of the blocker on the enhancement of a second stress by epinephrine which lasted at least 4 hours, (unpublished observations). The time sequence in this series of experiments was therefore; MJ-1999 followed 2.5 hours later by the injection of the neurohumor, followed 30 minutes later by one

Table 2. ACTH concentration in the plasma of saline-, phentolamine-, or MJ-1999-pretreated rats, 2.5 minutes after ether (1 minute), ether/sham ULA (1 minute ether and sham unilateral adrenalectomy) or 10 minutes after an intraperitoneal injection of histamine. Doses are expressed as μg free base per rat. Each value represents pooled plasma samples from at least ten rats. (Fiducial limits at P = 0.95 are given in parentheses.)

| Treatment | Stress | Plasma ACTH mU/100 ml | | |
|---|---|---|---|---|
| | | Saline (0.2 ml/rat) | Phentolamine (1 mg/rat) | MJ-1999 (4 mg/rat) |
| A. Saline | Ether (1 min) | 4.2 (3.7-5.1) | 5.0 (4.1-5.8) | 4.3 (3.8-4.8) |
| Isoproterenol 80 μg | Ether (1 min) | 9.1 (8.0-10.2) | 7.6 (6.9-8.1) | 3.8 (3.0-4.3) |
| Epinephrine 80 μg | Ether (1 min) | 7.0 (6.2-8.1) | 10.0 (9.6-10.9) | 5.3 (4.6-6.0) |
| Norepinephrine 80 μg | Ether (1 min) | 3.7 (3.1-4.3) | 3.4 (2.8-4.1) | 4.0 (3.1-4.9) |
| Dopamine 1 mg | Ether (1 min) | 4.8 (4.0-5.4) | 4.8 (4.1-5.6) | 4.4 (3.8-5.0) |
| Vasopressin 500 mU | Ether (1 min) | 7.1 (6.9-7.8) | 7.5 (6.5-8.3) | 5.8 (4.9-6.3) |
| 250 mU | Ether (1 min) | 5.8 (5.3-6.6) | 6.0 (5.0-7.0) | 4.5 (3.9-5.0) |
| Histamine 300 μg | Ether (1 min) | 4.0 (3.4-4.9) | 4.7 (4.0-5.4) | 4.3 (3.7-5.0) |
| B. Saline | Histamine (300 μg) | 3.9 (3.1-4.6) | - | - |
| Epinephrine 80 μg | Histamine (300 μg) | 5.7 (5.0-6.5) | - | - |
| Vasopressin 500 mU | Histamine (300 μg) | 3.1 (2.6-3.9) | - | - |
| C. Saline | Ether/Sham ULA | 10.1 (9.2-12.0) | - | - |
| Norepinephrine 160 μg | Ether/Sham ULA | 14.2 (12.9-15.9) | - | 11.1 (9.9-12.0) |

minute ether and decapitation at 2.5 minutes after the beginning of the ether stress. Table 2 shows that epinephrine, isoproterenol, and vasopressin were all effective in enhancing the ACTH secretion in response to ether stress. Histamine, dopamine, and 80 µg of norepinephrine were without effect; section C of Table 2 shows however, that increasing the dose of norepinephrine to 160 µg and the intensity of the stress to ether followed by laparotomy resulted in a significant enhancement of ACTH secretion. Section B of Table 2 shows that if histamine stress is substituted for the ether, epinephrine pretreatment still enhanced this stress response while vasopressin did not. In contrast to the ACTH stimulating activity of the catecholamines the order to potency of these substances in enhancing a second stress was: isoproterenol > epinephrine > norepinephrine. Pretreatment with phentolamine did not alter the potentiating ability of these substances and in fact increased that of epinephrine. Pretreatment with the β-blocker, MJ-1999, abolished the enhancing properties of all the substances tested including vasopressin.

In order to determine the site of action of these two differing effects of peripherally administered catecholamines, the corticotropin releasing activity (as measured by the increase in adrenal corticosterone) of a crude acid extract of rat median eminence (MEE) was compared in rats in which the endogenous secretion of CRF was effectively inhibited by a single subcutaneous injection of prednisolone (2.5 mg/100 g. 4 hours earlier) with animals given phentolamine only, steroid plus phentolamine, or normal saline only (see Table 3). Phentolamine injected 3-1/2 hours after the steroid and 30 minutes before use of the animals was without effect in preventing the stress of ether and intravenous saline and the response of the pituitary to MEE, nor did it affect the sensitivity of the adrenal cortex to injected ACTH.

Table 4 shows a similar series of experiments using MJ-1999 instead of phentolamine. Once more the blocker was administered 30 minutes before use of the animals. MJ-1999 markedly depressed not only the increase in adrenal corticosterone resulting from the stress of ether and intravenous saline but also that reflecting the corticotropin releasing activity of the median eminence extract both in the presence and in the absence of the steroid. In order to eliminate the possibility that this inhibition was exerted at the adrenal level, the increase in adrenal corticosterone in response to two doses of standard ACTH was compared in steroid blocked rats with and without MJ-1999. The adrenergic blocker had no direct effect on the responsiveness of the adrenal cortex to ACTH indicating that its site of action in depressing the corticotropin releasing activity of MEE was exerted primarily at the pituitary.

Table 3. Effect of ACTH, MEE or the stress of intravenous saline under ether anesthesia on the adrenal corticosterone concentration of rats pretreated with saline only, saline plus phentolamine (1 mg/rat), prednisolone (2.5 mg/100 g. body weight) plus saline or prednisolone plus phentolamine. Number of animals given in parentheses.

| | Adrenal Corticosterone Concentration (μg/100 mg tissue ±S.E.) | | | |
|---|---|---|---|---|
| I.V. Injection | Saline 4 hr + saline 30 min | Saline 4 hr + phentolamine 30 min | Prednisolone 4 hr + saline 30 min | Prednisolone 4 hr + phentolamine 30 min |
| None | 3.02±0.10(5) | 4.88±0.22(5) | 1.09±0.07(5) | 1.21±0.08(5) |
| Saline | 6.65±0.32(5) | 5.80±0.31(5) | 1.24±0.15(5) | 1.14±0.11(5) |
| 1 ME | - | 6.72±0.48(5) | 3.01±0.20(5) | 3.51±0.28(5) |
| ACTH 33.3 μU | - | - | 1.87±0.19(5) | 1.99±0.09(5) |
| ACTH 100 μU | - | - | 6.39±0.41(5) | 6.12±9.38(5) |

Table 4. Effect of ACTH, MEE, or the stress of intravenous saline under ether anesthesia on the adrenal corticosterone concentration of rats pretreated with saline only, saline plus MJ-1999 (4 mg/rat), prednisolone (2.5 mg/100 g. body weight) plus saline, or prednisolone plus MJ-1999. Number of animals given in parentheses.

| | Adrenal Corticosterone Concentration (μg/100 mg tissue ±S.E.) | | | |
|---|---|---|---|---|
| I.V. Injection | Saline 4 hr + Saline 30 min | Saline 4 hr + MJ-1999 30 min | Prednisolone 4 hr + saline 30 min | Prednisolone 4 hr + MJ-1999 30 min |
| None | 3.09±0.22(5) | 3.16±0.24(5) | 1.78±0.09(5) | 1.86±0.13(5) |
| Saline | 7.57±0.49(5) | 4.38±0.38(5) | 1.47±0.10(5) | 1.57±0.08(5) |
| 1 ME | - | 4.83±0.42(5) | 3.15±0.19(5) | 2.11±0.09(5) |
| ACTH 33.3 μU | - | - | 2.32±0.17(5) | 2.10±0.09(5) |
| ACTH 100 μU | - | - | 6.24±0.51(5) | 6.25±0.45(5) |

These findings suggested the hypothesis that catecholamines secreted peripherally during exposure to various stress situations affect the hypothalamic pituitary unit by at least two distinct mechanisms: direct stimulation of ACTH secretion related to their α-adrenergic hypertensive properties, mediated by the hypothalamus and possibly reflexly by higher centers in the central nervous system, and a β-adrenergic-receptor-mediated property of enhancing the ACTH secretion to a subsequent stress by an action primarily on the adenohypophysis. This latter property of the catecholamines appears to be shared by vasopressin; whether this is an intrinsic effect of this peptide or one mediated in some way by catecholamines remains to be determined. On the other hand the mechanism by which histamine stimulates pituitary ACTH secretion does not fall within either of these categories. It is of interest to point out in this respect the recent observation of Dallman (1967) that prior exposure to the stress of scalding enhances the increase in circulating corticosterone in response to histamine whereas prior administration of histamine has no effect on the response to scalding.

The presence and distribution of catecholamines in the central nervous system has been well documented (Vogt, 1954; Fuxe, 1963; Carlsson et al., 1962). However, studies with peripherally administered tritiated epinephrine and norepinephrine have shown that these substances do not cross the blood brain barrier although the anterior pituitary takes up large amounts of the labelled amines; to a lesser extent so does the median eminence and radioautographic studies demonstrate that they penetrate only a short distance into the hypothalamus (Axelrod et al., 1959; Weil-Malherbe et al., 1961; Samorajski and Marks, 1962). Attempts to study the role of brain amines on pituitary ACTH secretion have proved confusing largely because of the lack of specificity and knowledge of the site and mechanism of action of the drugs used to deplete or inhibit synthesis of these amines. Interest in α-methyl p-tyrosine (α-MT) stems from its ability to inhibit tyrosine hydroxylase and thereby to interfere with the synthesis of catecholamines (Spector et al., 1965). This agent depletes norepinephrine stores in peripheral sympathetic nerve endings and reduces the concentrations of norepinephrine and dopamine in the brain. It is especially useful for studying the role of brain catecholamines since (1) it is more effective in lowering the concentrations of these amines in the central nervous system than in the periphery (Torchiana et al., 1965), (2) it does not influence 5-hydroxytryptamine levels, and (3) unlike α-methyl-m-tyrosine and α-methyldihydroxyphenyl alanine, it is not converted into "false transmitters" (Spector et al., 1965). It therefore became of interest to use this drug as a tool in the study of the role of brain norepinephrine on pituitary ACTH secretion particularly since the long lag between maximal inhibition of brain and adrenal medullary amines because of the slower catecholamine turnover in the adrenal lent itself to the investigation of the relative contribution of these two components to the functional integrity of the hypothalamic-pituitary unit.

In this series of experiments the animals were used six hours after the intraperitoneal injection of a saline suspension of α-MT in a dose of 200 mg/Kg. body weight, unless otherwise specified. A similar suspension of l-tyrosine was injected into those animals that served as controls. At this time brain amines are markedly depressed (Rech et al., 1966).

Figure 2 shows the plasma corticosterone concentrations before and 15 minutes following the stress of ether (one minute) in rats that were either uninjected or had received an intraperitoneal injection of l-tyrosine or α-MT. The results show that the stress-induced increase in plasma steroids was reduced in the α-MT treated rats. Figure 3 shows that this reduced steroid response reflects reduced ACTH secretion in response to stress. Approximately 50% inhibition of the ACTH secretion in response to the stress of ether and sham adrenalectomy was found and increasing the dose of α-MT did not inhibit this response further. Since this phenomenon could have been due to the availability to the median eminence and pituitary of catecholamines originating from the adrenal, a similar experiment was performed in rats 24 hours after adrenalectomy. At this time period after removal of the adrenal glands female rats show a marked increase in the sensitivity to stress, secreting greater amounts of ACTH in response to the relatively mild stress of one minute ether than do intact animals (Hodges and Vernikos, 1959). Figure 4 shows that under these conditions one-tenth of the dose of α-MT sufficed to cause a 50% inhibition of the stress response and the usual dose of

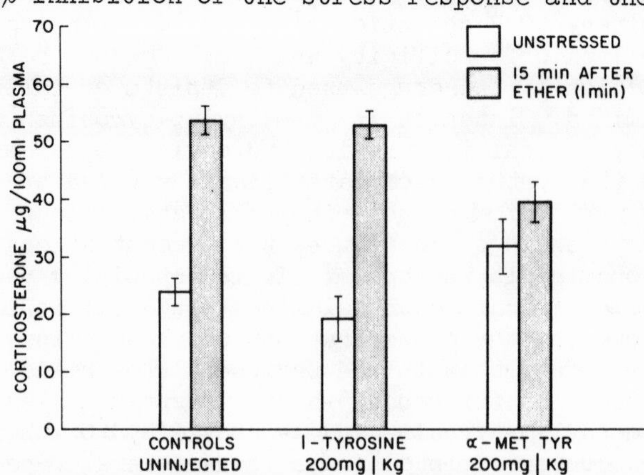


Figure 2. Plasma corticosterone concentrations before and 15 minutes following stress (ether 1 min) in uninjected rats or animals that had received six hours earlier an intraperitoneal injection of a saline suspension of l-tyrosine or α-methyl tyrosine in a dose of 200 mg/Kg. Number of animals given in parentheses.

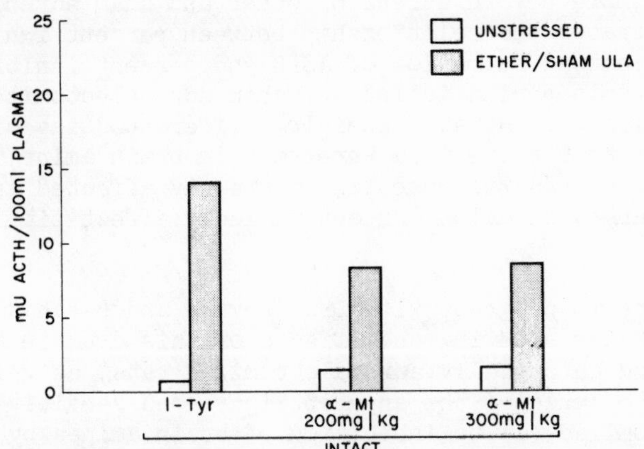


Figure 3. Plasma ACTH concentrations before and 2.5 minutes after stress (ether 1 min. and sham unilateral adrenalectomy) in rats given six hours earlier an intraperitoneal injection of a saline suspension of 200 mg/Kg. l-tyrosine or α-methyl tyrosine in dose of 200 mg or 300 mg/Kg.

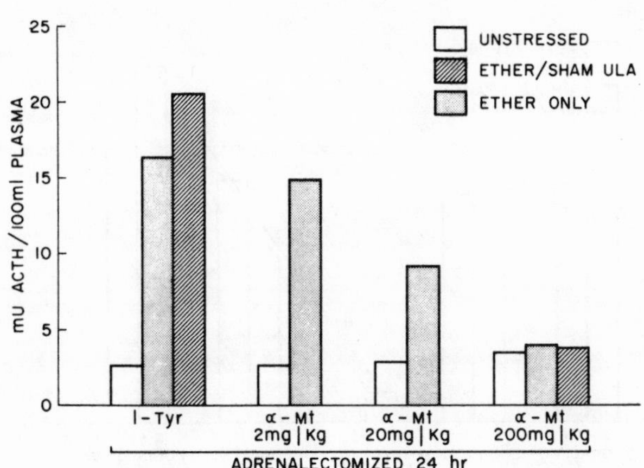


Figure 4. Plasma ACTH concentrations before and 2.5 minutes after stress (ether 1 min. or ether 1 min. and sham unilateral adrenalectomy) in rats 24 hours after the removal of their adrenal glands and six hours after an intraperitoneal injection of 200 mg/Kg. body weight l-tyrosine or 2, 20, or 200 mg/Kg. of α-methyl tyrosine.

200 mg/Kg. body weight now produced 96% inhibition of both ether stress and the more severe stress of ether and sham adrenalectomy. Figure 5 illustrates the relationship between percent inhibition of the stress-induced secretion of ACTH and percent inhibition of whole brain norepinephrine in intact, sham adrenalectomized and 24 hour adrenalectomized rats receiving different doses of α-MT. It would appear that there is a threshold in brain amine levels above which the stress response is not greatly affected and below which small changes in amine content markedly affect pituitary ACTH secretion.

Experiments in progress with Drs. Levine and Barchas of Stanford University studying the effects of this drug in adrenal demedullated and chronically adrenalectomized rats, as well as looking into the cause of the apparent increased sensitivity of the adrenalectomized rat to inhibition of brain amines by α-MT are too preliminary to discuss at the present time. Nevertheless, together with this rather elementary exercise in experimental pharmacological design presented here they help to point out certain well known but neglected facts, and suggest new cautions in the interpretation of results, the building of models and the art of hypothesizing, with regard to the role of catecholamines in regulating pituitary ACTH secretion.

The word catecholamines merely denotes structural relationship of those biogenic amines, possessing an o-dihydroxybenzene ring.

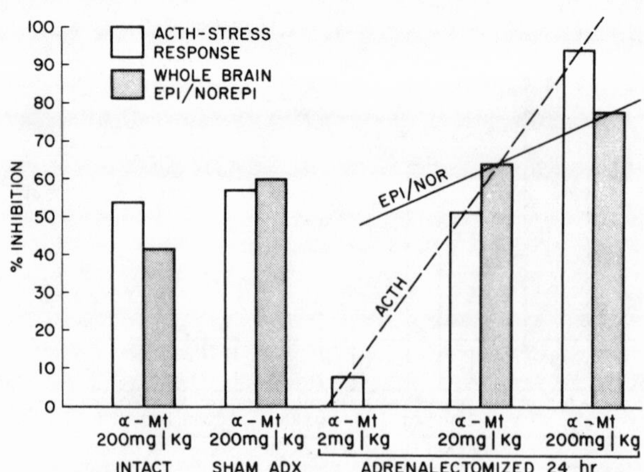


Figure 5. Comparison of the percent inhibition of the stress-induced secretion of ACTH and whole brain epinephrine/ norepinephrine content in intact, sham adrenalectomized, and 24 hour adrenalectomized rats given six hours earlier an intraperitoneal injection of different doses of α-methyl tyrosine.

It does not denote any kind of uniformity of physiological or pharmacological mechanism or site of action. In fact the members of this group have little in common other than coming through the same biosynthetic pathway and are characterized by an extraordinary ability to exert similar or differing effects at the same site, similar effects at different sites, or differing effects at different sites. Yet in the study of their role on pituitary corticotropic secretion they have been used almost interchangeably and often their role as a group emphasized or dismissed on the results of experiments obtained with a single member. The experiments described in this paper perhaps help to emphasize that (a) catecholamines do play a role in regulating ACTH secretion, (b) that each member of the catecholamine group exerts distinct, varied and often overlapping effects, (c) that there is an intricate relationship and balance between the actions of circulating catecholamines exerted at pituitary and hypothalamic level and those exerted by centrally located amines on the various structures of the central nervous system, both stimulant and inhibitory that appear to be involved in maintaining the functional state of the hypothalamic-pituitary ACTH secreting system.

Finally the physiological significance of the close association between the glandular tissues of the adrenal cortex and medulla has been recently emphasized by the work of Wurtman (1966) and of Jost and his associates (Margolies et al., 1966) by the demonstration that the synthesis of the N-methyl-transferase enzyme, which forms epinephrine, is under the control of adrenal corticosteroid hormones. Together with the present findings, it is tempting to suggest in addition to the well known negative feedback loop between adrenal cortex and central CRF controlling centers, the existence of a positive feedback loop between adrenal cortex and anterior pituitary mediated by epinephrine from the adrenal medulla.

Acknowledgements

My thanks are due to Dr. J. D. Barchas of the Department of Psychiatry, Stanford University, for the brain norepinephrine determinations, and to Dr. J. D. Fisher of Armour Laboratories, Kankakee, Illinois, for generous gifts of ACTH.

## REFERENCES

Axelrod, J., Weil-Malherbe, H. and Tomchik, R., (1959), J. Pharmacol. Exp. Therap. 127:251.

Butcher, R. W. and Sutherland, E. W. (1962), J. Biol. Chem. 237:1244.

Carlsson, A., Falck, B., and Hillarp, N. (1962) Acte Physiol. Scand. Suppl. 196,28.

Dallman, M. F. (1967) Ph.D. Thesis, Stanford University.

De Wied, D. (1967) Pharmacol. Rev. 19:251.

Fortier, C., Harris, G. W., and McDonald, I. R. (1957) J. Physiol. (London) 136:344.

Fuxe, K. (1963) Acte Physiol. Scand. 58:383.

George, R. and Way, E. L. (1957) J. Pharmacol. Exp. Therap. 119:310.

Guillemin, R. (1955) Endocrinology 56:248.

Hess, M. C., Hottenstein, D., Shanfeld, J., and Haugaard, N. (1963) J. Pharmacol. Exp. Therap. 141:274.

Hodges, J. R. and Vernikos, J. (1959) Acte Endocrinologica 30:188.

Hynie, S., Krishna, G., and Brodie, B. B. (1966) J. Pharmacol. Exp. Therap. 153:90.

Kvam, D. C., Riggilo, D. A., and Lish, P. M. (1965) J. Pharmacol. Exp. Therap. 149:183.

Long, C. N. H. (1952) Ciba Foundation Collogine in Endocrinology 4:139.

Margolies, F. L., Roffi, J., and Jost, A. (1966) Science 154:275.

Murad, F., Chi, Y. M., Rall, T. W., and Sutherland, E. W. (1962) J. Biol. Chem. 237:1233.

Rech, R. H., Borys, H. D., and Moore, K. E. (1966) J. Pharmacol. Exp. Therap. 153:412.

Robison, G. A., Butcher, R. W., and Sutherland, E. W. (1967) Annals N.Y. Acad. Sci. 139:703.

Samorajski, T. and Marks, B. H. (1962) J. Histochem. Cytochem. 10:392.

Smelik, P. G. (1959) Ph.D. Thesis, University of Groningen.

Spector, S., Sjoerdsma, A., and Udenfriend, S. (1965) J. Pharmacol. Exp. Therap. 147:86.

Sutherland, E. W. and Rall, T. W., (1958) J. Biol. Chem. 232:1077.

Tepperman, J. and Bogardus, J. S. (1948) Endocrinology 43:448.

Torchiana, M. L., Stone, C. A., Porter, C. C., and Halpern, L. M. (1965) Fed. Proc. 24:265.

Van Peenen, P. F. and Way, E. L. (1957) J. Pharmacol. Exp. Therap. 120:261.

Vernikos-Danellis, J. (1965) Vitamins and Hormones 23:97.

Vernikos-Danellis, J. (1966) Abstr. 48th Ann. Endocrine Soc. Meeting, Chicago, Ill., page 24.

Vernikos-Danellis, J., Anderson, E., and Trigg, L. (1966) Endocrinology 79:624.

Vogt, M. (1954) J. Physiol. (London) 123:451.

Weil-Malherbe, H., Whitby, L. G., and Axelrod, J. (1961) J. Neurochem. 8:55.

Wurtman, R. (1966) Endocrinology 79:608.

# EFFECTS OF ACTH- AND MSH-PEPTIDES ON CENTRAL NERVOUS SYSTEM

G.L. Gessa and W. Ferrari

Dept. of Pharmacology - University of Modena (Italy)

The unusual symptomatology produced by ACTH- and MSH-peptides injected into the cerebro-spinal fluid (CSF) of mammals: stretching and/or yawning movements, frequently repeated, characterize a syndrome, we have named stretching yawning-syndrome (SYS).

The most important features of the SYS so far recognised are the following:

1) only peptides possessing adrenocorticotrophic and/or melanocyte-stimulating activity are able to induce it;

2) SYS is shown only when the ACTH- and MSH-peptides are introduced by the cerebro-spinal route or are injected into selected brain areas;

3) the effective doses of the most active peptides are in the order of microgram fractions;

4) natural hormones are among the most active agents;

5) SYS does not appear immediately after the injection of ACTH- or MSH-peptides: onset is delayed, according to the site of the injection;

6) SYS lasts for a long period, as much as 48 to 72 hours - at least in the dog;

7) animals showing SYS remain in good health; they do not show overt signs of bodily or mental impairment;

8) SYS can be induced repeatedly without the appearance of any resistance or sensitization toward the inducer;

9) SYS appears also in anaesthetized animals (dog, rabbit and

cat);
10) SYS induced by intracisternal injection of ACTH- and MSH-peptides in anaesthetized dog causes breathing changes;
11) chlorpromazine, atropine, morphine, diparcol act antagonistically to ACTH- and MSH-peptides. That is, their injection into an animal exhibiting SYS causes the disappearance of SYS for some time. However, this antagonistic effect vanishes sooner or later and SYS reappears;
12) ACTH- and MSH peptides also act in dogs after adrenalectomy; moreover, ACTH also retains its full activity when boiled in NaOH N/10. On the other hand, there is no clear correlation between the SYS and the other known extra-adrenal effects of ACTH.
13) different animal species show different behavioural patterns: stretching predominates in dogs and cats, yawning in monkeys.

ACTH- and MSH-peptides clearly produce SYS by acting on brain structures. In order to locate the central nervous sites affected by MSH- and ACTH- we observe how the latency of the effects varies when the peptides are injected at different levels of the CSF system, that is cisterna magna, lateral and

TABLE 1 - $\beta^{(1-24)}$ACTH activity in producing SYS after intracisternal injection

| μg/cat | symptomatology (*) | latency min (mean and range) | % of animals with SYS (**) |
|---|---|---|---|
| 250 | +++ | 15 (13-37) | 100 (10) |
| 20 | +++ | 27 (21-70) | 100 (30) |
| 5 | ++ | 40 (30-104) | 60 (30) |
| 2.5 | ++ | 55 (43-101) | 18,7 (16) |
| 1,25 | - | $\infty$ | 0 (10) |

(*) +++ = at least one stretching act every 2 min, for more than one hour.
++ = at least one stretching act every 5 min, for more than one hour.

(**) in parenthesis the number of animals used.

third ventricle. These experiments were performed in adult conscious cats prepared with stereotaxically implanted cannulae. We have used synthetic 1-24 ACTH (Ciba 30-920 Ba) dissolved in saline.

The minimal dose of 1-24 ACTH capable of inducing SYS after intracisternal injection in the cat is 5 $\gamma$ per animal (table 1).

Using 20 $\gamma$ of 1-24 ACTH per animal the shortest delay in the effect was obtained after injection into the third ventricle . Inasmuch as the symptomatology evoked using this route was more pronounced than after injection into the lateral ventricle or into the cisterna magna (table 2), these results seem to suggest that the site(s) of action of ACTH are closer to the third ventricle than to other CSF sites.

However, the studies of the effect of intracerebral injection of 1-24 ACTH indicate that the delay in the onset of SYS is not attributable solely to the time the peptide requires to reach the sensitive areas. For these experiments we used adult cats of both sexes . Usually four cannulae (needles with 0.5 mm outer diameter and 0.3 mm tamping needles) were chronically implanted by mean of a stereotaxic apparatus.

The experiments were begun 5-7 days after surgery. 5 µl of saline containing different amounts of 1-24 ACTH were injected

TABLE 2 - Latency and intensity of SYS in cats after injection of $\beta^{(1-24)}$ACTH into different areas of cerebro-spinal fluid.

| Site of injection | symptomatology (*) | latency min (mean and range) |
|---|---|---|
| Cisterna magna | +++ | 27 (21-70) |
| Lateral ventricle | +++ | 20 (15-32) |
| Third ventricle | ++++ | 13 (11-20) |

(*) +++ = at least one stretching act every 2 min, for more than one hour.

++++ = at least one stretching act every minute for more than one hour.

into the cannulae, using only one cannula per cat in each session. A 5-day interval was kept among sessions and each brain area was tested in at least 3 cats.

Preliminary experiments were made employing a dose of 20 γ of 1-24 ACTH per cannula.

These experiments showed that the hypothalamic structures lining the third ventricle are the most sensitive area. Among the other brain areas tested the SYS could be elicited from the caudate nucleus and the red nucleus (table 3).

TABLE 3 - SYS induced by $\beta^{(1-24)}$ACTH (20 μg) injected into different areas of the cat brain.

| Site of injection | Animals with SYS/ treated animals | Latency min (mean and range) | Intensity of SYS (*) |
|---|---|---|---|
| Cerebellum (anterior lobe) | 0/5 | - | - |
| Nucleus centralis thalami | 0/3 | - | - |
| Nucleus lateralis thalami | 0/3 | - | - |
| Nucleus medialis thalami | 0/3 | - | - |
| Formatio reticularis (mesencephalica) | 0/5 | - | - |
| Amigdala | 1/5 | 35 | ++ |
| Putamen | 1/5 | 25 | ++ |
| Globus pallidus | 1/6 | 22 | ++ |
| Substantia nigra | 1/9 | 35 | ++ |
| N. ruber | 3/8 | 30 (25-45) | ++ |
| N. caudatus | 2/5 | 24 - 30 | ++ |
| Hypothalamus lateralis | 3/5 | 30 (22-40) | ++ |
| Hypothalamus posterior | 4/5 | 20 (12-27) | ++ |
| Hypothalamus ventromedialis | 5/5 | 14 (10-21) | ++++ |
| Hypothalamus anterior | 5/5 | 12 (9-25) | ++++ |
| Corpus mammillare | 5/5 | 12 (7-23) | ++++ |

(*) ++ = at least one stretching act every 5 min, for more than one hour.
+++ = at least one stretching act every 2 min, for more than one hour.

Only a single positive response was obtained out of 9 cats after injection into putamen, globus pallidus and substantia nigra.

There is the possibility of a peptide diffusion from the site of injection. Accordingly we evaluated the minimal dose effective when injected into the brain areas seen to be responsive. Table 4 confirms that hypothalamic areas closer to the hypophysis and the nucleus caudatus are the most sensitive.

We do not know if SYS has a physiological significance. Relevant to this problem are the following facts: a) SYS might be considered as an exaggeration of physiological acts, namely stretching and yawning; b) it can be specifically induced by low quantities of two physiological hormonal peptides; c) the animals showing SYS remain in good mental and bodily health; d) the hormones act only when they are directly applied to the brain, where only selected areas appear to be sensitive; e) the most sensitive area is very near to hypophysis.

Moreover, ACTH and MSH have been shown to be present in the hypothalamus. Finally, MSH modifies the electrical activity of certain spinal neurones.

The significance of SYS is not clear.

Owing to the antagonistic effect of chlorpromazine, atropine, morphine and other observations we are inclined to think that SYS might play a part in arousal mechanisms. Previously we said that spontaneous stretching movements appear in anae-

TABLE 4 - Minimal active dose of $\beta^{(1-24)}$ACTH injected into different areas of the cat brain.

| $\beta^{(1-24)}$ACTH μg/animal | No. of animals with SYS / No. of animals treated | | | | | | | | | |
|---|---|---|---|---|---|---|---|---|---|---|
| | Hypothalamus a. | Corpus mammillare | Hypothalamus v.-m. | Hypothalamus p. | N. caudatus | N. ruber | Globus pallidus | Putamen | Subst. nigra | F.Retic. mesenceph. |
| 5.0 | 5/5 | 5/5 | 5/5 | 5/5 | 6/10 | 0/5 | 0/5 | 0/5 | 0/5 | 0/5 |
| 2.5 | 5/5 | 5/5 | 5/5 | 2/5 | 2/5 | 0/2 | - | - | - | - |
| 0.5 | 2/2 | 2/3 | 0/2 | 0/3 | - | - | | | | |
| 0.25 | 2/5 | 1/3 | - | - | - | - | | | | |

sthetized dogs after i.c. ACTH.

However, a sensorial stimulus applied during the interval between two spontaneous stretching acts precipitates a stretching response. Moreover, head shakes precede the beginning of each act of stretching. They subside when the stretching movements end.

Some cats, carrying permanently implanted scalp electrodes, were placed in a large sound-proofed screened chamber with a one-way glass window and electromyogram from the neck muscles and EEG were recorded: 20 $\gamma$ of 1-24 ACTH were injected into the third ventricle. These cats showed about the same tendency as the controls (injected with saline) to light sleep with EEG synchronization. However, EEG was seen to be periodically interrupted by short periods of EEG arousals in coincidence with each stretching, indicated by a discharge of the EMG.

Finally, as pharmacologists, we emphasize the CNS effects of ACTH- and MSH peptides as a rare example of drugs possessing a highly selective action on the CNS.

This selectivity concerns the chemical structure of the agents, the CNS sites affected, and the symptomatology induced.

# PHARMACOLOGY OF SYNTHETIC ACTH PEPTIDES OF DIFFERENT CHAIN LENGTHS.

L. Szporny, Gy. T. Hajos, S. Szeberenyi and G. Fekete

Chemical Works of Gedeon Richter Ltd., Budapest

Highly purified polypeptides with corticotropic activity have been isolated from the pituitaries of different species including man. All these corticotropines contain a straight chain of 39 amino acides and species differences in chemical structure occur between positions 25 to 33 of the polypeptide chain. The exact structures of corticotropines from different species have been described by different authors (1, 2, 3, 4).

In previous reports the synthesis and potencies of ACTH peptides of chain lengths up to 39 amino acides of the porcine sequence have been described (5). Detailed pharmacology and clinical data are available of the 1-24 peptide (6, 7, 8).

It seemed of interest to investigate whether peptides of different chain lengths between 1-24 and 1-39 as well as between porcine and human sequences would have a different corticotropic effect. In order to elucidate this problem the following peptides have been synthetised by a Hungarian group of chemists (9, 10, 11).

1. 1-28 porcine
2. 1-28 human
3. 1-32 human
4. 1-39 human

The chemical structures of the peptides are shown in Table I.

For comparison of the biological activities of these compounds the following methods were used.

1. Subcutaneous Sayers' method (12) as modified by Hamburger (13).

Table 1. Chemical structures of the investigated peptides.

| | |
|---|---|
| 1-28 PORCINE | 1 - 24 - ASP - GLY - ALA - GLU |
| 1-28 HUMAN | 1 - 24 - ASP - ALA - GLY - GLU |
| 1-32 HUMAN | 1 - 24 - ASP - ALA - GLY - GLU - ASP - GLN - SER - ALA |
| 1-39 HUMAN | 1 - 24 - ASP - ALA - GLY - GLU - ASP - GLN - SER - ALA - - GLU - ALA - PHE - PRO - LEU - GLU - PHE |

2. Saffran's in vitro assay (14).

3. Assay of corticosteroidogenic potencies in peripheral blood of the rat (six point assays each) after subcutaneous and intravenous administration (15). Blood samples were taken 18 and 40 minutes after intravenous and subcutaneous administration respectively.

4. Subcutaneous and intravenous time curves of corticosterone in peripheral blood.

The 3rd International Working Standard of Corticotropine and a synthetic 1-24 peptide served as the bases of comparison.

Values of the Sayers' and Saffran's assays are shown in Table 2. The activities of the different compounds were practically the same, ranging between 88 and 119 I.U./mg.

Table 3 demonstrates I.U./mg values based on corticosterone production. In this assay the compounds had activities between 86.6 and 131.4 I.U./mg with the only exception of the 1-28 human sequence which only had a potency of 42.8 I.U./mg in the subcutaneous assay.

The time curves of blood corticosterone level after a single dose of the peptides are shown on Figures 1-10. Biologically equivalent doses were chosen on the basis of the assays based on corticosterone production. In the intravenous test peaks and shapes of curves show a very similar pattern, similar at the same time to those of the standard. On the other hand atendency

Table 2. Potencies based on depletion of adrenal ascorbic acid and in vitro corticosterone production.

| PEPTIDE | SAYERS S.C. I.U./mg | SAFFRAN I.U./mg |
|---|---|---|
| 1-24 | | 96 /72-119/ |
| 1-28 P | 107 /80-119/ | 119 /61-155/ |
| 1-28 H | 92 /69-115/ | 91 /76-120/ |
| 1-32 H | 104 /70-129/ | 88 /71-132/ |
| 1-39 H | 109 /78-153/ | 107 /87-136/ |

Table 3. Potencies based on elevation of corticosterone level in peripheral blood.

| PEPTIDE | I.U./mg BASED ON CORTICOSTERONE PRODUCTION | |
|---|---|---|
| | i.v. | s.c. |
| 1-24 | 94,7 ± 5,2 | 86,6 ± 7,4 |
| 1-28 P | 131,4 ± 8,3 | 99,8 ± 6,1 |
| 1-28 H | 120,0 ± 9,4 | 42,8 ± 10,6 |
| 1-32 H | 131,3 ± 9,7 | 90,6 ± 6,0 |
| 1-39 H | 106,0 ± 13,8 | 107,6 ± 8,2 |

of differences may be observed in case of the subcutaneous curves. All the peptides proved to possess an effect on blood corticosterone level of more or less shorter duration than the standard. In order to obtain a rough estimation of total corticosterone production numerical values of corticosterone levels at different time intervals were summed and compared. (Table 4). These "cumulated corticosterone values" seem to justify the conclusions drawn from the curves themselves. In the intravenous test these

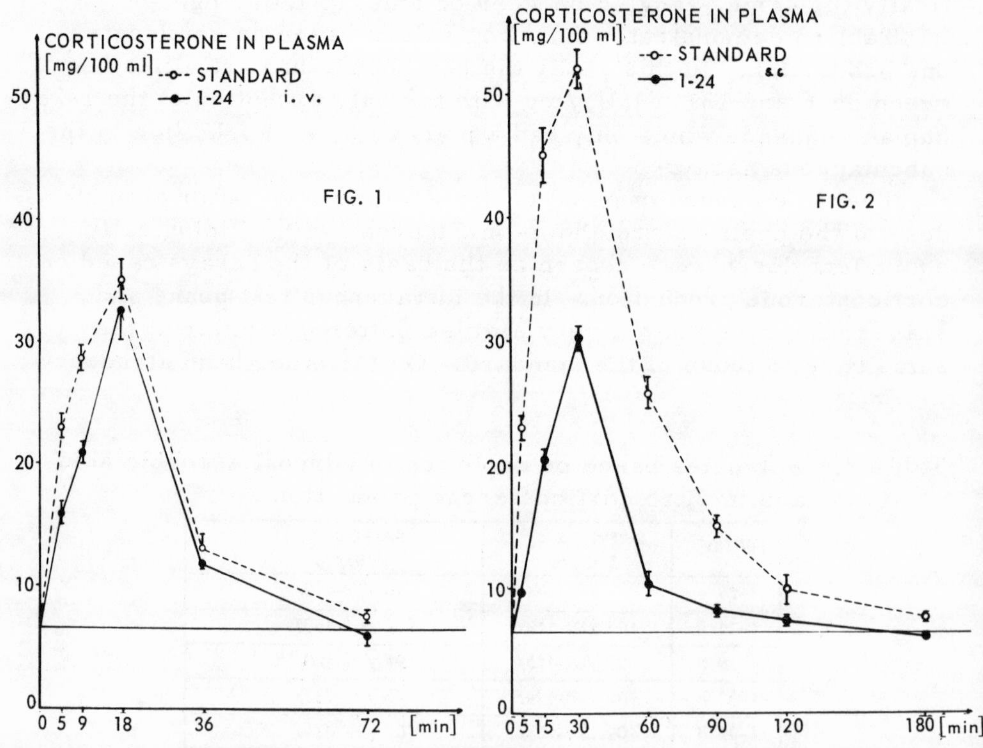

Table 4. Sum of numerical values of single corticosterone levels at different time intervals.

| PEPTIDE | CUMULATED CORTICOSTERONE VALUES | |
|---|---|---|
| | i. v. | s.c. |
| 3rd I.W.S. | 107,2 | 178,6 |
| 1-24 | 87,5 | 92,4 |
| 1-28 P | 118,8 | 153,2 |
| 1-28 H | 91,8 | 147,1 |
| 1-32 H | 85,2 | 126,0 |
| 1-39 H | 86,0 | 143,8 |

values range between 85. 2 and 118. 8 that is differences are almost negligible compared to the accuracy of such type of tests.

Values in case of subcutaneous administration on the other hand show a much greater tendency to differ from each other, their values ranging between 92. 4 and 178. 6.

Summing up our results obtained with ACTH peptides of different chain lengths it may be stated that very little difference

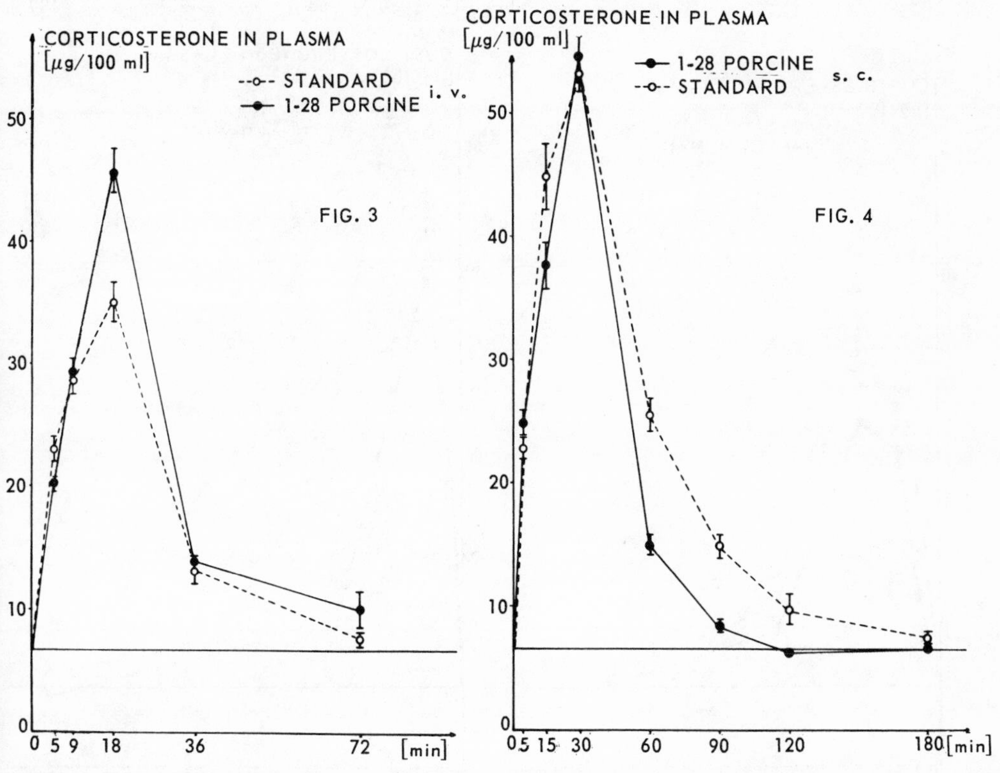

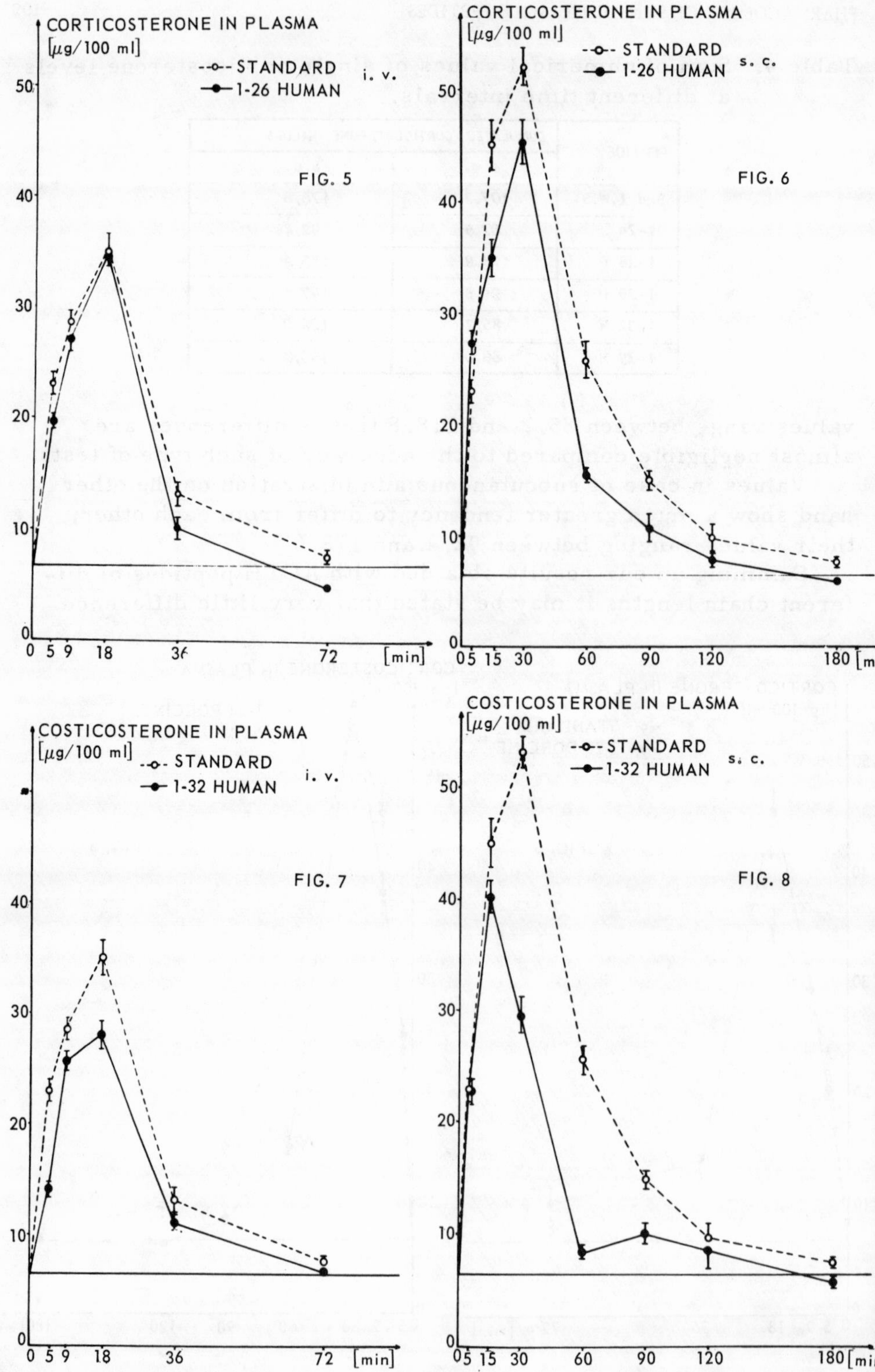
CORTICOSTERONE IN PLASMA
[μg/100 ml]
STANDARD
1-26 HUMAN
i. v.
FIG. 5
0 5 9 18 36 72 [min]
CORTICOSTERONE IN PLASMA
[μg/100 ml]
STANDARD
1-26 HUMAN
s. c.
FIG. 6
0 5 15 30 60 90 120 180 [min]
COSTICOSTERONE IN PLASMA
[μg/100 ml]
STANDARD
1-32 HUMAN
i. v.
FIG. 7
0 5 9 18 36 72 [min]
COSTICOSTERONE IN PLASMA
[μg/100 ml]
STANDARD
1-32 HUMAN
s. c.
FIG. 8
0 5 15 30 60 90 120 180 [min]

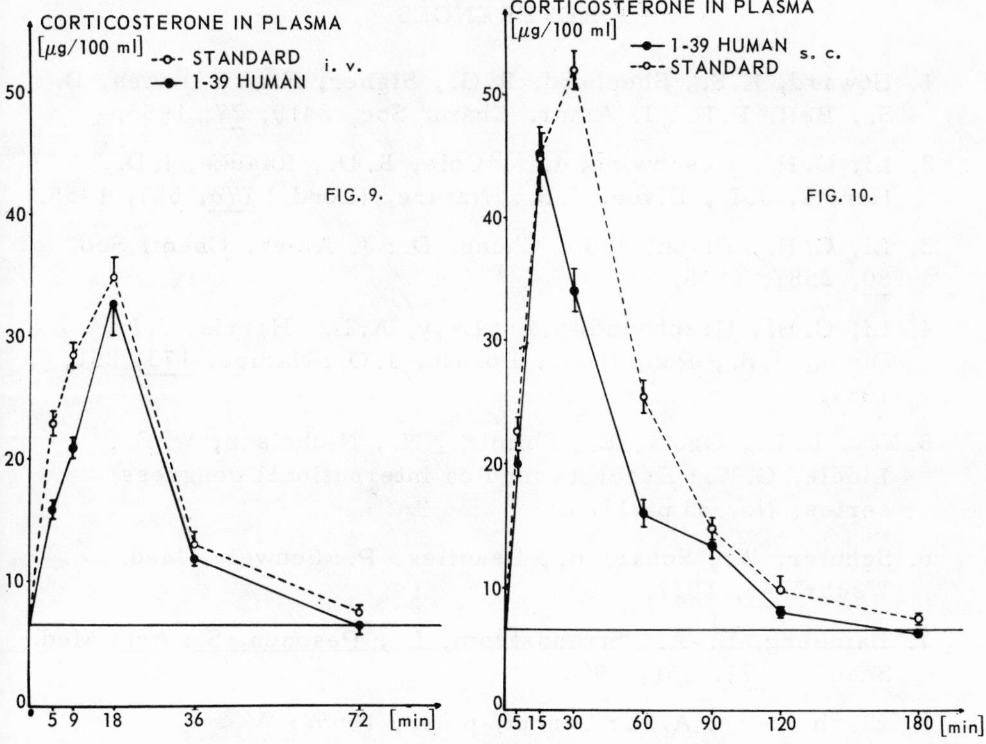


FIG. 9.

FIG. 10.

can be found between them in the usual types of assay. When comparing however their corticosterone blood curves after a simple subcutaneous dose, the 1-24 peptide is likely to produce a smaller and shorter elevation of blood corticosterone than the peptides with a longer chain. Should this phenomenon prove to be valid in the human assay as well it should speak in favour of the practical use of synthetic ACTH peptides of longer chain lengths. Hardly any difference could be demonstrated between the 1-28 procine and human peptides. The 1-39 porcine peptide has been found to possess an activity of 90 I. U. /mg by Ney et al. (5). This value corresponds very well to that found by us for the 1-39 human peptide.

On the basis of immunobiological findings obtained with highly purified corticotropines of animal origin in man however (16, 17) a significant preference should be given to the clinical use of human sequences.

## REFERENCES

1. Howard, K.S., Shepherd, R.G., Signer, E.A., Davies, D. S., Bell. P.H.: J. Amer. Cham. Soc. 3419, 77, 1955.

2. Li, C.H., Geschwind, I.I., Cole, R.D., Raacke, I.D., Harris, J.I., Dixon, J.S.: Nature, (Lond.) 176. 687, 1955.

3. Li, C.H., Dixon, J.S., Chung, D.: J. Amer. Chem. Sco. 80, 2587, 1958.

4. Li, C.H., Geschwind, I.I., Levy, A.L., Harris, J.I., Dixon, J.S., Pon, N.G., Porath, J.O.: Nature, 173, 251, 1954.

5. Ney, R.L., Ogota, E., Shimizu, N., Nicholson, W.E., Liddle, G.W.: Excerpta medica international congress series. No. 83 p. 1184.

6. Schuler, W., Schar, B., Dsaulles, P.: Schweiz, Med. Wschr. 93, 1027, 1963.

7. Lamberg, B.-A., Strandstrom, L., Pesonen, S.: Acta Med. Scand., 179, 551, 1966.

8. El-Shaboury, A.H.: Lancet, p:298, 1965.

9. Bajusz, S., Medzihradszky, K., Paulay, Z., Lang, Zs.: Acta Chim. Acad. Sci. Hung. 52, 335, 1967.

10. Bruckner, V., Medzihradszky, K., Bajusz, S., Kisfaludy, L., Low, M., Paulay, Z., Szporny, L., Hajos, Gy. T.: Hungarian Patent. Prov. No. RI. 306.

11. Kisfaludy, L., Low, M.: Unpublished data.

12. Sayers, M.A., Sayers, G., Woodbury, L.A.: Endocrinology, 42, 379, 1948.

13. Hamburger, C.: Acta Endocrinol. 35, 594, 1960.

14. Saffran, M., Schally, A.V.: Endocrinology 56, 523, 1955.

15. Purves, H.D., Sirett, N.E.: Endocrinology, 77, 366, 1965.

16. Buytendijk, H.J., Maesen, Fr.: Acta Endocr., 47, 613, 1964.

17. Felberg, J.P., Aschroft, S.H.J., Villanueav, R., Vanotti, A.: Nature, 211, 654, 1966.

# PURIFICATION OF PORCINE AND HUMAN ACTH

Aaron B. Lerner, G. Virginia Upton and Saul Lande

Departments of Medicine and Biochemistry

Yale University School of Medicine, New Haven, Conn.

Homogeneous ACTH from porcine and human sources can now be prepared in moderate quantities because suitable starting material is available, and fractionation on large Sephadex and carboxymethyl cellulose columns is readily reproducible. The purified peptides are necessary for immunologic, metabolic and chemical studies. To prepare porcine adrenocorticotropic hormone, the starting material is the fraction made for clinical use by carrying the extraction of whole pituitaries through an oxycellulose adsorption step. For human adrenocorticotropic hormone, by-products from preparations of human growth hormone are used. Detailed procedures for purifying adrenocorticotropic hormone isolated from human and porcine sources are described below.

## MATERIALS, PROCEDURES AND RESULTS

Assays for melanocyte-stimulating hormone were carried out by Mr. S. Kulovich using the in vitro frog skin assay (1). Subcutaneous assay for adrenocorticotropic activity (2), based on ascorbic acid depletion in hypophysectomized rats, was performed by Dr. J. D. Fisher of the Armour Laboratories. Acid hydrolysates (constant boiling HCl, deaerated, 22 hours, 110°) of these fractions were characterized by automatic amino acid analysis with a Spinco model 120B analyzer. The number of tryptophan residues in the intact peptide was estimated from the ultraviolet absorption curves made with a Cary model 15 spectrophotometer. Electrophoresis was carried out on Whatman paper No. 1 with pyridine-acetate buffer, pH 6.5, and 4 molar urea for 3 hours at 26 volts per cm. Peptides were detected with bromphenol blue(3).

## Porcine ACTH

The Armour Laboratories prepared a concentrate of adrenocorticotropic hormone from whole pituitaries of pigs by extraction of the glands with an acetone hydrochloric acid mixture (4) followed by adsorption and elution of adrenocorticotropic hormone - active material from oxycellulose (5). This product, similar to that marketed as "ACTH" for clinical use, contains about 35 per cent of the hormone. 3.53g of the concentrated adrenocorticotropic hormone, representing material from about 8,000 glands, was dissolved in 50 ml 5 per cent acetic acid and chromatographed on a 10 x 300 cm column of Sephadex G-25, fine grade. The flow rate was 300 ml per hour. The effluent, monitored continuously for absorbance at 2800 Å with a Cary Model 14 spectrophotometer, was divided into 10 fractions as shown in Fig. 1. Most adrenocorticotropic hormone activity was in fractions 3 and 4 which yielded 1.020 and 0.766 g, respectively. Each of these two fractions was dissolved in 4 ml of a 0.012 M sodium succinate-0.015 M acetic acid buffer at pH 4.9 (6) and chromatographed separately on a 2 x 92 cm carboxymethyl cellulose column at 60 ml per hour. As before, the development of the column was monitored by continuous, spectrophotometric measurements at 2800 Å. The chromatographic patterns from Fractions 3 and 4 were identical. The eluates were divided into subfractions A through G (Fig. 2) and the parts common to both runs were combined. Yields were as follows: A, 67 mg; B, 71 mg; C, 67 mg; D, 72 mg; E, 200 mg; F, 508 mg; and G, 80 mg. The small, unretarded Fraction A was a mixture of several peptides and was not fractionated further. Fractions B, C, D and E were each dissolved in 1 to 5 ml of the 0.012 M sodium succinate-0.015 M acetic acid buffer and rerun on a carboxymethyl cellulose column of 0.8 x 140 cm. Flow rate was 6 ml per hour. Eluates from each of these runs were labeled 1, 2 and 3 (Fig. 3). Fraction F, containing adrenocorticotropic hormone, and Fraction G, containing a moderate amount of this hormone, were not purified further. On electrophoresis Fraction F was nearly homogeneous. Amino acid analyses of the peptides are shown in Table I.

From 3.53 g starting material, 508 mg of purified adrenocorticotropic hormone was obtained. Fraction E3, 35 mg undoubtedly was adrenocorticotropic hormone also. Thus the overall yield was 543 mg from 3.53 g or 15 per cent. In a previous experiment when 6 g of starting material was chromatographed through a carboxymethyl cellulose column and the adrenocorticotropic hormone peak was rerun on another carboxymethyl cellulose column to give 1.2 g of highly purified adrenocorticotropic hormone, the yield was 20 per cent (7). Our laboratory facilities were such that for the present study chromatography on Sephadex was carried out before that on carboxymethyl cellulose.

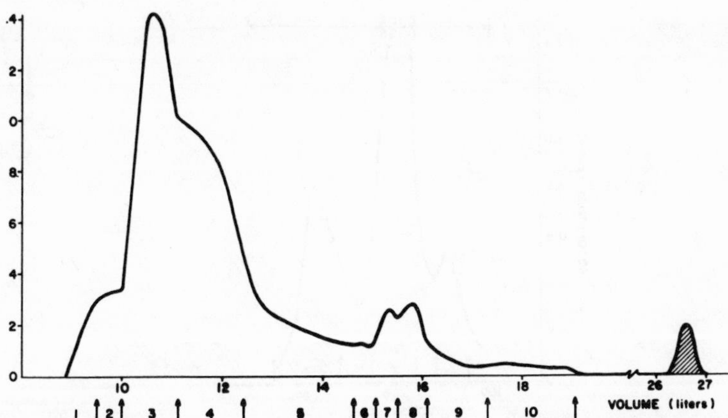


Figure 1: Gel filtration of commercial porcine ACTH on Sephadex G-25, fine grade. The column (10 x 300 cm) was developed with 5 per cent acetic acid at a flow rate of 300 ml per hour and monitored continuously for absorbancy at 2800 A°. The eluate was divided into 10 fractions as indicated by the arrows.

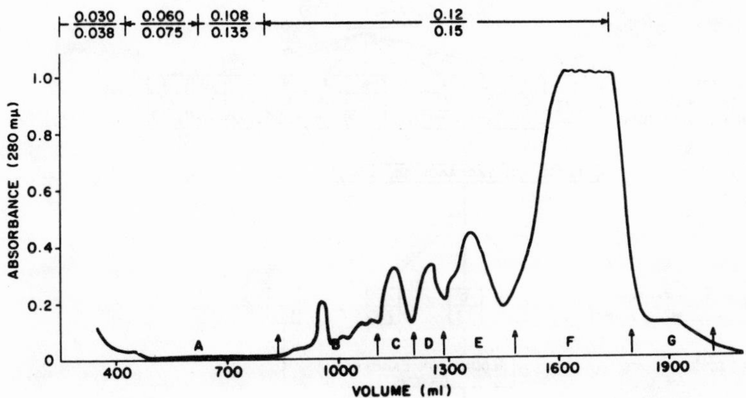


Figure 2: Chromatography of Fraction 3 on carboxymethyl cellulose. Development of a column 2 x 92 cm was by stepwise elution with disodium succinate-acetic acid buffer at a flow rate of 60 ml per hour. The molar concentrations of sodium succinate to acetic acid are given at the top of the graph. Fraction 4 gave an identical elution pattern. Subfractions A - G were collected as indicated by the arrows.

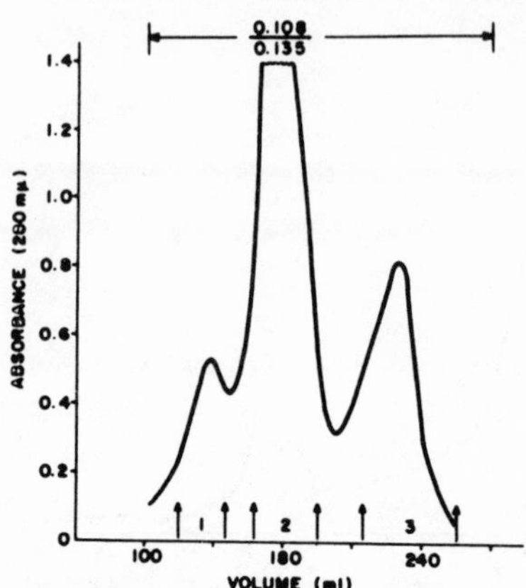


Figure 3: Rechromatograph of porcine fraction E on carboxymethyl cellulose. A column 0.8 x 140 cm was developed by a stepwise concentration gradient of disodium succinate-acetic acid buffer at a flow rate of 6 ml per hour. In each case effluents were pooled into three subfractions 1, 2 and 3. Fractions B, C and D gave similar patterns.

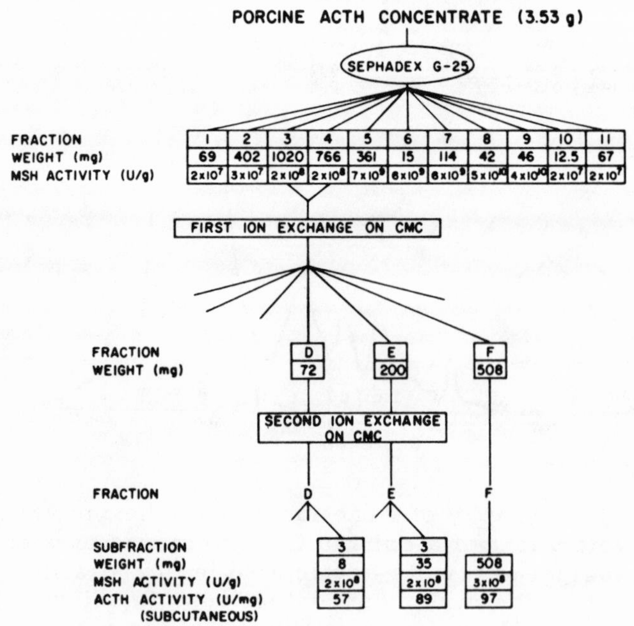


Figure 4: Flow diagram of purification of porcine ACTH. Melanotropic activities of subfractions are included.

Molar Ratios of Amino Acids in Subfractions

| Amino Acid | Porcine | | | | Human | | | | | | | |
|---|---|---|---|---|---|---|---|---|---|---|---|---|
| | $D_3$ | $E_3$ | F | Theory | 3A | 4A | 5B | 6B | 7A | 8B | 9A | Theory |
| Lys | 3.97 | 4.05 | 3.80 | (4) | 4.00 | 3.77 | 3.90 | 3.81 | 3.75 | 4.12 | 4.29 | (4) |
| His | 0.95 | 0.91 | 0.93 | (1) | 0.89 | 1.13 | 0.90 | 0.96 | 0.99 | 0.93 | 0.88 | (1) |
| Arg | 2.54 | 2.68 | 2.95 | (3) | 3.00 | 3.00 | 3.40 | 3.21 | 2.90 | 3.09 | 3.30 | (3) |
| Asp | 2.06 | 2.15 | 2.13 | (2) | 2.33 | 2.26 | 2.30 | 2.26 | 2.02 | 2.11 | 2.41 | (2) |
| Thr | 0.38 | 0.41 | Trace | (0) | 0.27 | Trace | 0.01 | 0.10 | 0.07 | Trace | 0.17 | (0) |
| Ser | 1.51 | 1.70 | 1.72 | (2) | 2.44 | 2.64 | 2.60 | 2.45 | 2.56 | 2.55 | 1.95 | (3) |
| Glu | 4.60 | 5.16 | 5.28 | (5) | 5.33 | 5.65 | 5.40 | 5.26 | 5.16 | 5.33 | 4.28 | (5) |
| Pro | 3.65 | 4.09 | 4.08 | (4) | 3.22 | 3.96 | 4.30 | 4.21 | 3.25 | 3.97 | 3.96 | (4) |
| Gly | 3.17 | 3.18 | 3.03 | (3) | 3.39 | 3.40 | 3.30 | 3.14 | 3.16 | 3.10 | 3.89 | (3) |
| Ala | 3.00 | 3.13 | 3.11 | (3) | 3.22 | 3.40 | 3.10 | 3.17 | 3.04 | 2.89 | 2.65 | (3) |
| ½ Cys | 0 | 0 | 0 | (0) | 0 | 0 | 0 | 0 | 0 | 0 | 0 | (0) |
| Val | 2.54 | 3.02 | 2.94 | (3) | 3.00 | 3.02 | 2.70 | 2.79 | 3.08 | 3.00 | 3.26 | (3) |
| Met | 0.43 | 0.51 | 0.80 | (1) | 0.52 | 0.47 | 0.70 | Trace | 0.35 | 0.78 | 0.18 | (1) |
| I Leu | 0.13 | 0.04 | 0.03 | (0) | 0.03 | Trace | Trace | Trace | Trace | Trace | 0.08 | (0) |
| Leu | 2.06 | 2.00 | 2.06 | (2) | 1.13 | 1.13 | 1.10 | 0.91 | 0.98 | 1.00 | 0.76 | (1) |
| Tyr | 1.37 | 1.71 | 1.83 | (2) | 1.78 | 2.08 | 1.90 | 1.82 | 1.62 | 1.97 | 1.66 | (2) |
| Phe | 2.38 | 2.85 | 2.95 | (3) | 2.78 | 3.00 | 3.07 | 2.91 | 2.74 | 3.00 | 2.00 | (3) |
| Trp | (1) | (1) | (1) | (1) | (1) | (1) | (1) | (1) | (2) | (1) | (1) | (1) |

Table I. Amino Acid Analyses of Porcine and Human Corticotropin Fractions. Methionine values represent the sum of methionine and methionine sulfoxide.

However, it is probably easier, when given a choice, to carry out carboxymethyl cellulose chromatography first.

By employing gel filtration first, we obtained in Fractions 3 and 4, peptides having molecular weights similar to that of adrenocorticotropic hormone. Greatest melanocyte-stimulating activities were found in Fractions 8 and 9. We do not know why Fraction 4 behaved as a shoulder of Fraction 3 on Sephadex as both gave identical elution curves from carboxymethyl cellulose. Furthermore, amino acid analyses of the major component (F in both cases) were the same and identical to that of porcine adrenocorticotropin. At least 6 components in addition to the major constituent obtained by carboxymethyl cellulose fractionation were similar to adrenocorticotropic hormone in both size and amino acid composition. However, adrenocorticotropic activity of these subfractions was low compared to that of purified adrenocorticotropic hormone (Fraction F) or of those peptides obviously made up largely of adrenocorticotropic hormone (i.e. E-3 and G). A flow diagram summarizing the fractionation procedure, weight yields and melanotropic and corticotropic activities is given in Figure 4.

## Human Adrenocorticotropic Hormone

A concentrate of human adrenocorticotropic hormone was obtained from Dr. Maurice Raben of Tufts University through the help of the National Pituitary Agency and the Endocrine Study Section of the National Institute of Arthritis and Metabolic Diseases. This material was prepared in a manner similar to but not identical with that used for porcine adrenocorticotropic hormone. Human pituitaries were collected and stored in cold acetone. Glands were accumulated for approximately six months, and an acetone dried powder was made. The powder was extracted with acetic acid at $60^{o}$ and treated as described earlier with acetone, sodium chloride and diethyl ether (8). The material obtained from ether precipitation was mixed with 20 per cent by weight of oxycellulose in 0.1 N acetic acid. Growth hormone, unlike adrenocorticotropic hormone is not adsorbed onto oxycellulose and can be separated in good yield by this procedure. Adrenocorticotropic hormone and melanocyte-stimulating hormone adsorbed to the oxycellulose was removed later by mixing with 0.1 N hydrochloric acid.

Two batches of the human adrenocorticotropic concentrate, one 800 mg and the other 730 mg, representing material from almost 7,000 glands, were dissolved separately in 5 ml 0.012 M sodium succinate-0.015 M acetic acid buffer. After centrifugation each clear supernatant solution was chromatographed on a 2 x 92 cm carboxymethyl cellulose column equilibrated with 0.012 M

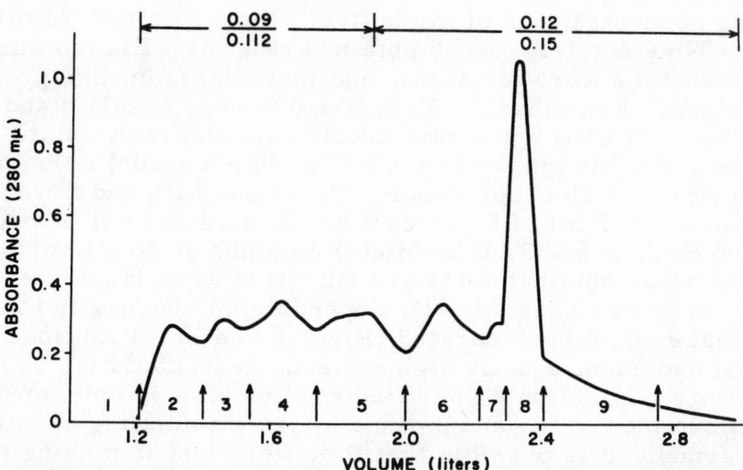


Figure 5: Chromatography of 600 mg human ACTH concentrate on carboxymethyl cellulose. The column 2 x 92 cm was developed by a stepwise concentration gradient of disodium succinate-acetic acid buffer in concentrations given at the top of the drawing. Flow rate was 75 ml per hour. A total of 1.53 g was run in this manner. Ten fractions were obtained as indicated by the arrows.

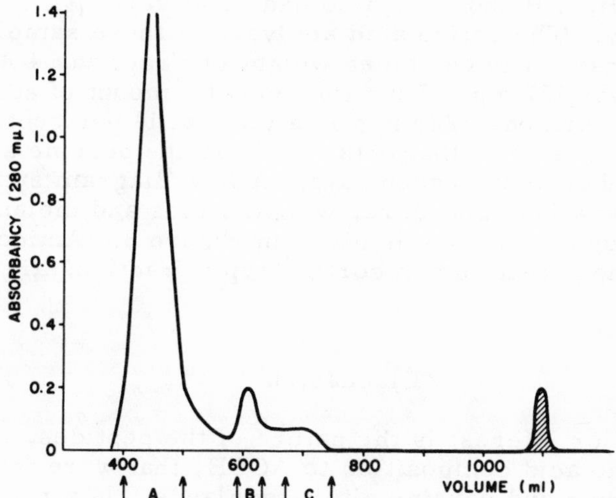


Figure 6: Gel filtration of human fraction 3 with Sephadex G-25 fine grade on a 2 x 246 cm column at 25 ml per hour. The columns were developed with 5 per cent acetic acid. Fractions A, B, C, etc. were collected as shown by the arrows. Similar elution patterns were obtained from chromatography of fractions 2, 4, 7 and 9.

sodium succinate - 0.015 M acetic acid buffer and developed with increasing concentrations of the buffer. Flow rate was 75 ml per hour. Nine fractions were obtained (Fig. 5). Elution curves from the two runs were identical, and material from like peaks was combined. Fractions 2, 3, 4, 7 and 9 were desalted and then dissolved in 2 ml 5 per cent acetic acid and rechromatographed on a 2 x 246 cm Sephadex - G 25 (Fine grade) column with a flow rate of 25 ml per hour. Fractions 5, 6 and 8 were each dissolved in 2 ml of 5 per cent acetic acid and put through 2 x 249 cm Sephadex - G 50 (medium) columns at 20 ml per hour. The subfractions obtained after gel filtration were labeled A, B and C or, in some cases, A + B, depending on how easily the various peaks could be separated (Figs. 6 and 7). Fractions 1 to 7 and 9 all had about equally high melanocyte-stimulating hormone activity. Fraction 8, purified adrenocorticotropic hormone, was the most active. $\alpha$- and $\beta$-melanocyte-stimulating hormones were not found. It is possible that they were lost in making the original, acetone-dried powder. These extractions will have to be repeated starting with fresh frozen glands.

From 1.53 g of starting material, 106 mg of adrenocorticotropic hormone, fraction 8 B, was obtained. This material was almost homogeneous on electrophoresis. The yield of almost 7 per cent was half that obtained with adrenocorticotropic hormone from pigs. On the other hand, several fractions, viz., 4 A, 5 B, 6 B and 7 A, also had relatively high corticotropin activity. The amino acid analyses of these samples were almost identical. The combined weight of Fractions 4 A, 5 B, 6 B and 7 A was 177 mg. Thus the overall amount of adrenocorticotropic hormone, 283 mg or a yield of 18 per cent, becomes comparable to that obtained from the porcine adrenocorticotropic hormone concentrate. A flow diagram summarizing the fractionation procedure, weight yields and melanotropic and corticotropic activities is given in Figure 8. Amino acid analyses of the potent human corticotropin fractions are given in Table 1.

## Discussion

Of major interest is the nature of the peptides, similar in size and amino acid composition to ACTH, that were isolated from both human and porcine pituitary glands. In a recent study involving extraction of ACTH from sheep pituitaries, Pickering et al (9) detected small amounts of a peptide with a sequence identical to that of porcine ACTH as well as other peptides closely related to ACTH. Our fractionation of porcine, oxycellulose-prepared ACTH gave at least 6 peptides similar in molecular weight to ACTH but somewhat different in composition. In fractionating the human ACTH concentrate we obtained

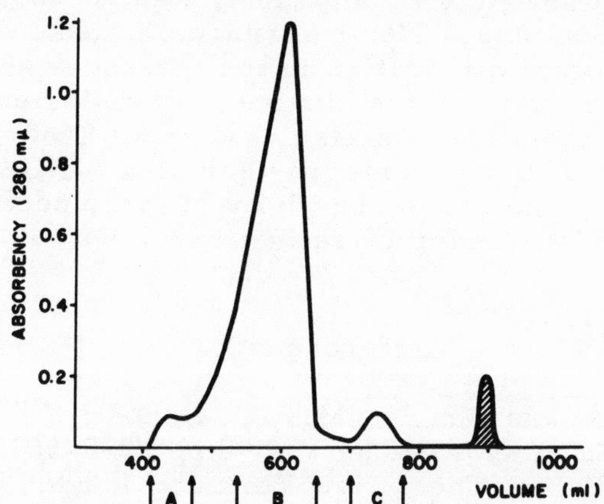


Figure 7: Gel filtration of human fraction 8 on Sephadex G-50, medium grade column 2 x 249 cm and at a flow rate of 20 ml per hour with 5 per cent acetic acid. Fractions 5 and 6 were purified in the same manner.

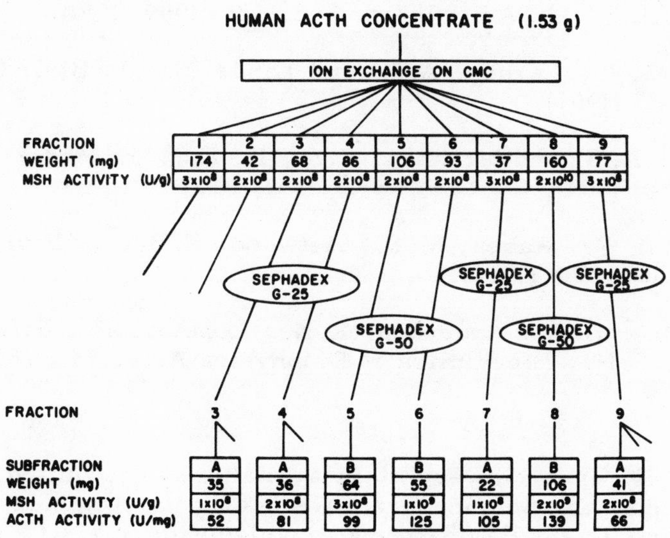


| FRACTION | 1 | 2 | 3 | 4 | 5 | 6 | 7 | 8 | 9 |
|---|---|---|---|---|---|---|---|---|---|
| WEIGHT (mg) | 174 | 42 | 68 | 86 | 106 | 93 | 37 | 160 | 77 |
| MSH ACTIVITY (U/g) | 3x10$^8$ | 2x10$^8$ | 2x10$^8$ | 2x10$^8$ | 2x10$^8$ | 2x10$^8$ | 3x10$^8$ | 2x10$^{10}$ | 3x10$^8$ |

| FRACTION | 3 | 4 | 5 | 6 | 7 | 8 | 9 |
|---|---|---|---|---|---|---|---|
| SUBFRACTION | A | A | B | B | A | B | A |
| WEIGHT (mg) | 35 | 36 | 64 | 55 | 22 | 106 | 41 |
| MSH ACTIVITY (U/g) | 1x10$^8$ | 2x10$^8$ | 3x10$^8$ | 1x10$^9$ | 1x10$^8$ | 2x10$^9$ | 2x10$^8$ |
| ACTH ACTIVITY (U/mg) | 52 | 81 | 99 | 125 | 105 | 139 | 66 |

Figure 8: Flow diagram of purification of human ACTH. Melanotropic activities of subfractions are included.

moderate yields of 4 additional peptides with the same amino acid composition as human ACTH. Biologic activity, although less than that of the major ACTH component, was, nevertheless, high in these related peptides. The differences in these substances may simply represent a variation in the content of amide groups or the presence of oxidized methionine, although some variation in amino acid composition was also observed. This problem deserves further study to decide whether or not different sequences occur. The cross reactivity of these peptides with antibodies to ACTH is under investigation.

## References


1. Lerner, A.B. and Wright, M.R., Methods of Biochemical Analysis. Ed. David Glick, Vol. 8 pp. 295-307, Interscience Publishers Inc., New York and London.

2. The United States Pharmacopoeia, Seventeenth Revision, 1965, pp. 147-149.

3. Durrum, E.L., J. Am. Chem. Soc. 72: 2943 (1950).

4. Lyons, W.R., Proc. Soc. Exp. Biol. Med. 35: 645 (1937).

5. Astwood, E.B., Raben, M.S., Payne, R.W. and Grady, A.B., J. Am. Chem. Soc. 73: 2969 (1951).

6. Upton, G.V., Lerner, A.B. and Lande, S., J. Biol. Chem. 241: 5585 (1966).

7. Lerner, A.B. and McGuire, J.S., N. Eng. J. Med. 270: 539 (1964).

8. Payne, R.W., Raben, M.S., Astwood, E.B., J. Biol. Chem. 187: 719 (1950).

9. Pickering, B.T., Anderson, R.N., Lohmar, P., Birk, Y. and Li, C.H., Biochimica et Biophysica Acta, 74: 763 (1963).


Funds for this research were made available by the United States Public Health Service Grant CA 04679; the American Cancer Society Grant P-168-B; and Development Award (S. Lande) 1-K3 AM 25757-01, United States Public Health Service.

# THE MECHANISM OF ACTION OF LUTEINIZING HORMONE ON STEROIDOGENESIS IN THE CORPUS LUTEUM *IN VITRO*

JOHN M. MARSH

University of Miami School of Medicine

Miami, Florida, U.S.A.

## INTRODUCTION

The interest of our laboratory for sometime, has been centered around the subject of ovarian steroidogenesis and its control by tropic hormones. I would like to present here some of the results we have obtained on one aspect of our work, namely, the study of the hormonal control of steroidogenesis in the corpus luteum. Historically the earliest demonstration of a hormonal influence on steroidogenesis in the corpus luteum was carried out using the rat, and it was found that prolactin was the luteotropic agent in this animal (1). Subsequently it has become increasingly clear that prolactin is probably not luteotropic in several other species, such as the pig, the rabbit, the cow, and the human. In regard to the cow, there is considerable evidence from *in vivo* and *in vitro* studies to suggest that luteinizing hormone (LH) is the luteotropic hormone which regulates steroidogenesis in the corpus luteum (2).

## THE EFFECT OF GONADOTROPINS *IN VITRO*

Our approach to this subject was to develop an *in vitro* model system in which we could study the effects of tropic hormones upon steroidogenesis in the corpus luteum under precisely controlled conditions. Essentially the system we worked out was to prepare slices of a single corpus luteum, obtained from a pregnant cow at slaughter, and to incubate these slices in Krebs-Ringer bicarbonate buffer in the presence or absence of test substances (3). After the incubation (usually 2 hours) progesterone was isolated from the tissue and medium, and measured spectrophotometrically.

Steroid synthesis was calculated by subtracting the amount of steroid present in an unincubated portion of the gland from that present after incubation. A second assessment of steroid synthesis used was the measuremen of the incorporation of radioactivity from acetate-1-$^{14}$C into the steroid.

It was found that if the slices of a corpus luteum were incubated in the presence of an LH preparation, there was a marked increase in the steroid synthesis, usually more than 100% over the control. Other substances such as prolactin, a preparation of LH inactivated by hydrogen peroxide, ACTH, glucagon, epinephrine, and bovine serum albumin, were completely ineffective. Follicle stimulating hormone and growth hormone were somewhat effective but their activity could be accounted for by the small amount of LH known to be present in these preparations. Thyroid stimulating hormone (TSH) has also been found to be effective in this syste but due to the fact that we have been unable to obtain a TSH preparation completely devoid of LH activity, it has been difficult to determine whether its activity is due to its LH content (4). With the exception of this possible effect by TSH, we have found that the stimulation of steroidogenesis in slices of bovine corpora lutea is specific for hormones with luteinizing hormone activity. This _in vitro_ model system is quite sensitive to small quantities of LH. The lowest effective dose is generally between 0.01 and 0.02 $\mu$g per gram of tissue. Using a molecular weight of 30,000 (5) this calculates to approximately 0.5 picomoles of LH per gram of tissue.

## THE ROLE OF CYCLIC AMP IN THE MECHANISM OF LH ACTION

After the initial phase of the work had been concluded, the investigation was divided into two parts. a) A study to determine what steps in the steroidogenic pathway were enhanced by the gonadotropin and b) an investigation to discover the mediators of this gonadotropic stimulation. The remainder of this presentation will be concerned with the second part; name the investigation of the mediators of LH action. We began this work by exploring the applicability of an hypothesis put forth by Haynes et al (6) to explain how ACTH stimulated steroidogenesis in the adrenal cortex. Briefly, this hypothesis proposed that ACTH acted initially by increasing the concentration of a substance called cyclic adenosine 3', 5'-monophosphate (cyclic AMP or 3', 5'-AMP) which in turn activated the enzyme phosphorylase. This increased phosphorylase activity lead to a breakdown of glycogen in the adrenal cortical cells, producing glucose-1-phosphate and glucose-6-phosphate. The latter substance when metabolized via the pentose pathway produces NADPH, and it was proposed that this nucleotide then stimulated corticosteroidogenesis through its role as a cofactor in many of the biosynthetic steps.

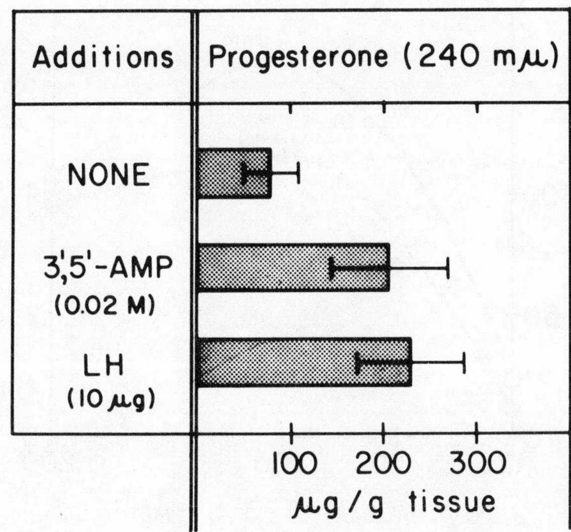


Fig. 1. Effect of cyclic AMP and LH progesterone synthesis. The height of each bar represents the mean, and the brackets the 95% confidence limits. Appropriate vessels contained 10 μg LH/5 ml or 100 μmoles of cyclic AMP/5 ml. Approximately 0.5 g. of slices was incubated in 5 ml of Krebs-Ringer bicarbonate medium in an atmosphere of 95% oxygen, 5% carbon dioxide for 2 hours.

In our studies, exogenous cyclic AMP was added to incubating slices of a corpus luteum and its effect on steroidogenesis was compared to that of LH (7). The results of 15 experiments are summarized in Fig. 1.

It can be seen that this cyclic nucleotide produced a significant stimulation of steroid synthesis (with a P value less than 0.001). In fact, the magnitude of this stimulation by cyclic AMP is about equal to that produced by saturating amounts of LH.

To determine the optimal concentration of cyclic AMP for maximum stimulation of steroidogenesis, graduated doses from 0.002M to 0.04M cyclic AMP were added to separate vessels containing acetate-1-$^{14}$C and slices from a single corpus luteum. Progesterone synthesis was assessed spectrophotometrically and by the incorporation of radioactivity. The results are illustrated in Fig. 2. As the concentration of cyclic AMP was increased, the synthesis of progesterone (assessed by both parameters) was increased. A maximum was reached at 100 μmoles per 5 ml of medium or 0.02M cyclic AMP.

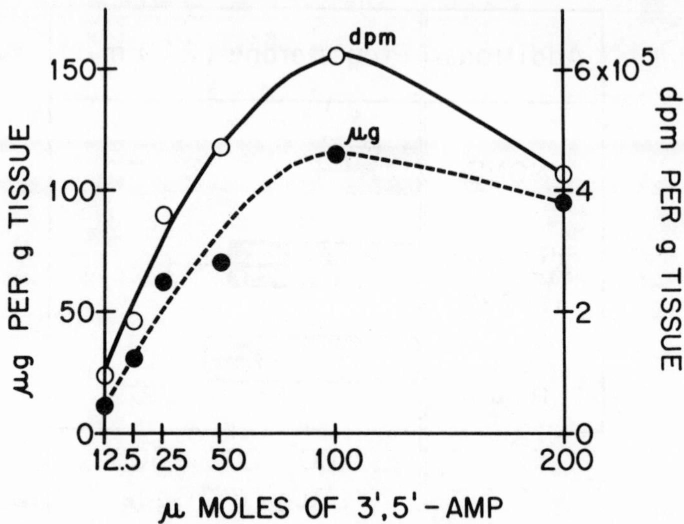


Fig. 2. Effect of cyclic AMP on progesterone synthesis. Graded amounts of cyclic AMP from 12.5 to 200 μmoles were added to separate vessels containing paired slices from a single corpus luteum, caffeine (15 μmoles) and acetate-1-$^{14}$C (10 μc 0.5 μmoles). The incubations were carried out as described in Fig. 1.

The concentration needed for maximal stimulation is relatively high and exceeds by far the endogenous concentration of this cyclic nucleotide usually found in the corpus luteum (8) and in other tissues (6, 9). Such a requirement for high concentrations of the nucleotide has, however, been observed previously by others in studies on the effect of exogenous cyclic AMP on phosphorylase and steroidogenesis in liver and adrenal cells (10). The interpretation given by these authors was that cyclic AMP did not easily penetrate the cell membrane, since much smaller concentrations were effective in homogenates of their tissues.

Other nucleotides, structurally related to cyclic AMP were tested for their effect on progesterone synthesis at the 0.02M concentration. In each experiment the effect of one or more related nucleotides was compared with the effects of cyclic AMP and LH, on slices from the same corpus luteum. Fig. 3 summarizes the results of these experiments, showing the mean change in synthesis above that of the control and the 95% confidence limit of the true mean for each treatment. None of the related nucleotides, 3'-AMP, 5'-AMP or ATP were effective in stimulating progesterone synthesis in corpora lutea which readily responded to cyclic AMP and LH.

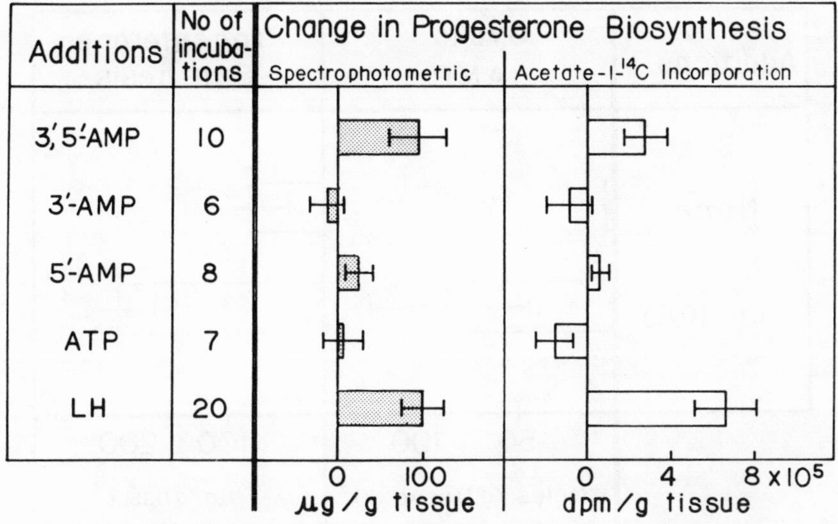


Fig. 3. Effect of various nucleotides and LH on progesterone synthesis. The height of each bar represents the mean change in progesterone synthesis over the control level, and the bracket the 95% confidence limit. The incubations were carried out as described in Fig. 1.

The ineffectiveness of these related nucleotides in stimulating progesterone synthesis serves to establish to a certain degree the specificity of the action of cyclic AMP on steroidogenesis in corpora lutea.

The data obtained with exogenous cyclic AMP were compatible with the hypothesis that cyclic AMP was a mediator of the action of LH. Such a consideration, however, should require that the addition of LH would bring about an increase in the endogenous concentration of cyclic AMP. An investigation was then undertaken in collaboration with Dr. R. W. Butcher and Dr. E. W. Sutherland to measure the amount of cyclic AMP in incubating luteal slices and to assess the effect of LH on the level of this substance (8). As shown in Fig. 4, it was found that the gonadotropin brought about a striking increase (as much as 100 fold) in the endogenous concentration of cyclic AMP measured after 15 minutes of incubation as well as producing its usual increase in steroidogenesis measured after 2 hours of incubation.

A number of other substances were tested for their effect *in vitro* on the accumulation of cyclic AMP in corpora lutea slices. These included LH inactivated by hydrogen peroxide treatment, prolactin and other substances such as ACTH, epinephrine and glucagon, which have been shown

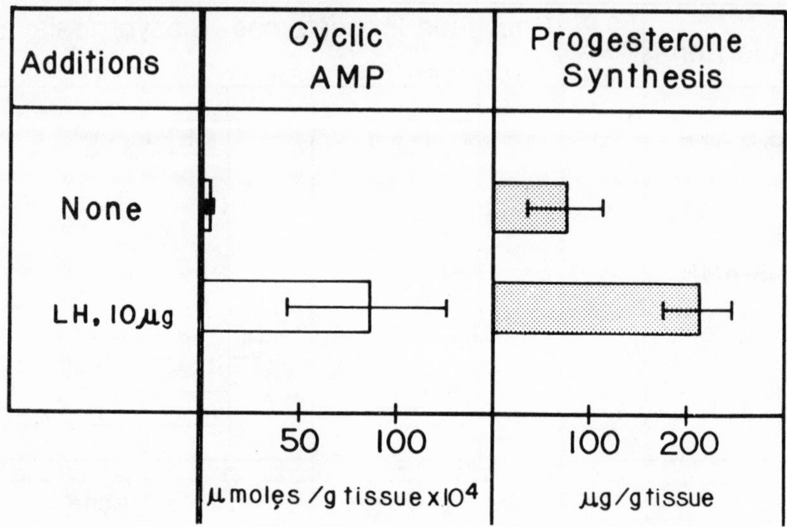


Fig. 4. Effect of LH on the concentration of cyclic AMP and the µg of steroid synthesized in 12 paired experiments. The height of each bar represents the mean and the brackets the 95% confidence limit. Each vessel contained 5 ml Krebs-Ringer bicarbonate buffer and approximately 0.3 to 0.5 g slices. The slices were incubated for 1 hour in buffer alone, and then transferred to fresh buffer alone or buffer containing 10 µg LH/5 ml for the second incubation. After 15 minutes, one set of slices was analyzed for cyclic AMP. The other set of slices was incubated for 2 hours and steroid synthesis measured.

to be effective in increasing the cyclic AMP concentration in other tissues (9). The latter preparations were tested at concentrations which were effective in the experiments on other tissues. All of these other substances were ineffective, while LH brought about a marked increase in the accumulation of cyclic AMP in slices from the same corpus luteum. It seems, therefore, that this response of cyclic AMP accumulation, like the response of increased steroidogenesis, is specific for substances with luteinizing hormone activity.

When the effects of LH on cyclic AMP accumulation and progesterone synthesis were compared at intervals throughout a two hour incubation, it was found that the rise in cyclic AMP occurred before the effect on steroidogenesis. As shown in Fig. 5, the stimulatory effect of LH on cyclic AMP accumulation occurred between 0 and 15 min. while the effect on the rate of steroidogenesis was not apparent until between 15 and 30 minutes. In

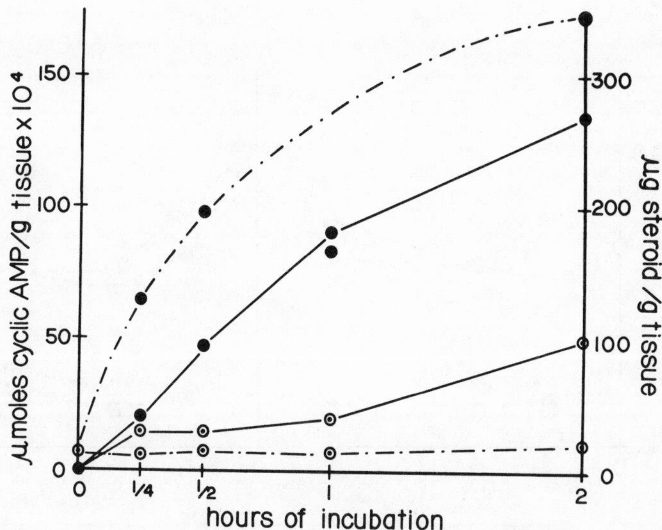


Fig. 5. Time course of the effect of 10 μg LH/5 ml on cyclic AMP accumulation and steroidogenesis. The changes in cyclic AMP concentration are shown as broken lines and the rates of steroidogenesis by solid lines. The control values are shown as open circles and the values obtained from LH treated tissue as solid circles. The incubation conditions were the same as in Fig. 4 except that the cyclic AMP and steroid determinations were made at various times during the second incubation.

other experiments, the time between the rise of cyclic AMP and the effect on steroidogenesis was even greater. This time relationship would be expected if cyclic AMP is the mediator of the action of LH on steroidogenesis.

On the basis of the above results it seems very likely that cyclic AMP acts as the mediator of the action of LH upon progesterone synthesis. Most of our current work has been directed toward determining how cyclic AMP mediates this effect. At this time our results are very preliminary and have been more informative in terms of indicating what mechanisms are probably not correct among the current hypotheses than in indicating the correct mechanism of action of cyclic AMP.

In earlier studies, we have tried to assess the possible role of NADPH in the mechanism action of LH in this model system. Our early results indicated that it might be part of the mechanism. The addition of exogenous NADPH led to increase in mass of the steroid produced in incubating corpora lutea slices (3). As the work progressed, however, results began to appear which indicated a difference in the action of exogenous NADPH and LH.

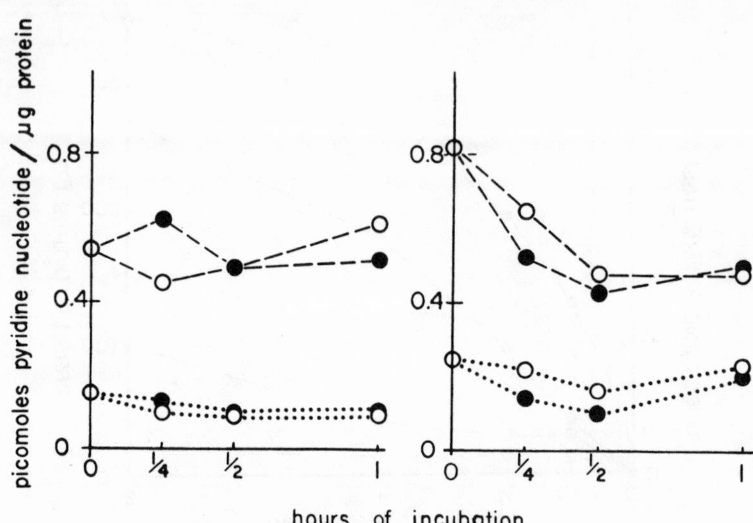


Fig. 6. The concentration of NADP (dotted lines) and NADPH (dashed lines) in control and LH (10 μg/5 ml) treated slices at various intervals during a one hour incubation. The control values are shown as open circles, and the values obtained from LH treated tissue as solid circles. The incubation was carried out as described in Fig. 1.

When both parameters of steroid synthesis were assessed (the increase in mass of progesterone and the incorporation of $^{14}$C), it was found that exogenous NADPH increased the mass of steroid produced without a concomitant increase in the incorporation of $^{14}$C. LH, on the other hand, increase both parameters (11).

Although these results suggest the LH does not act via NADPH, they are not conclusive since it is not known if the addition of exogenous NADPH is equivalent to increasing the endogenous concentration of this cofactor. A more conclusive test of whether or not LH acts by increasing limiting quantities of NADPH is to measure the endogenous concentration of this cofactor directly. Using the very sensitive enzyme cycling method developed by Lowry and Passonneau (12) we have measured both NADP and NADPH in control slices of corpus luteum and of slices treated with LH (13). Our preliminary results, shown in Fig. 6, indicate that there is no significant or consistent effect of LH on the level of either form of this pyridine nucleotide, when measured at several times during a one hour incubation. Slices from the same tissue, however, showed their usual response to LH (an increase in steroidogenesis). More experiments of this type are in progress

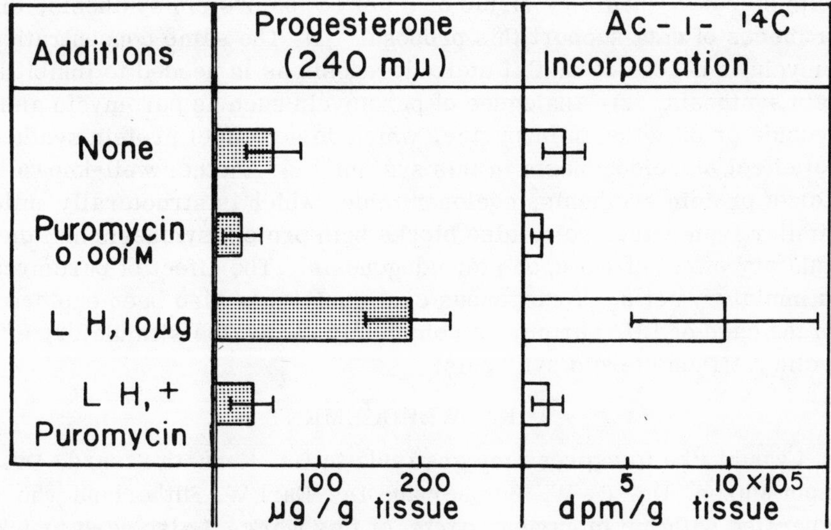


Fig. 7. Effect of puromycin on the stimulation of progesterone synthesis by LH. The height of each bar represents the mean and the brackets the 95% confidence limits. Appropriate vessels contained 10 $\mu$g LH and 5 $\mu$moles of puromycin. Acetate-1-$^{14}$C (10 $\mu$C, 0.5 $\mu$moles) was added to each vessel. The incubations were carried out as in Fig. 1.

Another approach we have taken in our study of the mechanism of action of LH and cyclic AMP on steroidogenesis in the corpus luteum is the exploration of the possible role of protein synthesis. In this work, we have used inhibitors of protein synthesis such as the antibiotic puromycin and assessed their effects on the action of LH in our model in vitro system. Slices of a corpus luteum were incubated in the usual fashion and the inhibitor of protein synthesis was added to both control incubations and incubations with LH. Fig. 7 summarizes the data of 10 experiments using puromycin, and it can be seen that this substance at the concentration of 0.001M completely blocked the gonadotropic effect on progesterone synthesis when it was added to incubations with LH. This inhibitory effect of the gonadotropic stimulation was highly significant with a P value less than 0.001. There is also some inhibition of the control steroid synthesis by the addition of puromycin alone, but this is a less significant effect with a P value of 0.05. Protein synthesis was simultaneously assessed in three of these experiments by the extent of incorporation of radioactive leucine into luteal protein. It was found that this antibiotic consistently produced its expected inhibition of protein synthesis in the controls and the incubations with LH.

While we do not wish to infer that it has been proven that puromycin inhibits steroidogenesis by virtue of its effect on protein synthesis, three other pieces of data support this proposal. 1) The same concentration of puromycin is needed to inhibit steroidogenesis as is needed to inhibit luteal protein synthesis. 2) Analogues of puromycin such as puromycin aminonucleoside or dimethylaminopurine, which do not effect protein synthesis, do not effect steroidogenesis in this system. 3) Another well-known inhibitor of protein synthesis, cycloheximide, which is structurally quite dissimilar from puromycin, also blocks both protein synthesis and the stimulatory effect of LH upon steroidogenesis. The effect of puromycin on the stimulatory action of exogenous cyclic AMP has also been explored. As in the case of LH, puromycin completely blocks the stimulatory effect of cyclic AMP on steroid synthesis.

## ACKNOWLEDGEMENT

I would like to express my gratitude to Dr. Kenneth Savard, Dr. Norman Mason, Dr. R. W. Butcher and Dr. Earl W. Sutherland who collaborated with me in various parts of this work. I also wish to acknowledge the excellent technical assistance of Mrs. Adalgisa Rojo.

This work was supported in part by Grant CA-04004 National Cancer Institute, United States Public Health Service.

## REFERENCES

1. Evans, H. M., Simpson, M. E., Lyons, W. R., and Turpeinen, K. Endocrinology, 28, 933 (1941).
2. Hansel, W., J. Reprod. Fert., Suppl. 1, 33 (1966).
3. Mason, N. R., Marsh, J. M., and Savard, K., J. Biol. Chem., 237, 1801 (1962).
4. This was first observed by P. Major and D. Armstrong. Since then it has been confirmed by us. (unpublished results)
5. Squire, P. G., and Li, C. H., J. Biol. Chem., 234, 520 (1959).
6. Haynes, R. C., Jr., Sutherland, E. W., and Rall, T. W., Recent Progress in Hormone Research, 16, 121 (1960).
7. Marsh, J. M., and Savard, K., Steroids, 8, 133 (1966).
8. Marsh, J. M., Butcher, R. W., Savard, K., and Sutherland, E. W. J. Biol. Chem., 241, 5436 (1966).
9. Sutherland, E. W., Oye, I., and Butcher, R. W., Recent Progress in Hormone Research, 21, 623 (1965).
10. Sutherland, E. W., and Rall, T. W., Pharmacol. Revs., 12, 265 (1960).
11. Savard, K., and Casey, P. J., Endocrinology, 74, 599 (1964).
12. Lowry, O. H., and Passonneau, J., Methods in Enzymology 4, 792 (1963).
13. Marsh, J. M., (unpublished results).

# FATE AND LOCALIZATION OF IODINE-LABELLED HCG IN MICE

Aliza Eshkol and B. Lunenfeld

Institute of Endocrinology, Tel-Hashomer Government Hospital, Israel

It has been shown in the past that labelling of human chorionic gonadotropin (HCG) with 1 - 2 atoms of radioactive iodine does not affect the biological, immunological and physicochemical behaviour of the hormone. (1). Furthermore, it has been established that the ovary, more than any other organ, has a capacity to concentrate HCG, and this is specific for the gonadotropic hormone. Labelled albumin or iodide$^{131}$ were not concentrated by the ovary even when the animals were stimulated simultaneously with unlabelled HCG (1, 2).

When increasing doses of HCG$^{131}$I, albumin$^{131}$I and sodium$^{131}$I (1.1-8.8 $\mu$c) were injected, a concomitant increase in ovarian radioactivity was observed only in the animals receiving the labelled HCG (3).

When antiserum to HCG was injected 5 minutes prior to the injection of labelled HCG, the concentration of radioactivity by the ovary was inhibited, being 56 cpm, as compared to those injected with the labelled HCG only, which was 1135 cpm. This observation indicated that the HCG is bound to the antiserum in the circulation and, as a hormone-antihormone complex, is not concentrated by the ovary.

Having demonstrated that the ovary has a specific capacity to concentrate HCG, the fate and some kinetics of HCG in infantile mice following intravenous injections were studied. Fig. 1 demonstrates the disappearance of labelled HCG from the circulation following a single intravenous injection. It can be noticed that the concentration fell to approximately 50% within the first 30 minutes and dropped to about 15% of injected material within the following hour; thereafter the concentration decreased slowly to about 7% after 6 hours.

During the period of rapid disappearance of radioactivity from the

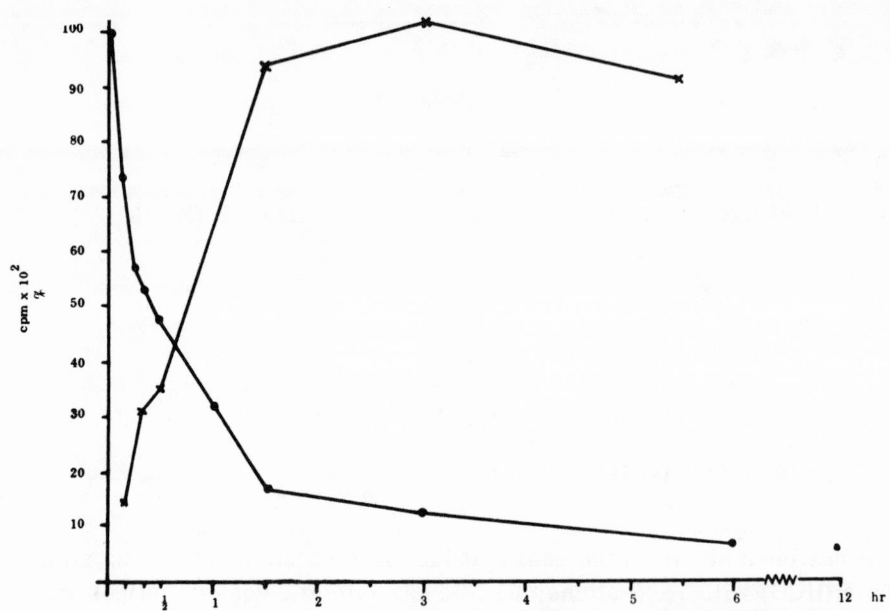


Fig. 1. Concentration of radioactivity ( ● expressed as %) and in ovaries ( ✖ given in cpm x $10^2$/ 100mg tissue) following a single i.v. injection of labelled HCG.

circulation (90 mins), the concentration in the ovary increased significantly (Fig. 1). It will be noticed that for a certain period of time the ovarian concentration remained almost constant and decreased later. The concentration of labelled HCG in the ovary at any given time is dependent on at least four different factors: (1) quantity accumulated in the ovary (by intra or intercellular trapping of HCG or adherence to cell membranes); (2) the level of HCG in the circulation; (3) rate of inflow; and (4) rate of outflow. The relationship between these events during different phases of ovarian uptake of labelled HCG was investigated.

Eliminating inflow of HCG to the ovary for specific periods of time would enable the assessment of rate of inflow and outflow during the same period. This was achieved by blocking circulating HCG with an antiserum at specific times after administration of the labelled HCG (the HCG-anti-HCG complex is not concentrated by the ovary).

The inflow during a given time interval is the difference between the concentration of ovarian radioactivity at the end of the desired period in animals which received HCG only and in animals in which the inflow of HCG to the ovary was blocked at the onset of the desired period.

The outflow during a given time interval is the difference between the concentration of ovarian radioactivity (following injections of HCG only) at the time chosen as the beginning of the desired period and the radioactivity at the end of the desired time interval in ovaries of animals in which the availability of HCG to the ovary was blocked at the onset of this period.

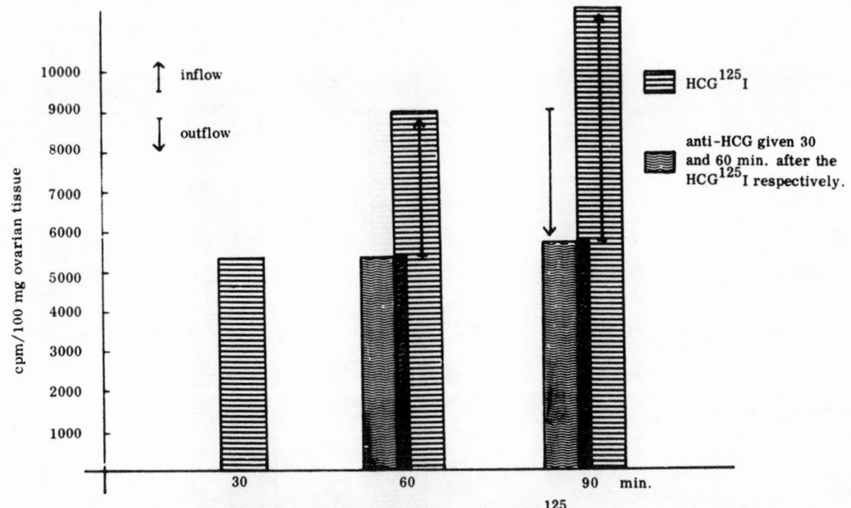


Fig. 2. Rate of inflow and outflow of radioactivity between 30 and 90 minutes following a single i.v. injection of $HCG^{125}I$.

Between 30 and 60 minutes a significant increase in ovarian concentration of labelled HCG was observed (Fig. 2). In this specific period the increased concentration reflects an inflow only. That there was no outflow becomes evident from the fact that when availability of $HCG^{125}I$ to the ovary was blocked 30 minutes after its administration (by anti-HCG), the ovarian concentration of radioactivity remained as it was at the time of the antiserum administration.

The changing concentration observed between 60 and 90 minutes is the result of both inflow and outflow. When antiserum was given 60 minutes following the injection of labelled HCG, the concentration of ovarian radioactivity at 90 minutes was lower than that of 60 minutes after the injection of HCG alone. This difference between the concentration of radioactivity reflects the outflow of labelled material from the ovary between 60 and 90 minutes. When no antiserum was given, then the concentration of radioactivity was increased, and the difference between radioactivity at 90 minutes between antiserum treated animals and those which received no antiserum reflects the inflow of labelled HCG from the circulation to the ovary within this period of time.

To summarize Fig. 3, it can therefore be concluded that within 30 to 60 minutes there was an inflow from the circulation to the ovary, but it seems that there is no outflow during this period, while between 60 to 90 minutes, concomitant with an inflow from the circulation to the ovary, there is also an outflow which is, however, lower than the inflow.

During the following 90 minutes, namely between 90 to 180 minutes following the administration of labelled HCG, ovarian concentration of

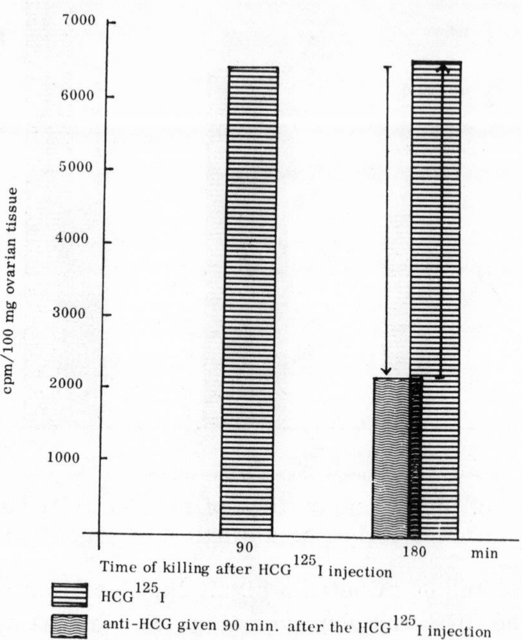


Fig. 3. Rate of inflow (↑) and outflow (↓) of radioactivity between 90 and 180 min. following a single i.v. injection of 1.2 IU $HCG^{125}I$.

radioactivity remained constant. It was shown that this was due to an equilibrium between inflow and outflow from the circulation to the ovary (Fig. 3). In this experiment the antiserum was injected intravenously 90 minutes after the injection of $HCG^{125}I$, when the ovarian concentration of radioactivity was already at its peak (6350 cpm). The animals were killed 90 minutes later (180 minutes after $HCG^{125}I$ administration), wherein the control animals - the ovarian concentration was still at its maximum (6430 cpm). In the animals receiving antiserum after 90 minutes, the radioactivity was significantly reduced (2200 cpm). The decrease in radioactivity which occurred during these 90 minutes reflects the quantity of radioactive material leaving the ovary during this time in the absence of newly available $HCG^{125}I$ from the circulation (4150 cpm).

Since the ovarian concentration of radioactivity of the animals which were not treated with antiserum remained constant during this period, 6350 and 6430 cpm respectively, it can be postulated that the inflow during this period was about 4200 cpm (difference in concentration of antiserum injected and control animals at 180 minutes).

During the phase in which ovarian concentration of radioactivity was

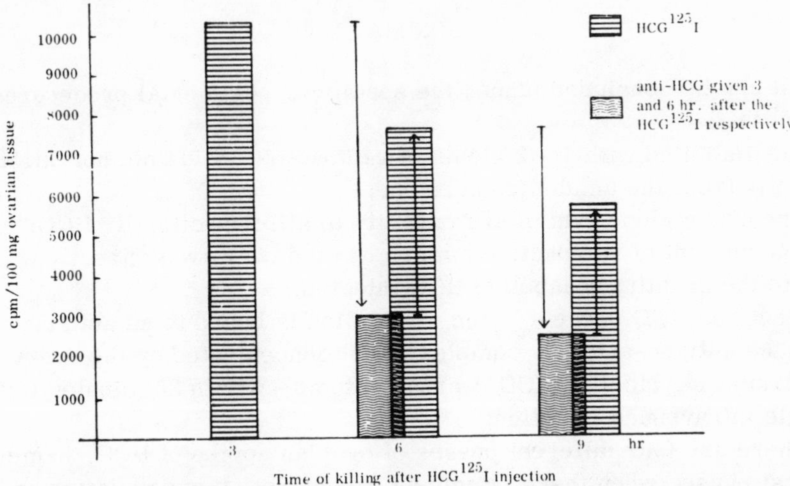


ig.4. Rate of inflow ( ↑ ) and outflow ( ↓ ) of radioactivity between 3 and 9 hr. following a single i.v. injection of 1.2 IU HCG$^{125}$I.

decreasing, it was demonstrated that this was not due to a lack of inflow but the result of a relatively higher outflow than inflow (Fig. 4).

In an attempt to localize the target cells which are responsible for the concentration of HCG, autoradiography of histological sections was performed. It was found that the radioactivity was confined mainly to the follicular envelopes, theca cells, and also to the stroma. It seems that the density of black grains reflecting concentration of radioactivity was proportional to the state of follicular development.

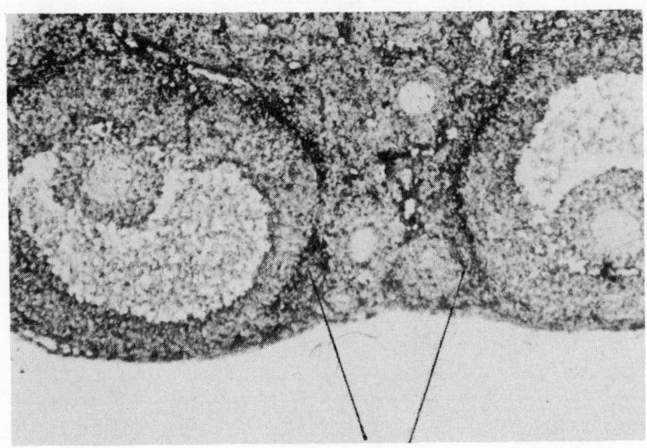

Fig. 5. Radioactivity in follicular envelopes.

## SUMMARY

It can be concluded (under the specific experimental procedures employed) that:-

1) HCG labelled with 1 - 2 atoms of radioactive iodine did not differ significantly from the unlabelled hormone;

2) The ovary alone exhibited a capacity to affix specifically HCG;

3) The amount of radioactive material in the ovary was directly proportional to the quantity of labelled HCG injected;

4) When the HCG present in the circulation is bound to an antiserum to HCG, the antigen-antibody complex is not concentrated by the ovary;

5) Circulating labelled HCG decreased to 50% within 30 minutes following a single intravenous injection;

6) There are four different phases of ovarian uptake of HCG, namely:- the first phase, when there is only an inflow from the circulation and storage mainly in the follicular envelopes; the second phase, when there is a greater inflow than outflow; the third phase, when the inflow is equal to the outflow; and the fourth period, when the outflow is bigger than the inflow.

## REFERENCES

1. Lunenfeld, B. and Eshkol, A. Vitamins and Hormones (1967) 25:165
2. Eshkol, A. In: Recent Research on Gonadotrophic Hormones, eds. E.T. Bell and J.A. Loraine, Edinburgh, Livingstone (1967), p.202.
3. Eshkol, A. and Lunenfeld, B. Proc. Tel-Hashomer Hosp. (1967) 6:4.

## ACKNOWLEDGMENTS

This work was supported in part by a grant from the Population Council, N.Y., U.S.A. and by Grant No. 67-470 from the Ford Foundation, N.Y., U.S.A.

# HUMAN FOLLICLE-STIMULATING HORMONE. PURIFICATION AND SOME BIOLOGICAL PROPERTIES

P.Donini, D.Puzzuoli, I.D'Alessio and S.Donini

Research Laboratories, Istituto Farmacologico

Serono, Rome (Italy)

In the last ten years increasing interest arose in the use of human FSH for therapeutic purposes and many attempts have been made to purify this hormone from human pituitaries and from post-menopausal urine. The results obtained by different experimental procedures in the purification of follicle-stimulating hormone (FSH) from human menopausal gonadotrophins (HMG) were reported by Donini et al. (1964; 1966a,b; 1967).

This report will deal with the preparation of highly purified urinary FSH and with some of its chemicophysical and biological properties.

## Materials and Methods

Materials. As staring material one batch of HMG (Pergonal, E146-2) was used. This material was prepared by a procedure previously described by us (Donini et al. 1964) and further purified by batchwise chromatography on DEAE-Cellulose (DEAE-C), according to the method of Roos & Gemzell (1964b).

Gel Filtration and DEAE-C Chromatography. Gel filtration on Sephadex G-100 was carried out according to the method of Roos and Gemzell (1964b). DEAE-C chromatography was carried out according to the method of Donini et al. (1966a).

Disc Electrophoresis. The preparative polyacrylamide gel electrophoresis was carried out according to the procedure of Stevens (1967). The instrument used was from Buchler Instruments, Inc., Fort Lee, N. J.

Immunoelectrophoretic Analysis. The immunoelectrophoretic studies were performed according to the method of Scheidegger (1955). The rabbit anti-HMG serum was prepared by immunizing the animals with roughly purified HMG. Rabbit anti-human albumin serum was supplied by Behringwerke A.G.

Biological Activity. The follicle-stimulating activity (FSH) was determined by the HCG-augmentation assay of Steelman & Pohley (1953). The luteinizing activity (LH) was assayed by the ovarian ascorbic acid depletion (OAAD) method of Parlow (1961). The FSH potency of the fractions obtained in the last step of purificiation was assayed also by the radioimmunological method. The FSH and LH potencies were expressed as international units (i.u.) of the 2nd IRP-HMG. The experimental design was 3+3. Statistical calculations were performed according to Borth et al. (1957).

## Purification

First Step: Chromatography on DEAE-C.
1200 mg of precursor (Pergonal, E146-2) were applied to 2x45 cm column packed with DEAE-C Whatman DE-50, previously equilibrated with borate-phosphate buffer, pH 8.0. The chromatography was carried out at 3°C by stepwise changing of solvents at flow rate of 25 ml/h. The collected fractions were the pooled tubes corresponding to each protein peak: fraction 1, 2, 3 and 4 corresponding to unadsorbed material and proteins eluted with buffer solution containing 0.05, 0.1 and 0.5 M NaCl respectively, were dialyzed against distilled watter at 3°C for ten hours and then lyophilized. After the whole desalting on Sephadex G-25, the fractions were lyophilized again. The chromatographic pattern and FSH activities are shown in Fig. 1.

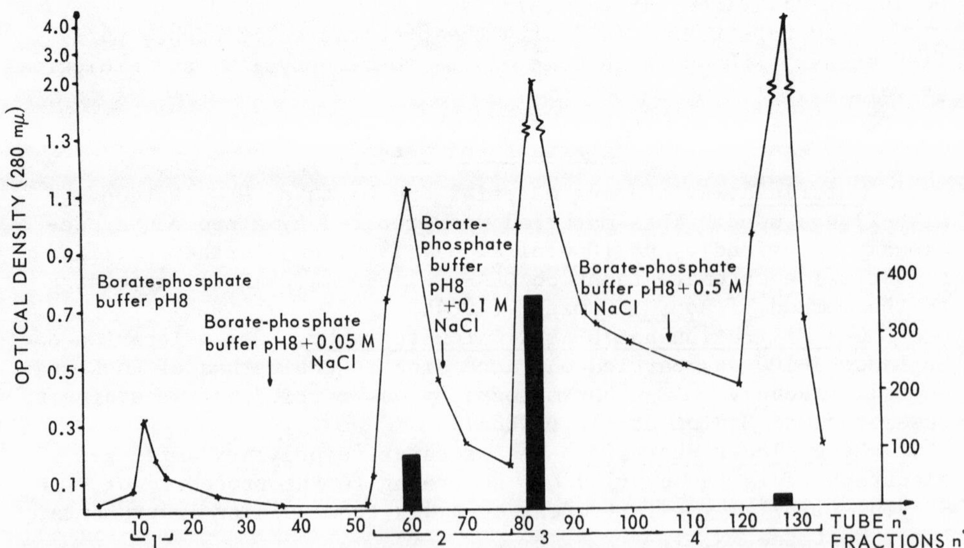


Fig.1: Chromatography of HMG (Pergonal-35 E146-2) on DEAE-Cellulose. 2x45 cm column equilibrated with 0.00213 M sodium borate - 0.00351 sodium phosphate buffer pH 8.0. Chromatography performed by stepwise changing of solvents. 1200 mg of protein applied to column; 6 ml/tube. Recovery total protein 82.8%, FSH 72.9%.

Table 1: Relative potencies and recoveries of the fractions obtained by chromatography on a DEAE-Cellulose column

| Fraction | Proteins mg.* | FSH | | LH | | FSH/LH ratio |
|---|---|---|---|---|---|---|
| | | i.u./mg | Fiducial limits (P = 0.95) | i.u./mg | Fiducial limits (P = 0.95) | |
| Precursor (Pergonal-35E146-2) | 1200 | 133.4 | 109.3-163.2 | 22.2 | 11.9-46.2 | 60.0 |
| 1 | 3 | — | — | — | — | — |
| 2 | 169.4 | 95.0 | 70.7-157.5 | — | — | — |
| 3 | 240.0 | 369.4 | 272.0-458.7 | — | — | — |
| 4 | 582.0 | 20.8 | 11.8-28.6 | — | — | — |
| | 994.4 | | | | | |

*Dry weight after desalting and lyophilization.

It should be noted that most of the FSH activity was recovered in fraction 3 obtained by pooling the tubes corresponding to the symmetric part of the peak. In Table 1 it should be noted that the FSH potency of fraction 3 was approximately three times as high as the precursor.

Second Step: Gel Filtration on Sephadex G-100. The most potent fraction obtained by the first step (E161-3) was used for the gel filtration on Sephadex G-100 column. 236 mg of this material were applied to the 3.4 x 150 cm Recychrom LKB 4900 A column

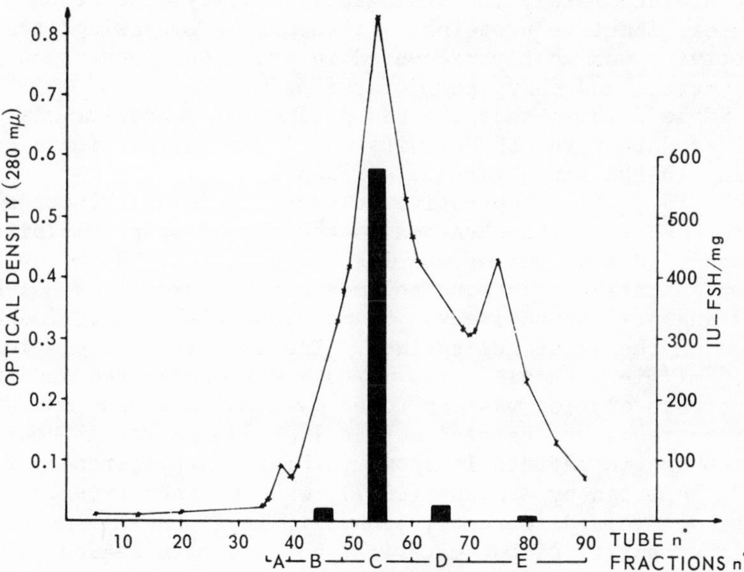


Fig.2: Gel filtration of partially purified urinary FSH (E161-3) on Sephadex G-100. 3.4x150 cm column equilibrated with 0.1 M potassium phosphate buffer pH 7.0 containing 0.3 M sodium chloride. 236 mg of protein applied to column; 10 ml/tube. Recovery total protein 87.5%, FSH 64.7%.

Table 2: Relative potencies and recoveries of the fractions obtained by gel filtration on Sephadex G-100

| Fraction | Proteins mg.* | FSH | | LH | | FSH/LH ratio |
|---|---|---|---|---|---|---|
| | | i.u./mg | Fiducial limits (P = 0.95) | i.u./mg | Fiducial limits (P = 0.95) | |
| Precursor (E161-3) | 236 | 369.4 | 272.0-458.7 | — | — | — |
| A | 5.0 | — | — | — | — | — |
| B | 18.8 | 19.1 | 12.7-24.8 | 90.0 | 42.5-161.1 | 0.21 |
| C | 94.2 | 582.8 | 481.3-702.0 | — | — | — |
| D | 41.5 | 25.6 | 12.4-37.3 | — | — | — |
| E | 47.2 | 4.3 | 2.5-5.9 | — | — | — |
| | 206.7 | | | | | |

*Dry weight after desalting and lyophilization.

packed with Sephadex G-100 previously equlibrated with the buffer solution pH 7.0. The flow rate was 34 ml/h and the samples 10 ml/tube. Before pooling, samples of the different tubes arbitrarily combined, were taken and the FSH activity was assessed by calculating the doses for bioassay on the basis of u.v. absorbancy at 280 mμ. The tubes were then combined into 5 fractions, which, after dialysis against distilled water at 3°C for 8 hours, were desalted on Sephadex G-25 and lyophilized. As illustrated in Fig. 2, the chromatographic pattern showed two peaks. The first peak contained mainly the biological activity, the second one biological inactive proteins. It should be emphasized that the FSH activity was mostly recovered in fraction C corresponding to the symmetric and sharp part of the peak.

Table 2 shows that the FSH potency of precursor was almost doubled and that the LH activity was found in fraction B corresponding to the first part of effluent.

Third Step: Preparative Polyacrylamide Gel Electrophoresis. The best FSH fraction obtained in the second step (E161bis-C) containing 582.8 i.u. per mg was further purified. Both the resolving and concentrating gels were polymerised by using Tris-HCl buffer at pH 8.7 and 6.7 respectively. 91 mg of partially purified FSH were layered on the upper gel surface. The flow rate of elution buffer was 0.6 ml/min. Samples of 4 ml/tube were collected and the protein pattern elution was monitored by reading at 280 mμ. The electrophoresis was carried out at 1°C, 300 mA for 17 hours. The electrophoretic pattern is shown in Fig. 3. Differently from the results reported by Stevens (1967), who obtained five peaks, we had two peaks. This difference, probably, is depending in part from the higher purity of our material, but the main reason should be the poor resolving capacity of our electrophoretic apparatus compared with Stevens' one. The first peak was unpolymerised acrylamide, which absorbed at 280 mμ, but contained a very small amount of protein. The second peak was arbitrarily divided in four fractions, which, after dialysis for 6 hours against distilled water

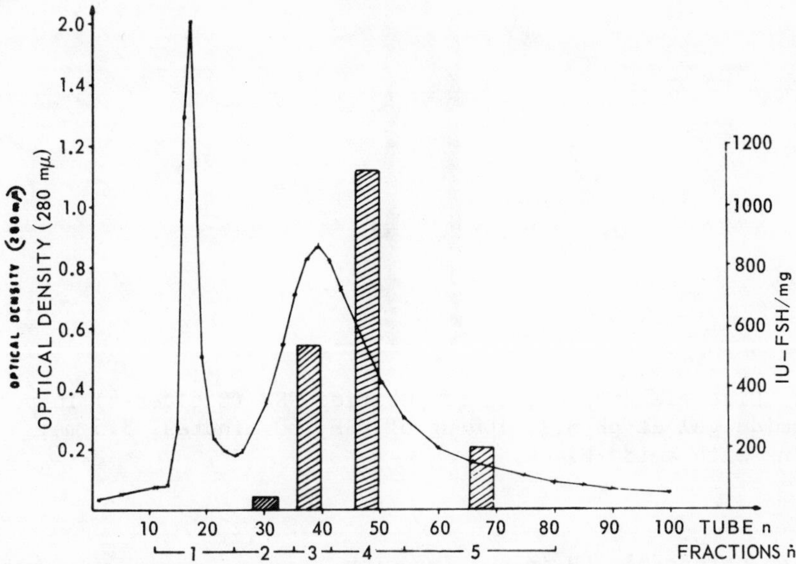


Fig.3: Preparative polyacrylamide gel electrophoresis of partially purified urinary FSH (E161bis-C). 91 mg of protein applied to gel column; 4 ml/tube. Recovery total protein 88.5%, FSH 95.3%.

at 3°C, were desalted on Sephadex G-25 and lyophilized.

Table 3 shows that the FSH activity was distributed in all fractions but mainly contained in fraction 4, which had 1115 i.u. and 1520 i.u. per mg, as determined by the biological and radioimmunological methods respectively. A low LH contamination was found in fraction 3 and 4 but the LH activity was mostly recovered in the slowest fraction, which had 99.2 i.u. LH per mg. These results are different from thos reported by Stevens (1967), who did

Table 3: Relative potencies and recoveries of the fractions obtained by preparative polyacrylamide gel electrophoresis of the partially purified FSH (E161bis-C)

| Fraction | Proteins mg.* | BIO-FSH | | BIO-LH | | FSH/LH ratio | Radioimmuno-FSH | |
|---|---|---|---|---|---|---|---|---|
| | | i.u./mg | F.L. (P = 0.95) | i.u./mg | F.L. (P = 0.95) | | i.u./mg | F.L. (P=0.95) |
| Precursor (E161bis-C) | 91.0 | 582.8 | 481.3-702.0 | — | — | — | — | — |
| 1 | 1.8 | — | — | — | — | — | — | — |
| 2 | 13.2 | 40.9 | 32.1-49.0 | — | — | — | — | — |
| 3 | 22.8 | 513.4 | 385.9-652.2 | 1.9 | 0.66-3.50 | 270.0 | 468.0 | |
| 4 | 32.5 | 1115.0 | 924-1397 | 4.3 | 1.7-7.8 | 259.3 | 1520.0 | |
| 5 | 10.3 | 199.7 | 130.7-263.6 | 99.2 | 55.0-261.1 | 2.0 | 418.0 | |
| | 80.6 | | | | | | | |

*Dry weight after desalting and lyophilization.

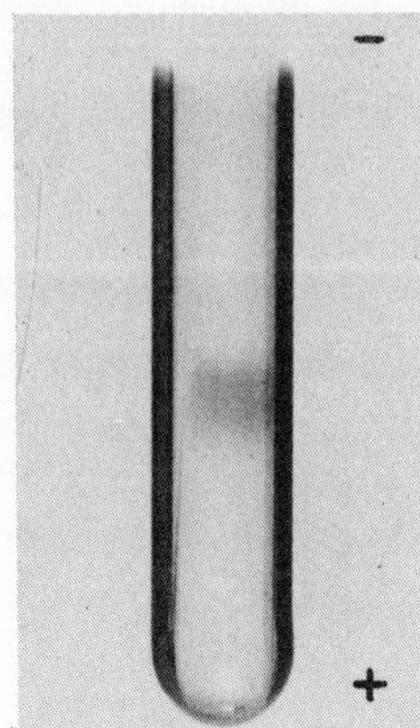


Fig.4: Disc electrophoresis of purified FSH (E161ter-4) in polyacrylamide gel at pH 9.1; 100 μg of FSH; 60 minutes, 3.5 mA; staining with amido-black.

not find detectable LH in any fraction.

Physicochemical and Immunological Characterization

The most potent FSH preparation (E161ter-4) was studied utilizing some of the phsicochemical and immunological methods commonly used to investigate the homogeneity of proteins.

Analytical Disc Electrophoresis. The FSH preparation subjected to polyacrylamide gel electrophoresis according to the method of Davies (1964) and Ornstein (1964) at pH 9.1 and 1°C showed two bands, as illustrated in Fig. 4.

Ultracentrifugation. The ultracentrifugation was carried out in the Spinco Mod.E ultracentrifuge at 59780 rpm, using 0.6% solution of FSH in 0.1 M phosphate buffer., pH 6.5. Fig. 5 shows that

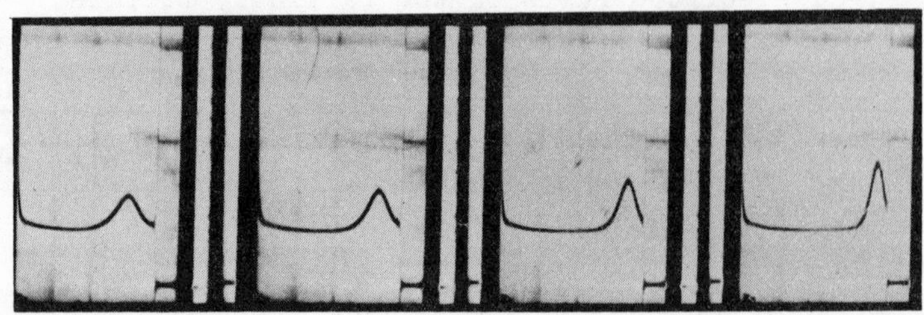

Fig. 5: Sedimentation pattern of purified FSH (E161ter-4) at a concentration of 0.6% in 0.1 M phosphate buffer, pH 6.5 at 59.780 rpm. The direction of sedimentation is toward the left.

the substance gave one rather symmetric peak. $S_{20,w}$ value was calculated to be 1.94 S. The constant diffusion ($D_{20,w}$) of the substance was found to be 5.98 x $10^{-7}$. The molecular weight calculated on the basis of these data was 31632.

Ultraviolet Absorption Spectrum. The optical densities of 0.0466% solution of FSH were 0.205 at 250 mμ, 0.237 at 260 mμ, 0.305 at 270 mu, 0.334 at 277 mμ, 0.331 at 280 mμ, and 0.231 at 290 mu. The absorbancy of 1% FSH solution at 277 mμ was calculated to be 7.17.

Immunolectrophoretic Analysis. The immunoelectrophoretic studies were carried out as described by Donini et al. (1964a). Unadsorbed and adsorbed anti-HMG sera were used. The absorption of the antiserum was performed with both HCG (3000 i.u./mg) and Kaolin-acetone extract of children urine, according to the method of Tamada et al. (1967). Any fraction obtained by the last step of purification did not show precipitin line when assayed with antiserum raised against human albumin.

Fig. 6 shows the immunoelectrophoretic pattern of some urinary extracts. Before absorption the anti-HMG gave some lines against a roughly purified Pergonal (40 i.u. FSH and 41 i.u. LH per mg) and children urinary extract, while one line was obtained against the

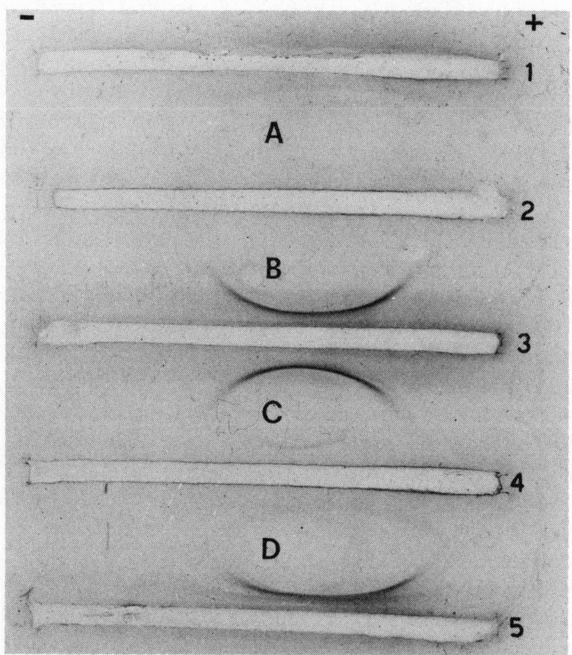


Fig. 6: Agar immunoelectrophoresis. Wells A,B,C and D filled with 100 μg of children urinary extract, HCG, purified FSH (E161ter-4) and crude HMG. Troughs filled with unadsorbed (1,3,5) and adsorbed (2,4) antiserum to HMG.

HCG (3000 i.u. per mg) and two lines against the purified FSH. After absorption two lines against the children urinary extract and HCG disappeared, the highly purified FSH gave one line and the less purified HMG two lines.

## Discussion

In recent years, several attempts to isolate a highly purified human FSH have been reported. We should like to mention some of the last papers and further references are given in a recent review by Harris and Donovan (1966). At present, the most potent FSH preparations from human pituitaries have been reported by Reichert & Parlow (1964) and by Roos & Gemzell (1965). Both the preparations were not isolated from the solution and their potencies have been calculated on the basis of u.v. absorption. The FSH of Roos & Gemzell was homogeneous in both the ultracentrifuge and the free-zone electrophoresis. The human pituitary FSH isolated by Butt et al. (1966) in the dry state contained approximately 1000 i. u. FSH and 180 i.u. LH per mg. This material, further purified by repeated chromatography on DEAE-Sephadex (Amir et al. 1966) behaves as a homogeneous protein on Sephadex filtration and the potency was presumed to be not more than 1500 i.u. FSH per mg.

Over the findings on the purification of urinary FSH reviewed by some of us (Donini et al. 1966a), Roos and Gemzell (1965) reported on the urinary FSH, which has not been isolated in the dry state. This preparation was claimed to be homogeneous in both the ultracentrifuge and in the free-zone electrophoresis. Donini (1967) and Stevens (1967) reported on their urinary FSH containing 597.4 and 789.1 i.u. FSH per mg respectively. These two preparations, which have been isolated in the dry state, were not tested for homogeneity.

Concerning the results of this study it should be noted that in the gel filtration on Sephadex we have used only the proteins corresponding to the symmetric part of the peak obtained in the first step of purification and eluted with the buffer containing 0.1 M NaCl. By this procedure, one run on Sephadex yielded material free from albumin and containing 582.8 i.u. FSH per mg. Furthermore, we have confirmed our results previously reported (Donini et al. 1966a, 1967) that LH moves faster than FSH during the gel filtration on Sephadex. Our chromatographic pattern differs from that of Stevens (1967), who found the LH activity in the last part of effluent. The preparative polyacrylamide gel electrophoresis yielded material containing 1115 i.u. FSH and 4.3 i.u. LH per mg, as determined by the biological methods, but the potency was 1520 i.u. FSH per mg, when determined by the radioimmunological method. We would like to point out that our FSH preparation was 12 times as potent as the starting material and that FSH/LH ratio arose from 60.0 to 259.3. The ultracentrifugal studies showed that the FSH was monodisperse and the $S_{20,w}$ value was calculated to be 1.94 S, which agrees very well with 1.95 S obtained by Roos and Gemzell (1965) for their urinary FSH. Although the ultracentrifugation indicates homogeneity of our FSH, its behaviour at the disc electrophoresis and at the immunoelectrophoresis demonstrates the inhomogeneity.

One of the most difficult problems in the purification of FSH is the complete separation from LH. A new approach, based on the immunological binding of LH and subsequent chromatography on DEAE-C, was attempted by some of us (Donini et al. 1966b). The apparently biological pure FSH, isolated in the dry state, did not produce uterine weight response when 128 mcg of this material (equivalent to 42 i.u. FSH) were injected into intact immature mice. Eshkol & Lunenfeld (1967) have also studied its effect on ovarian and uterine hystology in the mice demonstrating the follicular growth, while the uterus remained infantile. Although the results of some tests used for stating the homogeneity showed that the urinary preparation described in this study was not homogeneous, it should be pointed out that its FSH potency was one of the highest reported till now.

The authors wish to thank Dr. B. Lunenfeld, Institute of Endocrinology, Tel-Hashomer Government Hospital (Israel), for the ultracentrifugal studies and for the diffusion constant determination.

We are indebted to Dr. A. R. Midgley, Ann Arbor, Michigan, for the radioimmunological assay of FSH.

## Bibliography

Amir, S.M., Barker, S.A. Butt, W.R., and Crooke, A.C. (1966). Nature, 209, 1092.

Borth, R., Diczfalusy, E., Heinrichs, H.D. (1957). Arch. Gynäk. 188, 377.

Butt, W.R., Crooke, A.C., and Wolf, A. (1965). Ciba Found. Study Group, 22, 85.

Davies, B.J. (1964) Ann. N.Y. Acad.Sci. 121, 404.

Donini, P., Montezemolo, R., and Puzzuoli, D. (1964). Acta Endocr. 45, 321.

Donini, P., Puzzuoli, D., D'Alessio, I., Lunenfeld, B., Eshkol, A., and Parlow, A.F. (1966a) Acta Endocr. 52, 169.

Donini, P., Puzzuoli, D., D'Alessio, I., Lunenfeld, B., Eshkol, A., and Parlow, A.F. (1966b) Acta Endocr. 52, 186.

Donini, P. (1967). In "Recent Research on gonadotrphic hormones". Ed. E.T.Bell and J.A.Loraine, Publ. E&S Livingstone, 1967, p. 134.

Eshkol, A. and Lunenfeld, B. (1967). Acta Endocr. 54, 91.

Harris, G.W. and Donovan, B.T. (1966). "The Pituitary Gland: Anterior Pituitary. Publ. University of California Press.

Ornstein, L. (1954). Ann. N.Y. Acad.Sci. 121, 321.

Parlow, A.F. (1961). In "Human Pituitary Gonadotrophins: A Workshop Conference. Ed. A. Albert, Springfield, Ill. p. 300.
Reichert, L.E. and Parlow, A.F. (1964). Proc. Soc. Exptl. Biol. Med., 115, 286.
Roos, P. and Gemzell, C.A. (1964a). Biochim. Biophys. Acta 82, 218.
Roos, P. and Gemzell, C.A. (1964b). Biochim. Biophys. Acta 93, 217.
Roos, P. and Gemzell, C.A. (1965). Ciba Found. Study Group 22, 11.
Scheidegger, J.J. (1955). Int. Arch. Allergy 1, 103.
Steelman, S.L. and Pohley, F.M. (1953). Endocrinology 53, 604.
Stevens, V.C. (1967). In "Recent Research on Gonadotrophic Hormones". Ed. E.T. Bell and J.A. Loraine, Publ. E&S Livingstone, 1967, p. 142.
Tamada, T., Super, M., and Taymor, M.L. (1967). J. Clin. Endocr. 27, 379.

# THE TREND OF ANDROGEN METABOLITES AFTER OVULATION INDUCED BY FSH AND LH FRACTIONS OR BY HCG ALONE

Renzo Grattarola

Istituto Nazionale dei Tumori-Milano-Italy

In our earlier study (C. R. Acad. Sc. Paris, 260:6698, 1965) we treated 3 patients suffering from Stein-Leventhal syndrome with partially purified FSH and LH fractions obtained from human postmenopausal urinary gonadotropin (HMG) in order to induce ovulation. In 2 cases ovulation was induced, as was shown by the presence of a recently formed corpus luteum in the ovary, and the urinary levels of 11-deoxy-17-ketosteroids decreased after the treatment.

On the other hand the urinary excretion of 11-deoxy-17-ketosteroids was steadily increased in the patient in whom ovulation was not induced and in the ovary a marked hyperplasia of the theca cells was observed.

More recently (J. Endocrin. 38:77, 1967) we observed that in women without ovulatory cycle and whose endometrium obtained at premenstrual period was hyperplastic, the urinary 11-deoxy 17-ketosteroid level was significantly higher than in control subjects with ovulatory cycle and whose premenstrual endometrium was showing secretory activity (Table 1).

TABLE I

| premenstrual endometrium | Il-deoxy-17-ketosteroids mgr./24 hr. | P value |
|---|---|---|
| progestational | 2.24 ± 0.14 | < 0.01 |
| hyperplastic | 3.77 ± 0.44 | |

There seems to be a correlation between the absence of progestational activity and the increased excretion of urinary 11-deoxy-17-ketosteroids (dehydroepiandrosterone, etiocholanolone, androsterone) in the absence of ovulation.

In this paper we study the behavior of urinary androgen metabolites of patients without ovulatory cycle after treatment with a combination of FSH and LH fractions prepared from urine of postmenopausal women (Jutisz M. et al. C. R. Acad. Sc. Paris 259:1195, 1964) following the scheme we used in patients with the Stein-Leventhal syndrome.

7 Patients were treated with this combination (2 patients suffering from sterility, 2 with Stein-Leventhal syndrome and 3 patients with breast carcinoma).

At the end of treatment in 3 patients a marked decrease in urinary level of 11-deoxy-17-ketosteroids was observed as well as an increase in pregnanediol and the appearance of a secretory endometrium.

1 patient with Stein-Leventhal syndrome 3 months after treatment became pregnant.

In 4 patients the endometrium showed a more marked hyperplastic activity together with an increased excretion of urinary 11-deoxy-17-ketosteroids, after the treatment (Table 2).

TABLE 2

Urinary II-deoxy-17-ketosteroids (dehydroepiandrosterone (D), etiocholanolone (E), androsterone (A) excretion after treatment with combination FSH and LH fractions (HMG)

| | Endometrium | Urinary excretion (mgr./24 hr.) | | | Total II-deoxy-17-ketosteroids (mgr./24 hr.) |
|---|---|---|---|---|---|
| | | D | E | A | |
| before treatment | hyperplastic | 1.26 | 2.40 | 2.10 | 5.76 |
| after treatment | progestational | 0.49 | 1.54 | 1.37 | 3.40 |
| before treatment | hyperplastic | 0.59 | 1.44 | 1.33 | 3.36 |
| after treatment | hyperplastic | 1.53 | 1.69 | 1.63 | 4.85 |

6 patients with hyperplastic pattern of premenstrual endometrium (5 patients with breast cancer. 1 patient with mastopathy) were treated with human chorionic gonadotropin given alone (HCG was given 10 times on alternate days, at daily dose of 2,000 i.u. starting the treatment 5 days following the menstrual period.

In all the treated patients there was together with the increase of pregnanediol excretion and the appearance of a secretory

TABLE 3

Urinary II-deoxy-17-ketosteroids (dehydroepiandrosterone (D), etiocholanolone (E) androsterone (A) in patients with endometrial hyperplasia treated with HCG (20.000 i.u.).

| | Endometrium | Urinary excretion (mgr./24 hr.) | | | Total II-deoxy-17-ketosteroids (mgr./24 hr.) |
|---|---|---|---|---|---|
| | | D | E | A | |
| before treatment | hyperplastic | 1.36 | 1.80 | 1.05 | 4.21 |
| after treatment | progestational | 0.68 | 1.11 | 1.19 | 2.98 |

endometrium, a decrease in urinary 17-ketosteroids (Table 3).

It seems that the induction of progestational activity results in a decrease in androgenic activity.

For further confirmation of this observation 10 breast carcinoma patients whose premenstrual endometrium was hyperplastic were treated with a gestagen (17α-acetoxy-progesterone cyclopentyl enol ether) which at the oral dose of 20 mgr. daily for 10 days, induces a complete secretory transformation of the endometrium.

After the treatment we observed a marked decrease in urinary androgen metabolites (Table 4).

TABLE 4

Urinary II-deoxy-17-ketosteroids (dehydroepiandrosterone (D), etiocholanolone (E), androsterone (A) excretion after treatment with 17α-acetoxyprogesterone cyclopentyl enol ether in 10 patients with breast cancer and hyperplastic premenstrual endometrium.

| | Endometrium | Urinary excretion (mgr./24 hr.) | | | Total II-deoxy-17-ketosteroids (mgr./24 hr.) |
|---|---|---|---|---|---|
| | | D | E | A | |
| before treatment | hyperplastic | 1.06 | 1.79 | 1.32 | 4.17 |
| after treatment | progestational | 0.83 | 1.32 | 0.80 | 2.95 |

# SOME FACTORS INFLUENCING SUCKLING-INDUCED PROLACTIN RELEASE IN THE LACTATING RAT

C.E.Grosvenor, F.Mena, D.A.Schaefgen, A.P.S.Dhariwal
J.Antunes-Rodriguez, S.M.McCann

Departments of Physiology, Univ. of Tenn. Medical School
Memphis, and Univ. of Texas South Western Medical School
Dallas

The association of the suckling stimulus with alterations in prolactin secretion stems from experiments with rats possessing both intact and duct-ligated or teat-excised mammary glands in which the suckling of intact glands retarded involution of the glands from which milk removal was prevented (1,2). The retarding effect of suckling upon mammary involution was mimicked by systemic injections of prolactin (3). This effect is not specifically due to prolactin, since STH, adrenal corticoids, and oxytocin also retard the rate of mammary gland involution (see 4).

More direct evidence that suckling influences prolactin secretion came from measurements of the prolactin content or concentration in the pituitary. The pituitary content of prolactin was shown by Meites and Turner (5,6) to be higher in rats and in rabbits which had been suckled for several days than in rats or rabbits which had their pups removed at birth. They showed also that the prolactin content fell to the prepartum level much more rapidly in non-suckled than in suckled mothers (6,7).

Reece and Turner (8) demonstrated that 3 hours of suckling following 12 hours of non-suckling lowered the pituitary prolactin content of rats by about one-half in comparison with that of rats non-suckled for 15 hours. Holst and Turner (9) obtained similar results in rabbits and in guinea pigs. Grosvenor and Turner (10) then demonstrated in postpartum day 14 rats that 30 minutes of suckling following 10 hours of non-suckling reduced the pituitary prolactin concentration from one-half to one-third of that of rats non-suckled for 10 hours; restoration to pre-suckled values, on the other hand, occurred much more slowly. The reduction in pituitary prolactin stores following short-term nursing in the rat on postpartum day 14 was confirmed (11-18) and extended to days

2, 6, and 10 of lactation (13).

Pasteels (19), using light microscopy, observed that the prolactin cells of anterior pituitary glands of lactating rats became engorged with granules when the mothers were separated from their litters for 10 hours. Thirty minutes of nursing caused a disappearance of the granules from the cells but did not affect the size or number of the prolactin cells.

Pasteels (19), using electron-microscopy, also observed that the excretion of prolactin granules into the perisinusoidal spaces occurred at the pole of the cell. The granules quickly lost their individuality once they were excreted. Smith (R.E. Smith, personal communication), using electron-microscopic techniques, found that a short period of suckling following a period of non-suckling in rats results in a quick extrusion of prolactin granules into the pericapillary and intercellular spaces of the pituitary gland. The granules continued to be extruded for about one hour after the cessation of suckling.

Thus it would appear that the <u>immediate</u> effect of a single suckling episode, i.e. acute suckling, is to reduce the prolactin content of the anterior pituitary gland with a release of the hormone into the circulation. The overall importance of suckling to lactation has been adequately reviewed (20-24).

## METHODS AND RESULTS

Primiparous rats of the Sprague Dawley-Rolfsmeyer strain were used in all studies. Except where noted the litter size was adjusted to 8 pups shortly after birth and to 6 pups on postpartum day 4. On the day of the test (postpartum day 7 or 14) the pups were separated from their mothers for 8, 10, or 16 hours after which they were replaced and allowed to nurse for periods up to 30 minutes. The mothers and pups were killed a few minutes after nursing was completed. The mothers' anterior pituitary glands were rapidly removed, weighed and frozen until assayed for prolactin. The contents of the pups' stomachs were weighed to the nearest 0.1 g to assess the extent of milk ejection. The extent of release of prolactin from the pituitary gland as a result of suckling was estimated by comparing the prolactin concentration in the pooled pituitary glands of rats which had been suckled with those which had been isolated but not suckled. Prolactin concentration was determined by 4-point assay using the intradermal crop-sac method as described in detail previously (12). The standard tests for parallelism, assay validity and precision were made (25) and the data were expressed either as relative potency or as a percent of the non-suckled control.

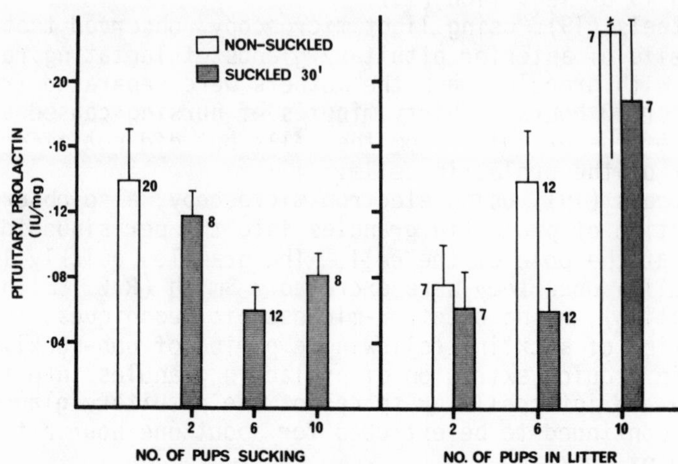


Fig. 1. Effect of number of pups suckling on pituitary prolactin concentration on postpartum day 7. Suckling for 30 min following 8 hrs of non-suckling. Left panel: Litters with 6 pups until time of suckling. Right panel: Litters adjusted to 2, 6, or 10 pups at parturition. Numbers refer to number of rats; vertical bars to standard error of the assay. (From Mena & Grosvenor, 17)

## Effects of Intensity and Duration of Suckling and Non-suckling Intervals upon Release of Prolactin

Litter size was adjusted shortly after parturition to contain 2, 6, or 10 pups. The pituitary prolactin concentration after 8½ hours of non-suckling on postpartum day 7 increased approximately 3-fold with an increase in the size of the litter from 2 to 10 pups (Fig. 1). This finding agrees well with cytological and ultrastructural evaluation of pituitary cells of the rat in relation to suckling strength (19) and suggests that suckling stimulation and perhaps the behavioral stimuli associated with it can modulate the secretory capacity of the prolactin cell in relation to the strength of the stimulation. Meites et. al. (26) found, however, that rabbits nursing large numbers of young (5-11) had no more prolactin in their pituitary glands than did those nursing only 2 young.

In two experiments significant reductions in prolactin concentration were provoked by 6 pups suckling whereas suckling by 2 pups provoked slight but insignificant falls (Fig. 1). We do not exclude the possibility, however, that a small, albeit physiologically important, quantity of prolactin was released by 2 pups suckling, for it is known (27) that nursing of small litters exerts effects on the reproductive tract which are clearly ascribable to prolactin. The effect of suckling by 10 pups upon the

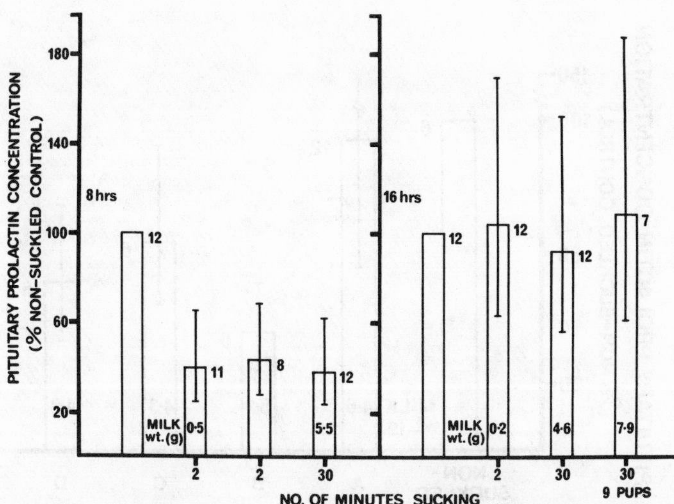


Fig. 2. Effect of duration of suckling and non-suckling interval upon pituitary prolactin concentration on postpartum day 7. Left panel: 6 pups not-suckled for 8 hrs prior to suckling. Right panel: 6 or 9 pups not-suckled for 16 hrs prior to suckling. Numbers refer to numbers of rats; vertical bars to 95% confidence limits. (From Grosvenor et. al., 15)

concentration of pituitary prolactin was equivocal. A significant fall occurred in one of two experiments. There is the possibility, however, that because of crowding and competition for nipples, an insufficient number of nipples may have been simultaneously suckled in the second experiment to have effectively released normal amounts of prolactin.

Two minutes suckling by 6 pups following 8 hours of non-suckling induced a fall in pituitary prolactin concentration of the same magnitude as that which occurred as a result of 30 minutes suckling by the same number of pups (Fig. 2). The similarity in the fall in prolactin concentration occurred, though much greater amounts of milk were obtained during the 30 minute suckling than during the 2 minute suckling period.

No fall in pituitary prolactin concentration occurred in response to either 2 or 30 minutes suckling by 6 pups or by 30 minutes suckling by 9 pups when the mothers were previously non-suckled for 16 hours (Fig. 2). The pups obtained normal quantities of milk which indicates that the suckling-generated stimulus for oxytocin release was reaching the central nervous system. The failure of suckling to alter prolactin concentration in the pituitary while eliciting normal milk ejection also provides additional evidence that prolactin is not released in response to oxytocin mediation as has been postulated (see 4 for review).

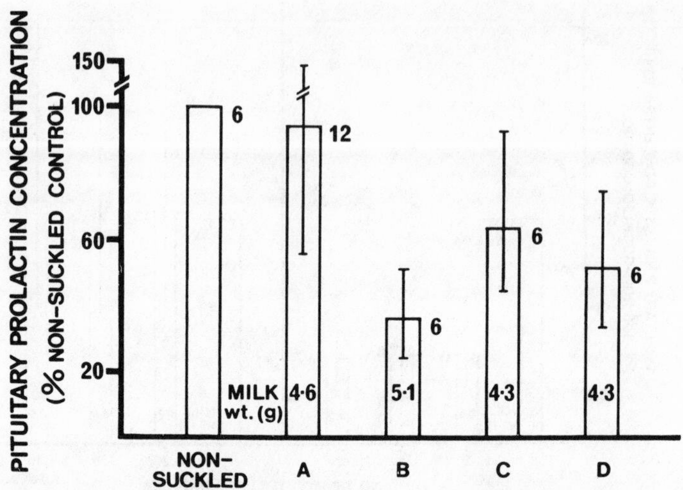


Fig. 3. Effect of suckling stimulation of the mammary gland during isolation of mother and 6 pups upon the ability of a terminal 30 min nursing to reduce pituitary prolactin concentration. 16 hr isolation was uninterrupted (A) or interrupted by a 30 min suckling period 8 hrs (B), by 5 min periods 4 & 8 hrs (C), or by 30 min periods 4 & 8 hrs after onset of isolation. Data obtained on postpartum day 7. Numbers refer to numbers of rats; vertical bars to 95% confidence limits. (From Grosvenor et. al., 15)

Interestingly, bleeding stress was able to reduce pituitary prolactin concentration in 14 day lactating rats following 16 hours of non-suckling (12).

The fall in pituitary prolactin concentration in response to suckling was restored when one or two 5 minute or 30 minute periods of suckling stimulation were provided to the mammary glands during the first 8 hours of the 16 hour isolation period (Fig. 3). On the basis of these data we propose that in the rat the mechanism involved in the release of prolactin in response to suckling operates efficiently only if the suckling stimulus is applied periodically and at short intervals. If this hypothesis is correct, the progressive reduction in the incidence of suckling in the rat as lactation proceeds, i.e. the progressive increase in the non-suckling interval (28) might induce a gradual alteration in the threshold for prolactin release in response to suckling to the point where re-establishment of the reflex is difficult. This event would occur as a concomitant to other factors, e.g. stagnation of milk with reabsorption of secretory tissue and also a reduction in the synthesis and storage of prolactin, which contribute to the cessation of lactation.

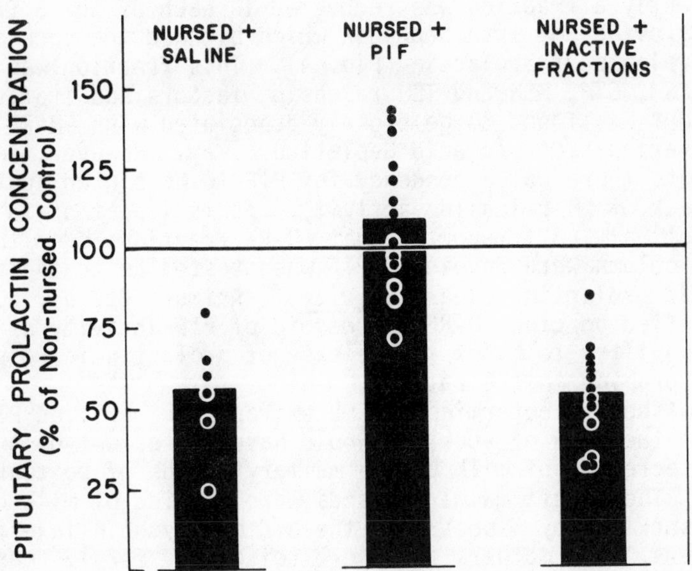


Fig. 4. Blocking effect of intraperitoneal injection of purified fractions of ovine SME upon suckling-induced fall in pituitary prolactin concentration on postpartum day 14. Suckling for 30 min by 6 pups following 10 hrs of non-suckling. Each point refers to an individual assay. (From Dhariwal et. al., 34)

## Effect of Hypothalamic Extracts Upon Suckling-Induced Prolactin Release

A considerable body of evidence from a variety of *in vivo* and *in vitro* experiments indicates that pituitary secretion of prolactin is unrestrained when the pituitary is disconnected from the hypothalamus (see 21, 29). These data have led to the hypothesis that the hypothalamus chronically holds prolactin release in check. Pasteels (30) and Talwalker et. al. (31) showed that a decreased release of prolactin from pituitary tissue incubated *in vitro* occurred following the addition of hypothalamic extracts to the medium. This effect was presumed to be due to a specific prolactin-inhibiting neurohumor (PIF) since known pharmacologically active agents were ineffective when added to the system (31). We provided *in vivo* support for the PIF concept by demonstrating that bovine stalk-median eminence (SME) but not cortex would inhibit the suckling-induced fall in pituitary concentration of prolactin if injected a few minutes before the onset of suckling (14). It was then found that the extracted PIF from as little as one-third of a rat SME would inhibit suckling-induced release of prolactin in lactating rats (12) suggesting that rat PIF was remarkably potent.

Recently a fraction was recovered in each of the 5 fractionations of ovine SME with Sephadex which blocked the suckling-induced release of prolactin (Fig. 4). This fraction was separated from FSH, STH, MSH and TSH releasing factors and from vasopressin but was found to be closely associated with LH-RF as judged by the ovarian ascorbic acid depletion assay. However, in 3 of 5 experiments there was a tendency for PIF to be eluted just prior to the peak of LH-releasing activity. It is important to note that Schally et. al. (32) reported that LH-RF fractions from their Sephadex column were devoid of PIF when tested as to their ability to inhibit prolactin release *in vitro*. Arimura et. al. (33) found that purified porcine LH-RF was devoid of PIF activity as judged by its inability to block the release of prolactin in response to cervical probing in the rat.

We wished to determine if SME extracts or purified PIF injected at the time of suckling would have any effect on the subsequent secretion of milk by the mammary glands of postpartum day 14 rats. The 6 left mammary glands were emptied of milk during 30 minutes suckling by 6 pups with the aid of oxytocin injected to the mothers. The mothers then were reisolated for 16 hours when the pups again were permitted to suckle. Milk yield was estimated from the weight loss of the mother and the weight gain of the young.

The injection of either crude rat SME extract or purified ovine PIF (T-135) at the time of suckling resulted in a significant reduction in the amount of milk secreted by the rat mammary gland during a subsequent 16 hour period of non-suckling, in comparison with that of glands of control rats injected with saline, ammonium acetate, rat cerebral cortical extracts or ovine hypothalamic fractions (T-153, T-155) outside the zone of PIF activity (Fig. 5). These data, in effect, imply that in control rats the prolactin discharged by short-term suckling (30 minutes) was sufficient to trigger nearly total reaccumulation of milk by the mammary gland.

The mechanism whereby milk secretion was inhibited by SME extracts is not known but probably is related, at least in part, to their demonstrated effect in blocking the suckling-induced release of prolactin (12, 14, 34). This supposition is supported by the observation that injection of prolactin to SME-treated rats restored milk reaccumulation to normal (Fig. 5). A peripheral interference by PIF with milk secretion is possible, of course, but is not deemed likely. The possibility that the inhibitory effect was related to other known releasing factors in the extract is rendered unlikely by the use of purified PIF in the one experiment. Although the purified fraction (T-135) employed did contain significant LH-releasing activity, any influence of LH- or FSH-releasing factor acting via release of LH and FSH was eliminated, since the rats were ovariectomized prior to the start of the experiment.

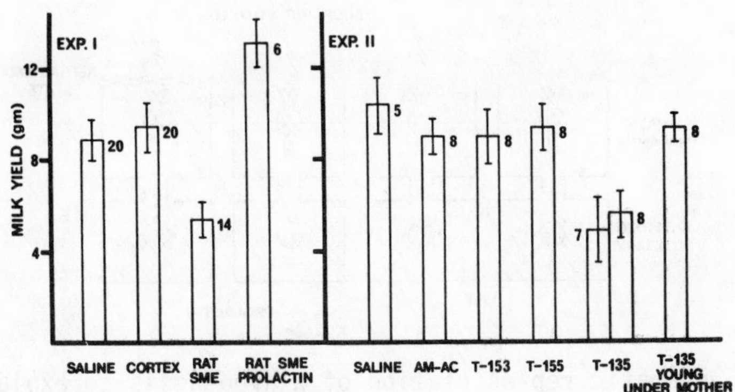


Fig. 5. Effect of rat SME (left panel) and of purified ovine PIF (T-135) (right panel) upon milk secreted 16 hrs subsequent to 30 min suckling and emptying of the mammary glands by pups on post-partum day 14. SME or PIF injected subcutaneously before suckling. Pups isolated from mothers in all groups but one during 16 hr period of milk secretion; in that group the pups were placed under but not in physical contact with their mother. Numbers refer to number of rats; vertical bars to standard error of assay. (From Grosvenor et. al., 35)

We were unable, however, to effect complete blockade of milk secretion following injection of either crude SME extract or purified PIF which suggests that mechanisms other than that of suckling may act to release prolactin in the lactating rat (see 35 for discussion).

We observed that normal secretion of milk took place in rats whose pituitary prolactin release had presumably been blocked at the time of suckling when the pups were placed under - but not in contact with - their mothers during the 16 hour period of milk reaccumulation (Fig. 5). These data suggest that the presence of the pups underneath the mothers stimulated the release of prolactin after the effect of the PIF had worn off. These results are in accord with bioassay data of other experiments (11) in which, in postpartum day 14 rats, the presence of hungry pups for 30 minutes under the mothers, after litter isolation, provoked a fall in pituitary prolactin concentration to approximately the same extent as 30 minutes actual suckling.

## Discussion

Though there is reasonably strong _in vitro_ and _in vivo_ evidence for the involvement of PIF in the mediation of prolactin release by suckling, the mechanism by which the suckling stimulus

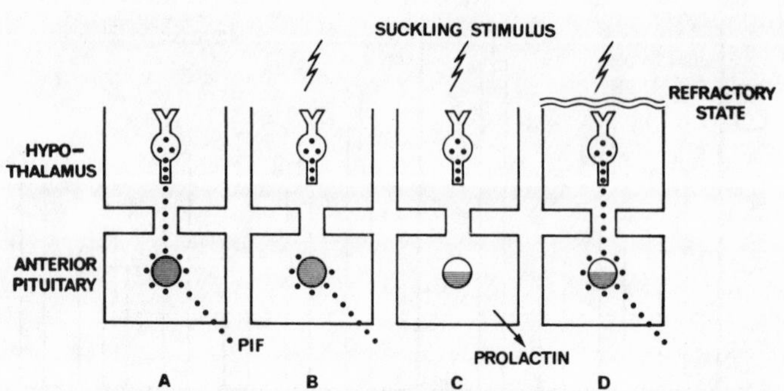


Fig. 6. Schematic representation of a hypothesis to explain the mechanism of suckling-induced prolactin release in the rat.

is able to erase the inhibitory control of PIF to permit prolactin release to occur is not known. Ratner and Meites (36) with an *in vitro* system could not detect PIF in hypothalami of lactating rats and concluded that depletion of hypothalamic PIF had occurred as a result of inhibition of synthesis of PIF by the suckling stimulus. Gala and Reece (37), also using an *in vitro* system, were able to detect PIF in hypothalami of lactating rats and in quantities which appeared to be comparable to that in hypothalami of non-lactating female rats. Lactating and non-lactating rat hypothalami, however, were not compared in the same experiment.

The PIF in the extract of as little as one-third of a hypothalamus prevented the suckling-induced fall in pituitary prolactin concentration regardless of whether the hypothalamus came from a rat whose pituitary prolactin level was high or from a rat whose prolactin level had been reduced by suckling or stress (12). In order to be fitted into the framework of action of a PIF, the data have been interpreted as indicating that short-term suckling suppresses the release rather than the synthesis of PIF. More recently, Minaguchi and Meites (38) found that anterior pituitaries incubated with SME from suckled rats released more prolactin than when incubated with those from cycling female rats and concluded that the release of PIF was depressed by suckling. However, it is possible that the initial effect of suckling upon PIF is one of inhibition of release with inhibition of synthesis occurring later on or perhaps after a more prolonged period of suckling. If this concept is true then it would be expected that inhibition of only the release phase would be evident if the hypothalami were harvested shortly after acute suckling. If hypothalami were obtained after more prolonged suckling, inhibition of synthesis more likely would be detected.

Our concept of one way in which suckling could effect the release of prolactin is shown schematically in Fig. 6. It is assumed firstly that a tonic secretion of PIF occurs in the absence of suckling and secondly, that PIF, because of its suspected small molecular size, probably passes rapidly through the pituitary. In effect, in the absence of suckling the pituitary prolactin cell would be tonically perfused with PIF-rich portal blood (Fig. 6). When the neural impulse of suckling reaches the hypothalamus, PIF release is reduced or perhaps shut off; the PIF released just prior to the suckling then quickly clears the pituitary; the prolactin cell, now free of restraint, discharges its contents into the bloodstream (Fig. 6 B,C). This scheme rather presupposes that once the level of PIF in the pituitary drops below the inhibitory threshold for prolactin release, the prolactin cell then discharges its contents into the circulation. It would logically be anticipated, however, that different threshold levels would exist in a population of prolactin cells. Operation of the system in the above manner would permit the rapid release of prolactin - within one minute after stress and within two minutes following suckling as have been demonstrated (14,15).

We suppose that the prolactin-releasing mechanism develops a refractory period following the initial rapid release of prolactin during which time further stimulation is ineffective and during which time the tonic release of PIF is restored (Fig. 6 D). However, it must be considered that suckling by more than 2 pups may constitute a sufficiently strong stimulus to release maximal quantities of prolactin from the pituitary. These suppositions follow from the observed inability of 30 minutes suckling by 6 pups to release more prolactin than 2 minutes suckling (24) and the inability of stress to release additional prolactin in the suckled rat (14), though in both cases large stores of assayable prolactin remained in the pituitary.

Exogenous PIF, in order to be effective, therefore, need only to be present in the pituitary in physiologically active quantities during the first few moments of suckling stimulation. (It is unlikely that a single systemic injection of PIF is physiologically effective for a very long time). This mechanism could explain the observed prolactin-blocking efficacy of a single injection of PIF for a 30 minute period of suckling (12,14,34) and for a 30 minute period of suckling followed by bleeding stress (14).

## References


1. Selye, H., Collip, J.B. and D.L. Thompson, Endocrinology, 18:237, 1934.
2. Selye, H., Amer. J. Physiol., 107:535, 1934.
3. Hooker, C.W. and W.L. Williams, Endocrinology, 28:42, 1941.
4. Meites, J., C.S. Nicoll and P.K. Talwalker, in Nalbandov, A.V. (ed), Advances in Neuroendocrinology, University of Illinois Press, Urbana, 1963, p. 238.
5. Meites, J. and C.W. Turner, Endocrinology, 30:726, 1942.

6. Meites, J. and C.W. Turner, Res. Bull. Mo. Agri. Exp. Sta. No. 416, 1948.
7. Meites, J. and C.W. Turner, Endocrinology, 31:340, 1942.
8. Reece, R.P. and C.W. Turner, Proc. Soc. Exp. Biol. and Med., 35:621, 1937.
9. Holst, S. and C.W. Turner, Proc. Soc. Exp. Biol. and Med., 42:479, 1939.
10. Grosvenor, C.E. and C.W. Turner, Proc. Soc. Exp. Biol. and Med. 96:723, 1957.
11. Grosvenor, C.E. Endocrinology, 76:340, 1965.
12. Grosvenor, C.E. Endocrinology, 77:1037, 1965.
13. Grosvenor, C.E. and C.W. Turner, Endocrinology, 63:535, 1958.
14. Grosvenor, C.E., S.M. McCann and R. Nallar, Endocrinology, 76:883, 1965.
15. Grosvenor, C.E., F. Mena and D.A. Schaefgen, Endocrinology, (in press) 1967.
16. Moon, R.C. and C.W. Turner, Proc. Soc. Exp. Biol. and Med., 101:332, 1959.
17. Mena, F. and C.E. Grosvenor, Endocrinology, (in press) 1967.
18. Grosvenor, C.E. and F. Mena, Endocrinology, 80:840, 1967.
19. Pasteels, J.L., Arch. Biol. Liege, 74:439, 1963.
20. Averill, R.L.W., Brit. Med. Bull. 22:261, 1966.
21. Meites, J. in Martini, L. and W.F. Ganong (eds), Neuroendocrinology, Academic Press, N.Y. 1966, ch. 16.
22. Lehrman, D.S. in Young, W.C. (ed), Sex and Internal Secretions Vol. II, Williams and Wilkins Co., Baltimore, 1961, ch. 21.
23. Cowie, A.T. and S.J. Folley, in Young, W.C. (ed), Sex and Internal Secretions, Vol. I, Williams and Wilkins Co., Baltimore, 1961, ch. 10.
24. Beyer, C. and F. Mena in Bajusz, E. (ed), The Physiology and Pathology of Adaptation Mechanisms. Neural and Endocrine Factors, Pergammon Press, Oxford, 1966, ch. 15.
25. Bliss, C.I., The Statistics of Bioassay, Academic Press, N.Y. 1952.
26. Meites, J., A.J. Bergman and C.W. Turner, Proc. Soc. Exp. Biol. and Med., 46:670, 1941.
27. Rothchild, I., Endocrinology, 67:9, 1960.
28. Grosvenor, C.E., Amer. J. Physiol., 186:211, 1956.
29. Everett, J.W., in Harris, G.W. and B.T. Donovan (eds), The Pituitary Gland, Butterworths, London, Vol. 2, 1966, p. 166.
30. Pasteels, J.L., C.R. Acad. Sci. (Paris), 352:3074, 1961.
31. Talwalker, P.K., A. Ratner and J. Meites, Amer. J. Physiol., 205:213, 1963.
32. Schally, A.V., A. Ratner and J. Meites, Proc. Soc. Exp. Biol. and Med., 117:252, 1964.
33. Arimura, A., T. Saito, E.E. Muller, C.Y. Bowers, S.Sawano and A.V. Schally, Endocrinology, 80:972, 1967.
34. Dhariwal, A.P.S., C.E. Grosvenor, J. Antunes-Rodriguez and S.M. McCann, Endocrinology (in press), 1967.

35. Grosvenor, C.E., F. Mena, A.P.S. Dhariwal and S.M. McCann, Endocrinology, (in press), 1967.
36. Ratner, A. and J. Meites, Endocrinology, 75:377, 1964.
37. Gala, R.R. and R.P. Reece, Proc. Soc. Exp. Biol. and Med., 117:883, 1964.
38. Minaguchi, H. and J. Meites, Endocrinology, 80:603, 1967.

PRODUCTION OF A THYROID STIMULATOR BY IMMUNIZING RABBITS WITH HUMAN THYROID

J.M. McKenzie, M.D.

McGill University Clinic, Royal Victoria Hospital

Montreal, Canada

With recognition of the IgG nature of the long-acting thyroid stimulator (1), the theory has arisen that it is an antibody to a component of human thyroid gland (1,2). An essential corollary of the theory is that it should be possible to produce the antibody - i.e., the long-acting thyroid stimulator - by immunizing animals with an appropriate antigen preparation. Preliminary reports indicating that this is possible have appeared (3,4,5,6); as pointed out (5) a potential artefact in these experiments is that immunization with thyroid preparations is likely to produce thyroiditis which may incite an enhanced secretion of thyrotropin in the immunized animal. Therefore it is necessary to ensure that a thyroid stimulator found in the blood in these animals is indeed a thyroid-stimulating immunoglobulin and not thyrotropin. This communication presents evidence disproving the latter possibility and showing that a thyroid-stimulating IgG can be produced in rabbits by immunization with human thyroid.

Rabbits weighing 2 - 3 Kg were immunized by a series of im injections of antigen preparations combined with equal volumes of Freund's (complete) adjuvant (total volume per injection, 1 - 2 ml). Three antigen suspensions were prepared from human tissue obtained at necropsy: 1) whole thyroid gland homogenate (filtered through gauze); 2) thyroid microsomes; 3) liver microsomes. An initial course of 4 weekly injections was followed by single "booster" injections at approximately monthly intervals; blood was taken at one and two weeks after the fourth and subsequent injections. Whole serum was used for immuno-electrophoresis studies and measurement of serum thyroxine and protein-bound iodine but a crude -globulin preparation was obtained by ammonium sulphate precipitation (4) to inject into mice for bioassay purposes; this was necessitated by the frequent lethality of iv injected rabbit serum in the mouse - an outcome avoided by use of the precipitate.

The bioassay referred to is that described for the measurement of thyrotropin and the long-acting thyroid stimulator (7). In this procedure mice are injected with $^{125}$I to label thyroid iodine and given thyroid to suppress endogenous thyrotropin. Test materials may be injected iv and thyroid stimulation indexed by an increase in radioactivity in the blood of the animal. Thus, 0.1 ml blood is taken immediately before and at 2 and 9 hr after the injection of the test substance and the contained $^{125}$I measured; the response may be expressed as a percentage increase in radioactivity or by a statistically superior procedure more recently developed (7). By computer program, the raw counts are converted to logarithms, adjusted by covariance analysis for variability in the zero hour radioactivity, and the adjusted logarithm is used in Duncan's (8) multiple range test to assess the significance of difference of a response from other responses and control values. The adjusted logarithm is usually expressed as the antilogarithm -- counts per minute (cpm) -- for convenience in presentation of data. Since the bioassay mice vary markedly both in their responsiveness to thyroid stimulation and to inert control materials (7), a response is taken as significant only if it differs ($P < 0.05$ or $< 0.01$ as indicated) from that seen with a variety of control materials tested concomitantly in the same batch of assay mice; routinely in the studies reported here groups of mice were given injections of 3 doses (0.05 mu, 0.2 mu and 0.8 mu) of thyrotropin (NIH S3 or B4) and 1% and 5% albumen solutions as well as other appropriate materials such as extracts of normal rabbit serum.

Characteristically (7) thyrotropin in this bioassay gives a response greater at 2 hr than at 9 hr and the reverse is seen with the long-acting thyroid stimulator. However, as previously reported (7), this is not always a clear-cut distinction -- hence the concern with the data obtained with some rabbit material reported below.

Thyroid function *in vivo* in the rabbits was assessed by injecting $^{131}$I and measuring thyroid gland radioactivity at intervals thereafter by holding the animal so that its neck was in a relatively constant geometric relationship to the open well of a $\gamma$-sensitive crystal. In this way radioiodine uptake and rate of release by the thyroid could be estimated. For a "suppression of release" test, sodium L-thyroxine was dissolved in 0.9% NaCl - 0.1% human serum albumen solution and injected every 2nd day ip in a dose of 100 μg (in 0.5 ml) per animal, during which time the $^{131}$I-release rate was measured.

## INVESTIGATION OF THE NATURE OF THE THYROID-STIMULATING MATERIAL IN ANTI-THYROID ANTISERA

It was reported previously (4) that most success - as judged by the production of a thyroid stimulator - was achieved by immunization with whole thyroid homogenate, rather than thyroid microsomes. [No thyroid stimulation was found with the injection of extracts of serum from either unimmunized rabbits or liver-microsome immunized controls]. The maximum response obtained to date has been 450% at 9 hr when the ammonium sulphate-precipitated gamma globulin fraction equivalent to 1 ml rabbit serum was injected per mouse. As illustrated in Fig. 1, the response to

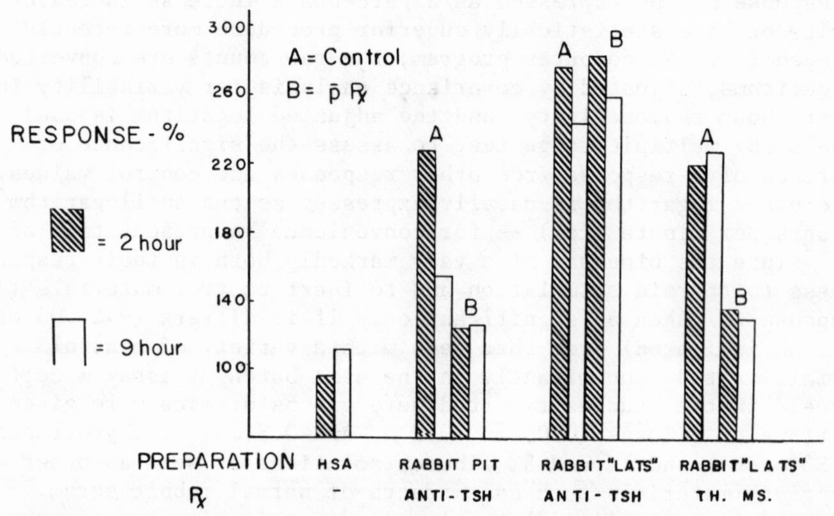


FIGURE 1. Effect of anti-thyrotropin antiserum and of thyroid microsomes on rabbit antiserum to thyroid homogenate.

Rabbit Pit. = homogenate of rabbit pituitary in 1% HSA, 1 mg:40 ml.
Rabbit "LATS" = antiserum to thyroid homogenate.
Anti-TSH = rabbit antiserum to bovine thyrotropin (USP Reference Standard).
Th.MS = human thyroid microsomes.

Microsomes were incubated at 4°C with antiserum overnight and then resedimented at 105000 G; the supernatant was assayed. Antiserum to thyrotropin was incubated with antiserum to thyroid homogenate for 1 hour at room temperature; the mixture was assayed without centrifugation. Control antiserum to thyroid homogenate was incubated with normal rabbit serum.

Each bar represents the mean results from the injection of 6 mice. Results "B" for "Rabbit Pit + Anti-TSH" and "Rabbit LATS + Th.MS" were not significantly different from control (HSA); all other values were significantly greater ($P < 0.01$).

the rabbit serum extract did not have a significant 9 hr maximum, when compared with the 2 hr value. Indeed, in approximately 50 such assays the mean response has sometimes been greater early, and in other instances greater at 9 hr but there has been no instance of a significant difference between them. This is in contradistinction to the typical 2 hr maximum response seen with rabbit pituitary extract (Fig. 1) i.e., thyrotropin, and the common [although, it should be stressed, not inevitable (7)] 9 hr maximal effect of the long-acting stimulator found in human serum.

It seemed probable that the immunization procedure with whole thyroid homogenate might lead to an immune-type thyroiditis (9). Although this was not systematically studied, the thyroids of two rabbits (both immunized with thyroid homogenate) which died were examined histologically; one showed a normal appearance and the other gross destruction with fibrosis and round-cell infiltration. It was recognized as probable that thyrotropin would be in supranormal concentration in the blood of any rabbit in which thyroid function was impaired as a result of thyroiditis, so that the failure to observe a clear-cut 9 hr maximal response in the assay of serum extracts could, in theory, be due to the presence of an excessive concentration of thyrotropin in the original serum. That this was potentially a source of artefact was confirmed by assaying the ammonium sulphate precipitate of rabbit serum to which rabbit pituitary extract had been added; approximately 50% of the added thyrotropin was found in the precipitate (which gave a 2 hr peak response in the bioassay).

The first step taken to confirm that the thyroid-stimulating activity in the serum of immunized rabbits was due to a gamma globulin, was the assay of the gamma globulin purified from the antisera. This was achieved by means of chromatography of the ammonium sulphate precipitate on diethylaminoethyl-Sephadex, in conditions described for use with diethylaminoethyl-cellulose (10) and the resulting $\gamma$-globulin was shown to produce but one (appropriate) band on immuno-electrophoresis, using antisera to rabbit whole serum. When this purified gamma globulin was assayed at twice the equivalent dosage, it was found to give a 2 - 9 hr assay response qualitatively similar to and quantitatively slightly greater than that seen with the crude extract (Table 1).

The data in Fig. 1 illustrate that the thyroid-stimulating activity of the immunized rabbit serum is not inhibited by a dose of antithyrotropin antiserum which inhibited the concentration of rabbit thyrotropin giving a similar 2 hr response in the bioassay. Further, the activity was apparently bound to thyroid microsomes; when the extract of rabbit serum was incubated with a human thyroid microsome preparation and the microsome pellet resedimented by ultracentrifugation, the supernatant contained no thyroid stimulating activity [distinct from results (not shown) found with a human liver microsome preparation which did not bind the thyroid stimulator]. The importance of the latter data is marginal since,

TABLE 1. Bioassay of extracts of serum from immunized rabbit.

| Test Material | Response (cpm) 2 hr | P vs. Control | 9 hr | P vs. Control |
|---|---|---|---|---|
| 5% Albumen (Control) | 1197 | - | 2148 | - |
| Ammonium Sulphate precipitate of serum | 3573[a] | $< 0.01$ | 3589[c] | $< 0.01$ |
| Purified $\gamma$-globulin | 3990[b] | $< 0.01$ | 4102[d] | $< 0.01$ |

The ammonium sulphate precipitate (4) of serum was assayed by injection of the equivalent of 1 ml per mouse, the purified $\gamma$-globulin at the equivalent of 2 ml of serum per mouse.

a vs. b
c vs. d = no significant difference ($P > 0.05$)

although there is one report (11) of thyroid microsomes binding the (human) long-acting thyroid stimulator but not thyrotropin, another group (12) found that both the stimulator and thyrotropin were bound. However, failure of antithyrotropin to prevent thyroid stimulation by the rabbit serum extract is one point in favour of the active serum factor's not being thyrotropin.

To date inhibition of the thyroid-stimulating activity in the serum of thyroid-immunized rabbits by antiserum to rabbit IgG has been inconstant and unconvincing. This is, however, not unexpected because of two factors; 1) if it is assumed that there is a thyroid stimulating IgG in the serum of the immunized rabbits, it presumably is a small proportion of the total IgG; 2) one is restricted in the total quantity of anti IgG antiserum (or gamma globulin) which may be added to the rabbit serum extract before iv injection into the mouse because of the artifactual effects seen with quantities of protein much in excess of the equivalent of 0.5 ml normal serum (unpublished data). Because of the failure of this approach, however, it was necessary to obtain additional data to exclude the possibility of artefact in the results.

If thyrotropin were causing, or playing any part in, the bioassay responses measured with the injection of serum extracts, then it should be possible to remove that influence by the administration of thyrotropin-suppressing doses of thyroid hormone. To attempt this, thyroxine was injected for 8 days in a dose of 100 $\mu$g every second day -- a "suppressing" dose as judged by experiments detailed below and by reports of Brown-Grant and his colleagues (13) Blood was taken from immunized rabbits (known to give positive responses in the bioassay) before and after the course of injections and serum was extracted as usual for assay purposes. The pre- and post-suppression serum extracts were then assayed concomitantly in a single batch of mice. With serum from 7 animals so treated the responses at 9 hr were all significantly greater than control data, and in no instance was there a significant difference when the responses obtained with the pre- and post-"suppression" sera from the individual rabbits were compared. Regarding the responses observed at 2 hr, again all values were significantly greater than the control value; in three instances the 2 hr response was significantly ($P < 0.05$) less when post-"suppression" serum was assayed, in three there was no change and in one there was a significant ($P < 0.05$) increase. It should be stressed that even with those pairs of serum extracts where there was an apparent

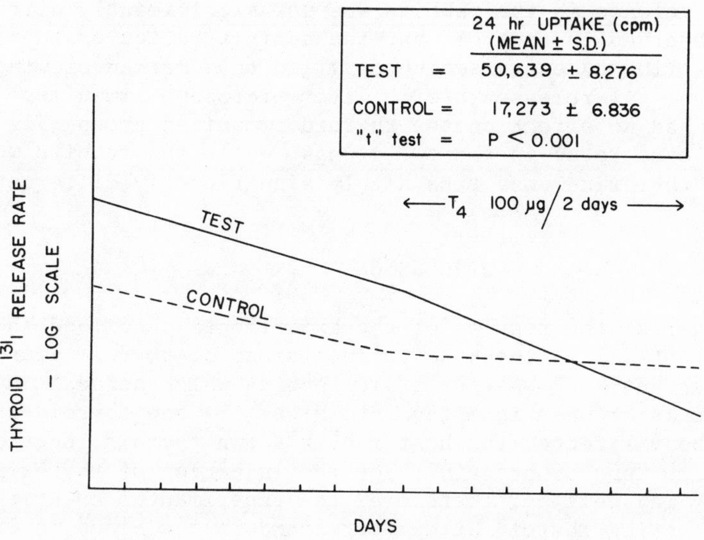


FIGURE 2. The $^{131}$I-release slopes were calculated as the means of daily observations made on 4 rabbits in the thyroid-immunized group and 7 rabbits immunized with liver microsomes. There was no significant difference between the pre-thyroxine slopes, but during thyroxine administration the slopes were highly significantly different ($P < 0.001$). Thyroxine was given ip, 100 μg every second day and was started 1 week before the recording of the release rate "on $T_4$". The release slopes are shown as continuous lines only for ease of presentation.

decrease in the response measured in the mice at 2 hr, there was no significant difference between the positive 9 hr values.

In conjunction with the preceding observations and in view of the additional data reported below, it seems reasonable to term the material being studied a thyroid-stimulating gamma-globulin.

## THYROID FUNCTION IN THYROID-IMMUNIZED RABBITS

The first report (4) of these data indicating the production of a thyroid-stimulator by thyroid-immunization described normal radioiodine uptake in the rabbits producing the thyroid stimulator but a strikingly elevated concentration of protein-bound iodine or total serum thyroxine in some. Since then it has been recognized (14) that the increased serum thyroxine was due to the production of antithyroglobulin antibody which bound thyroxine, and the concentration of free thyroxine was normal. However, in a few of the animals which survived 6 months or more of monthly "booster" injections of thyroid homogenate and repeated venesections, the radioiodine uptake was found to be increased significantly in comparison with that in control rabbits (Fig. 2). The rate of radioiodine release in the rabbits was not significantly different from that observed in controls but the administration of thyroxine for one week (100 µg every second day) led to a marked slowing effect on the $^{131}$I-release rate in liver microsome immunized animals but had no effect on the thyroid immunized group (Fig. 2); the mean release rates in the two groups, while the rabbits were being given thyroxine, was thus highly significantly different.

## DISCUSSION

To summarize the results of the experiments described above, it appears that 1) in response to injections of thyroid gland homogenate an antibody was formed in rabbits which acted as a thyroid stimulator when injected into mice; 2) the thyroid-stimulating antibody affected the host rabbit's own thyroid function so that it became "non-suppressible" by the injection of thyroxine. The antibody may well be, therefore, the experimental counterpart of the long-acting thyroid stimulator found in the blood in Graves' disease. At least the data may be seen as further confirming the concept that the human material is indeed an antibody to a thyroid gland component, although they shed no light on how that antibody might arise in the human.

The concern expressed in the introduction to this paper that a thyroiditis-induced increased secretion of thyrotropin might introduce an artefact into the experiments appears to have been largely allayed. Nonetheless it is reasonable to presume that thyroiditis (confirmed histologically in one instance) occurred although immunization with heterologous material, as in this study, appears to be a less efficient way of inducing thyroiditis than when homologous or autologous thyroid homogenate is used (9). The decrease in the 2 hr response in the bioassay of rabbit serum extract in 3 of the 7 instances where blood was taken after administration of thyrotropin-suppressive doses of thyroxine might indicate that the pituitary hormone had been present in the blood taken previously; however any such influence is clearly minor.

There is no evidence that the rabbits with thyroid-stimulating antibody in their blood were hyperthyroid and, indeed, the normal release rate of thyroid radioiodine probably indicates they were not. It would therefore be reasonable to question why they were not hyperthyroid or, in a similar vein, if they are likely to become so. The following points may be relevant to these considerations.

1) It is perhaps to be wondered at that an antibody to a human antigen should cross-react with (presumably) the same component in the rabbit thyroid -- although this wonder is surpassed by the surprise that it apparently has sufficient trans-species cross-reactivity to act in the mouse. Consequently, although such cross-reactivity appears to exist, it is only to be expected that there is loss of sensitivity (in terms of thyroid stimulation) in the various crossings from species to species so that a potent effect on the rabbit thyroid (or on the mouse gland) might require an inordinate titre of antibody.

2) The data presented indicate that over a period of 6 months there was an increased influence of the antibody on the hosts' thyroids in that initially, after 2 months of immunization, $^{131}I$ uptake was normal (4) whereas, after 3 months, it was increased; judging from the relevant bioassay data (unpublished) this would not likely be on the basis of increase in antibody titre. It may be, therefore, that hyperthyroidism might occur in time as an effect of continued stimulation by the antibody, and it is worth bearing in mind that nothing is known of how long the long-acting thyroid stimulator is active in man before overt hyperthyroidism is observed.

3) The thyroid-suppressing influence of iodide might be operative in the experimental rabbits since the standard rabbit laboratory chow is high in iodine content, which is reflected in a basically low percentage radioiodine uptake in these animals. Whether or not feeding the animals a low-iodine diet would precipitate hyperthyroidism is currently under study.

## ACKNOWLEDGEMENTS

I am grateful to Miss A.M. Williamson, B.Sc., for statistical analyses and other aid, to Mrs. R. Schnarch, Mrs. G. Kounoupi and Mr. D. Myles, B.Sc. for assistance with the bioassays and to Miss L. Hickey for expert manuscript preparation.

The work was financially supported by funds from the Medical Research Council of Canada (MT-884), the U.S. Public Health Service (A-04121) and the Foundation's Fund for Research in Psychiatry (65-318).

## REFERENCES

1. Kriss, J.P., V. Pleshakov, and J.R. Chien. J. Clin. Endocrinol. and Metab. 24: 1005, 1964.
2. McKenzie, J.M. Trans. Assoc. Am. Physicians 78: 174, 1965.
3. Pinchera, A., P. Liberti, and G. Badalamenti. Lancet 1: 374, 1966.
4. McKenzie, J.M. Recent Prog. Hormone Res. 23: 1, 1967.
5. McKenzie, J.M. J. Clin. Invest. 46: 1044, 1967 (abstract)
6. Solomon, D.H., and G.N. Beall. Clin. Res. 15: 127, 1967 (abstract).
7. McKenzie, J.M., and A.M. Williamson. J. Clin. Endocrinol. and Metab. 26: 518, 1966.
8. Duncan, D.B. Biometrics 11: 1, 1955.
9. Terplan, K.L., E. Witebsky, N.R. Rose, J.R. Paine, and R.W. Egan. Am. J. Path. 36: 213, 1960.
10. Levy, H.B., and H.A. Sober. Proc. Soc. Exp. Biol. Med. 103: 250, 1960.
11. Dorrington, K.J., L. Carneiro, and D.S. Munro. J. Endocrinol. 34: 133, 1966.
12. Beall, G.N., and D.H. Solomon. J. Clin. Invest. 45: 552, 1966.
13. Brown-Grant, K., C. von Euler, G.W. Harris, and S. Reichlin. J. Physiol. 126: 1, 1954.
14. McKenzie, J.M., and H. Haibach. Endocrinology 80: 1097, 1967.

# A COMPARATIVE STUDY ON THE MOUSE THYROID ULTRASTRUCTURE AFTER STIMULATION WITH TSH AND LATS.

G. Tonietti, P. Liberti, L. Grasso, A. Pinchera

I' Istituto di Patologia Medica, University of Rome

Italy

The biochemical structure and the immunological properties of the long-acting thyroid stimulator (LATS) and thyrotropin (TSH) are different (1), but their effects on the thyroidal iodine metabolism are similar, differing only in their time course (2, 3). Histologic studies have so far failed to show any major qualitative difference between the TSH and LATS stimulated thyroid glands. Both substances appear to affect in a similar manner the height of the thyroid follicular cells, the staining properties of the colloid (4) and the formation of intracellular colloid droplets (5). The similarities described above may suggest that TSH and LATS act on the thyroid gland through the same mechanism (6). In order to elucidate this problem, we have compared the ultrastructural changes induced by LATS and TSH in the thyroid of mice having their endogenous TSH secretion suppressed by the administration of thyroxine.

## MATERIAL AND METHODS

Female albino mice of the Swiss Webster strain, each weighing approximately 20 g, were used. TSH and LATS bioassays were performed according to a modification of the method of McKenzie (7). The animals received I-131 (10 μc) intraperitoneally and I-thyroxine (20 μg) subcutaneously. The thyroxine injection was repeated 20 to 24 hours prior to assay. The assays were carried out 3 days after the administration of radioiodine; 10-12 animals were used for each sample. Changes in the blood radioactivity were measured at graded intervals (10 and 45 min., and 3, 10, 24, 48 and 72 hours) after the intravenous injection of 0. 5 ml of the test material and expressed as percentage of the initial value.

The results were corrected by subtracting the control response. Control animals received either normal human serum or 0.9% NaCl solution containing 1% human serum albumin. Thyrotropin (Thytropar, Armour, or Ambinon, N. V. Organon) was dissolved in the saline-albumin solution and injected at the dose of 6.0 mU. LATS was constituted by a pool of sera eliciting high LATS responses (3200-3500% per 0.5 ml). The changes in blood radioactivity observed after the single injection of 6 mU of TSH or 0.5 ml of LATS are shown in fig. 1.

The mice used for the electron microscopic study received the same treatment as the assay animals, with the exception that no radioiodine was given. Specimens of the thyroid were obtained at the intervals described above from 2-3 animals for each sample. In vivo fixation was performed by dripping 1% osmium tetroxide in phosphate buffer (8) on the thyroid gland exposed under light aether anesthesia. After 5 min., the thyroid lobes were removed, fixed in the same solution for 1 hr, dehydrated in graded solutions of acetone and finally embedded in Araldite. Thin sections were cut on a Leitz ultramicrotome, stained with lead hydroxide and observed in the electron microscope (Siemens Elmiskop I).

## RESULTS

The thyroxine-treated control animals did not show any significant increase in the blood radioactivity when tested by the method of McKenzie. Electron microscopic examination of the thyroid specimens revealed that, with a few exceptions, the follicle cells

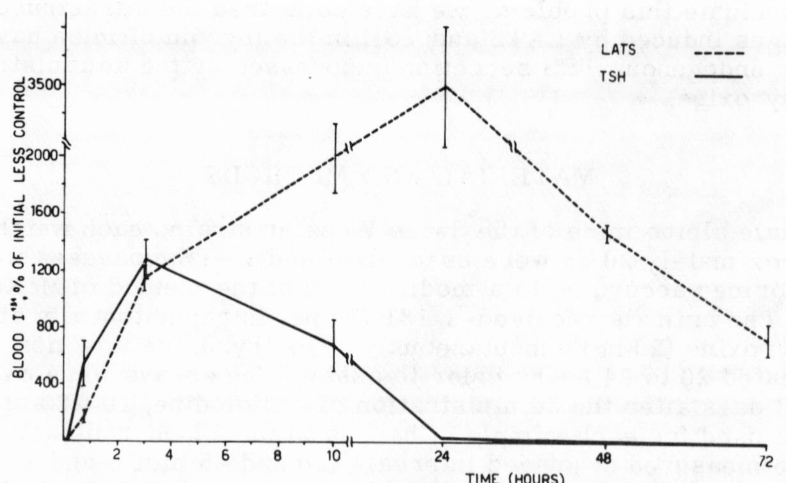


Fig. 1. Changes in blood radioactivity after injection of LATS (0.5 ml) and TSH (6mU) to the assay animals.

were devoid of colloid droplets (fig. 2). Both the number and the morphology of the other cytoplasmic components did not differ from those commonly described in normal glands. The assay ani mals treated with a single injection of 6 mU of TSH showed the expected rise in blood radioactivity, reflecting the discharge of thyroidal I-131. As illustrated in fig. 1, this effect could not be detected at the interval of 10 min., but was clearly evident after 45 min., and reached a maximal value at the 3rd hr, decreasing thereafter down to undetectable levels at the 24 th hr. The electron microscopic study of the thyroid specimens obtained 10 min. after the administration of thyrotropin indicated the presence of thyroid stimulation, as evidenced by the appearance of large colloid droplets in the apical region (fig. 3) and the formation of pseu dopod processes, projecting into the follicular lumen. The number and the size of the colloid droplets further increased in the thyroid specimens obtained after 45 min. and 3 hrs (fig. 4). An augmented number of the apical dense bodies was also observed. The ultrastructure of the thyroid glands excised 10 or more hrs after the TSH administration did not differ from that of the controls (fig. 5). Preliminary assays allowed the selection of appropriate doses of LATS, eliciting responses quantitatively similar to those produced by 6 mU of TSH, either at a given time interval or at their peak of action. The effects of the higher LATS dose was detectable 45 minutes after the injection, reached the highest values at the 24th hr and remained clearly evident after 72 hrs (fig. 1). Electron microscopic evidence of thyroid stimulation was absent in the specimens taken 10 minutes after LATS administration (fig. 6), became detectable after 45 min., progressively increased up to the 10th hr and was still marked after 72 hrs. The ultrastructure of the LATS stimulated thyroid gland was virtually indistinguishable from that observed after TSH treatment. The major changes included the appearance of large colloid droplets (fig. 7) and the formation of pseudopod processes (figg. 8, 9) morphologically identical to those commonly described in the TSH stimulated glands. The smaller doses of LATS elicited lower assay responses and correspondingly produced less marked ultrastructural changes in the thyroid.

Key to Figures (2-9). CO = follicular cavity; CD = colloid droplet; mv = microvilli; N = nucleus; m = mitochondrium; er = endoplasmic reticulum; BM = follicular basement membrane; PS = pseudopod; Cy = cytosegregosome; cn = centriole; END = endothelium of perifollicular capillary; CAP = lumen of perifollicular capillary. All figures were reduced 20% for reproduction.

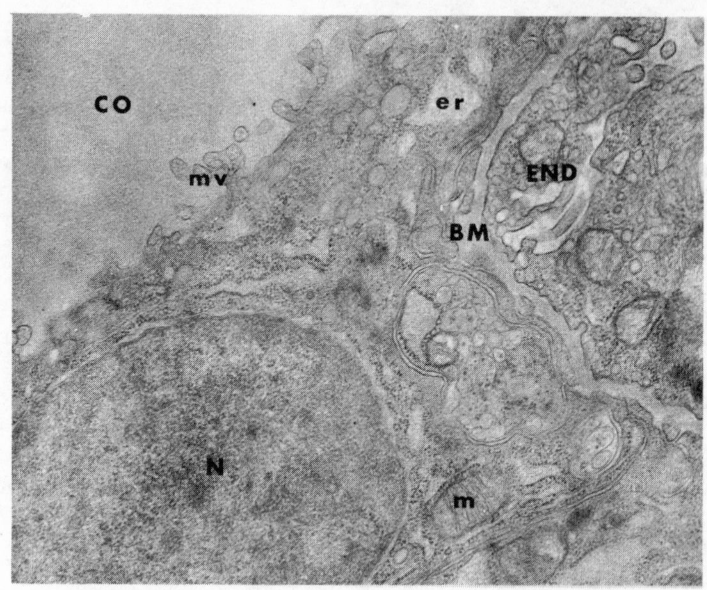


Fig. 2 Thyroid of a thyroxine-treated control mouse. No colloid droplet is visible at the cell apex. x 18. 000.

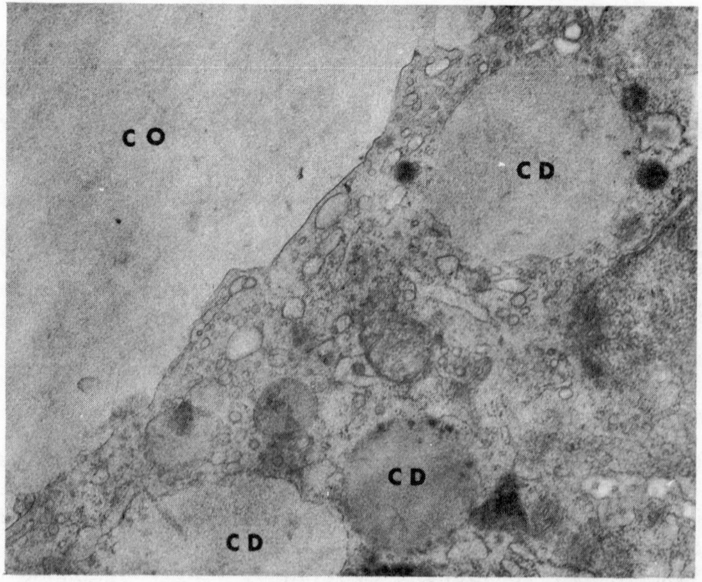


Fig. 3 Mouse thyroid gland 10 min. after injection of TSH. Colloid droplets of different size and electron density are visible in the apical part of a follicular cell. x 20. 000.

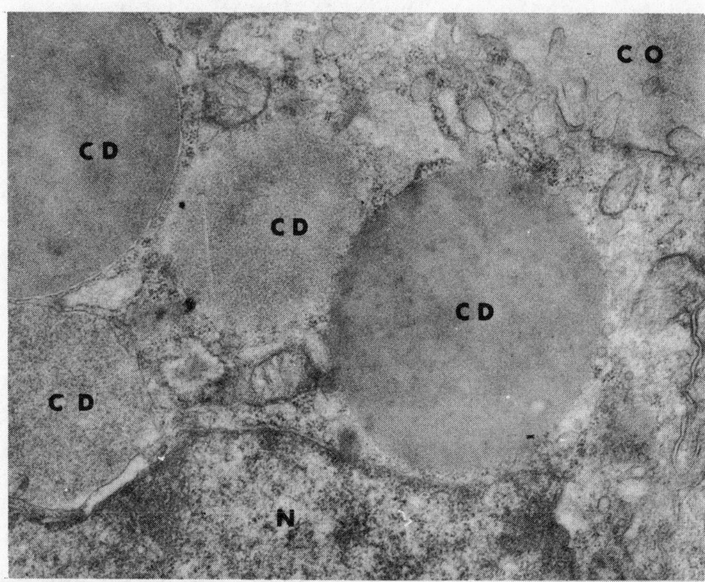


Fig. 4 Mouse thyroid gland 3 hrs after injection of TSH. Numerous large colloid droplets are present in the apical part of the cell. x 20. 000.

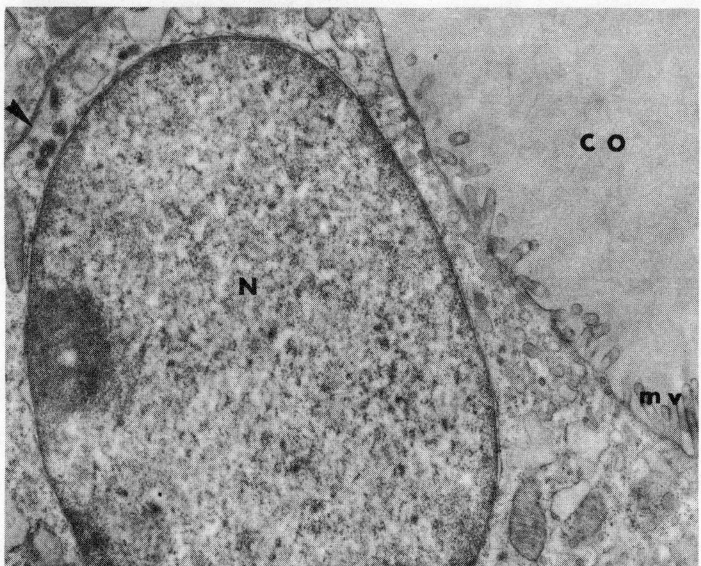


Fig. 5 Mouse thyroid gland 10 hrs after injection of TSH. Colloid droplets are no more visible within the follicular cell. The arrow indicates some dense bodies. x 15. 000.

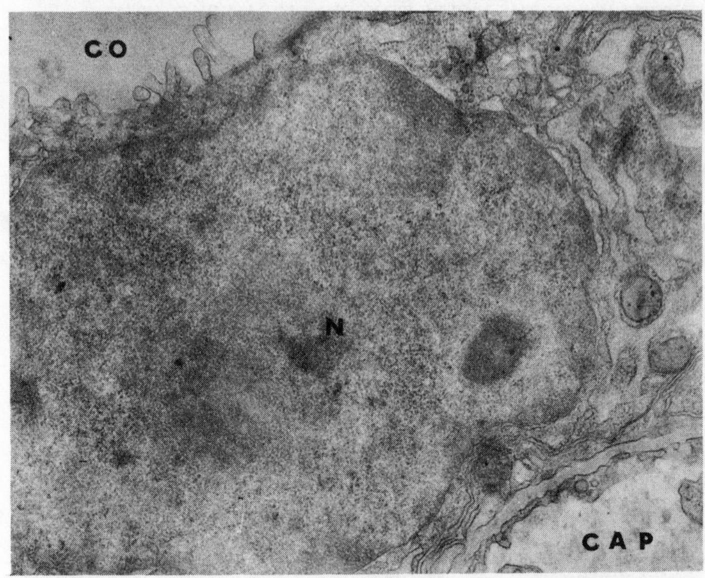


Fig. 6 Mouse thyroid gland 10 min. after LATS injection. The ultrastructural aspect of the follicular cells does not differ from that of control animals. x 16. 000.

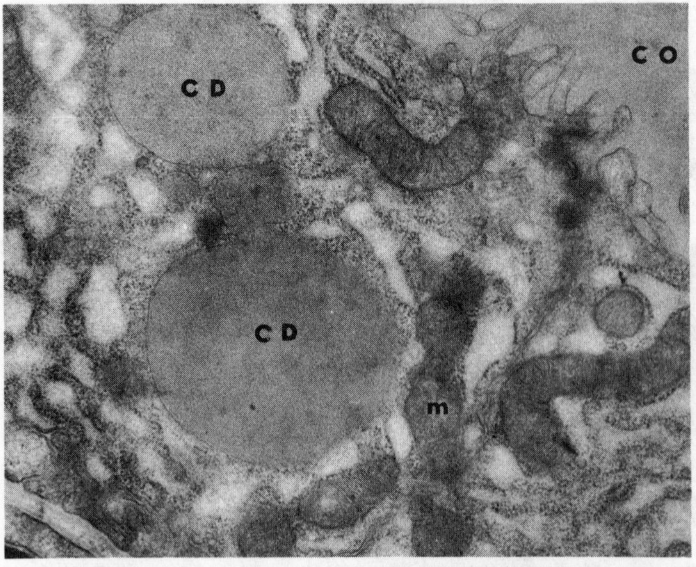


Fig. 7 Mouse thyroid gland 3 hrs after injection of LATS. Large colloid droplets are present in the apical part of a follicular cell. The endoplasmic reticulum cavities are slightly dilated. On the lower left corner, part of the follicular basement membrane and of a perifollicular capillary is visible. x 20. 000.

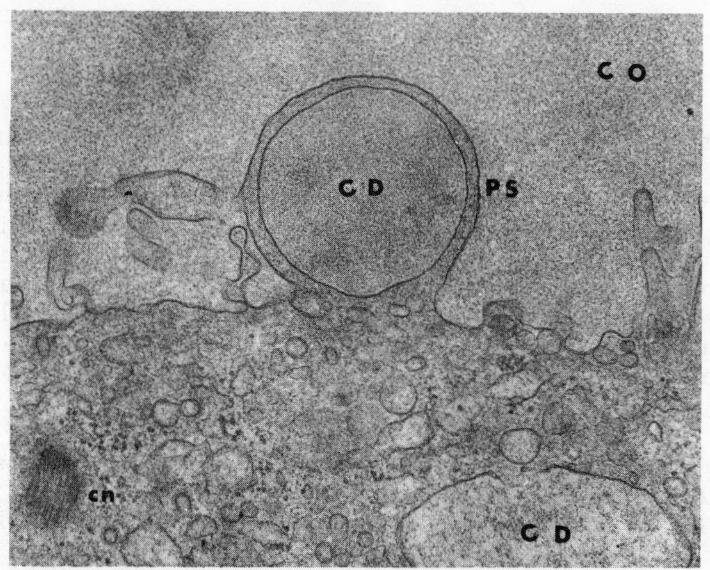


Fig. 8 Mouse thyroid gland 48 hrs after injection of LATS. A large pseudopod containing a colloid droplet is visible at the cell apex. Another colloid droplet of different content is visible in the apical cytoplasm. x 24. 000.

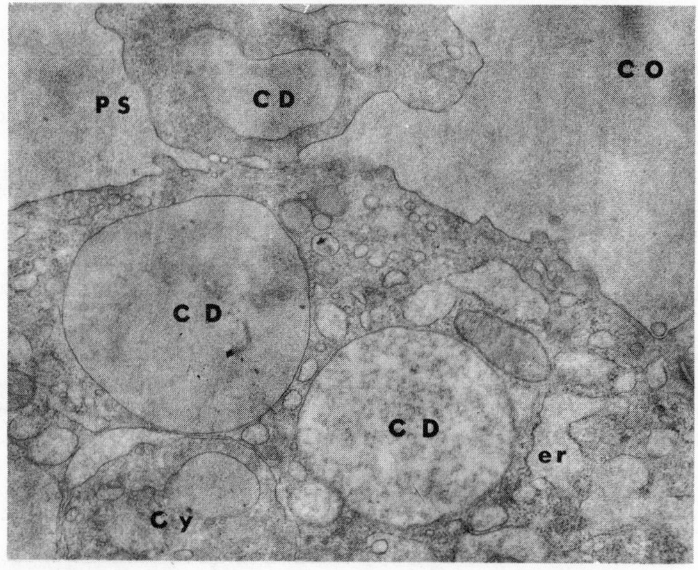


Fig. 9 Mouse thyroid gland 72 hrs after LATS injection. The persistence of stimulation is demonstrated by the presence of a pseudopod containing some colloid droplets and of large colloid droplets in the apical cytoplasm. A cytosegregosome is visible. x 27. 500.

## COMMENTS AND CONCLUSION

The electron microscopic study performed in the control animals has confirmed that the suppression of the endogenous TSH secretion is associated with the disappearance of the colloid droplets from the follicle cells (9). The usual ultrastructural modifications induced by TSH (9, 10) were evident before any significant change in blood radioactivity could be detected and were no more visible when the blood radioactivity was still increased. This time sequence is in keeping with the interpretation that the morphological changes reflect resorption of follicular colloid (10) and the radioactive modifications result from the release of colloid-stored I-131. The effects of LATS on the ultrastructure of the thyroid gland could not be distinguished from those observed in the animals treated with TSH, with the only exception that the signs of LATS stimulation appeared later and lasted for a longer period of time. The presumed slower diffusion rate of the larger LATS globulin molecule may account for these differences in the time course. Our results are consistent with the view that LATS and TSH stimulate the thyroid gland through the same mechanism at least as far as their effects on the discharge of the thyroidal colloid-stored iodine are concerned. This does not exclude the existence of more subtle differences, which cannot be detected in a morphologic study.

## REFERENCES

1) Kriss J. P., Pleshakov V. and Chien J. R. J. Clin. Endocrinol 24: 1005, 1964.
2) Pinchera A., Pinchera M. G. and Stanbury J. B. J. Clin. Endocrinol 25: 189, 1965.
3) Pinchera A. Folia Endocrinologica 18, 629, 1965.
4) McKenzie J. M. J. Clin. Endocrinol 20: 380, 1960.
5) Shishiba Y., Solomon D. H. and Beall G. N. Endocrinology 80: 957, 1967.
6) Adams D. D. Brit. Med. J. 1: 1015, 1965.
7) McKenzie J. M. Endocrinology 63: 372, 1958.
8) Millonig G. Proc. 5th Intern. Congr. Electron Microscopy, Philadelphia, 1962, vol. II, p. 8 Academic Press, New York 1962.
9) Seljelid R. J. Ultrastructural Research 17: 195, 1965.
10) Wetsel B. K., Spicer S. and Woolman S. M. J. Cell Biol 25: 593, 1965.

This work was supported by USPHS Grants AM 11030 and TW 00184 and by the Consiglio Nazionale delle Ricerche Impresa di Endocrinologia, Gruppo degli Ormoni Tiroidei.

# VASOPRESSIN EFFECT ON CORTISOL INCRETION IN MAN

G. Giordano, R. Balestreri, G.E. Jacopino and E. Foppiani

University of Genoa, Institute of Medical Semeiology
Genoa, Italy

The release of ACTH is triggered by a variety of polypeptidic substances (corticotropin-releasing factors or CRF's), elaborated at the level of the hypothalamic median eminence and delivered into pituitary portal vessel blood (1,2,3,4). Between the substances exhibiting CRF-like activity, beta-CRF, alpha1-CRF and alpha2-CRF are not chemically defined (1,4,5), whereas the structure of vasopressin and other synthetic analogues, also exhibiting CRF-like activity is established (2). The vasopressin triggers ACTH release in normal (2), corticosteroid-pretreated (2,6, 7) and, at variable degree, morphine or morphine-pentobarbital (6,7,9,10,11) pretreated animals. In man the studies with vasopressin are scanty. The lysine vasopressin enhances ACTH (8) and cortisol (12,13,14) plasma levels in normal subjects; adrenal response is inhibited by morphine pretreatment (14,15). Conflicting results have been reported in corticosteroid—pretreated subjects (failure (14) or raised threshold (13) for the response of ACTH release); to define the problem we have studied the effects of i.v. vasopressin on plasma cortisol levels after dexamethasone pretreatment in man.

## MATERIALS AND METHODS

Ten male subjects, free of pituitary or adrenal disorders and not receiving analgesic or sedative drugs, were employed. Each subject, at 8.30 A.M., received i.v. 5 U of synthetic lysine-8-vasopressin (Sandoz) in 5 ml of normal saline. Venous blood was drawn before and at 30, 60 and 120 minutes after vasopressin administration. Subsequently each subject was treated for three days with 0.5 mg of dexamethasone (9alpha-fluoro-11beta,17,21-

trihydroxy-16alpha-methyl-$\Delta^{1,4}$-pregnadiene-3,20-dione) every six hours (2 mg/die); the subsequent day, at 8.30 A.M., the treatment with vasopressin was repeated and blood drawn as previously described. Two control groups received i.v. at 8.30 A.M. 5 ml of normal saline before and after dexamethasone treatment; the blood was drawn as described. The plasma cortisol was determined according to Gantt and coworkers (16).

## RESULTS

In normal subjects (Table 1) the mean base line cortisol concentration of 16.7±3.1 increases rapidly after vasopressin to 32.2±6.1 and 27.6±4.9 µg/100 ml plasma 30 and 60 minutes following the injection ($P<0.05$); 120 minutes later the plasma cortisol is reduced to the control values.

Table 1.- Response of plasma cortisol to 5 U of lysine vasopressin i.v. in normal subjects

| µg/100 ml | | | |
|---|---|---|---|
| 0' | 30' | 60' | 120' |
| 16.7±3.1(°) | 32.2±6.1(°) | 27.6±4.9(°) | 16.4±3.1(°) |

(°) Mean±standard deviation

In dexamethasone pretreated subjects (Table 2) the zero plasma cortisol levels of 8.1±1.1 µg/100 ml are lower than without pretreatment ($P < 0.05$), but the rise 30 and 60 minutes after vasopressin is percentually not different ($P>0.05$) from that obtained without dexamethasone premedication (Table 3).

Table 2.- Response of plasma cortisol to 5 U of lysine vasopressin i.v. following dexamethasone

| µg/100 ml | | | |
|---|---|---|---|
| 0' | 30' | 60' | 120' |
| 8.1±1.1(°) | 15.7±2.5(°) | 13.3±1.9(°) | 8.2±1.1(°) |

(°) Mean±standard deviation

Table 3.- Increment per cent of plasma cortisol following 5 U of lysine vasopressin i.v. before and after dexamethasone pretreatment

| Increment % | | | |
|---|---|---|---|
| before | | after | |
| 30' | 60' | 30' | 60' |
| 93.1±10.3(°) | 65.5±4.6(°) | 93.7±8.7(°) | 64.0±4.0(°) |

(°) Mean ±standard deviation

The injection of 5 ml of normal saline i.v. at 8.30 A.M. is without effect on plasma cortisol both in normal and dexamethasone-blocked subjects.

## DISCUSSION

From the above data appears that dexamethasone does not impair ACTH release following i.v. administration of 5 U of lysine vasopressin. The results obtained in basal conditions are in agreement with those previously reported (12,13,14,15); but the response in dexamethasone-blocked subjects, in which the vasopressin exhibites the same degree of ACTH releasing activity as in untreated subjects, is quite different from that obtained by others, who have reported either an inhibition (14) or a diminution (13) of the vasopressin-induced ACTH secretion; the differences are not easy to explain because the injected amount of vasopressin is similar to (13) or greater (14) than that used in the present study. However it is necessary to take into account that the subjects were acutely blocked in the previously reported experiments and subchronically treated with dexamethasone in this study. Therefore the differences can be tentatively explained taking into account the circadian variations of ACTH release and cortisol secretion, with inherent, different sensitivity of the hypothalamic-hypophysial-adrenal system to the blocking effect of same doses of dexamethasone when administered at different hours of the day (3,4,6).

In order to ascertain the site and mode of action of vasopressin in man with hypophysial-adrenal axis blocked by corticosteroids, it is necessary to recall that morphine is always effective for the prevention of ACTH release after vasopressin (14,15); therefore it is conceivable that the site of corticosteroid inhibition is proximal to the site of vasopressin stimulation, whereas the morphine acts distally to or at the site of vasopressin stimulation (6,7,13).

## REFERENCES

1) Guillemin R.: Ann. Rev. Physiol., 29,313,1967
2) Martini L.: Neurohypophysis and anterior pituitary activity in: The Pituitary Gland, Harris G.W. and Donovan B.T. Eds., Vol. 3, Butterworths, London, 1966, pg. 535
3) Fortier C.: Nervous control of ACTH secretion, in: The Pituitary Gland, Harris G.W. and Donovan B.T. Eds., Vol. 2, Butterworths, London, 1966, pg. 195
4) Vernikos-Danellis J.: Vitamins and Hormones, 23,97,1965
5) Guillemin R. and Schally A., in: Advances in Neuroendocrinology, Nalbandov A.V. Ed. ,University of Illinois Press, 1963, pg. 314

6) Hedge G.A., Yates N.B., Marcus R. and Yates F.E.: Endocrinology, 79,328,1966
7) Leeman S.E., Glenister D.W. and Yates F.E.: Endocrinology, 70,249,1962
8) Gwinup G., Steinberg T., King C.G. and Vernikos-Danellis J.: J. Clin. Endocr., 27,927,1967
9) Oliver J.J. and Troop R.C.: Steroids, 1,670,1963
10) Rerup C.: Acta Endocr., 46,387,1964
11) Casentini S., De Poli A., Hukovic S. and Martini L.:Endocrinology, 64,483,1959
12) Clayton G.W., Librick L., Gardner R.L. and Guillemin R.: J. Clin. Endocr., 23,975,1963
13) Clayton G.W., Librick L., Horan A. and Sussman L.: J. Clin. Endocr., 25,1156,1965
14) Gwinup G.: Metabolism, 14,1282,1965
15) McDonald R.K., Evans F.T., Weise V.K. and Patrick R.W.: J. Pharmacol. Exptl. Ther., 125,241,1959
16) Gantt C.L., Maynard D.E. and Hamwi G.G.: Metabolism, 11, 1327,1964

# PSEUDOPREGNANCY AFTER MONOAMINE DEPLETION IN THE MEDIAN EMINENCE OF THE RAT

P.G. Smelik and J.H. van Maanen

Department of Pharmacology, Medical Faculty

University of Utrecht, UTRECHT,The Netherlands

Pseudopregnancy can be induced by a number of drugs (see Meites et al., 1963), among which brain catecholamine depletors as reserpine or α-methyl-dopa (Ratner et al., 1965; Coppola et al., 1965). This led to the assumption that brain catecholamines might be obligatory for the synthesis or release of the hypothalamic prolactin-inhibiting factor (PIF) (Coppola et al., 1965). In fact, in the ventral portion of the hypothalamus a high content of catecholamines could be demonstrated by Carlsson et al. (1962) and Fuxe (1964) by means of a histochemical fluorescence technique.

The present study was undertaken in order to obtain more direct information on the role of hypothalamic monoamines in the control of prolactin secretion. In case of systemic administration of drugs like reserpine the locus of action would remain unknown, and, moreover, effects other than on monoamine depletion would not be excluded. Therefore, an amount of approximately 2 μg reserpine was implanted into the hypothalamus of female rats, and the distribution of fluorescent material, indicating the presence of monoamines, was studied in the median eminence. It appeared that such implantations could cause a complete disappearance of the fluorescence in this area. The animals were otherwise completely normal, indicating that the implanted reserpine had no systemic effects.

In reserpine-implanted and sham-implanted rats daily vaginal smears were taken, and 4 days after implantation the left uterus horn was traumatized in order to provoke deciduoma formation. At autopsy 5 days later the uterine horns were dissected out and weighed. Table 1 shows that in reserpine-implanted rats a high incidence of pseudopregnancy was observed, whereas sham-implantation was virtually ineffective.

TABLE 1

| Number of rats with: | Sham implant | Reserpine implant |
|---|---|---|
| Normal cycle | 19 | 7 |
| Cycle disturbed | 5 | 22 |
| Idem + deciduomata | 1 | 17 |
| Weight in mg ± S.D. of | | |
| Intact uterus horn | 153 ± 16 | 139 ± 13 |
| Traumatized uterus horn | 218 ± 19 | 662 ± 92 |

In another experiment the MAO-inhibitor Iproniazide (150 mg/kg) was administered intraperitoneally on the day before and on the day after reserpine implantation. As shown in table 2, this treatment prevented the reserpine-induced pseudopregnancy.

TABLE 2

| Number of rats with: | Reserpine implant | Reserpine implant + MAO inhibitor |
|---|---|---|
| Normal cycle | 3 | 11 |
| Cycle disturbed | 10 | 0 |
| Idem + deciduomata | 7 | 0 |
| Weight in mg ± S.D. of | | |
| Intact uterus horn | 154 ± 18 | 161 ± 14 |
| Traumatized uterus horn | 471 ± 84 | 192 ± 23 |

These data indicate that reserpine induces prolactin secretion by depleting catecholamines in the median eminence. These amines have been found to be produced by cells in the infundibular and periventricular area, which send their axons to the outer zone of the median eminence (Fuxe, 1964). They appear to be dopaminergic (Fuxe and Hökfelt, 1966), and their nerve endings come into close contact with the pericapillary space of the capillary loops (Kobayashi et al., 1966). This strongly suggests that they may release dopamine directly into the portal vessel system (Sano et al., 1967). This arrangement would permit the idea that dopamine itself may act as a PIF.

## REFERENCES

CARLSSON, A., FALCK, B. and HILLARP, N.A. Cellular localization of brain monoamines. Acta physiol. scand. 56, suppl. 196, 1-28, 1962.

COPPOLA, J.A., LEONARDI, R.G., LIPPMANN, W., PERRINE, J.W. and RINGLER, I. Induction of pseudopregnancy in rats by depletors of endogenous catecholamines. Endocrinology 77, 485-490, 1965.

FUXE, K. Cellular localization of monoamines in the median eminence and the infundibular stem of some mammals. Z. Zellforsch. 61, 710-724, 1964.

FUXE, K. and Hökfelt, T. Further evidence for the existence of tubero-infundibular dopamine neurons. Acta physiol. scand. 66, 245-246, 1966.

KOBAYASHI, H., OOTA, Y., UEMURA, H. and HIRANO, T. Electron microscopic and pharmacological studies on the rat median eminence. Z. Zellforsch. 71, 387-404, 1966.

MEITES, J., NICOLL, C.S. and TALWALKER, P.K. The central nervous system and the secretion and release of prolactin. In: Advances in Neuroendocrinology, pp. 238-277. Ed. A.V. Nalbandov. Urbana: University of Illinois Press, 1963.

RATNER, A., TALWALKER, P.K. and MEITES, J. Effect of reserpine on prolactin-inhibiting activity of rat hypothalamus. Endocrinology 77, 315-319, 1965.

SANO, Y., ODAKE, G. and TAKETOMO, S. Fluorescence microscopic and electron microscopic observations on the tubero-hypophyseal tract. Neuroendocrinology 2, 30-42, 1967.

# THE SYNTHESIS OF INSULIN

Panayotis G. Katsoyannis

Brookhaven National Laboratory

Upton, New York

Pioneering studies by Sanger and co-workers in the period 1945-1955 led to the determination of the amino acid sequence and subsequently the overall structure of insulin from various species (1). This achievement, in conjunction with the earlier work of Banting and Best and of Abel, placed insulin in a unique place among the other proteins. It was one of the first proteins to be recognized as a hormone and thus forced the acceptance of the fact that a protein could be a hormone. It was the first protein whose structure has been elucidated and thus became the vanguard of protein structural analysis; and finally, from the discussion that will follow, it becomes apparent that insulin is the first protein to be chemically synthesized.

The structure of sheep insulin as proposed by Sanger is shown in Figure 1. We undertook the synthesis of this protein with the belief that, if chemically synthesized A and B chains with the sulfhydryl groups free were available, it might be possible to obtain insulin by air oxidation of a mixture of the sulfhydryl forms of these chains. While the work on synthesis toward that goal was in progress, Dixon and Wardlaw in 1960 confirmed that assumption with natural insulin chains (2). These investigators cleaved insulin to its two chains by oxidative sulfitolysis, namely by treatment with sodium sulfite in the presence of a mild oxidizing agent. As a result of that treatment, as shown in Figure 2, the insulin chains are converted to the S-sulfonated derivatives, which in turn, on exposure to a thiol, are converted to their sulfhydryl forms. Air-oxidation of a mixture of the reduced chains generated insulin. This is essentially the principle underlying procedures worked out subsequently in other laboratories, i.e., Du et al. (3,4) and Zahn et al. (5). Based on the amount of

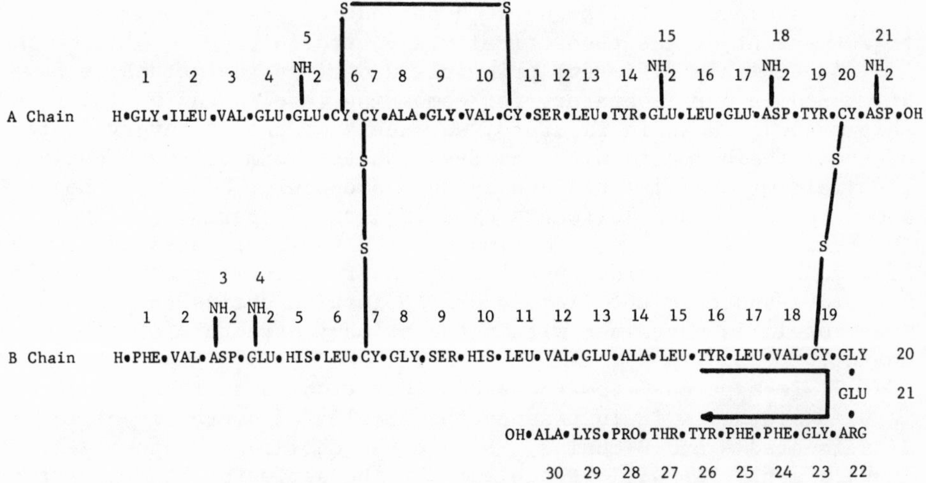


Fig. 1. Structure of sheep insulin.

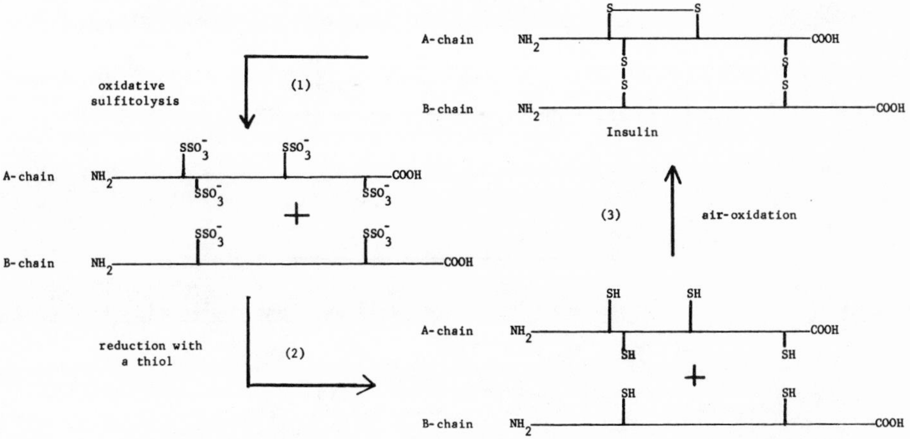


Fig. 2. Cleavage and resynthesis of insulin: (1) oxidative sulfitolysis and isolation of the S-sulfonated A and B chains; (2) conversion of the S-sulfonated chains to their sulfhydryl form; (3) insulin formation by air-oxidation of a mixture of the reduced chains.

the S-sulfonate of the B chain used, the yield of insulin produced in our laboratory, following such procedures was approximately 15 to 20 percent of the theoretically expected (6).

An entirely different principle underlies the method we have developed in our laboratory for combining insulin chains (6). In this method, shown in Figure 3, an excess of the sulfhydryl form of the A chain reacts with the S-sulfonated form of the B chain. The yield of insulin obtained by this procedure, based on the amount of B chain S-sulfonate used is about 70 percent of the theoretical prediction. The implication therefore arises from the high combination yields that the necessary information for complementarity and covalent linking of the insulin chains to produce the protein is contained within the primary structure of the chains.

It is apparent from these considerations that the problem of insulin synthesis is in essence the problem of synthesis of the S-sulfonated A and B chains. Should the chains be synthesized, insulin synthesis will be achieved by the series of reactions that I have discussed. The synthesis of the S-sulfonated A and B chains of sheep insulin was accomplished (7-11) in our laboratory in 1963 and the synthesis of the protected derivatives of these chains was reported at about the same time by Zahn and co-workers (12). The synthesis of the S-sulfonated A and B chains of bovine insulin (the bovine and sheep insulin B chains are identical) were reported by Wang and Niu and co-workers in 1965 and 1966, respectively (13,14). The synthesis, finally, of the S-sulfonated human insulin chains was reported (15-17) from our laboratory in 1966.

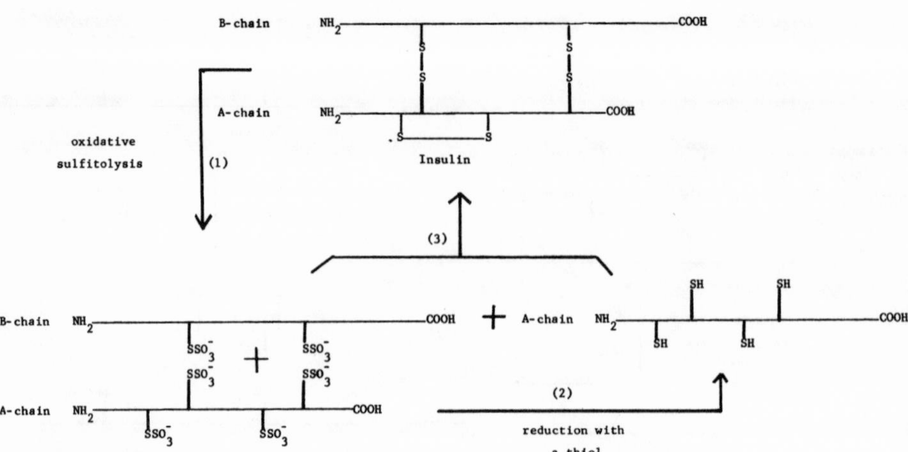


Fig. 3. Resynthesis of insulin by the improved method (Katsoyannis and Tometsko, 1966): (1) oxidative sulfitolysis; (2) conversion of the S-sulfonated A chain to its sulfhydryl form; (3) interaction of the reduced A chain with the S-sulfonated B chain to form insulin.

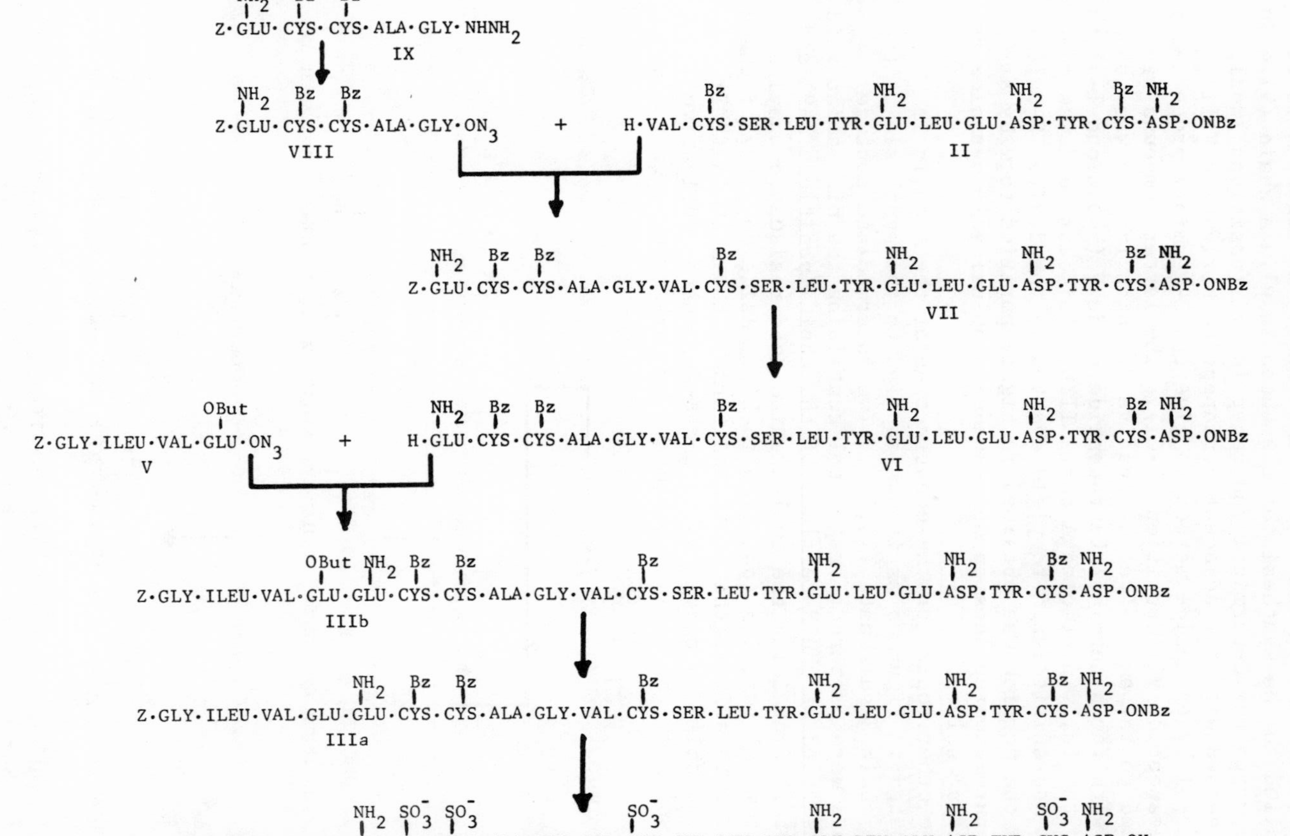


Fig. 4. Synthetic route to the S-sulfonated A chain of sheep insulin.

Implementing classical methods of peptide chemistry we have synthesized the S-sulfonated A chain of sheep and human insulin by two routes. One of these routes is showin in Figure 4 as it was applied to the synthesis of the sheep insulin A chain (9). In this approach the C-terminal dodecapeptide II (positions 10-21) was condensed with the adjacent pentapeptide derivative VIII (position 5-9) by the azide method to give the C-terminal protected heptadecapeptide VII (positions 5-21). The latter compound was converted to the amino-free derivative VI which in turn was condensed with the N-terminal tetrapeptide azide V (positions 1-4) to produce the fully protected A chain IIIb. This compound was deblocked and sulfitolyzed to yield eventually the desired S-sulfonated A chain which was obtained in highly purified form by a single chromatographic step on Sephadex G-50 using 1M pyridine as the eluting solvent.

The other route for the construction of the A chain is illustrated in Figure 5 as it was applied to the synthesis of the human A chain S-sulfonate (16,17). Two intermediate peptide fragments were prepared; fragment I containing the nine amino acid residues found at the N-terminus of the chain (positions 1-9) and fragment II containing the twelve amino acid residues occupying the C-terminal position (position 10-21). Azide coupling of these two fragments affords the protected form of the A chain (III) which eventually is converted to the S-sulfonated derivative IV. Purification of the chain constructed by this route required two

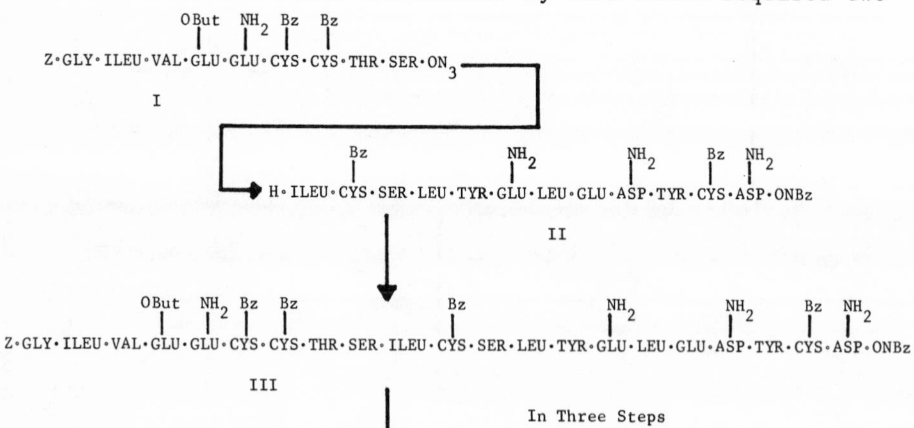

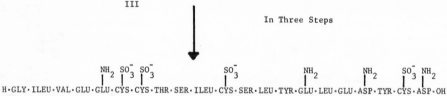

Fig. 5. Synthetic route to the S-sulfonated A chain of human insulin. Abbreviations: Z, carbobenzoxy group (amino protector); But, t-butyl group (carboxyl protector); Bz, benzyl group (sulfhydryl protector); NBz, p-nitrobenzyl group (carboxyl protector).

chromatographic steps on Sephadex G-25. One of the steps involves 1M pyridine and the other 1% acetic acid as the eluting solvents.

The synthetic chains prepared by either route exhibited an identical behavior with the natural compounds when compared as to amino acid composition after acid and enzymatic hydrolysis, electrophoretic mobility on paper and thin layer electrophoresis at two pH values, specific rotation and infrared pattern. The amino acid analysis of acid and enzymatic (LAP: leucine aminopeptidase) hydrolysates of the synthetic S-sulfonated sheep A chain is shown in Table I and of human A chain in Table II. As it can be seen from these tables the amino acid compositions of the synthetic chains expressed in molar ratios are in excellent agreement with theoretically expected values. Furthermore, complete digestibility of these chains by LAP establishes their stereochemical homogeneity.

The synthesis of the S-sulfonated B chain of sheep and human insulin which differ at the C-terminal amino acid residue was accomplished by the same overall approach. Figure 6 illustrates this approach as was applied to the synthesis of the S-sulfonated human B chain (15). Using again the classical procedures of

Table I

Amino acid composition of the synthetic S-sulfonate of the A chain of sheep insulin determined by the Stein-Moore procedure

Number of amino acid residues per molecule

| Amino Acid | Acid hydrolysate | | Amino Acid | LAP digest | |
|---|---|---|---|---|---|
| | Theory | Found | | Theory | Found |
| Aspartic acid | 2.00 | 2.00 | Asparagine | 2.00 | Emerge on the same position, and not determined. |
| Serine | 1.00 | 0.90 | Glutamine | 2.00 | |
| Glutamic acid | 4.00 | 4.20 | Serine | 1.00 | 0.90** |
| Glycine | 2.00 | 2.00 | Glutamic acid | 2.00 | 2.00 |
| Alanine | 1.00 | 1.10 | S-Sulfocysteine | 4.00 | 4.30*** |
| Cysteine | 4.00 | 4.00* | Glycine | 2.00 | 2.00 |
| Valine | 2.00 | 1.80 | Alanine | 1.00 | 1.00 |
| Isoleucine | 1.00 | 0.70 | Valine | 2.00 | 2.00 |
| Leucine | 2.00 | 2.00 | Isoleucine | 1.00 | 1.00 |
| Tyrosine | 2.00 | 2.00 | Leucine | 2.00 | 2.00 |
| Ammonia | 4.00 | 4.00 | Tyrosine | 2.00 | 2.00 |

* Determined as cysteic acid, S. Moore, J. Biol. Chem., 238, 235 (1963).

** Separated from glutamine and asparagine in a 30° chromatographic run.

*** Eluted from the long column of the Beckman-Spinco analyzer after 26 ml. of effluent.

## Table II

Amino acid composition of the synthetic S-sulfonate of the A chain of human insulin determined by the Stein-Moore procedure

Number of amino acid residues per molecule

| Amino Acid | Acid hydrolysate | | Amino Acid | LAP digest | |
|---|---|---|---|---|---|
| | Theory | Found | | Theory | Found |
| Aspartic acid | 2.00 | 1.90 | Asparagine | 2.00 | ) Emerge on ) the same ) position |
| Threonine | 1.00 | 1.00 | Glutamine | 2.00 | |
| Serine | 2.00 | 1.80 | Threonine | 1.00 | 1.00 |
| Glutamic acid | 4.00 | 4.20 | Glutamic acid | 2.00 | 2.00 |
| Glycine | 1.00 | 1.00 | Glycine | 1.00 | 1.10 |
| Cysteine | 4.00 | 3.70 | S-Sulfocysteine | 4.00 | 4.10* |
| Valine | 1.00 | 0.80 | Valine | 1.00 | 1.00 |
| Isoleucine | 2.00 | 1.80 | Isoleucine | 2.00 | 2.10 |
| Leucine | 2.00 | 2.00 | Leucine | 2.00 | 1.80 |
| Tyrosine | 2.00 | 1.90 | Tyrosine | 2.00 | 1.80 |
| Ammonia | 4.00 | 4.20 | Serine | 2.00 | 1.80** |

* Eluted from the long column of the Beckman-Spinco analyzer after 26 ml. of effluent.

** Separated from glutamine and asparagine in a 30° chromatographic run.

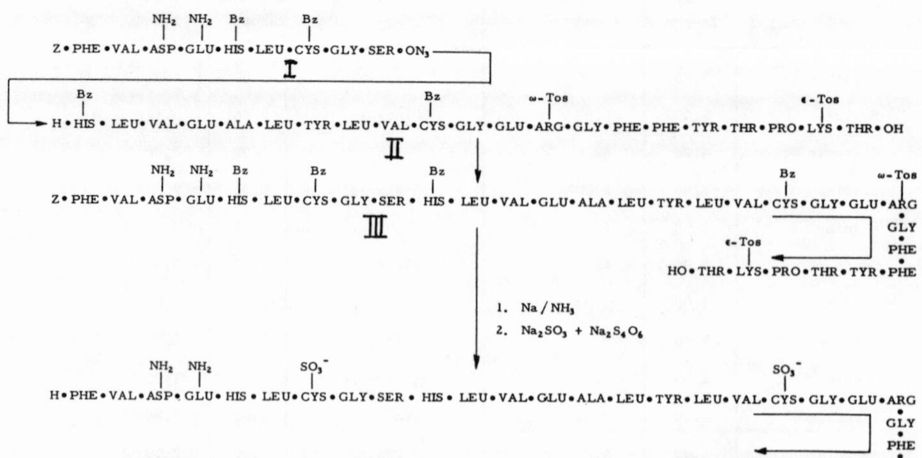


Fig. 6. Synthetic route to the S-sulfonated B chain of human insulin. Abbreviations: Z, carbobenzoxy group (amino protector); Bz, benzyl group (sulfhydryl and imidazole protector); Tos, p-toluene-sulfonyl group ($\varepsilon$-amino and guanido protector).

peptide chemistry we have prepared two peptide subunits. One of these subunits (I) contains the nine amino acid residues found at amino terminus of the chain (positions 1-9) and the other (II), the twenty one amino acid residues occupying the C-terminal position of that chain (position 10-30). Condensation of these subunits by the azide method afforded the protected triacontapeptide III which contains the entire amino acid sequence of the B chain. Removal of the blocking groups from III and sulfitolysis of the ensuing product yielded the S-sulfonated B chain of human insulin. Chromatography of the crude material on carboxymethylcellulose gave the S-sulfonated B chain in highly purified form as judged by amino acid analysis after acid and enzymatic hydrolysis, electrophoretic mobility on thin layer electrophoresis at two pH values, infrared pattern and optical rotation. The amino acid analysis of acid and enzymatic (APM aminopeptidase M) hydrolysates of the synthetic S-sulfonated B chain of sheep insulin is shown in

Table III

Amino acid composition of the synthetic S-sulfonate of the B-chain of sheep insulin determined by the Stein-Moore procedure

| Amino acid | Acid hydrolysate: Theory | Acid hydrolysate: Found | Amino acid | APM digest: Theory | APM digest: Found |
|---|---|---|---|---|---|
| Lysine | 1.00 | 1.00 | Lysine | 1.00 | 1.00 |
| Histidine | 2.00 | 2.00 | Histidine | 2.00 | 1.50 |
| Arginine | 1.00 | 1.00 | Arginine | 1.00 | 1.00 |
| Aspartic acid | 1.00 | 1.00 | Asparagine | 1.00 | Emerge on the same position. Not determined. |
| Threonine | 1.00 | 0.90* | Glutamine | 1.00 | |
| Serine | 1.00 | 0.80* | Serine | 1.00 | |
| Glutamic acid | 3.00 | 3.00 | Glutamic acid | 2.00 | 2.20 |
| Proline | 1.00 | 0.90 | S-Sulfocysteine | 2.00 | 1.80 |
| Glycine | 3.00 | 3.10 | Threonine | 1.00 | 1.00 |
| Alanine | 2.00 | 1.80 | Proline | 1.00 | 0.90 |
| Cysteine | 2.00 | 1.80 | Glycine | 3.00 | 3.00 |
| Valine | 3.00 | 3.00 | Alanine | 2.00 | 2.00 |
| Leucine | 4.00 | 3.90 | Valine | 3.00 | 2.70 |
| Tyrosine | 2.00 | 1.70 | Leucine | 4.00 | 4.00 |
| Phenylalanine | 3.00 | 2.80 | Tyrosine | 2.00 | 2.20 |
| | | | Phenylalanine | 3.00 | 3.00 |

Number of amino acid residues per molecule

* Uncorrected for destruction.

Table III and of human insulin in Table IV. As it can be seen, the amino acid compositions of the synthetic chains expressed in molar ratios are in excellent agreement with the theoretically expected values. Furthermore, complete digestibility of these chains by APM proves their stereochemical homogeneity.

Combination of the synthetic chains to produce insulin was originally carried out by the methods of Dixon and Du. Under these conditions, based on the starting amounts of the chains, all-synthetic insulin was produced in approximately 2 percent yield and half-synthetic insulins, consisting of one synthetic sheep or human insulin chain (A or B) and one natural bovine insulin chain, were produced in yields ranging from 4 to 8 percent. A most dramatic increase, however, in the yield of synthetic insulins is materialized when our method for combining insulin chains is employed (18,19). Using synthetic sheep and human insulin chains and natural bovine and porcine insulin chains, we were able

Table IV

Amino acid composition of the synthetic S-sulfonate of the B-chain of human insulin determined by the Stein-Moore procedure

| Number of amino acid residues per molecule | | | | | |
|---|---|---|---|---|---|
| Amino acid | Acid hydrolysate | | Amino acid | APM digest | |
| | Theory | Found | | Theory | Found |
| Lysine | 1,00 | 0,90 | Lysine | 1.00 | 1.00 |
| Histidine | 2,00 | 1.70 | Histidine | 2.00 | 1.50 |
| Arginine | 1,00 | 1.00 | Arginine | 1,00 | 1.00 |
| Aspartic acid | 1.00 | 0,70 | Asparagine | 1,00 | } Emerge on the |
| Threonine | 2,00 | 1.60* | Glutamine | 1.00 | same position. |
| Serine | 1.00 | 0.60* | Serine | 1.00 | Not determined. |
| Glutamic acid | 3.00 | 2.90 | Glutamic acid | 2.00 | 2,00 |
| Proline | 1.00 | 0.70 | S-Sulfocysteine | 2.00 | 2.00 |
| Glycine | 3,00 | 3.00 | Threonine | 2.00 | 2.10 |
| Alanine | 1.00 | 1.10 | Proline | 1.00 | 1.00 |
| Cysteine | 2,00 | 1.50 | Glycine | 3,00 | 2,70 |
| Valine | 3.00 | 2.70 | Alanine | 1,00 | 0.90 |
| Leucine | 4.00 | 4.00 | Valine | 3.00 | 2.80 |
| Tyrosine | 2,00 | 2.00 | Leucine | 4.00 | 3.70 |
| Phenylalanine | 3.00 | 3.00 | Tyrosine | 2.00 | 2.20 |
| | | | Phenylalanine | 3,00 | 3,10 |

* Uncorrected for destruction.

to produce several all-synthetic and half-synthetic insulins. Table V records the various insulins we have synthesized and the yields of their production (19). These insulins include the all-synthetic sheep, all-synthetic human, half-synthetic sheep, bovine, and porcine insulins and an insulin which has not been found in any of the species examined and which we tentatively designated as insulin $B_AH_B$. The latter insulin was produced by combining synthetic human B and natural bovine A chains. It might be pointed out that insulin $B_AH_B$ possessed the full biological activity of the naturally occurring insulin and was obtained in crystalline form.

Isolation of the synthetic insulins from the combination mixtures of the A and B chains was originally undertaken by methods devised to isolate insulin from natural sources. We were thus able to isolate (20) synthetic insulins by applying a modified version of the method suggested by Smith (21) which, like all the existing methods of isolation, included an acid-alcohol extraction as the initial step. The low recovery, however, of

Table V

Yields of insulins produced by combination of A and B chains

by the method of Katsoyannis and Tometsko

(Proc. Nat. Acad. Sci. USA 55, 1554 (1966))

| Type of chains used for combination | Insulin produced | Yields of insulin* produced by combination of A and B chains % (based on B chain used) |
|---|---|---|
| Synthetic sheep A + Natural bovine B | Sheep (half-synthetic) | 30-38 |
| Natural bovine A + synthetic sheep B | Bovine (half-synthetic) | 17-19 |
| Synthetic sheep A + synthetic sheep B | Sheep (all-synthetic) | 12-16 |
| Synthetic human A + natural bovine B | Porcine (half-synthetic) | 30-42 |
| Natural bovine A + synthetic human B | Insulin $B_AH_B$** | 18-23 |
| Synthetic human A + synthetic human B | Human (all-synthetic) | 15-22 |

* Biological assays were performed by the mouse convulsion method.

** This insulin has not been found in any of the species examined, and it was tentatively designated as insulin $B_AH_B$ from bovine A chain and human B chain.

the highly purified material made such isolation procedures impractical. Kung and co-workers (22), using the method proposed by Du, which is actually a modified version of the original alcohol extraction reported by Banting et al., have also isolated synthetic bovine insulin. The overall recovery of the pure material, however, based on the amounts of the starting A and B chains, was of the order of 0.05 to 0.1% and in terms of actual amount of material produced per experiment, it was of the order of a few hundreths of a milligram. Such a low recovery of material, however, not only precludes reliable calculation of the specific activity and the attainment of other important analytical data but also makes the synthesis of insulin for any practical purpose highly unrealistic.

This problem too has now been solved. We have developed simple isolation procedures (23) by which we can produce highly purified synthetic insulins (19) with overall recoveries, based on the amounts of the starting chains, at least 100-fold higher than the yields reported by Kung et al. Time limitations will not allow me to discuss the very interesting data we have obtained while working to devise these isolation techniques. I will, therefore, confine myself to summarizing, very briefly, some of our findings. From our data, it becomes clear that insulin is the exclusive product formed, among the many possible isomers, by combination of the A and B chains. Other products of the combination mixture are components derived from unreacted A chain and components derived from unreacted B chain. We have further found (23) that, under certain conditions, undesirable interactions of the components of the combination mixture occur which lead to insulin alteration or insulin destruction. From these studies, we have developed a simple isolation procedure which involves as the initial step the conversion of the insulin and of other components of the combination mixture to the picric acid and eventually to the hydrochloric acid salts. Separation of the insulin hydrochloride from the other products of the mixture was finally accomplished by chromatography on a carboxymethylcellulose column with an exponential sodium chloride gradient. The insulin was recovered from the effluent as the hydrochloride and eventually, if desirable, was crystallized. The recovery of the highly purified hormone, based on the activity present in the combination mixture, ranges from 50 to 65% in the chromatographic effluent and from 35 to 52% at the hydrochloride stage. Figure 7-I shows the chromatogram of natural bovine insulin in this chromatographic system and Figure 7-II depicts the chromatographic pattern obtained from a recombination mixture of natural bovine insulin chains. The regenerated insulin isolated by this procedure was identical with the natural hormone in respect to amino acid composition, specific activity, electrophoretic mobility on thin layer electrophoresis, mobility on carboxymethylcellulose columns, infrared pattern and

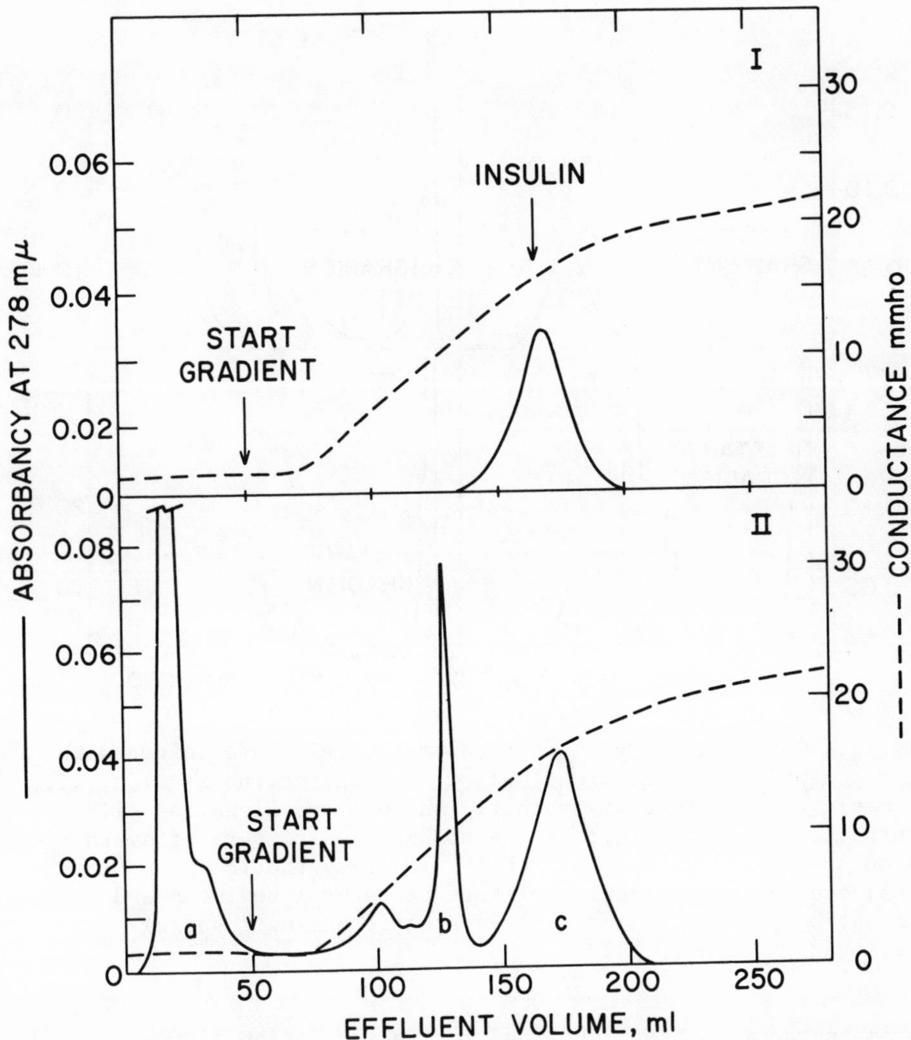


Fig. 7. Chromatography of natural bovine insulin (I) and of a recombination mixture of natural A and B chains of bovine insulin (II) on a carboxymethylcellulose column with an exponential salt gradient (Katsoyannis et al., 1967).

crystalline form.

This isolation procedure was applied with equally satisfactory results to the isolation of a number of regenerated natural insulins and to the isolation of several all-synthetic and half-synthetic insulins (19). Figure 8 illustrates the separation of the all-synthetic sheep (II), half-synthetic sheep (III), and half-synthetic bovine (IV) insulins. Figure 9 illustrates the

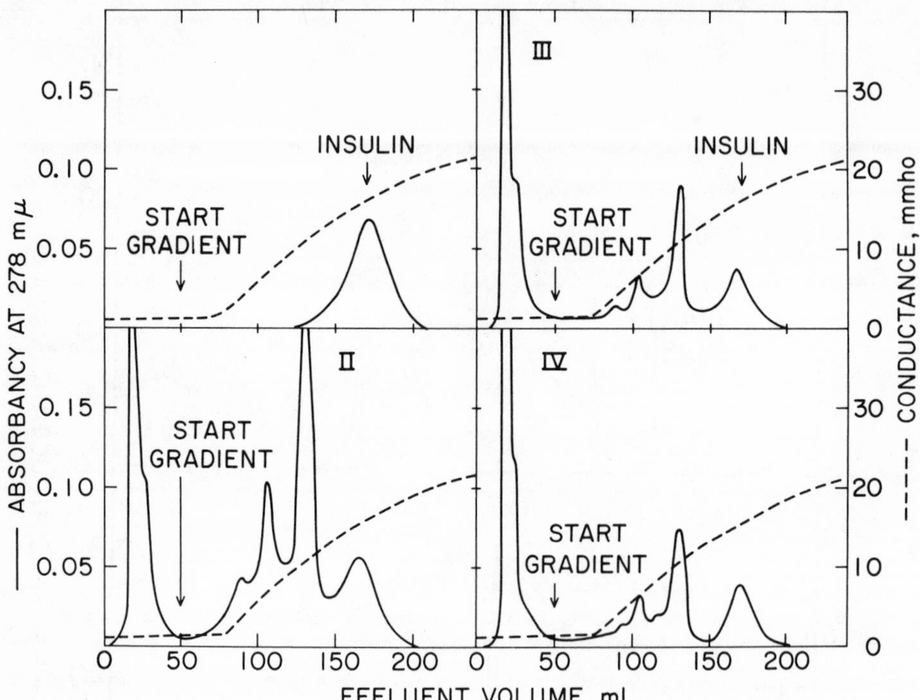


Fig. 8. Chromatography on a carboxymethylcellulose column with an exponential salt gradient of: I, natural bovine insulin; II, a combination mixture of synthetic sheep A and B chains (all-synthetic sheep insulin); III, a combination mixture of synthetic sheep A and natural bovine B chains (half-synthetic sheep insulin); and IV, a combination mixture of natural bovine A and synthetic bovine B chains (half-synthetic bovine insulin). (Katsoyannis et al., 1967)

chromatographic pattern obtained in the separation of regenerated porcine (II), regenerated bovine (III), half-synthetic porcine (IV), half-synthetic insulin $B_AH_B$ (V), and all-synthetic human (VI) insulins. As it can be seen from Figures 8 and 9 the synthetic insulins have the same mobility as the natural hormone on the carboxymethylcellulose column.

The recoveries of the regenerated, half-synthetic and all-synthetic insulins and their specific activities are given in Table VI. It is clear from this table that the regenerated natural, the half-synthetic and the all-synthetic insulins possessed specific activities identical to that of the natural hormone. The synthetic insulins were further compared with their natural counterparts with respect to amino acid composition and electrophoretic mobility on thin layer electrophoresis, a highly

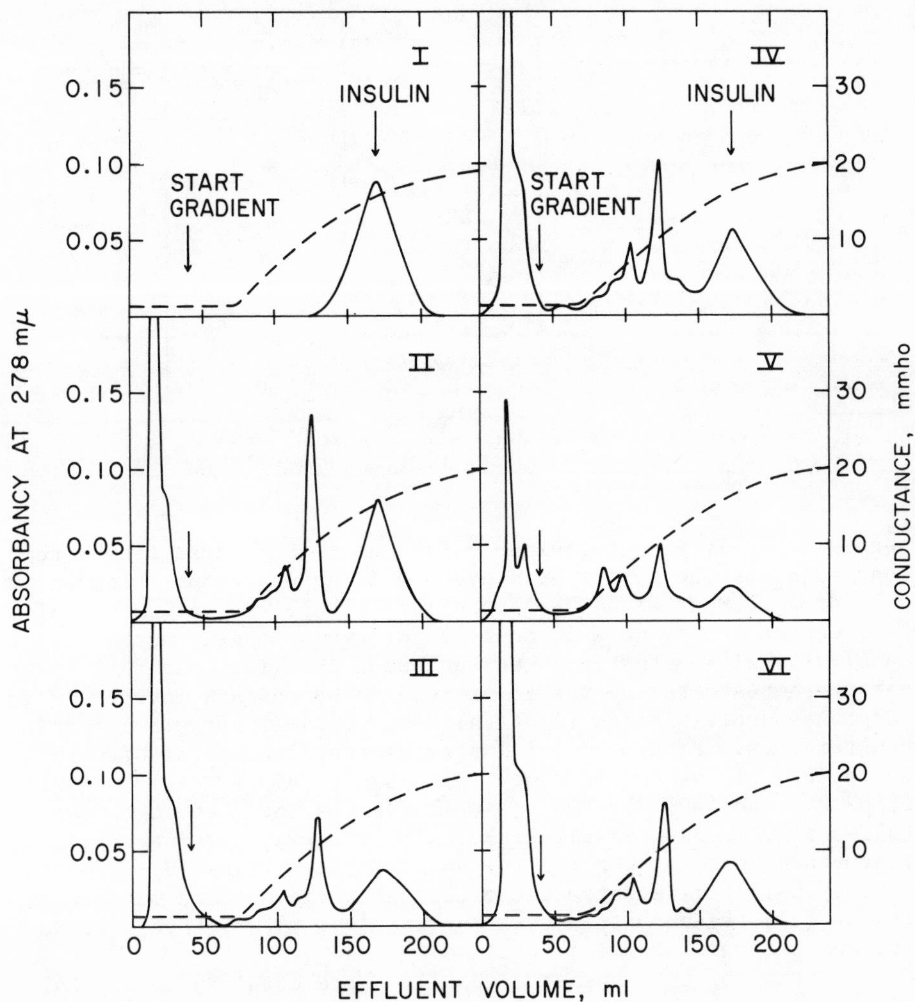


Fig. 9. Chromatography on a carboxylmethylcellulose column with an exponential salt gradient of: I, natural bovine insulin; II, a combination mixture of natural porcine A and natural bovine B chains (regenerated porcine insulin); III, a combination mixture of natural bovine A and natural porcine B chains (regenerated bovine insulin); IV, combination mixture of synthetic human A and natural bovine B chains (half-synthetic porcine insulin); V, a combination mixture of natural bovine A and synthetic human B chains (half-synthetic $B_AH_B$ insulin); VI, a combination mixture of synthetic human A and B chains (all-synthetic human insulin) (Katsoyannis et al., 1967).

Table VI

Over-all Recoveries and Specific Activities of Isolated Insulins
Synthesized by Combination of A and B Chains

| TYPE OF CHAINS USED FOR COMBINATION | INSULIN PRODUCED | OVER-ALL RECOVERY AS INSULIN HYDROCHLORIDE* % | SPECIFIC ACTIVITY OF ISOLATED INSULIN** IU/mg |
|---|---|---|---|
| Natural Insulin (Bovine, Sheep, Porcine, Human) | | | 23 - 25 |
| Natural Bovine A + Natural Bovine B | Bovine | 50 | 25 (Crystalline) |
| Synthetic Sheep A + Natural Bovine B | Sheep (Half-Synthetic) | 39 | 25 (Crystalline) |
| Natural Bovine A + Synthetic Sheep B | Bovine (Half Synthetic) | 37 | 22 (Crystalline) |
| Synthetic Sheep A + Synthetic Sheep B | Sheep (All-Synthetic) | 43 | 25 (Crystalline) |
| Natural Porcine A + Natural Bovine B | Porcine | 42 | 25 |
| Natural Bovine A + Natural Porcine B | Bovine | 39 | 23 (Crystalline) |
| Synthetic Human A + Natural Bovine B | Porcine (Half-Synthetic) | 52 | 22 |
| Natural Bovine A + Synthetic Human B | Insulin $B_AH_B$** | 35 | 22 (Crystalline) |
| Synthetic Human A + Synthetic Human B | Human (All-Synthetic) | 41 | 24 |

* The insulin activity present in the combination mixture taken as 100%.

** Calculated from determinations of protein content (Folin method) and of biological activity (mouse convulsion assay method).

sensitive analytical technique. Thus amino acid analysis of the synthetic hormones after acid hydrolysis gave a composition in excellent agreement with the theoretically expected values. Table VII records the amino acid analyses of two synthetic hormones, namely the all-synthetic sheep and human insulins. On thin layer electrophoresis the synthetic proteins behaved as homogeneous compounds and exhibited identical mobilities with their natural counterparts. Figure 10-I illustrates the electrogram of natural sheep (a), all-synthetic sheep (b) and natural bovine (c) insulins at pH 2.9 and 3400 v. Figure 10-II depicts the electrogram of half-synthetic $B_AH_B$ insulin (a), half-synthetic porcine (b), regenerated natural porcine (c) and natural porcine (d) insulins. Figure 10-III illustrates the electrophoretic pattern of natural bovine (a), regenerated natural bovine (b), all-synthetic human (c) and natural human (d) insulins.

Several of the synthetic insulins were obtained in crystalline form using the crystallization method suggested by Randall (24) or the method of Epstein and Anfinsen (25). Figure 11 illustrates crystals of regenerated natural bovine (I), half-synthetic sheep (II), all-synthetic sheep (III) half-synthetic bovine (IV), and half-synthetic $B_AH_B$ (V) insulins.

All these comparisons of the physical, chemical and biological properties of the all-synthetic and half-synthetic insulins with those of the natural counterparts justify, in our estimation, the conclusion that the synthetic materials are identical with the natural hormones and that the structures proposed for these proteins are correct.

In conclusion the synthesis of several insulins has been

achieved. It is true that a large number of chemical steps are required for the synthesis of the individual chains. However, the availability of highly efficient methods, both for the combination of the chains and the isolation of the insulin thus produced, makes the synthesis of this protein for practical purposes a possibility. No matter, however, how practical the use of synthetic insulin is at present, undoubtedly its synthesis marked the beginning of a new chapter in the history of this hormone. We have now reached the stage where the problem of the relationship between chemical structure, biological activity and antigenicity of insulin can be pursued in an unlimited way. Toward that goal, we have started the synthesis of selected analogues of insulin. I have already mentioned such an analogue, the insulin $B_AH_B$, and we have just completed the synthesis of another analogue from natural

Table VII

Amino acid composition* of the all-synthetic sheep and human insulins determined by the Stein-Moore procedure

| AMINO ACID | SHEEP INSULIN | ALL-SYNTHETIC SHEEP INSULIN FOUND | HUMAN INSULIN THEORY | ALL-SYNTHETIC HUMAN INSULIN FOUND |
|---|---|---|---|---|
| Lysine | 1 | 1.0 | 1 | 0.8 |
| Histidine | 2 | 1.9 | 2 | 1.9 |
| Arginine | 1 | 1.0 | 1 | 1.0 |
| Aspartic Acid | 3 | 2.6 | 3 | 2.8 |
| Threonine | 1 | 0.9 | 3 | 2.5 |
| Serine | 2 | 1.8 | 3 | 2.6 |
| Glutamic Acid | 7 | 7.2 | 7 | 7.1 |
| Proline | 1 | 1.1 | 1 | 0.9 |
| Glycine | 5 | 5.1 | 4 | 4.1 |
| Alanine | 3 | 3.1 | 1 | 1.1 |
| Valine | 5 | 4.7 | 4 | 3.9 |
| Isoleucine | 1 | 0.8 | 2 | 1.9 |
| Leucine | 6 | 6.2 | 6 | 6.1 |
| Tyrosine | 4 | ** | 4 | ** |
| Phenylalanine | 3 | 3.1 | 3 | 3.0 |
| Half-Cystine | 6 | ** | 6 | ** |

* Number of amino acid residues per molecule.

** Not determined.

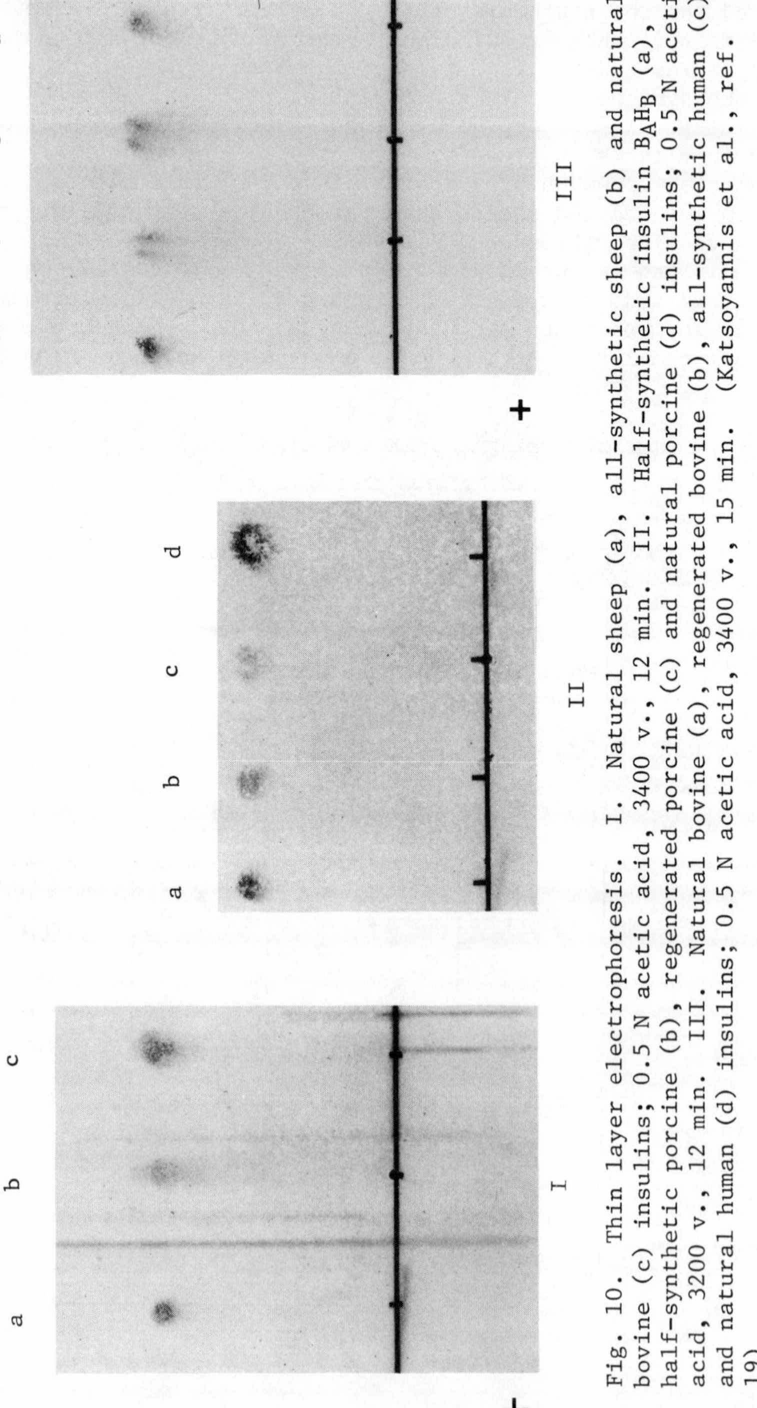


Fig. 10. Thin layer electrophoresis. I. Natural sheep (a), all-synthetic sheep (b) and natural bovine (c) insulins; 0.5 N acetic acid, 3400 v., 12 min. II. Half-synthetic insulin $B_AH_B$ (a), half-synthetic porcine (b), regenerated porcine (c) and natural porcine (d) insulins; 0.5 N acetic acid, 3200 v., 12 min. III. Natural bovine (a), regenerated bovine (b), all-synthetic human (c) and natural human (d) insulins; 0.5 N acetic acid, 3400 v., 15 min. (Katsoyannis et al., ref. 19)

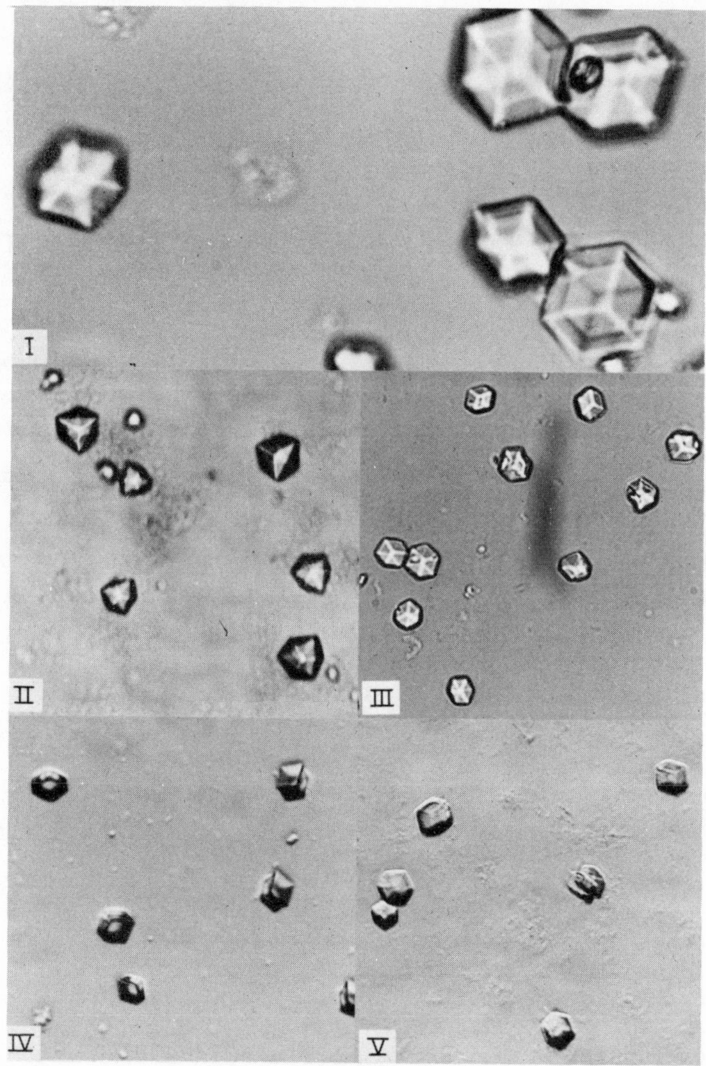


Fig. 11. Crystalline zinc insulins: I, regenerated bovine; II, half-synthetic sheep; III, all-synthetic sheep; IV, half-synthetic bovine; and V, insulin $B_AH_B$.

chains. This new compound is an insulin molecule missing the tripeptide fragment prolyl-lysyl-alanine from the carboxyl terminus of the B chain. When assayed by the mouse convulsion method, this analogue was found to have a potency ranging from 20 to 22 IU/mg. Finally, the synthesis of biologically active and doubly-labeled insulin, a task that can be now readily achieved, will undoubtedly open the way for studying the site of action and the metabolic fate of this protein.

## References


1. Sanger, F. (1959), Science 129, 1340 (Nobel Lecture).
2. Dixon, G. H., and Wardlaw, A. C. (1960), Nature 188, 721.
3. Du, Y.-C., Zhang, Y.-S., Lu. Z.-X., and Tsou, C.-L. (1961) Scientia Sinica 10, 84.
4. Du, Y.-C., Jiang, R.-Q., and Tsou, C.-L. (1965), Scientia Sinica 14, 229.
5. Zahn, H., Gutte, B., Pfeiffer, E.P., and Ammon, J. (1966), Ann. 691, 225.
6. Katsoyannis, P. G., and Tometsko, A. (1966), Proc. Natl. Acad. Sci. U.S.A. 55, 1554.
7. Katsoyannis, P. G., Tometsko, A., and Fukuda, K. (1963), J. Am. Chem. Soc. 85, 2863.
8. Katsoyannis, P. G., Fukuda, K., Tometsko, A., Suzuki, K., and Tilak, M. (1964), J. Am. Chem. Soc. 86, 930.
9. Katsoyannis, P. G., Tometsko, A., Zalut, C., and Fukuda, K. (1966), J. Am. Chem. Soc. 88, 5625.
10. Katsoyannis, P. G., (1964), Vox Sanguinis 9, 227.
11. Katsoyannis, P. G., (1964), Excerpta Medica Inter. Congr. Series No. 83, 1216.
12. Meienhofer, J., Schnabel, E., Bremer H., Brinkhoff, O., Zabel, R., Sroka, W. Klostermeyer, H., Brandenburg, D., Okuda, T., and Zahn, H. (1963), Z. Naturforsch. 18b, 1120.
13. Wang, Y., Hsu, J., Chang, W., Cheng, L., Li, H., Hsing, C., Shi, P., Loh, T., Chi, A., Li, C., Yieh, Y., and Tang, K. (1965), Scientia Sinica 14, 1887.
14. Niu, C., Kung, Y., Huang W., Ke, L., Chen, C., Chen, Y., Du, Y., Jiang, R., Tsou, C., Hu, S., Chu, S., and Wang, K. (1966), Scientia Sinica 15, 231.
15. Katsoyannis, P. G., Tometsko, A., Ginos, J., and Tilak, M. (1966), J. Am. Chem. Soc. 88, 164.
16. Katsoyannis, P. G., Tometsko, A., and Zalut, C. (1966), J. Am. Chem. Soc. 88, 166.
17. Katsoyannis, P. G., Tometsko, A., and Zalut, C. (1967), J. Am. Chem. 89, 4505.
18. Katsoyannis, P. G. (1966), Science 154, 1509.
19. Katsoyannis, P. G., Trakatellis, A. C., Zalut, C., Johnson, S., Tometsko, A., Schwartz, G., and Ginos, J. (1967), Biochemistry 6, 2656.
20. Katsoyannis, P. G. (1966), Am. J. Med. 40, 652.
21. Smith, L. F. (1964), Biochim. Biophys. Acta, 82, 231.
22. Kung, Y., Du, Y., Huang, W., Chen, C., Ke, L., Hu, S., Jiang, R., Chu, S., Niu, C., Hsu, J., Chang, W., Chen, L., Li, H., Wang, Y., Loh, T., Chi, A., Li, C., Shi, P., Yieh, Y., Tang, K., and Hsing, C. (1966), Scientia Sinica 15, 544.
23. Katsoyannis, P. G., Trakatellis, A. C., Johnson, S., Zalut, C., and Schwartz, G. (1967), Biochemistry 6, 2642.
24. Randall, S. S. (1964), Biochim. Biophys. Acta 90, 472.
25. Epstein, C. J., and Anfinsen, C. B. (1963), Biochemistry2,461

# EFFECTS OF DIGESTIVE SECRETAGOGUES ON THE ENDOCRINE PANCREAS IN MAN.

J. Dupré, J.D. Curtis, R. Waddell and J.C. Beck

Fraser Laboratory for Research in Diabetes, McGill University Clinic, Royal Victoria Hospital, Montreal

The rate of disposal of glucose administered to normal man by way of the gastrointestinal tract is greater than that of glucose given intravenously (1,2,3). This phenomenon is associated with potentiation of the increase in blood insulin in response to a rise in blood glucose concentration (4,5). It has also been shown that ingestion of glucose or infusion of glucose into the duodenum in man provokes a rise in plasma glucagon-like immunoreactivity (GLI), whereas intravenous infusion of glucose results in no change or a fall in this activity (6,7). The effect of alimentary glucose on glucose disposal (3) and blood insulin (8) are not impaired by portacaval anastomosis with ligation of the portal vein. Peripheral hyperglycaemia produced by infusions of glucose into peripheral or portal veins has been found to have equivalent effects on blood immunoreactive insulin (IRI) and GLI in man (9). It therefore appears that features distinguishing the response to ingested glucose from that to glucose administered intravenously, are not related to portal hyperglycaemia and probably depend upon alimentary function. These findings have revived the question of the possible role of humoral agents derived from the intestine in the regulation of the endocrine function of the pancreas, a problem that was the subject of extensive but inconclusive investigation in the past (10).

We have previously demonstrated stimulation of increments in blood insulin concentration with crude and purified preparations of porcine secretin (11, 12). We report here further investigation of the effects of hormones known to regulate digestive secretions, on the endocrine response of the pancreas to the intravenous administration of glucose or arginine in man. These studies have been carried out with preparations of porcine secretin and pancreozymin of high specific activity, supplied by

Jorpes and Mutt of the Karolinska Institute, Stockholm; and with synthetic human gastrin obtained from Gregory and Tracy of Liverpool University. In these studies we have had the help of Dr. R. H. Unger and his colleagues at the University of Texas Southwestern Medical School in Dallas, where assays for GLI were carried out.

The secretagogues were administered intravenously in doses shown to have little or no effect on peripheral blood glucose, IRI, or GLI in the fasting state. Each subject received infusions of glucose or arginine with and without one of the secretagogues, and experiments were carried out in random order. When glucose was given, glucose disappearance rates were derived from blood glucose concentrations at intervals during 50 minutes after completion of the infusion.

When secretin was administered intravenously (3.5 u/min, 0.25 µgm/min) together with glucose (7 g/min for 40 mins), the rise in peripheral venous serum IRI in response to hyperglycaemia was enhanced. The mean peak insulin concentration increased to 300% of that observed when glucose was given alone. In all of five subjects, the concentrations attained at corresponding times during infusion of glucose together with secretin exceeded those attained when glucose was given alone. The significance of these results and that of the other experiments to be described was tested by deriving the mean difference between paired values for corresponding times. The standard error of this mean for the whole set showed that the difference was significant ($P<.001$). The validity of this treatment was confirmed by analysis of variance. The half-time for glucose disappearance was significantly reduced when secretin was administered ($P<.02$). In these experiments, no significant change in peripheral serum GLI was observed. Similar studies with pancreozymin (7-21 u/min, 1-3 µgm/min) in six subjects showed that the rise in peripheral serum IRI in response to hyperglycaemia was increased when pancreozymin was given. The mean peak IRI increased to 240% of that observed when glucose was given alone. The paired values differed significantly ($P<.001$). The half-time for glucose disappearance after the infusion was significantly reduced ($P<.01$). There was no significant change in peripheral serum GLI.

The effect of synthetic human gastrin on the response to similar infusions of glucose was examined in a further group of six normal subjects. In these experiments, 25 micrograms of gastrin was given intravenously at a constant rate during the first five minutes of the glucose infusion. This dose was shown to produce an increase in gastric acid secretion that persisted for more than one hour. The rise in peripheral serum IRI in response to hyperglycaemia was increased when gastrin was given. The mean peak IRI concentration increased to 162% of that observed when glucose was given alone. The paired values differed significantly ($P<.001$), and it was also found that the half-time for disappearance of glucose after the infusion was significantly reduced when gastrin was given together with glucose ($P<.05$).

The effects of the same preparations of secretin and pancreozymin in the same doses on the response to the intravenous infusion of arginine were also studied. Arginine was infused at the rate of 0.375 g/min to a total dose of 15 grams in 40 mins. The addition of pancreozymin to the infusate was associated with an increase in the rise of peripheral serum IRI that accompanied the arginine infusions in all of five subjects. The paired values for peripheral serum IRI, treated as before, differed significantly ($P<.005$). In similar experiments with secretin, enhancement of the early rise in IRI was again observed. However, this difference was not maintained throughout the infusions, and in contrast with the effect of pancreozymin, increments in serum IRI at 30 and 40 minutes were consistently smaller than those attained when arginine was infused alone.

The infusion of arginine in all subjects was associated with a rise in blood glucose concentration which was rapid in onset and which was maintained for a variable period. This increase in blood glucose was slightly but consistently prolonged when pancreozymin was given together with arginine, and this effect was not seen with secretin.

In three subjects who received arginine infusions with and without pancreozymin, serum GLI was assayed. Significant changes in peripheral GLI were not observed. It has been shown that the administration of pancreozymin to the dog causes a sharp increase in the output of insulin and of glucagon into the pancreatic duodenal vein (13). It seemed that failure to observe similar effects of arginine infusions in man in studies on peripheral blood might be attributable to dilution of pancreatic venous effluent, and to hepatic uptake of glucagon. Further experiments were therefore carried out in which arginine was administered to normal subjects at a higher rate. The amino acid was delivered at 5 grams/min for 2 minutes with or without the addition of 150 units (20 micrograms) of pancreozymin. It was again found that the rise in peripheral serum IRI that followed injection of arginine alone, was increased significantly when pancreozymin was added. However, there was no demonstrable effect of pancreozymin on the transient rise in blood glucose concentration that occurred in these experiments. Thus it appears that enhancement of the rise in blood IRI when pancreozymin was given in these experiments was not dependent on changes in blood glucose concentration. Moreover, in all of four subjects, a rise in peripheral serum GLI followed the injection of arginine together with pancreozymin and treatment of the paired values for these increments showed that the rise was significantly greater than that observed after infusion of arginine alone ($P<.01$). This effect of pancreozymin on peripheral plasma GLI was apparent only on consideration of the increment in GLI above the values obtained immediately before the infusion was given, and it was not seen in comparisons of total blood GLI. Since there is evidence that total blood GLI consists in at least two components with different origins and different ranges of biological effects, it is suggested that the physiological significance of blood GLI should be assessed in relation to changes in concentration and not to total concentration. Thus with regard to the overall response to glucose administered by way of the gastrointestinal tract, there is accumulating evidence in favour of the suggestion that a humoral agent secreted by the intestine and detectable in the immunoassay for glucagon mediates the potentiation of insulin secretion associated with this particular stimulus (14, 15). Unger and his colleagues have reported experiments with GLI extracted from the intestine of dogs which show that this agent is capable of causing release of insulin from the pancreas in doses which fail to stimulate glycogenolysis (16). On the other hand, changes in blood GLI in response to administration of amino acids appear to depend on the release of glucagon from the pancreas. Thus Unger and his colleagues have found that increments in blood GLI observed in association with administration of mixed amino acids by intravenous infusion or by way of the intestinal tract, were associated with increased rates of release of glucagon into the pancreaticoduodenal vein in dogs (17). It is therefore suggested that the change in peripheral plasma GLI observed in the present experiments in man in response to administration of arginine intravenously and its potentiation by the administration of pancreozymin, is the result of an effect on the release of glucagon from the pancreas. This agent may have local effects on insulin secretion within the islets, and its glycogenolytic effect may be responsible for the rise in blood glucose that occurs in association with administration of amino acids.

Another problem is the question of the physiological significance of effects of endogenous digestive secretagogues on the endocrine functions of the pancreas. Until this can be approached directly by means of assays for these activities in blood, a judgment will have to be based on studies of the effects of procedures which provoke their endogenous secretion. Infusion of isotonic glucose into the small bowel in man does not provoke the exocrine effects of secretin or pancreozymin (18), but it appears that this procedure alone can lead to effects that are qualitatively the same as those of ingested glucose. However, in man the addition of protein to ingested carbohydrate is associated with improvement in glucose tolerance associated with the development of relatively high blood insulin concentrations (19), and this suggests that insulin release can be further potentiated by stimuli which may effect the release of other intestinal hormones in addition to GLI.

The most potent stimulus to the release of secretin appears to be acidification of the duodenal mucosa. We have studied the effect of delivery of hydrochloric acid into the upper gastrointestinal tract during intravenous infusions of glucose. In these studies, glucose was given according to the procedure described above. On one occasion, each subject received an infusion of 30 m.e. of hydrochloric acid in 0.1 normal solution, which was delivered at a constant rate into the duodenum or into the stomach during the intravenous infusion of glucose. In four experiments, acid was infused into the duodenum, and in two it was infused into the stomach. Two of these subjects were studied on two occasions, once with infusion of acid into the stomach and once with infusion of acid into the duodenum. The results obtained on the two occasions were very closely similar. The increase in peripheral serum IRI in each subject was greater when acid was infused and the paired differences were significant ($P<.01$). The mean peak IRI concentration increased to 161% value obtained when glucose was given alone. There was no change in blood GLI. In four out of five of these experiments, the half-time for glucose disappearance following the end of the infusion, was reduced when acid was administered but in the fifth no change was observed. The paired differences in the half-time for glucose disappearance were not statistically significant. However, the mean half-time for glucose disappearance after infusions with acid was very close to that obtained after infusion of glucose with secretin or gastrin. The dose of acid differed from that administered by previous workers who observed no response to delivery of citric acid into the duodenum of normal subjects (20). Since the amount of acid used in the present experiments was well within estimates of acid secretion in response to a meal in man (21), we believe that its effects represent physiological events. It is possible therefore that the effects of gastrin are mediated by stimulation of release of secretin as a result of gastric acid production, although the studies of Unger and his colleagues have demonstrated an apparently direct response to intravenous gastrin in the dog (12).

With regard to the possible physiological role of pancreozymin, there is also some indirect evidence of its importance. The most potent stimulus to the release of this hormone is probably the presence of protein digest in the small intestine. It has been mentioned that the addition of protein to carbohydrate administered by mouth leads to further enhancement of insulin secretion and glucose tolerance. Potentiation of the response to intravenous glucose by the intravenous administration of amino acid might be taken to account for observed effects of mixed feeds (22). However, the effects of ingested protein on insulin secretion in normal man are seen when blood levels of amino acid are much lower than those attained in experiments in which effects of intravenous infusion of amino acids were demonstrated (23). Moreover, the demonstration of enhanced insulin secretion in response to ingested protein in maturity onset diabetes contrasts with the impaired response of similar patients to intravenous arginine (24,25). Taking these findings together with the present evidence, it is suggested that pancreozymin is secreted in response to the protein component of mixed meals and contributes its potentiating effect to the actions of rising blood glucose and amino acid concentrations. On the basis of the effects of intravenous infusions of arginine and pancreozymin, it is suggested that pancreatic glucagon participates in the response to ingested protein. However, when carbohydrate is added to protein, it may be that the rise in blood glucose concentration suppresses the release of pancreatic glucagon, and that the effects of intestinal hormones, including GLI, on insulin secretion are largely direct.

In conclusion, it has been shown that secretin causes insulin release in man. This effect is exaggerated during hyperglycaemia, and is associated with accelerated disposal of glucose. Similar effects result from the infusion of hydrochloric acid in physiological amounts into the upper gastrointestinal tract. The same effect obtained with gastrin may therefore be at least partly mediated by secretin. Pancreozymin also has these effects. In addition pancreozymin, unlike secretin, provokes maintained enhancement of secretion of insulin during intravenous infusion of arginine and causes enhancement of the rise in blood GLI that occurs after rapid intravenous injection of arginine. Arguments have been put forward in support of the suggestion that these effects of the digestive secretagogues are physiologically important. Their relative magnitude and their interactions have yet to be assessed, and their place in the pathophysiology of gastrointestinal or metabolic diseases remains to be explored.

## BIBLIOGRAPHY

1. O. Sommersalo (1950). Acta Paediat. (Helsinki), Supp. 78
2. V. Conard (1955). Acta Gastroenterol., Belg. 18:655.
3. J. Dupré (1964). Lancet ii:672.
4. N. McIntyre, C.D. Holdsworth, D.S. Turner (1964). Lancet ii: 20.
5. H. Elrick, L. Stimmler, C.J. Hlad, Y. Arai (1964). J. Clin. Endocrinol. 24:1076.
6. E. Samols, J. Tyler, A. Marri, V. Marks (1965). Lancet ii: 1257.
7. A.M. Lawrence (1966). Proc. Nat. Acad. Sci. (U.S.A.) 55:316.
8. N. McIntyre, C.D. Holdsworth, D.S. Turner (1965). J. Clin. Endocrinol. 25:1317.
9. J. Dupré, L. Rojas, R.H. Unger, J.J. White, J.C. Beck (1965). Programme 26th Annual Meeting, American Diabetes Association, p. 29.
10. E.R. Loew, J.S. Gray, A.C. Ivy (1940). Am. J. Physiol. 129: 659.
11. J. Dupré, J.C. Beck (1965). Diabetes 15:555.
12. J. Dupré, L. Rojas, J.J. White, R.H. Unger, J.C. Beck (1966). Lancet ii:26.

13. R.H. Unger, M. Ketterer, J. Dupré, A.M. Eisentraut (1967). J. Clin. Invest. 46:630.
14. R.H. Unger et al. (1967). Proceedings of 6th Congress of International Diabetes Federation (in press).
15. E. Samols, V. Marks (1967). Proceedings of 6th Congress of International Diabetes Federation (in press).
16. R.H. Unger et al. (1967). Proceedings of 6th Congress of International Diabetes Federation (in press).
17. ibid.
18. P. Sum and R. Preshaw (1967). Lancet ii:340.
19. L. Kinsell (1967). Proceedings of 6th Congress of International Diabetes Federation (in press).
20. D.R. Boyns, R.J. Jarrett and H. Keen (1967). British Med. J. i:676.
21. S.J. Rune (1967). Clin. Sci. 32:443.
22. J.C. Floyd, S.S. Fajans, S. Pek, C.A. Thiffault, R.F. Knopf, J.W. Conn (1967). Excerpta Medica Int. Cong. Series No. 140, p. 73.
23. J.C. Floyd, S.S. Fajans, J.W. Conn, R.F. Knopf, J. Rull (1966). J. Clin. Invest. 45:1479 and 1487.
24. S. Berger and S. Vargaraya (1966). Diabetes 15:303.
25. T.J. Merimee, J.A. Burgess, D. Rabinowitz (1966). Lancet i:1300.

# SERUM INSULIN RESPONSE TO GLUCAGON AS AN INDEX OF INSULIN RESERVE

A. Benedetti,* R.C. Simpson, G.M. Grodsky, and P.H. Forsham

Metabolic Research Unit, University of California Medical Center, San Francisco, U.S.A.

An attempt to evaluate the endocrine function of the islet cells of the pancreas in man has been possible since reliable methods of measuring serum insulin concentration became available to the clinician.

An approach largely employed in clinical endocrinology is based upon the measurement of the secretory response of a strongly stimulated gland. Following this line, our group has studied the plasma insulin rise after an intravenous acute injection of glucose in normal subjects, as well as in several conditions related to diabetes mellitus (Simpson et al., 1967).

Since in vitro studies with perfused rat pancreas (Fig. 1) have shown that immunoreactive insulin (IRI) is released rather instantaneously when glucose or glucagon are added to the perfusate (Grodsky and Bennett, 1966), an attempt was made to investigate in man the very early phase of insulin release during the first 5 minutes after stimulation (Simpson et al., 1967). We characterized the early insulin response to intravenous glucose in normal human subjects, in subjects with diabetic heritage, lean noninsulin dependent diabetics (Simpson et al., 1966), and nondiabetic obese patients (Bendetti et al., 1967).

Normal controls showed an immediate release of insulin following glucose; potential diabetics (offspring of two diabetics) had, as a group average, a decreased IRI rise even though glucose tolerance was normal; some apparently healthy individuals with normal glucose tolerance showed a very small rise of IRI, in the same range as subjects with potential or overt diabetes and lower K values.

* Present address: Clinica Medica, Universita di Padova, Italy

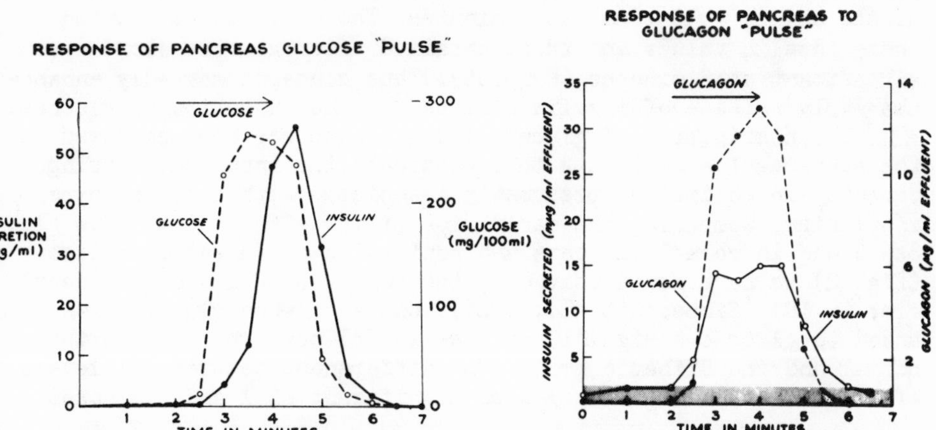


Fig. 1 - Effect of glucose and glucagon on insulin release in vitro.

This observation called for a search for a stronger stimulus, capable of promoting a maximal discharge of insulin.

The specific mechanisms by which insulin is released are still unknown. In addition to glucose, other metabolizable sugars, aminoacids, drugs (such as isoproterenol and sulfanylureas) and hormonal polypeptides (such as glucagon, secretin, pancrezymin, ACTH) stimulate insulin secretion. Despite the accumulation of data it is not clear whether all these substances induce the release of insulin through more than one biochemical signal (Kilo et al., 1967). A considerable interest is connected with the action of glucagon because of the physiological implications due to the glucoregulatory activity of both peptide hormones and because of the proximity of the respective hormone-producing cells. In view of the ability of glucagon to further raise serum IRI when added to a large load of glucose, a different mechanism has been postulated for glucagon.

So far, several investigators (Samols et al., 1965; Porte et al., 1965; Karam et al., 1965; Crockford et al., 1966) noticed that the intravenous injection of glucagon was followed by a rapid significant rise of plasma insulin and that this response was enhanced by simultaneous infusion of glucose.

## RESULTS

We injected glucagon (1 mg) intravenously in 9 normal subjects obtaining an appreciable rise of IRI in 6, with an average peak in nine subjects of 30 and 25 $\mu$U/ml at 1 and 5 minutes respectively. With the addition of glucagon to a 25 g of glucose injection, the mean K value increased from 2.10 $\pm$ 0.07 (mean $\pm$ S.E.M.) to 3.14 $\pm$ 0.32, a statistically significant difference ($P < 0.001$).

Serum IRI concentration reached a mean peak of 106 ± 28 μU/ml at 1 minute and 82 ± 13 μU/ml at 5 minutes. The differences between these insulin values and those obtained after glucose alone were significant at 5 minutes ($P < 0.05$). Thus glucagon markedly enhanced the early release of insulin promoted by the injection of glucose.

The combination of glucose and glucagon has been employed in the subsequent experiments to investigate the insulin-releasing capacity in conditions presumably associated with hypo- or hyper-insulinism. Comparing the early rise of IRI after intravenous glucose alone in normal subjects and noninsulin dependent diabetics (Fig. 2), we have found that only the former showed a significant rise in IRI (Simpson et al., 1966). However, when glucagon was added to glucose a significant rise in IRI occurred in both the normal and the diabetic group. The differences between IRI levels after glucose plus glucagon were significant at 1 and 5 minutes.

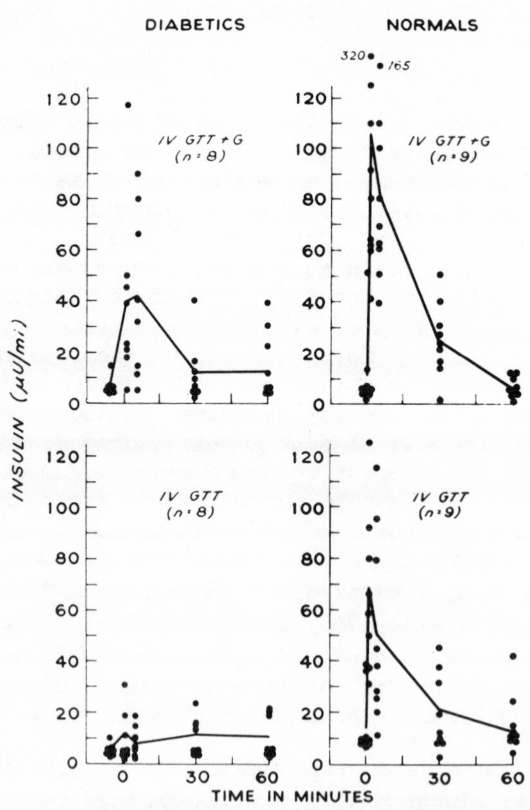


Fig. 2 - Serum IRI responses to intravenous glucose (IV GTT) and glucose plus glucagon (IV GTT G) in normal subjects are contrasted to the responses of nonobese diabetic subjects.

Blood glucose levels at 5 minutes were comparable when either glucose alone or glucose plus glucagon were injected, thus suggesting a direct effect of glucagon on the beta cells and a responsiveness of the diabetic pancreas to such combined stimuli.

On the other hand, when glucagon alone was injected into markedly obese nondiabetic subjects (Benedetti et al., 1967), the early insulin release was significantly higher than in normal subjects ($P < 0.0025$) at 1 and 5 minutes. After the injection of the mixture of glucose plus glucagon, the obese subjects showed a more striking rise in serum IRI than normal controls (Fig. 3). Again, the difference in insulin release between the normal and the obese group is statistically significant.

The results of glucagon plus glucose administration to normal, diabetic and obese subjects are summarized in Table 1.

In acromegalic patients the synergistic effect of glucagon and glucose in promoting insulin secretion was equally evident.

A serum IRI rise was promoted in a patient with longstanding acromegaly and overt diabetes who did not show any IRI response after glucose alone. The most striking insulin rise was observed in a young woman with active acromegaly and obesity, with a peak up to 850 μU/ml following glucose plus glucagon. Again, IRI response was higher than after glucose.

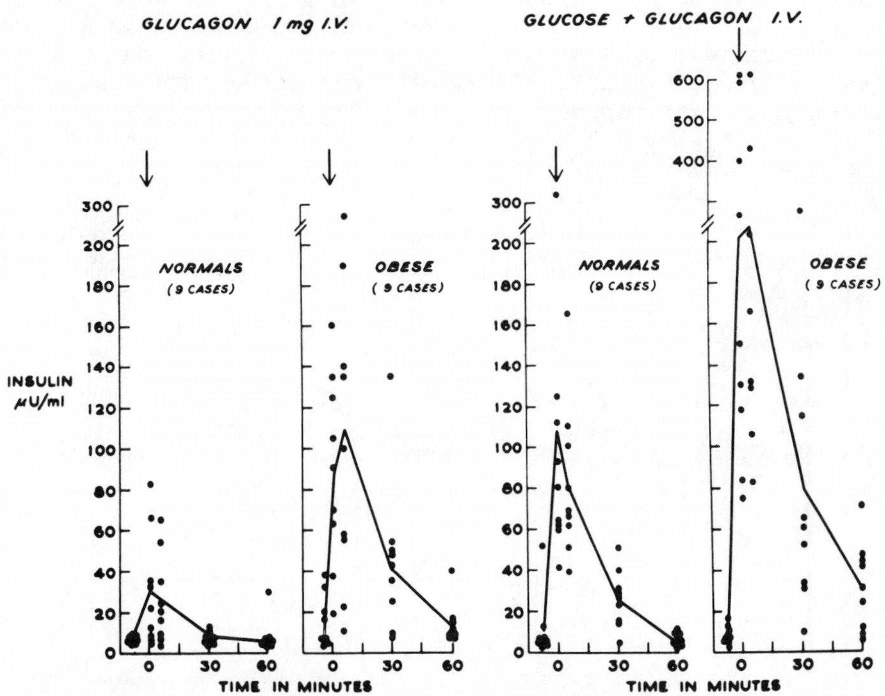


Fig. 3 - Serum IRI responses to intravenous glucose or glucose plus glucagon in normal subjects are contrasted to the responses of markedly obese subjects.

Table 1 - Serum insulin response to glucose plus glucagon in normal, lean diabetic and obese nondiabetic subjects.

| | | Serum insulin ( μU/ml) | | | | | K value |
|---|---|---|---|---|---|---|---|
| | Time | 0' | 1' | 5' | 30' | 60' | |
| Normal subjects (10) | Mean | 11 | 106 | 82 | 25 | 7 | 3.14 |
| | S.E.M. | 5 | 28 | 13 | 5 | 1 | 0.32 |
| Lean diabetics (10) | Mean | 6 | 40 | 42 | 12 | 15 | 0.90 |
| | S.E.M. | 1 | 12 | 12 | 4 | 5 | 0.15 |
| Obese subjects (10) | Mean | 8 | 278 | 272 | 99 | 34 | 1.51 |
| | S.E.M. | 1 | 88 | 71 | 28 | 7 | 0.22 |

We have compared serum IRI responses to various insulin releasing substances (Benedetti, 1967) injected intravenously into the same subjects on separate occasions including glucose, mannose, glucagon, glucose plus glucagon, secretin, glucose plus secretin, secretin plus glucagon. We have found that the combination glucose plus glucagon was followed by the most marked rise of IRI in all patients tested, including individuals with either subnormal or supranormal insulin reserve (Fig. 4).

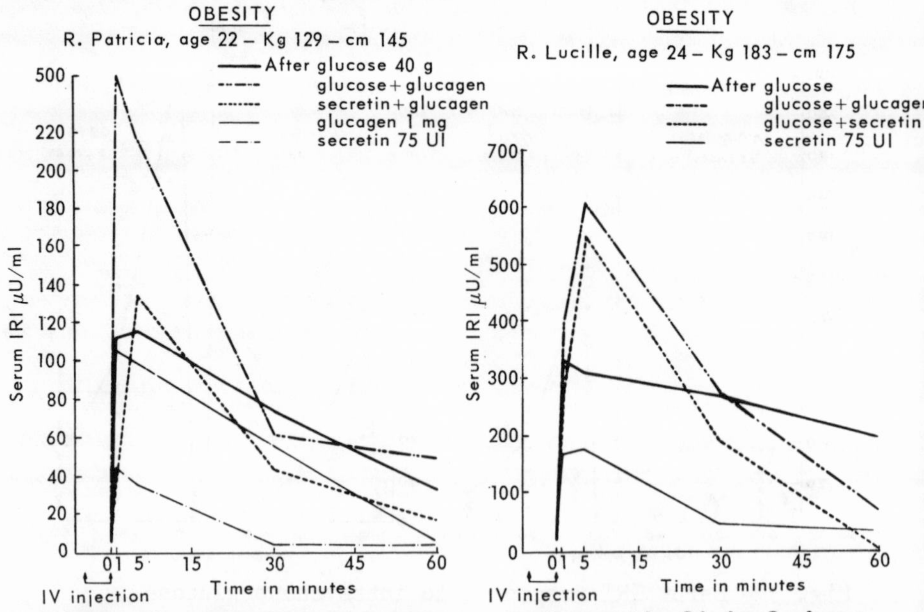


Fig. 4 - Serum IRI response to various stimuli in 2 obese subjects.

Finally, it has been shown that serum IRI rises after glucagon alone are occasionally more pronounced than after glucose injection in patients with mild diabetes, either in acromegalic or obese subjects.

## DISCUSSION AND LITERATURE REVIEW

Glucagon has been shown to increase the artério-venous blood glucose difference in man (Campanacci and Butturini, 1957).

When added to an intravenous glucose load, glucagon increases significantly the glucose disappearance rate in normal subjects.

In obese nondiabetic subjects, the K value rise is not significant. In mild diabetics the K value changes variably, remaining unchanged as a mean (table 1). The increased disposal of glucose found in normal subjects is related to the release of larger amounts of insulin. In the obese the peripheral effect of insulin is less prominent; in diabetics, the insulin response is not adequate to overcome the hyperglycemic effect of glucagon (Benedetti, 1966).

It is known that the stimulation of the vagus nerve promotes insulin secretion. However, glucagon stimulation of insulin has been shown not to be vagus mediated (Frohman et al., 1967).

Insulin secretion has been recently demonstrated in man with beta adrenergic stimulation using isoproterenol (Porte, 1967); this agent has no effect on IRI stimulation by glucagon. The rise in IRI elicited by isoproterenol is prevented by the beta-adrenergic receptor blocking agent, propranolol, which has no effect on glucagon stimulation. Thus the ability of glucagon to elevate serum insulin appears to be mediated through a different mechanism.

Studies employing rabbit or rat pancreas preparations in vitro (Turner and McIntyre, 1966; Vecchio et al., 1966; Sussman and Vaughan, 1967) have shown that glucagon acts directly on the beta cells. An increased production of cyclic 3',5'-AMP within the beta cells with subsequent activation of phosphorylase activity, and thereby increased intracellular glycogenolysis and increased concentration of glucose, has been suggested (Samols et al., 1966; Vecchio et al., 1966); this hypothesis needs further experimental support (Keen and Jarrett, 1967).

We emphasized that in some circumstances glucagon is more effective than glucose in promoting insulin release acutely, as in obese subjects and acromegalics with chemical diabetes.

In premature and newborn infants, Grasso (1967) observed that the infusion of glucose is followed by only a slight and sluggish rise of serum IRI. However, the administration of glucagon to the babies was followed by a marked rise of IRI, up to 206 $\pm$ 14 $\mu$U/ml. This is one more example of a selective sensitivity to glucagon of the insulin-releasing mechanisms.

Tolbutamide promotes insulin secretion within 5 minutes when injected intravenously alone or with glucose, just as glucagon does, in normals as well as in many maturity onset diabetics (Maingay et al., 1967). The rapid sequential administration of three beta cytotropic substances (glucose, glucagon, tolbutamide) appeared to provide an acute maximal stimulus to insulin secretion that could be used to quantify the beta cells reserve (Ryan et al., 1967).

A synergistic effect of certain aminoacids (arginine, leucine, histidine) and glucose upon insulin secretion in man has been reported (Floyd et al., 1967).

The quantity of glucagon employed in our experiments (a standard dose of 1 mg) was above the physiological range; however, Ketterer et al. (1967) obtained in dogs a rise in pancreaticoduodenal vein insulin levels, injecting rapidly as little as 1 $\mu$g of glucagon or even less via the portal vein, confirming that glucagon in physiological doses possesses the qualification of a potentiator of insulin secretion.

Very little is known about the physiological role of glucagon on insulin release. According to Unger (1966), it is tempting to suggest that the food load results in the release of either glucagon from the small intestine or of a hormone which stimulates the release of pancreatic glucagon which in turn potentiates the stimulatory action of hyperglycemia upon insulin release.

A direct action of the alfa cell hormone on the beta cells cannot be excluded because of the contiguity of the cell membranes in the islets.

Several hormonal polypeptides with different structure and physiological properties share the ability of promoting insulin release. Either a relatively simple aminoacid sequence retains this ability, or a single biochemical system is influenced by a number of different polypeptides. More work is necessary to develop this point which holds many important physiological and therapeutic implications.

To further elucidate the mechanism of the beta cytotropic action of glucagon, it would be of importance to determine whether all 29 aminoacids in the molecule are required or if fragments do retain activity. Working on this line, a possible dissociation between the hyperglycemic-glycogenolytic action and the beta-cytotropic activity may be detected, and thus lead to new developments in this field.

## SUMMARY AND CONCLUSION

Our own work has confirmed that the acute injection of glucagon promotes an immediate release of insulin in normal subjects, and we have shown an exaggerated secretory response in nondiabetic obese and acromegalic patients.

The hyperinsulinemic effect occurs within 5 minutes and it is independent of glucagon-induced hyperglycemia.

When glucagon was added to the glucose injection, it potentiated the stimulatory effect of glucose on insulin release in all the groups of patients tested, including those with a low insulin releasing capacity which show poor or no insulin response to intravenous glucose (nonobese insulin-independent diabetics), and those we consider to have secondary hyperinsulinism due to peripheral resistance or antagonism to endogenous insulin (obese and acromegalic patients). The latter gave an exaggerated rise in serum IRI after glucagon suggesting a local effect on the pancreatic beta cells, known to be hyperactive in both conditions.

Our findings suggest a selective sensitivity to glucagon of an insulin-releasing mechanism. Some possible modes of action have been discussed.

ACKNOWLEDGMENT

We are indebted to Miss Graziella Tedeschini for her excellent secretarial assistance.

## REFERENCES

Benedetti, A., 1966, Acta Med. Patav. 26:475.
Benedetti, A., 1967, (abstract), Excerpta Med. International Congress Series No. 140, p. 67.
Benedetti, A., Simpson, R. G., Grodsky, G. M., and Forsham, P. H., 1967, Diabetes, in press.
Campanacci, D., and Butturini, U., 1957, "Il Glucagone in Biologia ed in Clinica", Pacini Mariotti, Pisa.
Crockford, P. M., Porte, D., Wood, F. C., and Williams, R. H., 1966 Metabolism, 15:114.
Floyd, J. C., Fajans, S. S., Pek, S., Thiffault, C. A., and Knopf, R. F., 1967, (abstract), Diabetes, 16:510.
Frohman, L. A., Ezdinli, E. Z., and Javid, R., 1967, Diabetes, 16:443.
Grasso, S. G., 1967, (personal communication).
Grodsky, G. M., and Bennett, L. I., 1966, Diabetes 15:521.
Karam, J. H., Grasso, S. G., Wegienka, L.C., Grodsky, G. M., and Forsham, P. H., 1965, Diabetes 14:444.
Keen, H., and Jarrett, R. J., 1967, (abstract), Diabetes 16:528.
Ketterer, H., Eisentraut, A. M., and Unger, R. H., 1967, Diabetes 16:283.
Kilo, Ch., Devrin, S., Balley, R., and Recant, L., 1967, Diabetes 16:377.
Maingay, D., Ruyter, H. A., Touber, J. L., Croughs, R. J. M., Schopman, W., and Lequin, R. M., 1967, Lancet 1:361.
Porte, D., 1967, Diabetes 16:150.
Porte, D., Graber, A., Kuzuya, T., and Williams, R. H. (abstract) 1965, J. Clin. Invest. 44:1087.
Ryan, G. W., Nibbe, A. F., and Schwartz, T. B., 1967, Lancet 2:1255.
Samols, E., Marri, G., and Marks, V., 1965, Lancet 1:415.
Samols, E., Marri, G., and Marks, V., 1966, Diabetes 15:855.
Simpson, R. G., Benedetti, A., Grodsky, G. M., Karam, J. H., and Forsham, P. H., 1966, Metabolism 15:1046.

Simpson, R. G., Benedetti, A., Grodsky, G. M., Karam, J. H., and Forsham, P. H., 1967, New England J. Med., in press.
Sussman, K. E., and Vaughan, G. D., 1967, Diabetes 16:449.
Turner, D. S., and McIntyre, N., 1966, Lancet 1:351.
Unger, R. H., 1966, Diabetes 15:500.
Vecchio, D., Luyckx, A., Zahnd, G. R., and Renold, A., 1966, Metabolism 15:577.

# THE EFFECT OF GLUCOSE, GLUCAGON, ARGININE, VASOPRESSIN AND ACTH ON PLASMA INSULIN AND GROWTH HORMONE CONCENTRATIONS IN NORMAL CHILDREN AND IN PITUITARY INSUFFICIENCY

Z. Laron, M. Karp, A. Pertzelan, M. Nitzan, M. Doron and T. Waks

Dept. of Pediatrics, Tel Aviv Univ. Med. School and Rogoff-Wellcome Med. Res. Inst., Beilinson Hospital Petah Tikva, Israel

The accumulating evidence that the action of insulin is connected with that of growth hormone (1,2) and the possibility that both hormones act synergistically on the process of growth, led to studies of the physiologic role of insulin in the pediatric age group. The aim of the present study was to investigate the insulin secretory capacity in newborns and children after various stimuli.

Methods. Plasma insulin and growth hormone were measured by radioimmunoassay (3,4). The standards were: Human insulin - Novo Batch 33-3865; HGH - Wilhelmi Lot HS 612B.

The first study was performed in newborns whose mothers received during delivery an infusion of 300 ml 5% glucose, which gave the babies an intravenous glucose load. Blood glucose, plasma free fatty acids (FFA), insulin and HGH were determined in the mother at delivery, in the cord blood and in the newborn at age of 2 hours. The various groups studied and part of the results are shown in Table 1.

TABLE 1

Blood Glucose and Insulin in Mother, Cord and Newborn

| Group | No. | Glucose mg/100 ml | | | Insulin µU/ml | | |
|---|---|---|---|---|---|---|---|
| | | M | C<br>m±SE | N | M | C<br>m±SE | N |
| I Normal | 9 | 88±9 | 80±14 | 46±4 | 16±3 | 28±7 | 38±8 |
| II Diabetic no insulin | 10 | 97±16 | 131±40 | 49±7 | 32±7 | 31±7 | 41±9 |
| III Diabetic with insulin | 8 | 151±35 | 85±7 | 37±10 | *143±23 | 113±23 | 147±30 |
| IV Toxemia | 12 | 106±36 | 98±12 | 28±7 | 23±4 | 29±3 | 29±3 |

M - Mother, C - Cord, N - Newborn

* - Results unreliable because of antibodies in serum

Comparing the glucose concentration in the cord blood and two hours later, it is seen that the glucose decreases, though the insulin concentration is unchanged. A similar decrease in blood glucose in the newborn has been observed without extra stimulation of pancreatic insulin (5). In contrast to babies whose pancreas was not stimulated (plasma insulin<10 µU/ml) all the newborns had elevated plasma insulin concentrations denoting active response to the intrauterine glucose load. There was no difference between babies of normal mothers and those with diabetes not receiving insulin. Serum HGH rose markedly during the first two hours of life to concentrations of Gr: I:65±16 ng/ml; II:53±27; III:56±11; IV:72± 26; plasma FFA rose as well.

Ten children of both sexes ranging in age from 2 to 12 years, were given intravenous injections of crystalline glucagon (Lilly) 50 µg/kg weight. The results obtained in two brothers are illustrated in Fig. 1. During two hours there were two peaks of plasma insulin; the first with the rise of glucose during the first 15 minutes, the second later, at the return of the glucose to fasting levels. The first rise may be a direct stimulation of glucagon on the pancreas in combination with the rise of glucose. The second rise is not understood. Serum growth hormone showed two peaks parallelling those of insulin, possibly due to a glucagon stimulating effect on GH secretion. The fall in plasma FFA may be secondary to the rise of the insulin. Intravenous arginine infusion tests

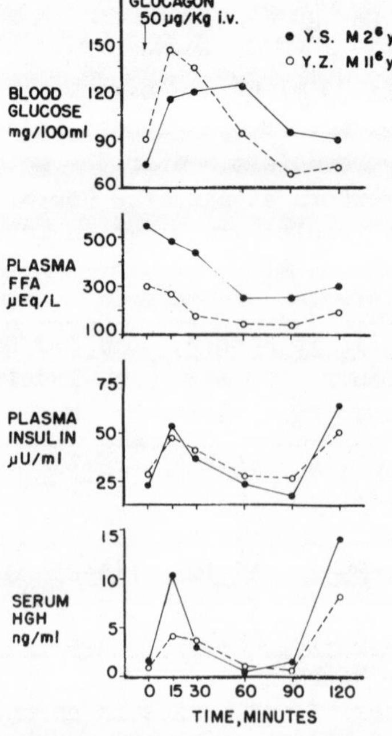


Fig. 1: Intravenous Glucagon Tolerance Test

(0.5 gm/kg, minimal dose 5 gm) were performed in 13 children of both sexes of prepubertal age, from 2 to 13 years old, and 2 pubertal boys. They included normal controls, idiopathic pituitary insufficiency, and genetic pituitary insufficiency (monohormone deficiency of HGH and children with the syndrome of pituitary dwarfism and high serum of inactive HGH (7)). The results are shown in Fig. 2. It is evident that only in the normal children, arginine induced a marked and sustained rise of plasma insulin. In children with pituitary insufficiency there was a slight rise in blood glucose. There was a slight decrease in plasma FFA. Serum GH rose in most of the normal children.

Vasopressin test. (L-8-V-P-Sandoz, 5 to 10 u i.m.) was performed in 28 children of both sexes, including the groups mentioned above and additional children. The results obtained are partly illustrated in Fig. 3. In normal children L-8-vasopressin induced a slight rise in plasma insulin; this was less marked and of shorter duration in idiopathic pituitary insufficiency and non-existent in genetic GH insufficiency. Blood glucose rose mostly in the control group. FFA decreased. There was little change in serum HGH.

Intravenous injection of ACTH (25 U cortrophine - Organon) resulted in a rise of plasma insulin in six normal children. Intravenous injection of tolbutamide induced marked rise in plasma insulin at any age.

In conclusion, the main findings in the present study revealed that: a. Normal children of any age and even newborns respond actively to pancreatic insulin secretion. b. The insulin secretory response in children with pituitary insufficiency of the idiopathic or genetic types with monodeficiency of GH or high plasma inactive GH, was diminished compared to children with normal GH reserve, proving the influence of growth hormone on insulin secretion. c. Vasopressin and ACTH are weaker stimulants of insulin secretion in normal children, as compared to arginine. d. Glucagon resulted in a slight rise in both insulin and growth hormone.

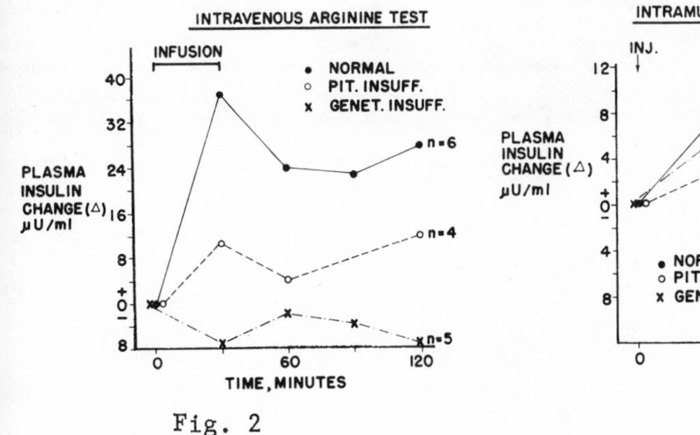


Fig. 2

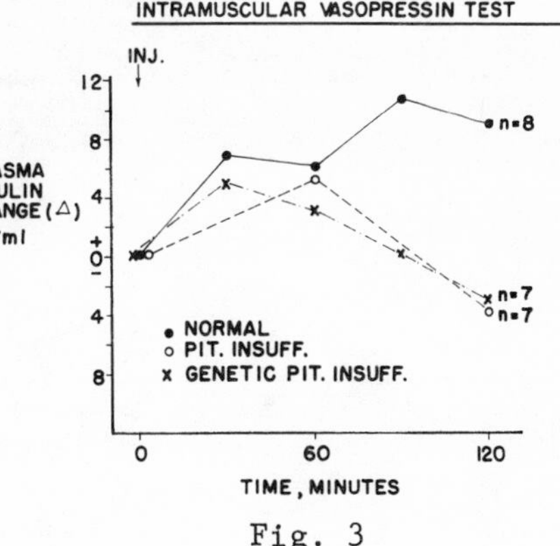


Fig. 3

References.

1. Rabinowitz, D., Merimee, T.J. and Burgess, J.A.
Growth hormone-insulin interaction. Fact and speculation.
Diabetes 15:905-910, 1966.

2. Daughaday, W.H. and Kipnis, D.M.
The growth-promoting and anti insulin actions of somatotropin.
Rec. Progr. Hormone Res. 22:49-99, 1966.

3. Hales, C.N. and Randle, R.J.
Immunoassay of insulin with insulin-antibody precipitate.
Biochem. J. 88:137-146, 1963.

4. Laron, Z. and Mannheimer, S.
Measurement of human growth hormone.
Israel J. Med. Sci. 2:115-119, 1966.

5. Laron, Z., Mannheimer, S., Nitzan, M. and Goldmann, J.
Growth hormone, glucose and free fatty acid levels in mother and infant in normal, diabetic and toxaemic pregnancies.
Arch. Dis. Childh. 42:24-28, 1967.

6. Samols, E., Marri, G. and Marks, V.
Interrelationship of glucagon, insulin and glucose. The insulinogenic effect of glucagon.
Diabetes 15:855-866, 1966.

7. Laron, Z., Pertzelan, A. and Mannheimer, S.
Genetic pituitary dwarfism with high serum concentration of growth hormone. A new inborn error of metabolism ?
Israel J. Med. Sci. 2:152-155, 1966.

INSULIN SECRETION "IN VITRO":
COMPARATIVE ASPECTS AND COMPARISONS WITH STUDIES ON RELEASE OF INSULIN "IN VIVO"

Ernst F. Pfeiffer

Dept.of Clinical Endocrinology and Metabolism

Center of Medicine, University of Ulm/Donau

## I) INTRODUCTION

In the preceding papers many studies were reported regarding the release of insulin "in vivo" in animals and men, mainly under the aspects of gastrointestinal control of insulin secretion. I will take a slightly different position. I shall describe experiments performed in collaboration with Drs. Telib, Ammon, Hildebrandt, Raptis, Schröder, Yeboah, Schwarz and Melani, in which

1) insulin secretion "in vitro" was studied in the presence of different sugars and amino acids, polypeptide hormones of gastrointestinal origin and sulfonylureas of different ß-cytotropic activities,
2) the response of the pancreatic tissue of various species of animals including amphibians was examined
3) the relation between the extractable pancreatic insulin content and the amount of insulin released following stimulation was investigated as well as the conceivable incorporation of glucose and amino acids into the insulin molecule.

Eventually, the problem of existent and non-existent correlations between data gained from "in vitro" and "in vivo" experiments was approached.

Our work was initiated by the consideration that the "true" "in vitro" preparation, i.e. the slices of pancreatic tissue incubated in buffer and stimulating substances, perhaps might offer certain advantages over all other "in vivo" studies concerned with the question of insulin secretion: The responses of the β-cells can be determined directly,independently from glucose and other substances also affecting insulin secretion, and apart from the possible indirect influences of the stimulators on pancreatic insulin secretion, by virtue of their effects on pancreatic blood circulation.

## II) METHODS

The details of the methods employed are given elsewhere (Pfeiffer et al (1965), Telib (1967), Yeboah (1967), Schwarz et al (1967), Hildebrandt (1967)).

1) In brief, the pancreatic glands of rats, rabbits, dogs, calves and frogs were cut in such a fashion that slices of 30-35 mg were collected from caput, corpus and cauda, for equalizing, by mixed incubation, the uneven distribution of the islets of Langerhans in the pancreatic gland. In one tube about 100 mg of pancreatic tissue was incubated in Krebs-Ringer-bicarbonate-buffer with additives of fumarate, pyruvate and glutamate. The use of isolated islets prepared according to Moskalewski (1965) was abandoned because of the clearly lower reactivity when compared with the response of the tissue slices. Viability was controlled over 6 hrs. by measurements of oxygen consumption, and responses to stimulation of insulin secretion by glucose. Best responses were obtained after incubation of 2 and 3 hours duration, respectively.
2) Extraction of pancreatic insulin was carried out according to Davoren (1962), the bioassay of the secreted insulin was performed on the basis of the transformation of 1d-14C-glucose to $^{14}CO_2$ by the rat epididymal fat pad (Martin et al (1958), Ditschuneit, Faulhaber and Pfeiffer (1962)), the immuno-assay following that of Melani et al (1965) with the slight modification of separation of free and bound insulin by charcoal-dextran (Telib (1967)). In some studies insulin secretion was measured by employing antibody-binding of the insulin released into the incubation medium according to Malaisse, Malaisse-Lagae and Wright (1967). The bound labelled insulin in the supernatant was counted 2 hours after further incubation. Under the constant presence of 1 and 100 λ antibodies and 10 μU of labelled

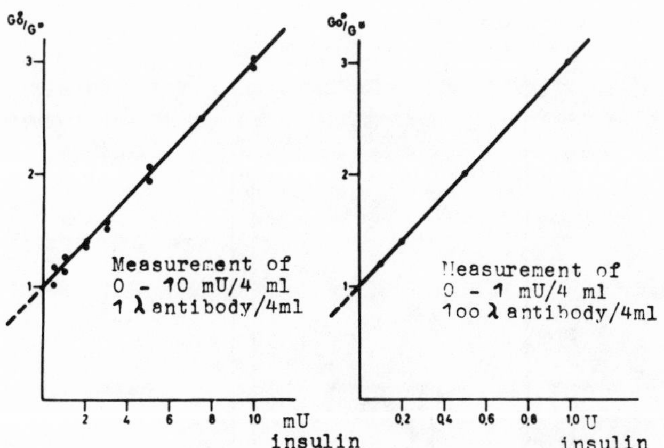


Fig. 1. Radioimmuno assay of insulin secreted in vitro.

insulin in the medium, increasing amounts of unlabelled insulin plotted on the abscissa resulted in standard curves corresponding to o.1 to 1.o and 1.o to 10.o mU insulin, respectively (Fig.1).

3) As shown by Hildebrandt (1967) in our laboratory, the addition of antibodies to the medium prevented insulin degradation by raw pancreatic homogenates, as well as by proteolytic enzymes of pancreatic and other origins. However, the presence of the insulin antibodies per se already produced a high basic rate of insulin secretion. Therefore, in these experiments, only the percentage increase following stimulation exceeding the basic secretion rate has been calculated.

4) Despite the marked conformity in the insulin binding curves of the insulins derived from different species and exposed to one anti-pig-insulin-antiserum, a marked discrepancy between immunological inhibition of biological activity and radio-immunologically measurable insulin was noted. In contrast to the well known incomplete inhibition of the biological activity of serum insulin effected by antibodies, insulin activity in the incubation medium of the tissue pieces was neutralized, before and after glucose stimulation, completely by the antiserum (Fig. 2). Identical suppression of biological activity by antibodies was recorded when pancreatic pieces of dogs, cats, rats and frogs were examined.

Because the ratio of the insulin concentrations measured by bio- and immuno-assay amounted constant-

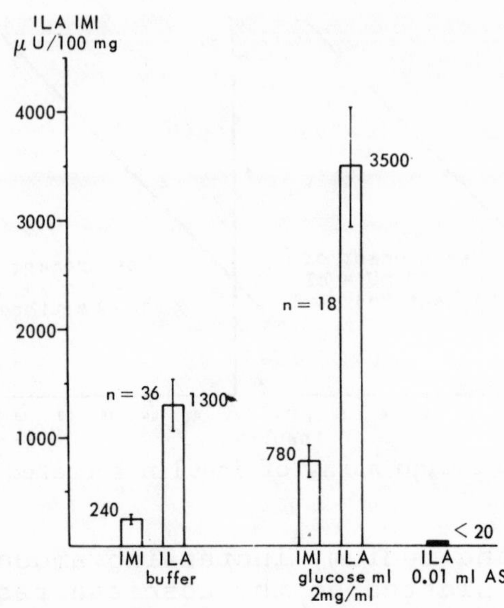


Fig. 2. Insulin secretion "in vitro".: Insulin release from pieces of rabbit pancreas effected by glucose before and after addition of guinea pig antiinsulin serum. Note complete inhibition of ILA by antibodies.

ly to about 5:1, this observation was taken as another example of the lack of correlation between immunologically measurable insulin, and suppressible insulin-like activity (see Pfeiffer (1966), Schöffling (1966)). Therefore, the usual expression IRI, i.e. immunologically reactive insulin, is substituted by IMI, i.e. immunologically measurable insulin, in the following.

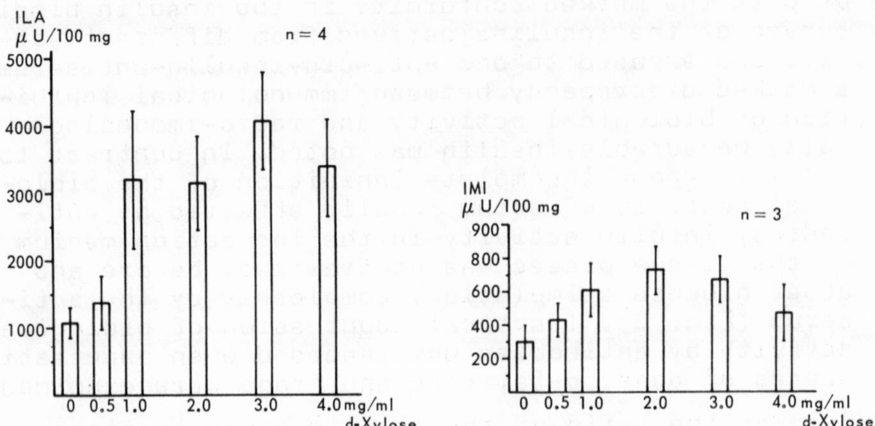


Fig. 3. Insulin secretion "in vitro": Effect of d-Xylose on insulin secretion in vitro (pieces of rabbit pancreas).

## III) RESULTS

1) As shown by Telib (1967) in our laboratory, stimulation of insulin secretion in pieces of rabbit pancreas resulted in similar increases following glucose, fructose, ribose and d-xylose, the most effective concentrations of the hexoses and pentoses being 2 mg/ml (Fig.3 as example). No effect was seen following both galactose and d- and l-arabinose (Fig. 4 as example).
Regarding glucose, fructose, galactose and d- and l-arabinose, our results correlate well with the "semi- in vitro" and "in vivo" data reported by others, using the perfused pancreas or the intact animal (cf. Candela and Coore (1967), Telib (1967)). The ineffectiveness of galactose might be understood on the basis that the intact galactose molecule is not able to enter cellular metabolism, whereas arabinose is known to be metabolically inert in most tissues. On the other hand, following oral ingestion galactose stimulates insulin secretion (Grodsky (1967)).
The positive action of ribose and xylose incubated with pieces of rabbit pancreas is in agreement with the "in vivo" studies of a number of workers (Pozza et al (1958), Biermann et al (1959), Williams (1966)) but at variance with Grodsky's (1963) observation on the perfused rat pancreas, and the studies of Coore and Randle (1964) who used the incubated rabbit pancreas likewise.
Without going into further details it can be stated here, that all sugars participating directly or indirectly in the glucose 6-phosphate-pool, the pentose-phosphate-cycle, the citric acid cycle or the metabolism of nucleic acids do stimulate insulin release from the ß-cells, whereas hexoses and pentoses

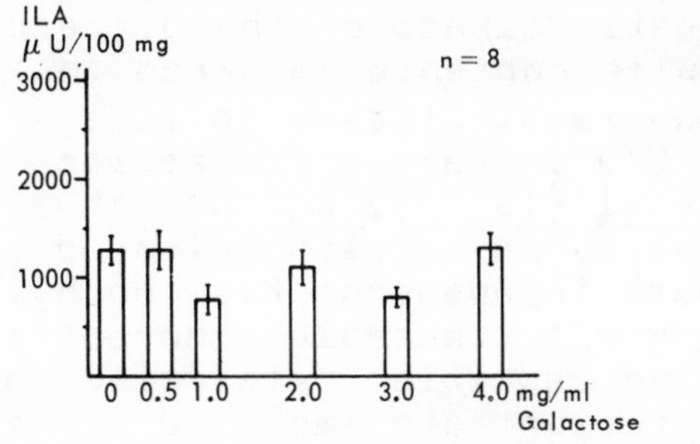


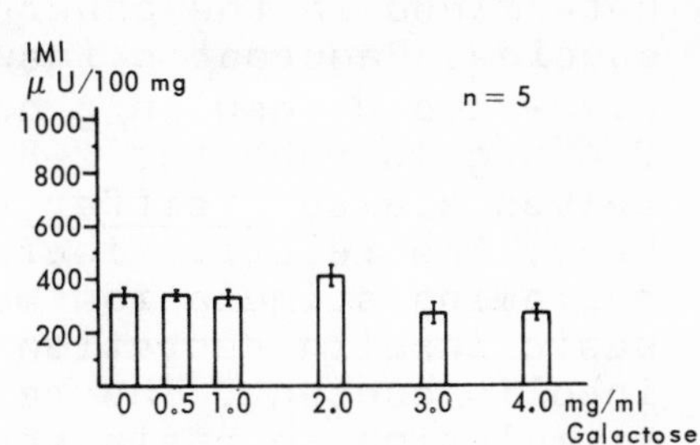


Fig. 4. Insulin secretion "in vitro": Ineffectiveness of galactose on insulin secretion in vitro (pieces of rabbit pancreas).

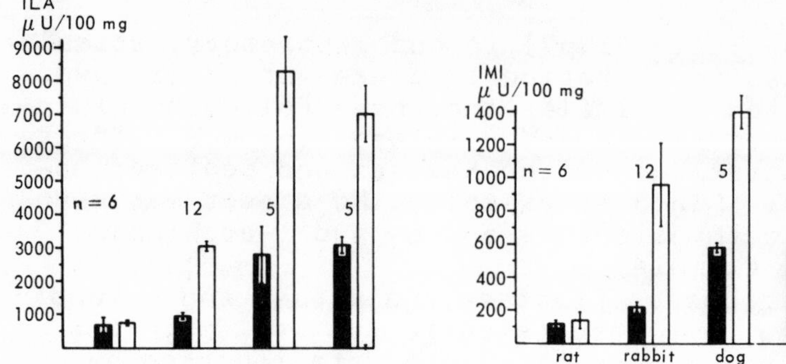


Fig. 5. Insulin secretion "in vitro": The effect of glucose on the release of insulin from pieces of rat, dog and calf pancreas, respectively.

not participating in cellular energy production are ineffective. Obviously, no relation exists between the insulin-mediated uptake of certain sugars by insulin-dependent tissues, and the capacity of hexoses and pentoses to influence insulin secretion, as suggested elsewhere.

Repetitive stimulation of insulin secretion by means of multiple addition of sugars to the same tissue pieces induced higher releases of insulin after each incubation than the single glucose additions (Pfeiffer and Telib (1967)). This pointed to increased sensitization of the ß-cells towards sugars, as established earlier by means of the perfused rat pancreas and "in vivo" in man (Grodsky et al (1967), Pfeiffer (1967)).

Fig. 5 shows the response of the pancreatic slices of rats, rabbits, dogs and calves to glucose stimulation, respectively. The absolute amount of insulin released increased according to the absolute rate of basic insulin secretion, which in turn, correlated well with the quantity of extractable insulin determined in the pancreatic glands of the different species. Pancreatic insulin contents amounted to o.7 – 1.o U insulin/g pancreatic tissue in rats, 2–4 U/g in rabbits, 2–8 U/g in calves (regarding calves s.also Pfeiffer etal (1957)). On the other hand, the relative increases in insulin released following stimulation were independent from both, basic insulin secretion and extractable pancreatic insulin content. The ratio secretion rate following stimulation to basic secretion rate amounted to 5:1 for rabbits and 3:1 for dogs and calves, the absolute loss in extractable pancreatic insulin content not

exceeding 1 to 5% (cf. Candela and Coore (1967), Telib (1967)).
The extractable pancreatic insulin content in frogs amounting to about 20-40 mU/g, was independent of the season in which the animals were caught. However, the pancreatic pieces of frogs caught in Winter and early Spring did not show any responses to glucose stimulation. Only at the beginning of the Summer period and lasting until Fall, were clearcut increases in insulin release and reductions in the extractable pancreatic insulin content induced by addition of glucose (Fig.6). This seasonal variation in pancreatic insulin secretion of amphibians correlated only to a certain extent with the basic insulin secretion, which showed differences in the 4 seasons, too. It might depend, to some degrees, on the consistency of the pancreatic gland in the different periods. Pancreatic glands of the winter frogs contained not much water and were of a harder consistency, while the glands of summer frogs contained much water, showing softer consistency (further details s. Telib (1967)).

2) Regarding stimulation of insulin secretion "in vitro" by amino acids, pieces of rat and rabbit pancreas were incubated in media consisting of the usual buffer plus o.1, 1.o and 10 mg amino acid/ml for 2 and 3 hours, respectively (Pfeiffer et al (1967), Yeboah (1967)). In Fig. 7 are shown the responses of rabbit and rat pancreas "in vitro" to l-leucine, the first amino acid shown to induce insulin secretion "in vivo" in insulinoma patients (Cochrane et al (1956)). Increases in IMI concentrations up to 300% of the initial values recorded as 100% were noted. Higher sensitivity of the rabbit than of the rat ß-

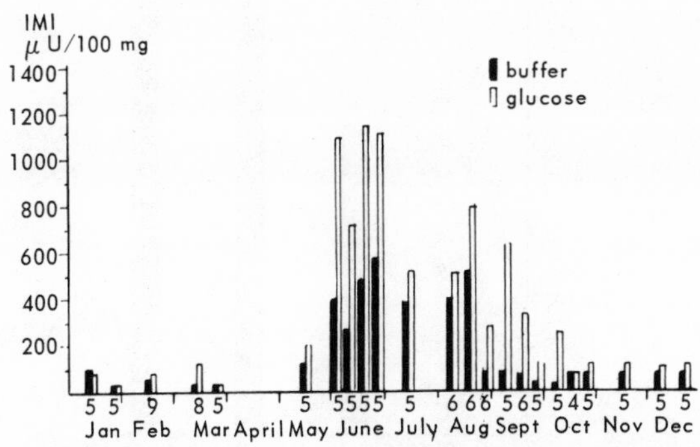


Fig. 6. Insulin secretion "in vitro": The effect of glucose (2.0 mg/ml) on the release of insulin from pieces of frog pancreas in relation to the seasons. Number of the frogs examined is given below the bars.

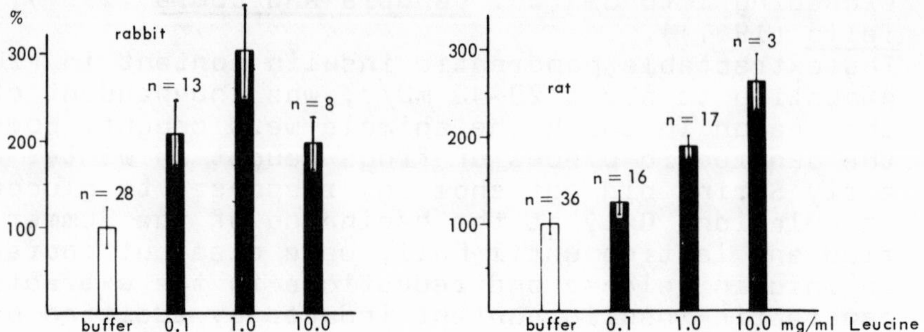


Fig. 7. Insulin secretion "in vitro": The effect of leucine on the release of insulin from pieces of rabbit and rat pancreas.

cell system may be inferred from the lower effective leucine concentration in the former species. This lower threshold of the rabbit pancreas went parallel with a lower or limited responsiveness to the 10.o mg/ml concentration which induced the greatest release of insulin from the rat pancreas. Addition of glucose to arginine did not produce further increases in IMI concentrations in the medium. Similar observations were made when other amino acids were examined, i.e.iso-leucine, arginine, methionine, tryptophane and valine.

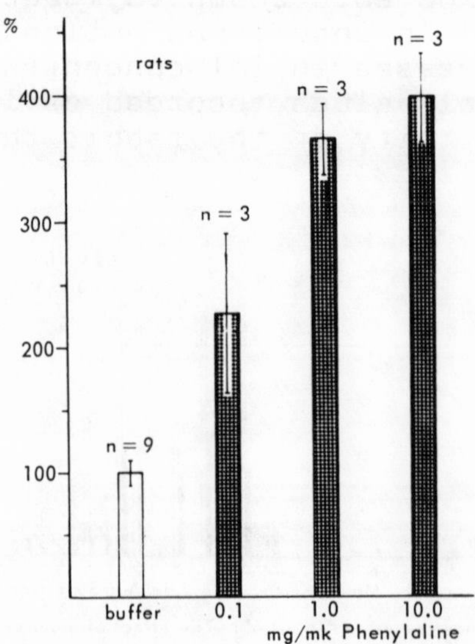


Fig. 8. Insulin secretion "in vitro": Insulin secretion by pieces of rat pancreas stimulated by different doses of phenylalanine.

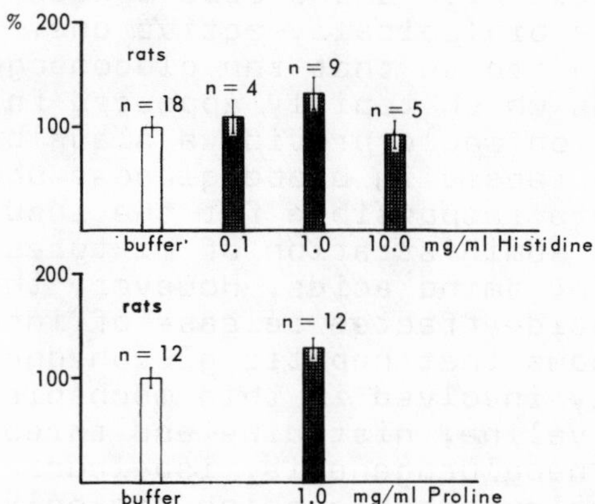


Fig. 9. Insulin secretion "in vitro": No insulin secretion from the rat pancreas induced by histidine and proline.

Insulin release from rat pancreas was most effectively stimulated by phenylalanine (Fig.8), least effectively by histidine and proline (Fig.9). A comparison of the results obtained with pieces of rat pancreas incubated in concentrations of 1.o and 10.o mg/ml, respectively, of different amino acids is given in Fig.10. The ineffectiveness of histidine with regard to stimulation of insulin secretion "in vitro" is in accordance with the observations made with the intravenous administration of essential l-amino acids to healthy human subjects (Floyd et al (1966)). On the other hand, the highest activity of leucine is somewhat at variance with these studies where arginine was the most active substance. However, in accord with the observations in humans is our own finding that the lowest effective concentrations of o.ol and 0.1 mg/ml were found when arginine, leucine and methionine, respectively, were added to the medium. Thus, the studies carried out with the delibe-

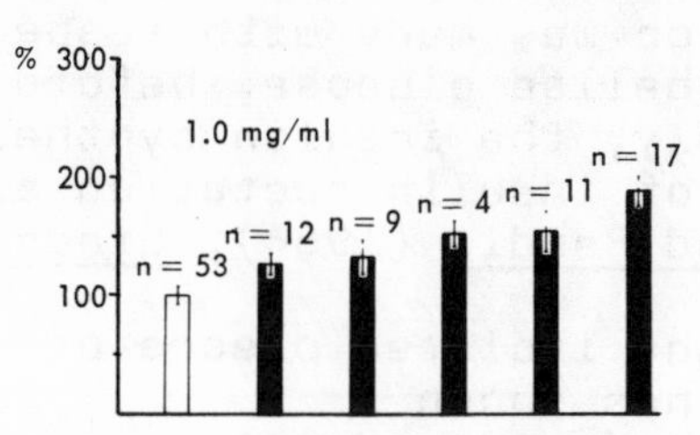


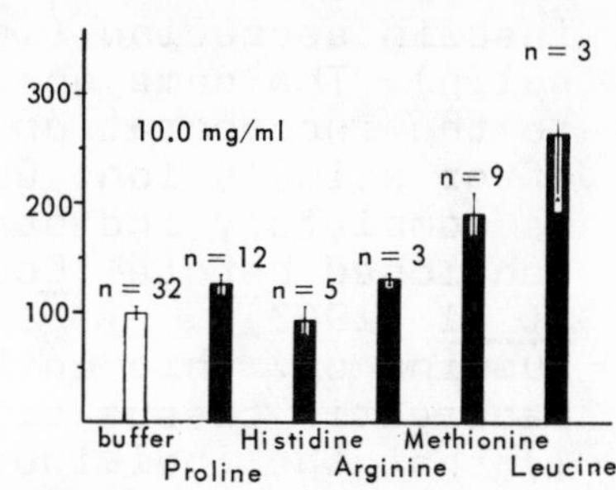


Fig. 10. Insulin secretion "in-vitro": Insulin release from pieces of rat pancreas effected by different aminoacids. Most effective stimulation induced by 1-leucine, least effective by histidine.

rately chosen high amino acid concentrations may not reflect the biologically active one.

It has been argued that the gluconeogenesis from amino acids which rapidly appeared following amino acid infusion would provide a plausible explanation for the increases in blood glucose observed, which in turn were responsible for the insulin release ensuing from administration of mixtures of or individually potent amino acids. However, the demonstration of amino acid-effected release of insulin "in vitro" clearly shows that hepatic gluconeogenesis is not necessarily involved in this mechanism. Furthermore, arginine, valine, histidine and threonine do participate in neoglucogenesis, iso-leucine is only weakly glucogenic, phenylalanine has only questionable glucogenic activity, and leucine, lysine, methionine and tryptophane are non-glucogenic. Thus, amino acids of very different capacity with regard to stimulation of insulin secretion are found in each of these groups.

The direct stimulation of insulin liberation "in vitro" by a number of individual amino acids indicates that secondary factors in metabolism of amino acids must be of minor importance. It certainly cannot be excluded that incorporation of the amino acids into the intracellular metabolic processes of the ß-cells is the effector mechanism. Arginine and lysine participate in the urea cycle, histidine, proline, leucine, isoleucine and valine enter the citric acid cycle. It remains completely open whether the metabolism of these substrates is operative in inducing insulin secretion.

All of the 9 amino acids examined, except methionine and tryptophane are present in the insulin molecule. However, as shown by <u>Meier</u> (1967) in our laboratory, none of the 14-C-amino acids found in insulin were incorporated at a higher rate during stimulation of insulin secretion (into the freshly synthesized insulin). The same observation was made with respect to the incorporation of labelled glucose, before and after stimulation. Obviously, the insulin synthesis is completely independent of insulin secretion as mentioned before (<u>Coore and Randle</u> (1964), <u>Grodsky et al</u> (1963)).

Summing up, this work on the isolated pieces of pancreatic tissue clearly has shown

1) that the individual amino acid and not a complete mixture of several amino acids is capable of inducing insulin secretion;

TABLE 1

Intestinal Hormones affecting Insulin Secretion in vitro

| Substance | Effect (Species)* | Reference |
|---|---|---|
| Glucagon | + (d) | Candela et al (1961) |
| | +(+) (R) | Turner & McIntyre (1966) |
| | + (r) | Vecchio et al (1966) |
| | + (R) | Milner & Hales (1967) |
| Secretin | +(+) (R) | Pfeiffer et al (1965) |
| | + (R) | McIntyre et al (1965) |
| | +(D,R,r,C,F) | Telib (1967) |
| Pancreozymin | + (R) | Schröder et al (1967) |
| Gastrin | + (R) | Schröder et al (1967) |
| Serotonin | + (R) | Pfeiffer (1967) |
| Duodenal Extract | + | Candela et al (1967) |

* R = Rabbit, D = Dog, r = rat, d= duck, F = Frog, C = Cat
\+ = Glucose independent stimulation
\- = no effect
(+)= Glucose supported stimulation

2) that in addition to leucine, also the other amino acids possess high insulin stimulating capacity;
3) that direct stimulation of insulin occurs which is glucose-independent.

3) In Table 1 are listed the studies relevant to the stimulation of insulin secretion "in vitro" by the different hormones of gastrointestinal origin. The early observations of Candela and his associates (1961) regarding the insulin stimulating action of glucagon by the "in vitro" preparation of duck pancreas were corroborated by several workers, among them Vecchio et al (1966) who worked with the cultured fetal rat pancreas supposedly free from $\alpha$-cells producing endogenous glucagon. This is the only experiment in which the participation of glucagon in the glucose-mediated insulin release "in vitro" was made unlikely. The possibility that endogenous pancreatic glucagon is the trigger of insulin release following addition of glucose certainly has to be considered with regard to all other "in vitro" studies using simple incubated pancreatic slices, including our own.

Following secretin, about the same increases in insulin secreted into the medium relative to the basic secretion were noted in our studies on the pancreatic glands of summer frogs, rabbits and dogs (Fig.11).

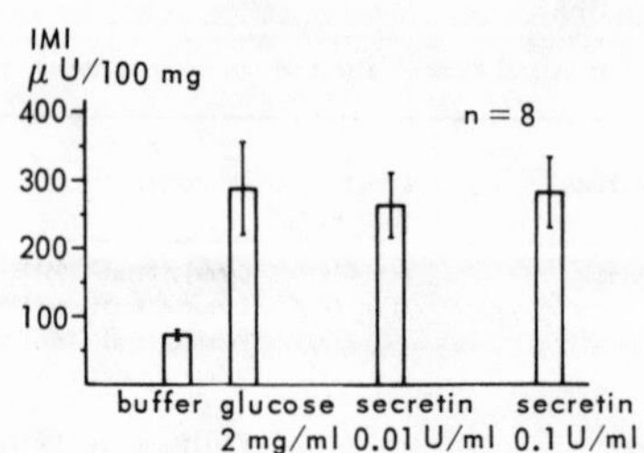


Fig. 11. Insulin secretion "in vitro": Insulin secretion from pieces of frog pancreas induced by secretion (0.01 and 0.1 U/ml). 0.1 U/ml required for producing effect comparable to that one effected by optimal glucose concentration.

Differences existed only with respect to the lowest effective secretin concentrations. Whereas in rabbits, maximum increases in insulin output were induced by o.l U secretin/ml (Pfeiffer et al (1965)), in frogs concentrations as low as o.ol U secretin/ml effected maximum insulin liberation (Telib, (1967)). Obviously, secretin-mediated insulin secretion phylogenetically belongs to the older endocrinological mechanisms.

Pancreozymine, obtained from "Karolinska Institute", Stockholm, was effective in concentrations of 1,5 and 10 U/ml on pieces of rabbit pancreas (Fig.12). Pancreozymine, obtained from "Boots", England, in concentrations of 1 U/ml, produced increases in insulin output similar to those obtained by the same concentration of pancreozymine "Karolinska".Following incubation with 5 U of the British preparation/

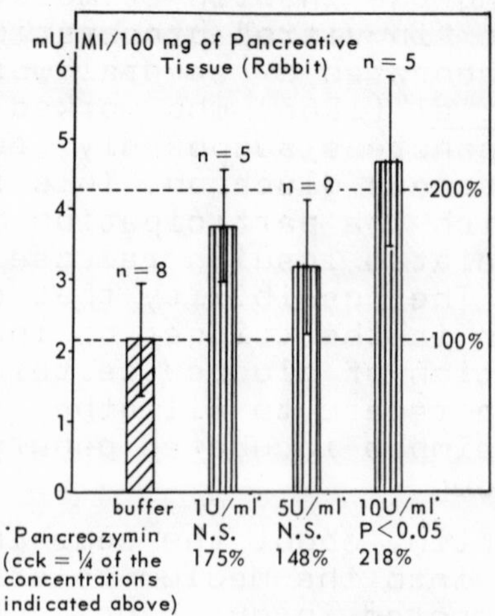


Fig. 12. Intestinal hormones affecting insulin secretion "in vitro:" Pancreozymin (CCK-PZ "Vitrum"), 1-10 U/ml. Release of insulin from rabbit pancreas.

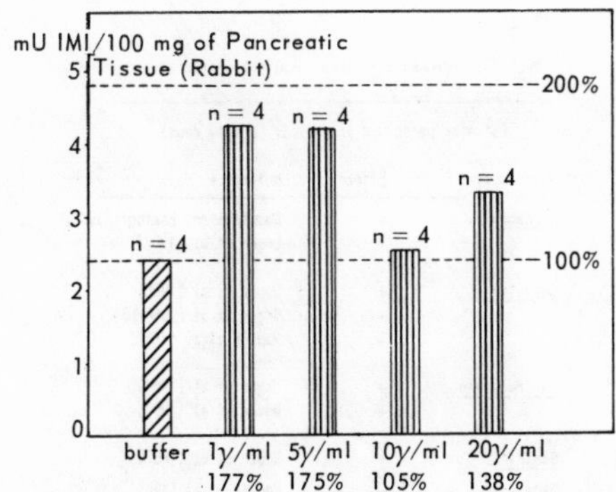


Fig. 13. Intestinal Hormones Affecting Insulin Secretion in Vitro Gastrin-Pentapeptide "ICI" 1-20 γ/ml. Release from Insulin from Rabbit Pancreas.

ml, however, a quite unexpected steep rise in immuno-insulin far exceeded the stimulation of insulin secretion obtained by the same concentration of the "Karolinska" preparation. The possibility of a simultaneous stimulation of glucagon secretion by the "Boots" pancreozymine has to be considered.

Gastrin-pentapeptide from Imperial Chemical Industries, England, though somewhat less potent clearly induced release of insulin by the rabbit pancreas "in vitro", too (Fig.13). More details on the insulin stimulating action of these intestinal hormones are given elsewhere (Schröder et al (1967)).

Finally, serotonin was also shown to effect insulin release from the rabbit pancreas in the "in vitro" preparation (Fig.14), whereas "in vivo" neither in animals nor in men did stimulation of insulin secretion occur. However, with the exception of this

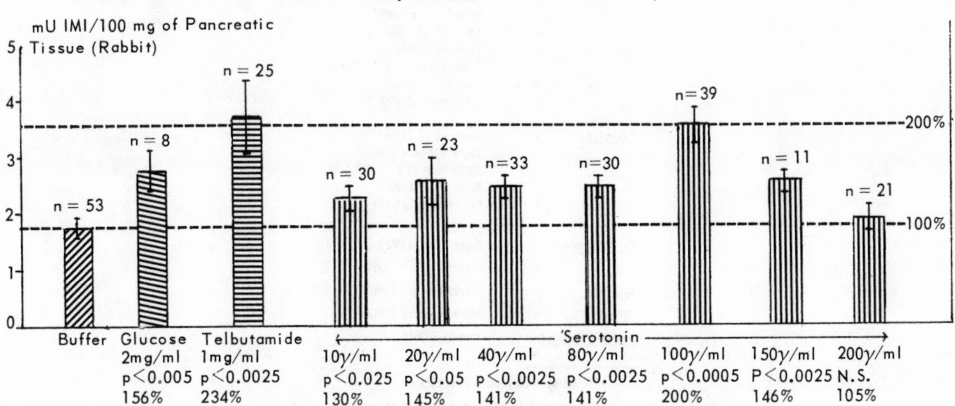


Fig. 14. Intestinal Hormones Affecting Insulin Secretion in Vitro: Serotonin 10-200 γ/ml. Release of Insulin from Rabbit Pancreas.

TABLE 2

Intestinal Hormones affecting Insulin Secretion in vivo

(Studies performed in animals, mostly dogs)

| Substance | Effect* | Reference |
|---|---|---|
| Glucagon | + | Campbell and Rastogi (1966) |
| | + | Unger et al (1967) |
| Secretin | + | Dupré et al (1966) |
| | + | Unger et al (1966/67) |
| | + | Raptis et al (1967) |
| Pancreozymin | + | Unger et al (1967) |
| | + | Meade et al (1967) |
| Gastrin | + | Unger et al (1967) |
| Serotonin | - | Raptis et al (1967) |
| Duodenal Extract | ∅ | |

* + = Stimulation
- = no Stimulation
∅ = no data available

lack of correlation between the "in vitro" and "in vivo" studies on serotonin, the results of the "in vitro" experiments concerned with gastrointestinal factors in control of insulin secretion, in general, were well in accordance with the "in vivo" observations in both, animals and men (Table 2 and 3).

One of the advantages of the use of the "in vitro" preparations for clarifying the action of enteric

TABLE 3

Intestinal Hormones affecting Insulin Secretion in Man

| Substance | Effect* | Reference |
|---|---|---|
| Glucagon | + | Langs and Friedberg (1965) |
| | + (+) | Samols et al (1965) |
| | + | Melani et al (1966) |
| | + | Karam et al (1966) |
| | + | Lawrence (1966) |
| | + | Grodsky and Bennett (1966) |
| | + (+) | Ryan et al (1966) |
| | + | Simpson et al (1967) |
| Secretin | + | Dupré et al (1966) |
| | + | Dupré and Beck (1966)* |
| | (+) | Bottermann et al (1967 a &B) |
| | + (+) | Boyns et al (1967) |
| | + (+) | Raptis et al (1967) |
| Pancreozymin | - | Boyns et al (1967) |
| | + | Pfeiffer (1967) |
| Gastrin | - | Pfeiffer (1967) |
| Serotonin | - | Schröder et al (1967) |
| Duodenal Extract | + | Vanotti (1967) |

* + = Glucose independent stimulation
(+) = Glucose or Tolbutamide supported stimulation
- = no effect
* = crude extract of Secretin

hormones on the β-cells is exemplified by the fact that in studies in dogs "in vivo" Delaney and Grim (1966) found increases in pancreatic perfusion following secretin and serotonin, whereas pancreozymine had no marked effect. An indirect method utilizing isotope clearances was employed for measuring pancreatic blood flow. This certainly permitted more correct estimates of pancreatic blood circulation than the surgical procedure which was used by others and ourselves (Raptis et al (1967)). Furthermore, in the fasting dog external pancreatic secretion is so low, that constant secretin infusion is needed for evaluating changes in the volume output. Without the results obtained by means of the incubated pancreatic tissue it seemed rather unlikely that such profound variations in the working activity and fluid production of the whole pancreatic gland, as effected by secretin perhaps via changes in the effective blood flow, were not primarily involved in the secretin-induced stimulation of insulin liberation.

4) Finally, an excellent quantitative correlation between the capacity of different substances of similar chemical configuration to stimulate insulin secretion in both, "in vivo" and "in vitro", was found in the β-cytotropic actions of tolbutamide and a new compound, HB 419, cyclohexysulfonylurea, respectively (Schwarz et al (1967)). "In vivo" in rats the new substance, HB 419, was about hundred times, and in rabbits about thousand times more effective than tolbutamide. "In vitro" in rats, HB 419, in concentrations of 10.o γ/ml, stimulated insulin secretion to the same extent as 1.o mg tolbutamide ml, i.e. about hundred times more effectively (Fig.15). Pieces of rabbit pancreas responded to 1.o γ HB 419/ml incubation medium by increasing the insulin secretion to about 175% of the controls, i.e. to the same percentage which was obtained by thousand times higher concentrations of tolbutamide (1.o mg/ml). Therefore, the high activity of the new compound permitting doses as small as o.1 mg/kg in treatment of diabetic subjects may be ascribed entirely to its direct pancreotropic effect.

## IV) SUMMARY AND CONCLUSIONS

Time does not permit to present further studies on the stimulating and inhibiting influences of other hormones

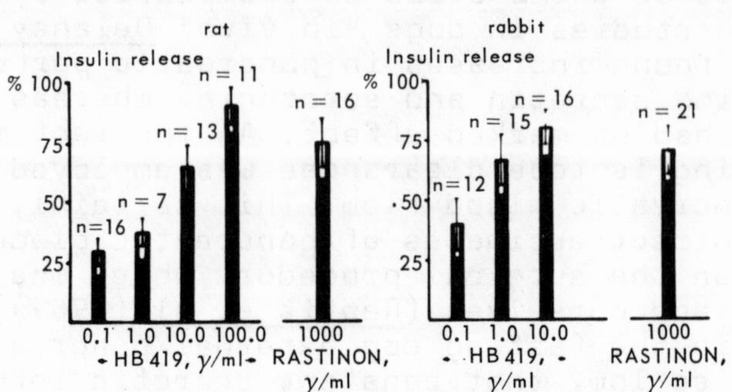


Fig. 15. Insulin secretion "in vitro": Insulin release from pieces of rat and rabbit pancreas, respectively, effected by Hb 419 and Tolbutamide in different concentrations. Note the approximately 100 and 1000 times more effective actions of Hb 419 on rat and rabbit pancreas, respectively.

and pharmacological substances on the "in vitro" preparation.

In conclusion, since in all species examined, including amphibians, similar relative increases in insulin secretion were observed following stimulation, the same basic mechanism operative in insulin secretion, perhaps concerning the same fraction of intracellular insulin, may be assumed. This common operator has to be found. It seems highly unlikely that any of the specific pathways of intracellular metabolism of glucose and amino acids mentioned is involved. The only fact established beyond doubt is the dominant influence of extracellular glucose and amino acid concentration. The effects of glucagon, adrenaline, theophylline, caffeine and cyclic 3' 5' AMP "in vitro" (cf. Candela and Coore (1967)) may indicate the primary role of this nucleotide in the insulin releasing and - as regards adrenaline - inhibiting mechanism.

As stated before by Grodsky and Bennett (1963), Coore and Randle (1964) no evidence was also provided by our studies that insulin secretion following stimulation is connected with insulin synthesis. However, this result of the acute experiment does not exclude any relationship between insulin secretion and total insulin production in the long term "in vivo" (Candela and Coore (1967)). The highest absolute quantity of insulin liberated following different stimuli was established in the incubation medium of pancreatic pieces of animals which showed the highest basic secretion and the highest content in extractable insuln (Telib (1967)).

However, this relationship was found only with regard to the pancreatic tissue of mammalians. In the studies in frogs, despite marked seasonal differences between insulin secretion in the fasting and in the stimulated state, the pancreatic insulin content remained unchanged (Telib (1967)). Obviously, in this species in certain annual periods the capacity of the ß-cells to respond to stimulators is blocked while the basic insulin secretion is sufficient for providing normal glycogen reserves in muscles and liver tissue (Hanke and Bergerhoff, unpubl.)

The minimal reduction in extractable pancreatic insulin content observed following stimulation "in vitro" is in contrast to the remarkable losses in pancreatic insulin determined "in vivo",as has been established, e.g. in calves following tolbutamide (Pfeiffer et al (1957)). This certainly is inducing one to draw only cautious conclusions regarding the biological significance of the studies carried out on the "in vitro" preparation. The more impressive is the close correlation which has been established between the "in vitro" and "in vivo" findings with regard to the insulin secretion induced by sugars, amino acids, enteric hormones and sulfonylureas.

## REFERENCES

Biermann E.L., E.M.Bakar, J.C.Plough and W.H.Hall: Metabolism of d-ribose in diabetes mellitus. Diabetes 8, 455, (1959)

Candela J.L.R., R.R.Candela, D.Martin and T.Hernandez and Castilla Costazar: Insulin secretion "in vitro" (prelim.comm.) Proc.4th Congr.I.D.F. Geneva (1961) p.629

---- and H.G. Coore: Insulin secretion "in vitro" in: Handbuch des Diabetes mellitus, Pathophysiologie und Klinik, Ed. E.F. Pfeiffer, Lehmanns Verlag, München (in press)

Coore G. and Randle P.J.: Regulation of insulin secretion studied with pieces of rabbit pancreas. Biochem. J. 93, 65, (1964)

Davoren P.R.: The isolation of insulin from a single rat pancreas. Biochem.Biophys. Acta 63, 150, (1962)

Delaney J.P. and E.Grim: Influence of hormones and drugs on canine pancreatic blood flow. Am.J.Physiol., 211, 1398-1402, 1966

Ditschuneit H., J.P.Faulhaber und E.F.Pfeiffer: Verbesserung der Methode zur Bestimmung von Insu-

lin im Blut mit Hilfe radioaktiver $^{14}$J-C-Glukose und dem epididymalen Rattenfettgewebe.
Atompraxis 8, 172 (1962)
Floyd J.C., S.S.Fajans, J.W.Conn, R.F.Knopf and J.Rull: Insulin secretion in response to protein ingestion J.Clin.Invest. 45, 1479 (1966)
Grodsky G.M. and L.L.Bennett: Insulin secretion from the isolated pancreas in absence of insulinogenesis: effect of glucose.
Proc.Soc.Exper.Biol.and Med. 114, 769 (1963)
---- and D.F.Smith: Effect of Carbohydrates on secretion of insulin from isolated rat pancreas.
Amer.J.Physiol. 205, 638 (1963)
----, L.L.Bennett, D.F.Smith and F.C.Schmid: Effect of pulse administration of glucose or glucagon on insulin secretion in vitro.
Metabolism 16, 222 (1967)
Hanke W. and K.Bergerhoff: (unpubl.)
Hildebrand H.E.: Contribution to discussion.
Communication to Stop-Press-Session, 6th Congr. IDF Stockholm, Sweden, Aug.1967
Malaisse W., F.Malaisse-Lagae and P.H.Wright: A new method for the measurement in vitro of pancreatic insulin secretion.
Endocrinology 80, 99 (1967)
Martin D.B., A.E.Renold and Y.M.Dagenais: An assay for insulin-like-activity using rat adipose tissue.
Lancet II, 76 (1958)
Meier J.M., J.Ammon, U.Gröschel-Stewart, F.Melani, J.E.Yeboah und E.F.Pfeiffer: Eine einfache, schnelle Methode zur Bestimmung von Insulin in kleinsten Gewebsmengen (bis 10 mg)
13.Symp.Dtsch.Ges.Endocrin. Würzburg 2.-4.3.1967
Abstr.Nr.57 (in press)
Melani F., J.Lawecki, K.M.Bartelt und E.F.Pfeiffer: Insulinspiegel bei Stoffwechselgesunden, Fettsüchtigen und Diabetikern nach intravenöser Gabe von Glukose, Tolbutamid und Glukagon.
Abstr.Nr.115, 2.Tg.Europ.Ges.f.Diabetol.1966, Diabetologia (in press)
Moskalewski S.: Isolation and culture of the islets of Langerhans of the guinea pig.
Gen.and Comp.Endoc. 5, 342 (1965)
Pfeiffer E.F.: Die Immunologie des Insulins.
Verh.Dtsch.Ges.Inn.Med. 72,811 (1966)
---- : Introduction to Panel Discussion "Intestinal Function in Relation to Insulin Secretion".
6th Intern.Congr.I.D.F.Stockholm 1967
---- und M.Telib: 1967 (in preparation)
----,H.Steigerwald, W.Sandritter, A.Bänder, A.Mager, U.Becker und K.Retiene: Vergleichende Untersuch-

ungen von Morphologie und Hormongehalt des Kälberpancreas nach Sulfonylharnstoffen (D 860).
Dtsch.Med.Wschr. 82, 1568 (1957)

----,M.Telib, J.Ammon, F.Melani und H.Ditschuneit: Direkte Stimulierung der Insulinsekretion in vitro durch Sekretin.
Dtsch. Med.Wschr. 90, 1663 (1965)

----,J.E.Yeboah, M.Telib, J.Ammon, H.E.Hildebrandt, Becker und Meier: 1967 (in preparation)

Pozza G., G.Galansino, H.Hoffield and P.Foa: Stimulation of insulin output by monosaccharides and monosaccharides derivates.
Amer.J.Physiol. 192, 497 (1958)

Raptis S., K.E.Schröder, F.Melani und E.F.Pfeiffer: Die Beeinflussung der Insulinsekretion durch Secretin in vivo.
Panel discussion "Intestinal Function in Relation to Insulin Secretion". - 6th Congr.I.D.F.Stockholm, Aug.1967

Schöffling K.: Der Insulinstoffwechsel des pancreaslosen Hundes. - 12.Symp.Dtsch.Ges.Endokrinologie, Wiesbaden (1966)

Schröder K.E., S.Raptis, M.Telib und E.F.Pfeiffer: Intestinal Hormones effecting insulin secretion in vitro: Secretin, Pancreocymin and Gastrin.
Panel Discussion "Intestinal Function in Relation to Insulin Secretion". - 6th Congr. I.D.F. Stockholm, Aug.1967

Schwarz H., J.Ammon, J.E.Yeboah, H.E.Hildebrandt und E.F.Pfeiffer: Förderung der Insulinsekretion in vitro durch ein neues, hochwirksames Antidiabeticum.
Diabetologia 1967 (in press)

Telib M.: Der Einfluß von Monosacchariden und Hormonen auf die Insulinsekretion verschiedener Wirbeltiere.
Inaug.Diss. Frankfurt (1967)

Vecchio D., A.Luyckx, C.R.Zahnd and A.E.Renold: Insulin release induced by glucagon in organ cultures of fetal rat pancreas.
Metabolism 15, 577 (1966)

Williams R.H.: Secretion, Tates and action of insulin and related products.
Diabetes 15, 623 (1966)

Yeboah J.E.: Insulin secretion in vitro: Stimulation der Insulinsekretion in vitro durch einzelne Aminosäuren.
Inaug.Diss. Ulm (1967)

# REACTIVITY OF -S-S- BONDS IN RELATION WITH INSULIN BIOLOGICAL ACTIVITY AND CATABOLISM

U. ROSA, C.A. ROSSI and L. DONATO

Joint Radiochemical Research Unit (CNR Group of Clinical Physiology,Pisa and SORIN, Saluggia) and Institute of Biochemistry,University of PISA

## Introduction

Disulphide bonds play a unique role in insulin in connection with the two chain structure of the hormone. Moreover, the suggestion that they are directly involved into the hormone mode of action, has been made (1, 2) and there is a large body of evidence indicating that the reductive cleavage of the sulphur bridges is one of the steps of the hormone catabolism (3).

Since specific substitutions or modifications of the various functional groups of insulin are known to lead to the loss of the biological activity, we have thought of interest to study the reactivity of the disulphide bonds in some insulin derivatives. The rationale of such an investigation was that the reactivity of the -S-S- bonds in simple organic disulphides is known to be strongly dependent on the surrounding chemical structure (4); on this basis one could expect the -S-S- bonds reactivity in insulin to be affected by those substitution processes leading to a modification of the native structure. The reaction with sodium sulphite at pH 7 is a realiable and simple tool to investigate the -S-S- bonds reactivity in insulin.

Disulphide bonds react with sulphite according to the equation (1)

$$RS.SR+SO_3^{2-} \rightleftharpoons RS.SO_3^- + RS^- \qquad (1$$

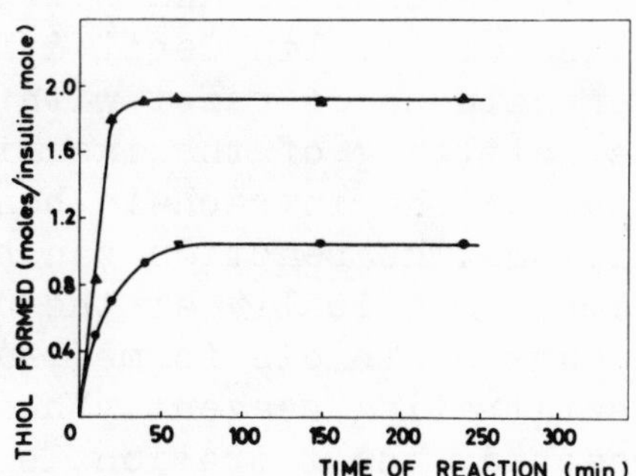


Fig.1-Time course (●) of the reaction between insulin iodinated at 6.4 ID and sodium sulphite at pH 7 and 37°C. The time course (▲) of native insulin is also reported. The concentration of insulin and $Na_2SO_3$ were 0.176 and 31 mM respectively. The titrations were carried out at pH 2, with mM HOPhHg (from Rosa et al., Biochem. J., 103, 407, 1967).

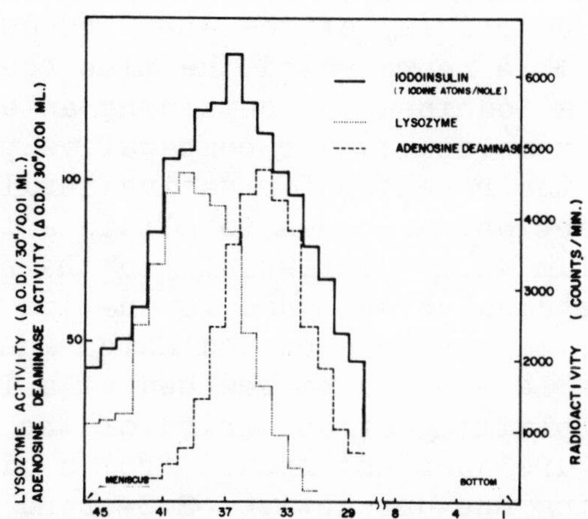


Fig.2-Centrifugation pattern for iodinated (7 ID-labelled with $^{125}I$ 2 μC/mg) insulin on sucrose gradient (——). Lysozyme (·······) and adenosine deaminase (-----) were used as standards. Assuming for lysozyme and adenosine deaminase a mol.wt. of 17200 and 35000 respectively, the mol. wt. of iodoinsulin results of about 26000. Under the same conditions native insulin gave a mol. wt. of about 30000.

The reaction is reversible and will go to completion at pH 7. In the case of insulin Cecil & Loening (5) found that the two interchain bonds react with sulphite at pH 7, leading to the splitting of the insulin molecule into its components chains; the intrachain bond does not react under these conditions. The reaction can be easily followed by titrating amperometrically at the dropping mercury electrode, the amount of thiols formed, using phenyl mercury hydroxide as titrating reagent. When the reaction equilibrium is reached, the titration is carried out at pH 2, since the reaction equilibrium is frozen at this pH value.

## Reactivity Towards the Sulphite of Some Insulin Derivatives

Iodination is one of the most specific substitution processes in insulin, since iodine reacts selectively with the tyrosyl groups unless substitution degrees higher than 7-8 atoms per molecule (M.W. 6000) are attained. Iodoinsulin was therefore the first insulin derivative on which the reactivity of the -S-S- bonds was studied (6). In Fig. 1 is represented the time course of the reaction for a iodoinsulin containing an average of 6 iodine atoms per molecule. For comparative purposes the time course of the reaction for native insulin is also given. Both reactions are seen to attain a true equilibrium; the equilibrium value corresponds to the splitting of two thirds of the disulphide bonds for native insulin. On the contrary, in the case of iodinated insulin, the equilibrium value is seen to be reached at a level corresponding to the splitting of one-third of the -S-S- bonds.

This figure does not result from a mixture of products averaging one unreactive -S-S- bond per molecule, since neither intact insulin nor soluble B chain can be identified among the products of the reaction between iodoinsulin and sulphite. Moreover, the state of aggregation of iodinated insulin does not differ from that of the native one (Fig. 2)so that one can exclude that the observed effect is due to a sort of masking of one-half of the monomer units caused by aggregation of the iodinated insulin. Further, fibrous insulin is completely spli

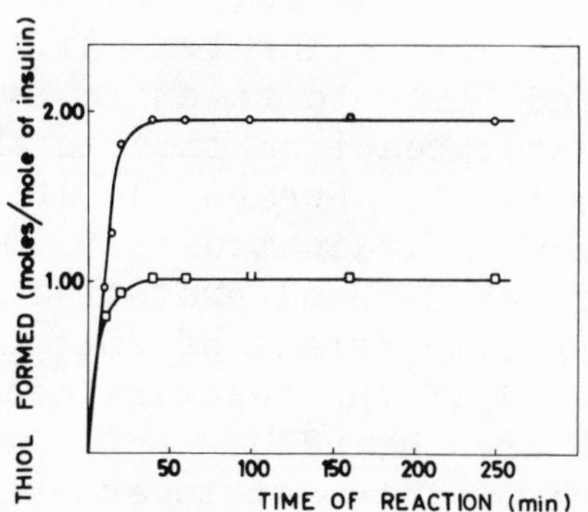


Fig.3-Time course of the reaction between fully methylated (□) (6-$OCH_3$/molecule) insulin and sodium sulphite at pH 7 and 37°C. The time course of native insulin (o) is also reported. The concentrations of insulin and $Na_2SO_3$ were 0.185 and 36 mM respectively. The titrations were carried out at pH 2, with mM HOPhHg (from Massaglia et al., Biochem. J., in press).

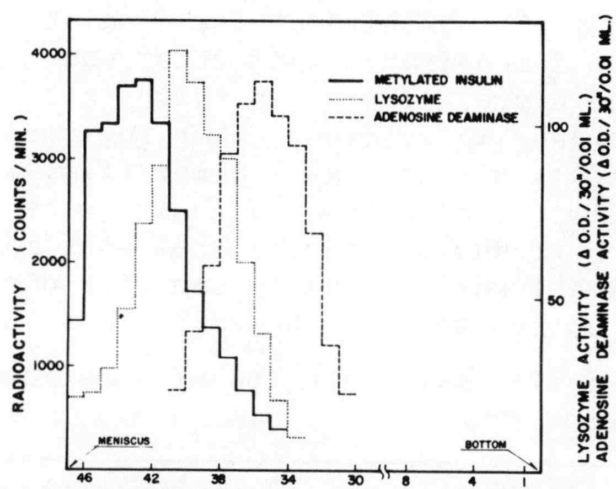


Fig.4-Centrifugation pattern for methylated (6-$OCH_3$/molecule-labelled with $^{14}C$ at 0.5 μC/mg) insulin on sucrose gradient (——).Lysozyme (⋯⋯)and adenosine deaminase (----) were used as standards. Assuming for lysozyme and adenosine deaminase a mol. wt. of 17200 and 35000 respectively, the mol. wt. of methyl insulin results of about 10000.Under the same conditions, native insulin, gave a mol. wt. of about 30000.

by sulphite at pH 7, and the reaction proceeds at the same rate as for the native insulin (3). On the other hand, one -S-S- bond fails to react in iodoinsulin, even after a tenfold longer reaction time in the presence of an excess of sulphite. It appears therefore that each monomer unit in insulin iodinated at 5 ID or more contains an interchain -S-S- bond protected in some way against the nucleophilic attack of $SO_3^{2-}$ ions at pH 7. The effect is removed if the reaction occurs at pH 9 or in presence of 8 M urea at pH 7.

It was thought of interest to establish whether the described effect is limited to iodination or whether other substitution processes are able to interfere on the disulphide bridges reactivity. Besides iodination, one of the most selective substitution processes is the esterification of the carboxyl groups of insulin with methanol. As reported in fig. 3 fully methylated insulin showed the same behaviour of the iodinated one. As one can see, the reaction between fully methylated insulin and sulphite attains a true equilibrium corresponding to the splitting of one-third of the disulphide bonds. Here again it was demonstrated that this result cannot be due to increased aggregation of the methylated insulin, which the sucrose density gradient centrifugation suggests to exist in solution mainly as a dimer (fig. 4).

In addition, no separated chains were found among the reaction products, proving that each monomer unit in esterified insulin actually contains one interchain bridge unreactive towards the sulphite at pH 7.

Table I.-Reaction with sulphite of some insulin derivatives

| Insulin derivative | -S-S- bonds reacted per molecule at equilibrium |
|---|---|
| Fully methylated | 1.00 ± 0.06 (20) |
| Fully acylated | 0.99 ± 0.05 (10) |
| Fully acetylated | 2.09 ± 0.14 (10) |
| Fully succinylated | 1.99 ± 0.15 (10) |
| DAA fully acetylated | 1.89 ± 0.06 (10) |

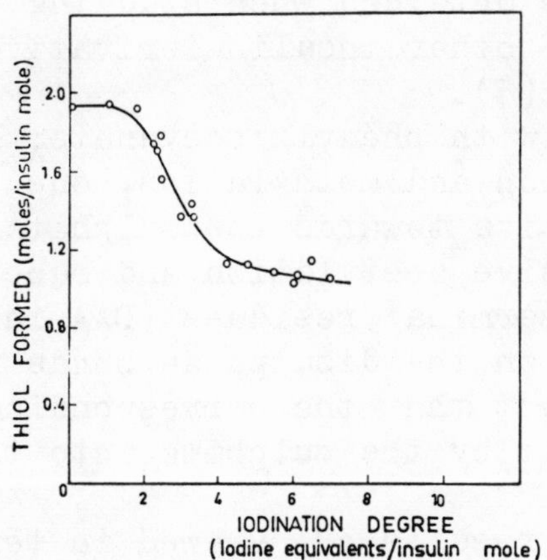


Fig.5-Extent of the reaction between iodoinsulins and sulphite. The time course of the reaction at pH 7 and 37°C was determined as shown in fig. 1 and the maximum extent of the reaction plotted against the iodination degree (from Rosa et al., Biochem. J., 103, 407, 1967).

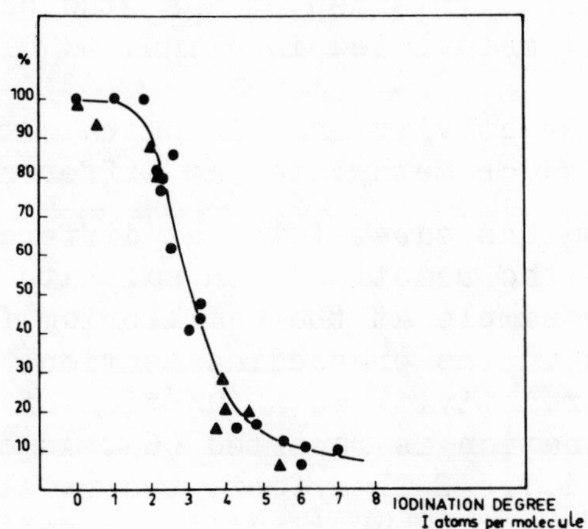


Fig.6-Biological activity (▲) and -S-S- reacted/mole (●) versus iodination degree. Titration results are expressed in %, taking as 100% the difference between the -S-S- titration value measured on native insulin and that measured on insulin iodinated at 5 ID.

The results obtained when studying the reaction with sulphite of other insulin derivatives are summarized in Table I. (7).

Acylation, with phenyl isocyanate, has the same effect of iodination and methylation, one interchain bond becoming unreactive towards the sulphite at pH 7. On the contrary, extensive acetylation and succinylation or removal of the C-terminal residues (DAA insulin) have no apparent effect on the disulphide bonds reactivity as proven by the fact that the corresponding insulin derivatives are split by the sulphite into the constituent chains.

When these results are viewed in terms of the hormonal activity of the substituted insulins, an interesting parallelism is noted between the effects of the substitutions on disulphide bonds reactivity and on the biological activity: the disulphide bonds reactivity appear to be affected only by those substitutions which are known to abolish the biological activity of the hormone, when carried beyond a certain level. It was therefore planned to study the changes of biological activity and of -S-S- bonds reactivity as a function of the degree of substitution. This aspect was studied in detail on iodinated and methylated insulins.

## -S-S- Bonds Reactivity and Biological Activity of Insulins Iodinated or Methylated at Different Extents

Iodoinsulins substituted at different extents were prepared and the reaction with sulphite at pH 7 was studied on each sample at the equilibrium following the procedure given in the preceeding section. In fig. 5 the -S-S- bonds reactivity as a function of the extent of insulin iodination is reported (6). As one can see, iodination beyond 1.5-2 ID affects the reaction of insulin with sulphite. At a substitution degree of 4-5 ID only one-third of the -S-S- bonds is split.

In fig. 6 the results are expressed in %, taking as 100% the difference between the -S-S- titration value measured on native insulin and that measured on insulin iodinated at 5 ID. On the same graph and with the same scale, the biological activity of the iodinated in-

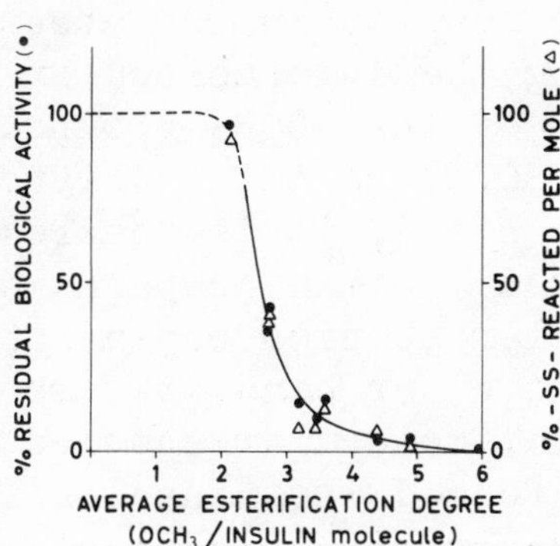


Fig.7-Biological activity (▲) and -S-S- reacted/mole (•) versus average number of methoxyl groups/molecule. Titration results are expressed in %, taking as 100% the difference between the -S-S- titration value measured on native insulin and that measured on insulin iodinated at 5 ID.

sulin samples is reported. Biological inactivation and unreactivity of the -S-S- bonds evolve following the same law. They occur when the average ID of the preparation reaches a value of the order of 2-3 iodine atoms. The study of the effect of progressive methylation gave essentially similar results, as shown in fig. 7, where the results of the -S-S- bonds titrations and the residual biological activity are reported on the same graph, as a function of the average methoxyl content of the protein (7). No data are available in the region between 0 and 2 methoxyl groups per molecule, since we were not able to prepare insulins methylated at such an extent. This must be ascribed to the fact that the terminal carboxyl groups on each chain are much more reactive towards methanol than the side chain carboxyls of the glutamic residues. This could imply that the effect on the -S-S- bonds reactivity results from the methylation of one of the side-chain carboxyl groups, since the phenomenon is already complete at a methylation degree of about 3.The biological activity follows the same course.

The actual substitution level at which the above

effects occur on the single molecule cannot be predicted on the basis of our results in the absence of any possibility of evaluating the intermolecular distribution of the iodine atoms or of the $-OCH_3$ groups. Since the insulin molecule (M.W. 6000) contains 4 tyrosine residues, i.e. 8 substitution sites, it is theoretically possible for a given iodinated insulin preparation to be composed of different molecular species containing from 0 to 8 iodine atoms. On the other hand, 6-carboxyl groups being available in insulin, molecular species containing from 0 to $6-OCH_3$ can be formed when insulin is methylated.

Furthermore the critical level of substitution cannot be only a consequence of the extent of the substitution on the single molecule, but it must correspond to the attainment of a critical intramolecular distribution of the substituent. In the case of iodination, evidence has been obtained (6) that the formation of di-iodinated tyrosine groups on the A chain represents the critical change at which the hormone loses its biological activity and one of the -S-S- bonds becomes unreactive.

The Reaction of Some Insulin Derivatives with GSH-Insulin Transhydrogenase

As it is well known glutathione-insulin-transhydrogenase splits native insulin into its component chains (8,9), and it can be considered as one of the possible mechanisms for disulphide cleavage at cellular level. Moreover it is well known that the enzyme does not require the native configuration of the substrate (10). The reaction of insulin with glutathione-insulin-transhydrogenase was choosen as a different kind of nucleophilic attack of the disulphide bonds, to test the behaviour of the substituted derivatives. Data from a typical experiment are shown in table II.

They demonstrate that the cleavage of insulin into the component chains is inhibited in the fully iodinated derivative. It has to be noted that DAA insulin is cleaved by the enzyme at the same rate of native insulin. These data are therefore consistent with those obtained with sulphite.

Table II.-Glutathione-insulin-transhydrogenase activity on modified insulins

| Substrate | Relative initial velocity |
|---|---|
| Native insulin | 100 |
| Iodinated insulin (7-8 ID) | 14 |
| DAA insulin | 117 |

## Disulphide Bonds Reactivity and Protein Conformation

The results of the studies reported above can be summarized as follows:
1) In iodinated, methylated and acylated insulin, one of the interchain -S-S- bonds becomes unreactive towards the sulphite at pH 7 at a critical level of substitution. 2) The appearence of one unreactive disulphide coincides with the loss of the hormonal activity. A critical intramolecular distribution of the substituent is responsible for both effects. 3) Substitution processes not affecting the biological activity do not lead to the formation of unreactive -S-S- bonds. 4) Enzymatic splitting of the interchain bonds of iodinated insulin is also inhibited.

Some possible explanations of the lack of reactivity of protein disulphide bonds have been discussed previously (4,11,12,13). It has been suggested that the important factors were steric hindrance and negative charges adjacent to the bonds.

We believe that neither mechanism can explain our results. In fact, acetylation and succinylation, which both increase the negative charge of the molecule, have no effects; whereas esterification, which decreases the negative charge affects the -S-S- bond reactivity. On the other hand the hypothesis of a direct steric effect is not tenable when considering that the substitution reactions which affect the -S-S- bond reactivity involve sites having many different locations inside the molecule.

It appears more conceivable that the substitution processes act on the behaviour of the interchain bonds "via" a conformational rearrangement of the molecule. This is consistent not only with the widely different location of the sites which entail the occurrence of the described effects, but also with the finding that the loss of hormonal activity and the failure to react with sulphite go together. Preliminary results of studies now in progress seem to confirm the hypothesis of a conformational change.

However, any attempt to suggest a plausible mechanism through which a conformational rearrangement could affect the -S-S- bonds reactivity is, at the present time, largely speculative.

## Catabolism of Insulin Derivatives in Normal Humans

Some experiments on the catabolism of various insulin derivatives in normal humans, gave results which ca be interpreted if one admits that the reactivity of the -S-S- bonds is an important determinant in catabolism of the protein.

We have shown that low iodinated insulin, labelled with radioiodine, behaves as a true tracer of the native one (13). When injected in normal humans, its rate of catabolism expresses as a fractional catabolic rate, is of the order of 4-6%/min.; as it is well known, the fractional catabolic rate expresses the fraction of the residual amount of the injected preparation which is degraded per unit time (table III) (15).

Table III.-Fractional catabolic rate of $^{131}$I-insulin preparations

| | Subjects | Mean±S.D. |
|---|---|---|
| Beef insulin (ID=0.2) | 16 | 4.06±1.23 |
| Beef insulin (ID=5.0) | 6 | 0.93±0.35 |
| Pig insulin (ID=0.2) | 14 | 4.33±0.82 |
| Pig insulin (ID=5.0) | 6 | 1.32±0.35 |
| Methylated pig insulin (ID=0.2) | 2 | 1.69 |
| A chain | 2 | 3.51 |
| B chain | 2 | 3.84 |
| DAA insulin | 2 | 3.17 |

While the fractional catabolic rate of low iodinated insulin is about 4-6%, that of high iodinated is of the order of 2%; fully methylated or acylated insulins are degraded at the same rate of about 2%/min. (16).The peculiarity of these results lays in the fact that degradative processes, such as excessive iodination, usually lead to an increase in rate of catabolism of the proteins: for instance, this behaviour is constantly observed in the case of plasma proteins. When the separated chains at different level of iodine substitution, either in the reduced or in the fully sulphonated form, were studied in man their rate of catabolism was found to be of the same order of that of native insulin. On the other hand DAA insulin is degraded at a rate of the same order of that of native insulin. It may be worth to note that DAA insulin is biologically inactive, but its disulphide bonds are fully reactive.

In our view our results on the reactivity of the -S-S- bonds in the different insulin derivatives, give the key to understand the biological measurements. If we accept the hypothesis that the splitting of the interchain bonds is the main step of the insulin degradation the biological results are well explained.

In fact such a step must be much faster than the following ones, represented by the hydrolitic demolition of the separated chains. Since the method of evaluating insulin catabolism is mainly affected by the rate of the slower step, it is perfectly understandable why the catabolic rates of native insulin and separated chains are in the same range: in both cases, in fact, we essentially measure the rate at which the chains are degraded.

It is also understandable why substituted insulins are degraded at a slower rate than the native one; in substituted insulins the splitting into the component chains is inhibited, so that the hydrolitic degradation occurs at a slower rate. On the other hand it is not surprising that DAA insulin, in which the -S-S- bonds are reactive, although biologically inactive be indistinguishable from the native one.

## Conclusions

The results we have reported prove that the reactivity of the disulphide bonds in insulin is affected by selective substitutions on functional groups located in different part of the molecule. It is postulated that a change in reactivity of the sulphur bridges reflects the occurrence of a conformational rearrangement of the protein molecule.

An immediate implication of the present results is that selective substitutions confined to a few definited sites, induce modifications of the chemical behaviour of groups located in different parts of the molecules. Moreover, the fact that the groups indirectly affected are the interchain -S-S- bonds, whose structural role is unique in insulin, implies the extension of the effects of localized substitutions to the molecule as a whole. Many attempts have been made in the past to consider the hormonal activity as depending on the integrity of certain key chemical groups, the "active sites". This concept was mainly based on the finding that only some specific substituent like iodine, methoxyl, phenylisocyanate lead to inactivation.

Our results prove that the effect of specific substitutions is not necessarily a direct consequence of the key role of the substituted groups.

Some aspects of our results can be discussed in the light of the pecualiar position that insulin occupies among the proteins, as far as the reactivity of its -S-S- bonds is concerned. In most native proteins the disulphide bonds are relatively unreactive towards the sulphite and thiols at pH 7 (4,14).

In this respect the behaviour of insulin appears peculiar, since the disulphide bonds are easily split by sulphite and thiols; this could be inherent to the insulin biological function, since a metabolic regulator is unlikely to be effective unless it is not only rapidly made available, but also rapidly destroyed (3). There is some evidence in favour of the hypothesis that the reductive cleavage of the disulphide bridges is the initial step of the insulin degradation, thus representing the rapid inactivating reaction required by a well working regulation mechanism. In addition it has been suggested that the mode of action of the hormone could involve attachment to the cell membranes by thiol-disulphide exchange reaction.

Viewed against this background our results could suggest that the high reactivity of the disulphide bonds in insulin strictly depends on the preservation of the native conformation, in the sense that limited substitutions are sufficient to induce in the protein structural conditions no longer compatible with such a high reactivity. In other words, the availability of the sulphur bridges to the nucleophilic attack could be viewed as one of the basic properties of the molecule that the native conformation as to secure. That the biological activity follows the same fate of the disulphide reactivity could imply that both properties depend at the same extent on the preservation of the native structure, adding further evidence to the hypothesis that the availability of the -S-S- bonds plays a role in the biological function of the hormone.

The present work has been carried out under the EURATOM contract 048-65/1 BIOI and 026-63-4 BIAC.

## References

1) Cadenas E., Kaji H., Park C.R. & Rasmussen M. (1961). J. Biol. Chem., 236, PC63.
2) Fong C.T.O., Silver L., Papenoe E.A. & Debous A.F. (1962). Biochim. Biophys. Acta, 56, 109.
3) Stanbury J.B., Wyngaarden J.B. & Fredrickson D.S. (1966). The Metabolic Basis of Inherited Disease. 2nd Ed., pg.69, McGraw Hill Book Co., New York.
4) Cecil R. & McPhee J.R. (1959) Adv. Prot. Chem. 14, 250.
5) Cecil R. & Loening U.E. (1960) Biochem.J. 76, 146.
6) Rosa U., Massaglia A., Pennisi F., Cozzani I. & Rossi C.A. (1967) Biochem.J. 103, 407.
7) Massaglia A., Pennisi F., Rosa U., Ronca-Testoni S., & Rossi C.A. (1967) Biochem.J. (in press).
8) Givol D., De Lorenzo F., Goldberger R.F., Anfinsen C.B. (1965). Proc. N.A.S., 53, 676,
9) Varandani P.T. (1966). Biochim. Biophys. Acta, 118, 198.
10) Katzen H.M. & Tietze F. (1966). J. Biol. Chem., 241, 3561.

11) Boyer P.O. (1960). Brookhaven Symp.Biol. 13, 1.
12) Hird F.J.R. (1962). Biochem. J. 85, 320.
13) Mancini P., Pennisi F., Rosa U. & Rossi C.A. (in preparation).
14) Davidson B.E. & Hird F.J.R. (1967). Biochem. J. 104, 473.
15) Bianchi R., Federighi G., Vitek F. & Donato L. (1966), EURATOM Conf. Labelled Proteins, EUR 2950. d,f,e, 371.
16) Federighi G., Giagnoni P., Giordani R., Novalesi R. Bianchi R. & Donato L. (in preparation).

# THE PHYSIOLOGIC ROLE OF GLUCAGON

Piero P. Foà

Sinai Hospital and Wayne State

University, Detroit, Michigan

Several very interesting papers dealing with the physiology of glucagon have been presented at this meeting and they bear witness to the revival of interest in glucagon which has taken place in the last few years. In order to summarize this work I recently found it necessary to read and, hopefully, digest more than 750 papers published since 1963 alone (Foà, 1967)*. Allow me then to follow the advice given by Professor Trabucchi in his opening address and attempt to give you a broad line summary of the problem, propose a few working hypotheses and suggest a possible physiologic role for glucagon, limiting to a minimum the presentation of new experimental data and the citation of old, although perhaps, in Dr. Guillemin's words, "prophetic literature". We have heard a great deal this morning about gastrointestinal glucagon. It seems now probable that the hyperglycemic glycogenolytic factor, which Sutherland and his collaborators (for references see Sutherland and Robison 1966) found in the mucosa of the gastrointestinal tract, probably consists of at least two types of substances: a glucagon-like material endowed with insulinogenic properties and a material or materials, possibly serotonin or the catecholamines, endowed with phosphorylase-activating and hyperglycemic properties. Histologic, biochemical and immunologic evidence strongly suggest that these substances are not identical with pancreatic glucagon and, while this conclusion opens new fields of investigation, it also simplifies our task for, hopefully, we may now again study glucagon deficiency in the relatively normal depancreatized animal, rather than being faced with the

---

* The reader will find a more complete bibliography in this reference.

impossible situation of studying the physiologic role of a hormone in the highly unphysiologic eviscerate preparation. The significance of this fact is emphasized by the data in Table I, which has been prepared using data from various laboratories, including our own. Please note that the total glucagon-like material contained in the gastrointestinal tract is comparable to that contained in the pancreas. Clearly, the recognition that these materials are not one and the same, makes a great deal of physiologic difference. By contrast with the relatively high glucagon concentration in pancreatic tissue, its concentration in the peripheral blood is very low and its precise measurement requires sensitive and specific methods. Recently we (Shima and Foà, 1967) have adapted the double-antibody method of Hales and Randle (1963) to the determination of serum glucagon, obtaining good recoveries of added glucagon (77-100; ave. 89.5%) and satisfactory standard curves in the physiologic range of concentrations (Fig. 1). Using this method the average fasting glucagon concentration of five normal individuals was found to vary between 0.44 and 0.46 (Ave. 0.44) mμg/ml and to decrease significantly following an intravenous load of glucose, reaching a minimum average value of 0.27 mμg/ml after 15 minutes. May I add that in a few experiments in man, we have observed a significant

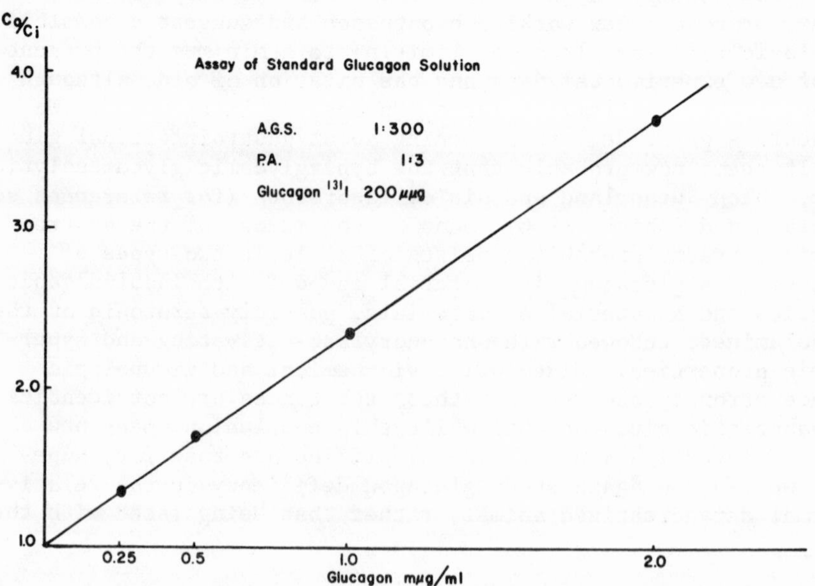


Fig. 1. Double antibody assay for glucagon. Standard curve.

increase in serum glucagon during insulin hypoglycemia. These findings confirm the results of our early perhaps "prophetic" cross-circulation experiments and the subsequent more precise experiments performed in Unger's laboratory (for references see Unger and Eisentraut 1964). More puzzling is the finding that the concentration of serum glucagon, which in the same individuals had decreased after intravenous glucose, rose slightly, but in a statistically significant manner after an oral glucose load, reaching an average value of 0.52 (0.46-0.62) mμg/ml thirty minutes after the ingestion of 100 g of glucose. A satisfactory explanation for this difference between the effects of oral and intravenous glucose is not available. Let us then proceed to the next topic and consider some of the <u>physiologic</u> effects of glucagon. <u>Hyperglycemia</u> is, of course, a readily demonstrable result of glucagon injections. It has been obtained in many animal species, including the depancreatized iguana, the hypophysectomized toad and the alligator: in the latter, under suitable conditions, it may last for two weeks (for references see Foà, 1967). In adequately fed animals, glucagon hyperglycemia occurs promptly after the injection of the hormone, suggesting that, at least its initial phase, is due primarily to hepatic glycogenolysis. Quantitative studies have contributed greatly to the clarification of this phenomenon. Thus, Kibler and collaborators (1964) have demonstrated that in normal human subjects glucagon raises hepatic glucose production from about 50 to about 150-200 mg/min per square meter of body surface, within five minutes. In the isolated rat liver, the production of glucose may be stimulated with a concentration of glucagon as low as $10^{-10}$M and with higher concentrations of the hormone, may approach the rate of 1 mg/min per gram of tissue, an amount near the maximum which can be supported by the glucose-6-phosphatase activity of the liver (Sokal, 1966).

The classic experiments of Sutherland and collaborators have demonstrated that this glycogenolytic action of glucagon is based on its ability to activate adenyl cyclase and thus, indirectly, hepatic phosphorylase. A great deal of work has elucidated many details of this fundamental action of glucagon (Sutherland and Robison 1966), while other work has clarified new aspects of the hormonal regulation of the enzyme reactions of glycogen synthesis and breakdown. A few facts will illustrate this statement: they have been selected arbitrarily because they contribute to the construction of a hopefully logical scheme: a) cyclic AMP decreases the total activity of synthetase (Sutherland and Robison 1966), and increases its glucose-6-phosphate dependent (D) fraction (Appleman <u>et al</u>., 1966); b) glucagon prevents glucokinase induction by glucose (Niemeyer <u>et al</u>., 1966) and therefore, the production of glucose-6-phosphate, necessary for the action of synthetase D; thus, c) glucagon not only promotes glycogen breakdown, but tends to inhibit its synthesis. On the other hand, insulin

TABLE I
Immunoreactive glucagon content of tissues

μg equivalents

| | Pancreas | | Stomach | | Duodenum | | Jejunum | | Ileum | | Colon |
|---|---|---|---|---|---|---|---|---|---|---|---|
| | per g wet wt. | Total | per g wet wt. | Total | per g wet wt. | Total | per g wet wt. | Total | per g wet wt. | Total | per g wet wt. |
| Rat | 2.4 | 2.0 | 0.0031 | 0.05 | 0.006 | -- | 0.14 | 0.5 | -- | 0.1 | -- |
| Dog | 4.5 | 127.0 | 0.2-0.47 | 44.5 | 0.043 | 1.2 | 2.1 | 67.0 | -- | 10.9 | -- |
| Beef | 8.0 | -- | -- | -- | -- | -- | 1.2 | -- | 0.24 | -- | -- |
| Man (biopsy) | 0.4-9.0 | -- | 0-0.004 | -- | -- | -- | 0.061 | -- | -- | -- | -- |
| Man (autopsy) | 0-0.8 (Ave. 0.24) | 23.0 | -- | -- | 0.004 | 0.7 | 0.009 | 2.2 | 0.032 | 0.94 | 0.006-0.01 |

induces glucokinase synthesis, stimulates synthetase activity, inhibits adenyl-cylase and phosphorylase activities (Niemeyer _et al_. 1966; Salas _et al_., 1963; Bishop and Larner, 1967; Fridland and Nigam, 1965). The conclusion that in liver tissue insulin and glucagon act antagonistically at the enzymatic level is tempting, indeed almost inescapable. In muscle the situation is not as clear, although glucagon, by stimulating lipolysis and increasing the flux of free fatty acids, may inhibit glucose utilization indirectly through the glucose-fatty acids cycle (Randle _et al_., 1966). Although glycogenolysis is the main mechanism for the prompt hyperglycemic effect of glucagon, _gluconeogenesis_ undoubtedly contributes significantly to it. Thus, for example, quantitative studies with perfused rat liver have shown that glucagon may increase the production of new glucose up to a maximum of 0.05 mg/min per gram of tissue, corresponding to approximately 700 mg/day per rat. This amount which is two to three times greater than the total liver glycogen of a fed rat, is sufficient to provide 10 to 15% of the animal's basal energy requirement (Sokal, 1966). Indeed, according to Schimassek (1967), in the presence of lactate, glycogenolysis accounts only for about one-third of the total glucose produced by the isolated perfused rat liver, while gluconeogenesis accounts for the remaining two-thirds. Glucagon promotes gluconeogenesis from both protein and non-protein sources. From protein by increasing the release of amino acids from skeletal muscle (Beatty _et al_., 1963), by inhibiting their incorporation into protein (Jarrett, 1963), and by stimulating the activities of hepatic transaminase (Greengard and Baker 1966) and of the major enzymes of the urea cycle (McLean and Novello 1965). Gluconeogenesis from non-protein sources requires bypassing of the energy barriers which make reversal of the glycolytic pathway impossible. Glucagon may accomplish this task in three possible ways: a) by a specific action on the glycolytic pathway, resulting in a greatly increased ratio of glucose-6-phosphate to fructose diphosphate (Schimassek and Mitzkat 1963); b) by increasing the activity of phosphoenolpyruvate carboxykinase (Exton and Park 1966; Lardy, 1964-65); c) by increasing pyruvate carboxylase indirectly. This enzyme has an absolute requirement for acetyl CoA which is made available in increased supply by the fact that glucagon stimulates _lipolysis_. This action of glucagon depends upon the activation of lipase by cyclic AMP and results in a breakdown of tissue triglycerides accompanied by an increase in serum free fatty acid concentration (Lefebvre, 1966; Whitty _et al_., 1967). The phenomenon is illustrated in Figure 2. It is hardly necessary to remind the reader that insulin has an effect opposite to that of glucagon both on gluconeogenesis and lipolysis. Figure 2 also illustrates the insulinogenic action of glucagon. Undoubtedly, in the intact organism, glucagon-induced hyperglycemia contributes to its insulinogenic action, however, under suitable experimental

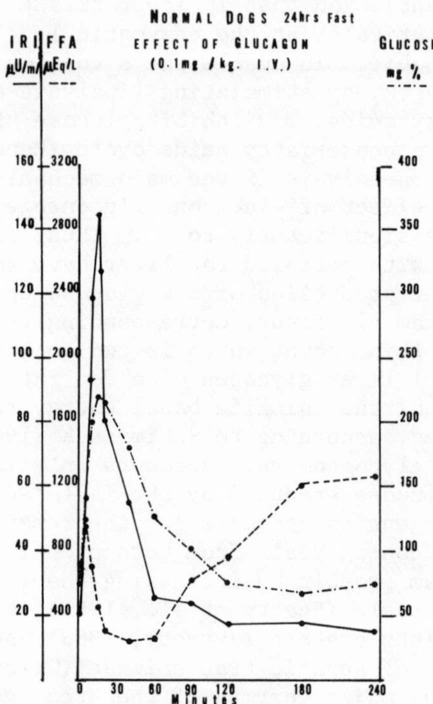


Fig. 2. Effect of glucagon on serum immunoreactive insulin (IRI), free fatty acid (FFA) and glucose in normal dogs.

conditions, a direct effect of glucagon on the pancreatic beta cell can be demonstrated. Thus, for example, following glucagon injections, the increase in serum immunoreactive insulin (IRI) concentration is greater than that caused by a comparable level of glucose-induced hyperglycemia and precedes the rise in serum glucose (Samols et al., 1965); the concentration of IRI in the serum or in the medium increases also when a rise in glucose concentration is not possible, such as in patients with von Gierke's disease (Benedetti and Kolb, 1966) and in the isolated pancreas perfused with a liquid of constant composition or incubated in a glucose-free medium (Grodsky et al., 1967; Devrim and Recant 1966). The enhancement of the insulinogenic effect of glucagon by caffeine or theophylline suggests that also this effect of glucagon is mediated by cyclic AMP. Indeed, cyclic AMP itself seems to be a potent insulinogenic stimulus (Sussman and Vaughan 1967). Another insulinogenic stimulus in the isolated pancreas preparation is the potassium ion (Grodsky and Bennett 1966), a fact of interest in view of the rapid and marked rise in serum potassium concentration induced by glucagon (Fig. 3; Galansino _et al._, 1960). Glucagon influences the serum concentration and urinary excretion of several other ions, among them calcium. Single injections of glucagon cause significant hypocalcemia and if the injections are repeated daily for several weeks, chronic hypocalcemia with secondary hyperparathyroidism may be produced in rabbits. It is of interest to note that animals with chronic experimental pancreatitis have elevated serum glucagon level, hypocalcemia and secondary hyperparathyroidism, suggesting that glucagon hypersecretion may be the cause of the parathyroid hyperplasia often noted in patients suffering from chronic pancreatic disease (Paloyan et al., 1967). On the basis of the evidence summarized in the preceding pages and of other observations and hypotheses (Foà, 1967), one may discuss the physiologic role of glucagon by assuming that a state of glucocytopenia exists, perhaps as the result of fasting, exercise or

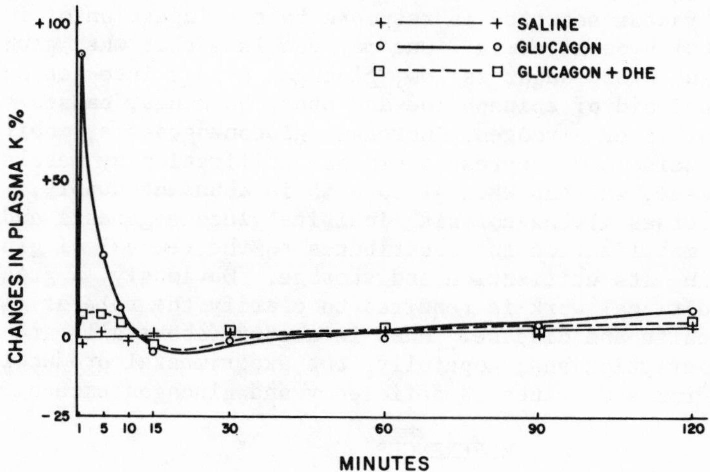


_Fig._ 3. Effect of intraportal injections of saline (5 exp.) and of glucagon on plasma potassium of normal anesthetized dogs (5 exp.) and of dogs pretreated with DHE (5 exp.). Initial concentrations 3.6-4.3 (av. 3.9) mEq.l.

excessive insulin administration. Under these conditions, insulin production would decrease as a result of fasting and hypoglycemia, or because an excess of circulating insulin inhibits further insulin secretion. At the same time, hypoglycemia would cause an immediate increase in glucagon secretion and the concentration of blood glucose would be restored to normal by the rapid increase in hepatic glycogenolysis and gluconeogenesis. In addition, the secretion of glucagon and the relative insufficiency of insulin would raise the level of blood sugar by promoting lipolysis, increasing the concentration of serum free fatty acids and ketone bodies and thus inhibiting glucose utilization. In this effort, glucagon would be aided by a simultaneous increase in the secretion of hormones such as STH, adrenal corticoids and epinephrine which tend to increase glucose production and fat mobilization and to inhibit insulin secretion, insulin action and glucose utilization by muscle and adipose tissue. Thus, the dwindling supplies of glucose would be diverted from the liver and the peripheral tissues to the essential function of the brain. Conversely, if a state of hyperglycemia or glucose abundance exists, perhaps as a result of food intake, inhibition of glucose utilization or excessive glucagon administration, insulin would be secreted, glucagon and STH secretion would be inhibited, glycogenolysis, gluconeogenesis and lipolysis would be suppressed, glucose utilization by muscle would be stimulated and the concentration of glucose in the blood would be restored to normal. Under these circumstances insulin secretion would be stimulated not only by the elevated levels of blood glucose and amino acids, but by glucagon itself and, possibly, by secretin or a gastrointestinal factor secreted in response to the ingestion of food and to portal hyperglycemia. One may conclude that when glucose is scarce and blood sugar is low, glucagon swings into action and, with the aid of epinephrine and other hormones, causes a breakdown of liver glycogen, increases gluconeogenesis, mobilizes free fatty acids and suppresses glucose utilization by muscle and adipose tissue, whereas when glucose is in abundant supply, insulin reverses glycogenolysis, inhibits gluconeogenesis and fatty acid mobilization and contributes to the removal of glucose by increasing its utilization and storage. Obviously, a great deal of additional work is required to clarify the role of glucagon in health and disease. What is needed, above all, are a precise description and, hopefully, the experimental production of the syndromes of glucagon deficiency and glucagon excess.

## REFERENCES

APPLEMAN, M. M., BIRNBAUMER, L. and TORRES, H. N. (1966): Factors affecting the activity of muscle glycogen synthetase. III. The reaction with adenosine triphosphate, $Mg^{++}$, and cyclic 3'5'-adenosine monophosphate. Arch. Biochem. Biophys. 116, 39.

BEATTY, C. H., PETERSON, C. H., BOCEK, R. M., CRAIG, N. C. and WELEBER, R. (1963): Effect of glucagon on incorporation of glycine -$C^{14}$ into protein of voluntary skeletal muscle. Endocrinology, 73, 721.

BENEDETTI, A. and KOLB, F. O. (1966): Metabolic effects of glucagon and epinephrine in four adults with type I glycogen storage disease. Diabetes, 15, 529.

BISHOP, J. S. and LARNER, J. (1967): Rapid activation-inactivation of liver uridine diphosphate glucose-glycogen transferase and phosphorylase by insulin and glucagon in vivo. J. Biol. Chem., 242, 1354.

DEVRIM, S. and RECANT, L. (1966): Effect of glucagon on insulin release in vitro. Lancet, 2, 1227.

EXTON, J. H. and PARK, C. R. (1966): The stimulation of gluconeogenesis from lactate by epinephrine, glucagon, and cyclic 3',5'-adenylate in the perfused rat liver. Pharmacol. Rev., 18, 181.

FOA, P. P. (1967): Glucagon. Erg. Physiol., in press.

FRIDLAND, A. and NIGAM, V. N. (1965): Effects of G6P, 2-deoxy-G6P and UMP on phosphorylase and glycogen synthetase of pigeon liver. Arch. Biochem. Biophys., 111, 477.

GALANSINO, G., D'AMICO, G., KANAMEISHI, D., BERLINGER, F. G. and FOÀ, P. P. (1960): Hyperglycemic substances originating in the pancreato-duodenal area. Am. J. Physiol., 198, 1059.

GREENGARD, O. and BAKER, G. T. (1966): Glucagon, starvation, and the induction of liver enzymes by hydrocortisone. Science, 154, 1461.

GRODSKY, G. M. and BENNETT, L. L. (1966): Cation requirements for insulin secretion in the isolated perfused pancreas. Diabetes, 15, 910.

GRODSKY, G. M., BENNETT, L. L., SMITH, D. F. and SCHMID, F. G. (1967): Effect of pulse administration of glucose or glucagon on insulin secretion in vitro. Metabolism, 16, 222.

HALES, C. N. and RANDLE, P. J. (1963): Immunoassay of insulin with insulin-antibody precipitate. Biochem. J., 88, 137.

JARETT, L. (1963): In vivo and in vitro effect of glucagon on DL-leucine-1-$C^{14}$ incorporation into protein of rat pancreas. Proc. Soc. Exp. Biol. Med. (N.Y.), 114, 550.

KIBLER, R. F., TAYLOR, W. J. and MYERS, J. D. (1964): The effect of glucagon on net splanchnic balances of glucose, amino acid nitrogen, urea, ketones, and oxygen in man. J. Clin. Invest., 43, 904.

LARDY, H. A. (1964-65): Gluconeogenesis - Pathways and hormonal regulation. Harvey Lecture Series, 60, 261.

LEFEBVRE, P. (1966): The physiological effect of glucagon on fat mobilization. Diabetologia, 2, 130.

MCLEAN, P. and NOVELLO, F. (1965): Influence of pancreatic hormones on enzymes concerned with urea synthesis in rat liver. Biochem. J., 94, 410.

NIEMEYER, H. N. PERÉZ and RABAJILLE, E. (1966): Interrelation of actions of glucose, insulin, and glucagon on induction of adenosine triphosphate: D-hexose phosphotransferase in rat liver. J. Biol. Chem., 241, 4055.

PALOYAN, E., PALOYAN, D. and HARPER, P. V. (1967): Glucagon-induced hypocalcemia. Metabolism, 16, 35.

RANDLE, P. J., GARLAND, P. B., HALES, C. N., NEWSHOLME, E. A., DENTON, R. M. and POGSON, C. I. (1966): Interactions of metabolism and the physiological role of insulin. Recent Progress Hormone Res., 22, 1.

SALAS, M., VINUELA, E. and SOLS, A. (1963): Insulin-dependent synthesis of liver glucokinase in the rat. J. Biol. Chem. 238, 3535.

SAMOLS, E., TYLER, J., MARRI. G. and MARKS, V. (1965): Stimulation of glucagon secretion by oral glucose. Lancet, ii, 1257.

SCHIMASSEK, H. (1967): Proceedings of this meeting.

SCHIMASSEK, H. und MITZKAT, H. J. (1963): Über eine spezifische Wirkung des Glucagon auf die Embden-Meyerhof-Kette der Leber. Versuche an der isoliert perfundierten Rattenleber. Biochem., 337, 510.

SHIMA, K. and FOÀ, P. P. (1967): A double-antibody immunoassay for glucagon. In press.

SOKAL, J. E. (1966): Glucagon - An essential hormone. Am. J. Med., 41, 331

SUSSMAN, K. E. and VAUGHAN, G. D. (1967): Insulin release after ACTH, glucagon and adenosine-3'-5'-phosphate (cyclic AMP) in the perfused isolated rat pancreas. Diabetes, 16, 449.

SUTHERLAND, E. W. and ROBISON, G. A. (1966): The role of cyclic-3',5'-AMP in responses to catecholamines and other hormones. Pharmacol. Rev., 18, 145.

UNGER, R. H. and EISENTRAUT, A. M. (1964): Studies of the physiologic role of glucagon. Diabetes, 13, 563.

WHITTY, A. J., SHIMA, K., TRUBOW, M. and FOA, P. P. (1967): Effect of glucagon on serum FFA in normal and depancreatized dogs. Federation Proc., 26, 257.

# SODIUM DEPENDENCY OF INSULIN-STIMULATED GLUCOSE TRANSPORT IN ISOLATED FAT CELLS.

J. Letarte and A. E. Renold

Institut de Biochimie Clinique, Universite de Geneve, Geneva Switzerland

The metabolic effects of insulin on muscle and adipose tissue are well established. However, its site and mode of action are still subjected to much discussion. Some tend to explain all its metabolic effects as a consequence of a stimulated transport of specific substrates while others would tend to explain the metabolic actions by a direct stimulation of the enzymatic processes. In fat tissues, it has been shown that glucose transport is a carrier mediated process having the characteristics of facilitated diffusion (1) on which insulin acts, in an unknown fashion, while also favoring $K^+$ retention and $Na^+$ efflux (2). In order to investigate the possible relationships between glucose transport and cations on insulin stimulated fat tissue, we have incubated isolated fat cells in the presence of various amounts of the different cations normally found in the body fluids and submitted them to insulin stimulation.

## MATERIAL AND METHOD

Cells were isolated from epididymal fat pads (3) of 5-7 weeks old Ivanovas mice and incubated in Krebs Ringer bicarbonate buffer containing various amounts of different cations and 3.5% human serum albumin. The media were kept isotonic and at pH 7.4. The incorporation of U $^{14}$C D-Glucose into $CO_2$ and total lipids was measured and used as an expression of glucose uptake (4). Results are expressed as uatoms glucose carbon incorporated into $CO_2$ or total lipids, per gram total lipid weight, per 2 hours of incubation.

Table 1. Glucose U $^{14}C$ metabolism in isolated fat cells: effect of the replacement of a single cation by $Na^+$ in the presence of insulin (1 mU/ml).

| Medium cations | | | Glucose U $^{14}C$ incorporated into | |
|---|---|---|---|---|
| $K^+$ | $Ca^{++}$ | $Mg^{++}$ | $CO_2$ | Total Lipids |
| + | + | + | 12.6* $\pm$ 1.20 | 29.8 $\pm$ 0.52 |
| - | + | + | 12.2 $\pm$ 0.88 | 28.5 $\pm$ 0.64 |
| + | + | - | 12.5 $\pm$ 0.90 | 28.9 $\pm$ 0.34 |
| + | - | + | 11.4 $\pm$ 0.98 | 27.8 $\pm$ 0.36 |
| | | | 12.8 $\pm$ 0.94 | 29.2 $\pm$ 0.40 |

* µAt. C metabolized/g. lipid/2h ; mean of 6 $\pm$ SEM.
Glucose 2.5 mM.

## RESULTS

In experiments reported earlier (5), we have described a possible link between glucose transport and cations in the isolated fat cells from mice. In order to further investigate these interactions we have incubated fat cells with insulin and studied the effects of various cations on their glucose metabolism.

Neither $Ca^{++}$ nor $Mg^{++}$ nor $K^+$ could be shown to have any modifying effect on the metabolic indices measured, namely $CO_2$ and total lipids (Table 1). As $K^+$ had been described as essential for insulin action by Rodbell (6) using similar techniques, but working with fat cells from rats, we have further investigated this cation.

Using maximal stimulating concentration of insulin (1mU/ml) repeated experiments never showed a depressive effect of $K^+$ lack. As this requisite for the insulin stimulation might have been present only at sub-maximal stimulation, we have also investigated low insulin concentrations. The stimulatory effect of $K^+$ lack, described previously (5), could again be seen in the absence of insulin. As insulin concentration was increased in the incubating medium, the activity of the cells, both in the $K^+$ containing and $K^+$ free media, reached a maximum value where they merge together. This experiment shows that, in those conditions, $K^+$ is not essential for the activity of the mice fat cells induced by insulin. Moreover, at less than maximal stimulation the cells in the $K^+$ free media were more active than similar cells incubated in $K^+$ containing media.

Table 2. Glucose U $^{14}C$ metabolism in isolated fat cells: effect of total $Na^+$ replacement by $K^+$ in the presence of insulin (1 mU/ml).

| Medium Na mM | Glucose $^{14}C$ incorporated into $CO_2$ | Total lipids |
|---|---|---|
| 144 | $12.2^* \pm 0.22$ | $21.9 \pm 0.14$ |
| 0 | $7.2 \pm 0.49$ | $10.4 \pm 0.68$ |
| 144 | $11.8 \pm 0.10$ | $36.8 \pm 0.18$ |
| 0 | $8.1 \pm 0.22$ | $26.6 \pm 0.41$ |
| 144 | $9.78 \pm 0.22$ | $17.4 \pm 0.12$ |
| 0 | $5.28 \pm 0.28$ | $12.1 \pm 0.65$ |

*µAT. C metabolized / g. lipid / 2h; mean of 6 $\pm$ SEM.
Glucose 2.5 mM.

To rule out a possible link between intra-cellular $K^+$ and the insulin effect, we did measure the total exchangeable $K^+$ using $K^{42}$ as a tracer agent. At different media $K^+$ concentrations, the exchangeable $K^+$ was not modified by the presence of insulin although the hormone had maximal stimulatory effect on the metabolism.

These facts, in our mind, tend to rule out that in insulin stimulated cells $K^+$ has any effect on the metabolic indices measured and used as an expression of total glucose uptake.

The situation is entirely different when it comes to the $Na^+$ dependence of insulin action on fat cells.

Using $K^+$ as replacing agent for $Na^+$ (Table 2)., it can be seen that, the metabolic activity of the insulin stimulated cell is repeatedly reduced to values which approach those of the cells incubated without insulin but whose metabolism is stimulated by the absence of $Na^+$.

It is readily evident that these effects could also be ascribed to the very high $K^+$ content of our incubating medium. To rule out this possibility, isolated cells were incubated in $Na^+$ free media in which the replacing cation was $Tris^+$, $Choline^+$ or $K^+$ (Table 3). Again, the same effect was registered: namely a marked depressive effect on the metabolism. It thus seems reasonable to link this diminution of transport to the $Na^+$ lack. In these experimental conditions measured exchangeable $K^+$ was not altered by the addition of insulin.

It could also be argued that these results are secondary to a breakage of the incubated cells which would then show a dimin-

Table 3. Glucose-U $^{14}C$ metabolism in isolated fat cells: effect of total $Na^+$ replacement by other ions in the presence of insulin (1 mU/ml).

| Medium | | Glucose-$^{14}C$ incorporated into | |
|---|---|---|---|
| $Na^+$(mM) | Replaced by | $CO_2$ | Total Lipids |
| 144 | | 16.9* ± 0.33 | 25.8 ± 0.46 |
| 0 | $Tris^+$ | 5.82 ± 0.10 | 8.49 ± 0.20 |
| 144 | | 16.1 ± 0.12 | 26.9 ± 0.46 |
| 0 | $Choline^+$ | 6.92 ± 0.14 | 10.2 ± 0.25 |
| 144 | | 12.2 ± 0.22 | 21.9 ± 0.14 |
| 0 | $K^+$ | 7.22 ± 0.49 | 10.4 ± 0.68 |

*uAt. C metabolized/g. lipid/2h; mean of 6± SEM.
Glucose 2.5 mM.

ished glucose uptake. It must be pointed out here, that the $Na^+$ lack, in the absence of insulin, has a marked stimulatory effect, the reverse of what is seen in the presence of the hormone. Moreover, this effect of $Na^+$ depletion, has been shown to be reversible. If there is a modification of the cell wall or of the cell metabolic integrity, it must be remembered that this situation produces a stimulation or a depression of the cell activity depending on the absence or the presence of insulin. This fact is hardly explainable by a non specific lesion induced, for example, by an osmotic trauma. Moreover, the reversibility of these effects is against brutal disruption or permanent leakage of the cell wall.

Table 4. Glucose-U-$^{14}C$ metabolism in isolated fat cells: effect of progressive $Na^+$ replacement by $Tris^+$ in the presence of insulin (1 mU/ml).

| Medium $Na^+$ mM | Glucose-$^{14}C$ incorporated into $CO_2$ | Total Lipids |
|---|---|---|
| 144 | 16.9 ± 0.33 | 25.8 ± 0.46 |
| 72 | 10.5 ± 0.27 | 15.6 ± 1.36 |
| 50 | 7.08 ± 0.26 | 10.2 ± 0.17 |
| 25 | 6.55 ± 0.22 | 9.07 ± 0.28 |
| 10 | 6.14 ± 0.36 | 8.32 ± 0.35 |
| 0 | 5.82 ± 0.10 | 8.40 ± 0.20 |

*µAt. C metabolized/g. lipid/2h; mean of 6± SEM.
Glucose 2.5 mM.

A stoichiiometric relationship between $Na^+$ concentration and glucose transport is suggested in experiments where the $Na^+$ ions were progressively replaced by $Tris^+$ ions (Table 4). The same response was observed using $Choline^+$ or $K^+$ as replacing cations.

The metabolic activity of the insulin stimulated cell is progressively decreased as the $Na^+$ content of the incubating medium is decreased. This fact speaks in favor of a direct link between $Na^+$ concentration and glucose transport in the isolated fat cell of mice. Moreover, kinetic studies of cells incubated in low $Na^+$ media, show a definite decrease in the maximal velocity of the glucose uptake without alteration of the calculated Km.

Another important point in these experiments, was to demonstrate that these effects could also be seen at various insulin concentrations. When partially depleted of $Na^+$ and in the absence of insulin, the fat cell shows an enhanced metabolism. The addition of insulin further increases the glucose uptake but to a lesser degree than the stimulation induced by the hormone in the presence of $Na^+$. In every case, the insulin effect is greater in the presence of normal $Na^+$ concentration than in the low $Na^+$ containing media.

## CONCLUSION

From the presented results, we would like to suggest a working hypothesis for the transport of glucose in insulin stimulated cells isolated from epididymal fat pads of mice. In accordance with other authors (7), we postulate that insulin activates an otherwise inactive carrier. This carrier had different characteristics than the one transporting glucose in the absence of the hormone. A major one is its absolute requirement for $Na^+$ to transport glucose through the cell membrane. $Na^+$ could be bound to the glucose carrier which then would cross the membrane on the inner side of which both $Na^+$ and glucose would be liberated. In so doing the carrier would be inactivated, needing a new insulin stimulation to cycle again.

In the absence of $Na^+$, this insulin activated carrier, would not be in a suitable form to bind glucose and it could not perform any inward transport of glucose. But, meanwhile, the insulin independent carrier would be subjected only to $K^+$ stimulation and would carry inward more glucose than it normally does in the presence of both $Na^+$ and $K^+$.

In the absence of glucose, the insulin stimulated carrier would still bind $Na^+$ and transport it inside of the cell. This would, maybe, permit to explain why insulin has still the same effect on the membrane resting electrical potential whether glucose is present or not (8).

## REFERENCES

1. O. Crofford, A.E. Renold, J. Biol. Chem. 240 (1965), 3237.
2. D.R.H. Gourley, M.D. Bethea, Proc. Soc. Exp. Biol. Med. 116 (1964), 31.
3. M. Rodbell, J. Biol. Chem. 239 (1964), 375.
4. H.G. Wood, J. Katz, B.R. Landau, Biochem. Z. 338 (1963), 809.
5. J. Letarte, A.E. Renold, Protides of the Biological Fluids Edit. H. Peeters, Elsevier, 1967, in press.
6. M. Rodbell, Handbook of Physiology, Section 5: Adipose tissue, Edit. A.E. Renold, G.F. Cahill, Jr., Am. Physiol. Soc. Washington, D.C. 1965, p. 475.
7. H.T. Narahara, P. Oxard, J. Biol. Chem. 238 (1963) 40.
8. P.M. Beigelman, P.B. Hollander, Proc. Soc. Exp. Biol. Med. 116 (1964) 821.

This work was supported by a grant from the Fonds National Suisse de la Recherche Scientifique and a research fellowship (to J.L.) from the Queen Elizabeth II Canadian Research Fund.

# The Role of Cyclic AMP in the Action of Insulin

J. G. T. Sneyd, J. D. Corbin and C. R. Park

Department of Physiology, Vanderbilt University

Nashville, Tennessee

It is now clear that not all the biological effects of insulin are caused by an increased rate of transport of sugars and amino acids across the cell membrane. One well documented effect that does not depend upon a stimulation of glucose metabolism is the antilipolytic action in adipose tissue which can be observed when rat epididymal fat pads are incubated in a glucose-free medium (1). Insulin has also been shown to increase the fraction in the independent (I) form of glycogen synthetase in rat diaphragm incubated in a medium without added substrate (2). Recently, Jungas (3) reported that insulin increased the percentage of the I form of glycogen synthetase in incubated adipose tissue and decreased the activity of adipose tissue glycogen phosphorylase assayed without added 5'-AMP.

Raising intracellular levels of cyclic AMP stimulates lipolysis in fat pads and isolated cells (4) probably by activating the triglyceride lipase. The nucleotide also stimulates the conversion of phosphorylase _b_ to phosphorylase _a_ in muscle and the conversion of the I form of glycogen synthetase to the D form (5, 6). It thus seemed possible that the inhibitory action of insulin on these processes might be mediated by a fall in tissue levels of cyclic AMP. Accordingly we have examined the effect of insulin on cyclic AMP levels in epididymal fat pads incubated with epinephrine and caffeine in the absence of glucose.

As shown in Table 1 the level of cyclic AMP in unstimulated

TABLE 1

Effect of insulin on level of cyclic AMP and lipolysis in epididymal fat pads.

Fat pads were incubated for 24 min in medium without glucose or albumin. Cyclic AMP was measured by the method of Butcher *et al.* (4) and glycerol by the enzymatic method of Wieland. Results are expressed per g of wet tissue.

| Additions | Cyclic AMP (mmolesx10$^{7}$/g) | Glycerol Release (μmoles/g) |
|---|---|---|
| None | 1.8 ± 0.2 | 0.06 ± 0.02 |
| Insulin 1 mU/ml | 1.5 ± 0.1 | 0.08 ± 0.02 |
| Epinephrine (1 μg/ml) | 2.8 ± 0.25 | 1.25 ± 0.14 |
| Epinephrine + insulin | 2.4 ± 0.41 | 0.66 ± 0.11 |
| Epinephrine + caffeine (1 mM) | 10.2 ± 1.1 | 1.38 ± 0.12 |
| Epinephrine + caffeine + insulin | 4.0 ± 0.7 | 1.32 ± 0.13 |

tissue was high and was not decreased by insulin. Epinephrine raised the level some 55% and stimulated lipolysis (as shown by the increase in glycerol release). This increase in cyclic AMP was not affected by insulin. Caffeine (which inhibits the phosphodiesterase that inactivates cyclic AMP) and epinephrine added together raised tissue cyclic AMP levels five-fold. Inhibition of this rise by insulin was readily observed (Table 1). Similar results have been seen with isolated fat cells i.e., this effect of insulin is on the fat cell itself.

In these experiments there was a poor correlation between the effects of insulin on cyclic AMP levels and lipolysis. When insulin had its most pronounced anti-lipolytic effects (i.e., in the presence of epinephrine alone) it did not appear to lower cyclic AMP, and when insulin lowered cyclic AMP (in tissue incubated with epinephrine and caffeine) it did not inhibit lipolysis. A small rise in the cyclic AMP level in adipose tissue (from about 1.5 x $10^{-7}$ mmoles/g to 3.5 x $10^{-7}$ mmoles/g) is enough to stimulate lipolysis maximally and cyclic AMP levels much higher than this have no further effect (4). Thus, in order to see an antilipolytic action of insulin at the same time as a fall in cyclic AMP levels we would have to work in this narrow range of cyclic AMP concentrations. We re-investigated this problem with isolated adipose tissue cells. This tissue preparation is much more uniform than intact fat pads and small changes in cyclic AMP levels due to experimental treatments are less likely to be masked by variation from one piece of tissue to another. When isolated cells were incubated with epinephrine, insulin both lowered cyclic AMP levels and inhibited lipolysis (Table 2). It thus seems likely that the effect of insulin on lipolysis and perhaps the effects on phosphorylase and glycogen synthetase in adipose tissue reported by Jungas (3) are mediated by a fall in cyclic AMP.

TABLE 2

Effect of insulin on level of cyclic AMP and lipolysis in isolated fat cells.

Isolated fat cells from fed rats were incubated for 30 min in medium containing 4% albumin but no glucose. Cyclic AMP and glycerol were measured as in Table 1. Results are expressed per g of dry tissue.

| Additions | Cyclic AMP (mmolesx$10^7$/g) | Glycerol Release (μmoles/g/hr) |
|---|---|---|
| None | 1.53 ± 0.17 | 1.2 ± 0.9 |
| Epinephrine (1 μg/ml) | 3.55 ± 0.47 | 39.4 ± 2.8 |
| Epinephrine + insulin (1 mU/ml) | 2.53 ± 0.32 | 28.4 ± 2.4 |
| P (Epi. vs Epi. + ins) | < 0.025 | < 0.001 |

An important action of insulin on muscle and adipose tissue is to stimulate transport of glucose into the cell (7) and we thought it interesting to see whether the action of insulin on sugar transport could be caused by changes in the cellular level of cyclic AMP. Adipose tissue rather than muscle was used in this study because in preliminary experiments we could not detect any effect of insulin on cyclic AMP levels in the perfused rat heart.

Before measuring the transport of non-metabolized sugars into the fat pad it is necessary to have a marker for the extra-cellular space. Crofford and Renold (7) found $^3H$ sorbitol satisfactory for this purpose and we used the same compound. The sorbitol space of incubated fat pads was 15.2 ± 1.7 ml/100g (8 expts) which is similar to the values of 13.9 to 15.4 reported by Crofford and Renold. It did not change significantly between 10 and 40 minutes' incubation so it seemed valid to use incubation periods of 15 min for measuring the uptake of non-metabolized sugars. The intracellular water space measured by the method of Crofford and Renold was 3.72 ± 0.25 ml/100g (16 expts). The test sugars used in these experiments were D-galactose and 3-0-methyl glucose. The intracellular galactose or 3-0-methyl glucose space was taken as a measure of the rate of membrane transport of the sugar.

Initially we decided to raise tissue cyclic AMP to high levels with epinephrine and caffeine to see if this would inhibit the action of insulin on galactose accumulation. The intracellular galactose space in control tissue was appreciable and was significantly increased by insulin (Table 3). Addition of epinephrine and caffeine to the control tissue produced a significant fall in the intracellular galactose space ($P < 0.05$), and in the presence of insulin, epinephrine and caffeine reduced the intracellular galactose space to low levels. Almost identical results were observed when 3-0-methyl glucose was used instead of galactose (Table 3, third column). In these experiments the sorbitol space was unaffected by epinephrine and caffeine but appeared to be increased slightly by insulin. Epinephrine or caffeine tested singly had no significant effect on galactose accumulation either in the presence or absence of insulin. (Table 3 parts B and C).

Galactose is metabolized slowly by adipose tissue and it thus became important to determine whether the effects of insulin, epinephrine and caffeine on the accumulation of galactose-$C^{14}$ were due to an effect on the transport of galactose into the cell or to effect on its subsequent metabolism. If, for example, insulin markedly stimulated the conversion of galactose-$C^{14}$ to other radioactive products without affecting the

TABLE 3

Effect of epinephrine and caffeine on sorbitol space, galactose space, and 3-0methyl glucose space of epididymal fat pads.

Epididymal fat pads were incubated for 10 min at 37° in 2 ml of Krebs-Ringer bicarbonate buffer containing 3 1/2% bovine serum albumin, transferred to 2 ml of fresh medium containing 3 H sorbitol (5 mM) and $^{14}C$ galactose (100 mM) or $^{14}C$ 3-0-methyl glucose (5 mM) and incubated for a further 15 min. Following incubation the fat pads and medium were analyzed for radioactive sugars. Hormones and caffeine (when present) were in both incubating media. Results are means ± S. E. M.

| | Additions | Sorbitol | Intracellular Galactose Space | Intracellular 3-0-MG Space | Cyclic AMP (mmoles x $10^7$/g) |
|---|---|---|---|---|---|
| A | None | 14.7±0.8 (20) | 1.29±.21 (20) | 1.22±.22 (4) | 1.4 ± 0.2 |
| | Insulin | 18.4±0.8 (20) | 2.71±.19 (20) | 2.74±.20 (4) | 1.8±0.3 |
| | Epinephrine + Caffeine | 15.6±0.6 (20) | 0.62±.25 (20) | 0.86±.34 (4) | 6.6 ± 2.0 |
| | Epi + caffeine + insulin | 19.0±1.1 (20) | 1.14±.19 (20) | 1.09±.36 (4) | 3.8 ± 0.9 |
| B | None | | 0.89±.18 (8) | | |
| | Insulin | | 2.21±.29 (8) | | |
| | Epinephrine | | 0.89±.23 (8) | | 2.3 ± 0.2 |
| | Epinephrine + insulin | | 2.46±.49 (7) | | 2.5 ± 0.5 |
| C | None | | 1.46±.28 (4) | | |
| | Caffeine | | 1.28±.11 (4) | | |
| | Insulin | | 3.48±.46 (4) | | |
| | Insulin + caffeine | | 2.73±.45 (4) | | |

rate of transport, the entry of galactose into the cell would not be offset by the exit of intracellular galactose, and the rate of transport into the cell would appear to increase. When adipose tissue was incubated with galactose-1-$C^{14}$ under the same conditions as those used for the transport experiments and the $C^{14}O_2$ trapped in hyamine and counted, no radioactivity could be detected. If 2% of the intracellular galactose had been oxidized to $C^{14}O_2$ it would have been detected without difficulty. In other experiments adipose tissue was incubated with galactose-$C^{14}$ and the tissue glycogen and lipid isolated and counted. Again no radioactivity was found; if 1% of the intracellular galactose had been converted to glycogen or lipid it would have been detected. These experiments indicate that under the conditions of the transport experiments very little of the intracellular galactose was converted to $CO_2$, glycogen or lipid. In a further series of experiments an attempt was made to precipitate galactose-$C^{14}$ with $Ba(OH)_2$ and $ZnSO_4$ from the perchloric acid extracts of incubated adipose tissue. If some of the radioactive galactose were present in the form of sugar phosphates it should be precipitated under these conditions. A small amount of radioactivity was removed by this procedure from perchloric acid extracts of adipose tissue incubated with galactose-$C^{14}$ and insulin, the amount lost corresponding to 0-11% of the intracellular galactose-$C^{14}$. No radioactivity was lost from adipose tissue incubated without insulin or from perchloric acid extracts of the incubating medium. Thus it appears that no more than 10% of the intracellular galactose was converted to other products. The loss of intracellular free galactose would not appreciably affect the determination of intracellular galactose space and has been neglected. Moreover, the fact that almost identical results were obtained when accumulation of the non-metabolized sugar, 3-0-methyl glucose, was measured suggests that the metabolism of galactose did not contribute appreciably to the observed effects of insulin, epinephrine and caffeine.

Although it would be desirable to measure sugar transport in isolated adipose tissue cells as well as in intact epididymal fat pads we have so far been unable to make satisfactory measurements of galactose space in the isolated cells. We have instead used the incorporation of radioactivity from glucose-U-$C^{14}$ into $CO_2$ and lipid as an indication of glucose transport. This approach is based on the assumption that transport of glucose across the cell membrane is the rate-limiting step in glucose metabolism in the absence of insulin.

When isolated fat cells were incubated with glucose-U-$C^{14}$ insulin produced the expected increase in incorporation of label into $CO_2$ and lipid. Epinephrine stimulated incorporation of label in the absence of insulin but had little effect in the presence of insulin. Theophylline (or caffeine) inhibited incorporation of label into $CO_2$ and lipid both in the presence and absence of insulin. When epinephrine and theophylline were added to cells incubated with insulin, the inhibition of glucose incorporation was much greater than that seen with theophylline alone (Table 4).

These results are very similar to those obtained with non-metabolized sugars in the intact fat pad. In both sets of experiments insulin stimulated sugar transport whereas epinephrine plus a methyl xanthine was inhibitory. Epinephrine alone stimulated transport in the isolated fat cells but the effect was small and the space measurements in the intact fat pads were not accurate enough to detect such a small change.

TABLE 4

Effect of insulin, epinephrine and theophylline on glucose metabolism by isolated fat cells.

Isolated fat cells were incubated for 1 hr at 37° with 0.3 mM glucose-U-$C^{14}$. $CO_2$ was trapped in hyamine and neutral lipid extracted with chloroform-methanol. The values are means ± S. E. M. of 8 experiments.

| Additions | Glucose converted to $CO_2$ and Lipid (μmoles/g/hr) |
|---|---|
| None | 0.25 ± 0.04 |
| Insulin 1 mμ/ml | 3.46 ± 0.19 |
| Epinephrine 0.3 μg/ml | 0.53 ± 0.04 |
| Theophylline 0.3 mM | 0.16 ± 0.03 |
| Epinephrine + Theophylline | 0.20 ± 0.02 |
| Insulin + epinephrine | 3.59 ± 0.22 |
| Insulin + theophylline | 2.80 ± 0.15 |
| Insulin + epinephrine + theophylline | 1.89 ± 0.15 |

If insulin stimulated transport by lowering cyclic AMP we would expect that raising cyclic AMP would inhibit transport. There are, however, at least two circumstances under which a high rate of transport and a high cyclic AMP level can be seen at the same time, i.e., in tissue stimulated with epinephrine or epinephrine plus insulin. It thus seems unlikely that glucose transport is controlled by cyclic AMP, at least in the range where it affects lipolysis and glycogen metabolism. The very high levels of cyclic AMP in tissue treated with epinephrine and caffeine or theophylline do appear to inhibit transport, but it is unlikely that such high cyclic AMP levels are ever reached *in vivo* and the physiological significance of this finding is not clear.

It remains possible that sugar transport in adipose tissue is controlled by lower levels of cyclic AMP. If, in fact, cyclic AMP is important in mediating the action of insulin on sugar transport we would have to postulate that the cyclic AMP level in control tissue is high enough to inhibit transport almost completely. If all the cyclic AMP in unstimulated adipose tissue were distributed evenly through the cell water the intracellular concentration would be about $4 \times 10^{-6}$M which is considerably higher than the concentration required to activate phosphorylase *in vitro*. Since phosphorylase is not in fact activated in unstimulated tissue, much of the cyclic AMP in the cell must be inactive, perhaps through binding or sequestration. If insulin produced a fall in a small active fraction the fall might not be detectable against the large background of inactive cyclic AMP.

Prostaglandin $E_1$ lowers cyclic AMP in isolated fat cells under conditions where insulin has no detectable effect (9). If insulin stimulates transport by lowering cyclic AMP levels we would expect prostaglandin $E_1$ to stimulate transport also.

Fat cells were incubated with prostaglandin $E_1$ epinephrine and insulin and the incorporation of glucose-U-$C^{14}$ into lipid and $CO_2$ and glycerol release measured. Prostaglandin $E_1$ inhibited epinephrine-induced lipolysis more than a maximally-effective concentration of insulin, and thus presumably lowered cyclic AMP levels more than insulin. Prostaglandin $E_1$ was on the other hand much less effective than insulin in stimulating glucose transport (Table 5). Thus it appears unlikely that the effect of insulin on glucose incorporation into $CO_2$ and lipid (i.e., on transport) can be caused by a fall in tissue cyclic AMP

TABLE 5

Action of prostaglandin $E_1$ on isolated fat cells.

Isolated fat cells were incubated for 30 min at 37° with 0.3 mM glucose and U-$^{14}$C-glucose. The figures are means ± S. E. M. of 6 experiments.

| Additions | Glucose converted to $CO_2$ and Lipid (μmoles/g/hr) | Glycerol Release (μmoles/g/hr) |
|---|---|---|
| None | .59 ± 0.14 | 2.0 |
| Insulin (mU/ml) | 3.80 ± 0.31 | 2.0 |
| $PGE_1$ (1 μg/ml) | .68 ± .11 | 2.0 |
| Epinephrine (1 μg/ml) | 1.17 ± .11 | 51.5 ± 1.7 |
| Epi. + $PGE_1$ | 1.96 ± .21 | 27.7 ± 0.6 |
| Epi. + insulin | 3.43 ± .32 | 45.9 ± 2.5 |

levels. Furthermore, we have never been able to show that insulin lowers the cyclic AMP level in tissue that has not been stimulated with lipolytic hormones with or without caffeine or theophylline.

Insulin exerts important metabolic effects on muscle and liver as well as on adipose tissue and it seemed possible that cyclic AMP might be important in mediating some of these actions, too. In the perfused rat liver insulin inhibits gluconeogenesis, glycogenolysis and potassium output whereas glucagon and epinephrine stimulate these processes and at the same time raise tissue cyclic AMP levels. Insulin has been found to lower cyclic AMP in livers perfused with glucagon, and it appears to lower the cyclic AMP in livers perfused with buffer alone. Furthermore, the cyclic AMP level in the livers of alloxan diabetic rats is considerably higher than in controls; 30 min treatment of the rats with PZI markedly lowers cyclic AMP. It seems likely that this lowering of liver cyclic AMP levels produced by insulin accounts for many of its actions (8).

In preliminary experiments we were unable to detect any changes in cyclic AMP in rat hearts perfused with insulin under various conditions. Whether cyclic AMP has any role in mediating the actions of insulin on muscle must remain doubtful.

In summary we can say that insulin lowers cyclic AMP levels in adipose tissue and liver when the cyclic AMP has first been raised by various hormonal manipulations. This fall in cyclic AMP seems to account for several important actions of insulin, in particular its action on lipolysis, glycogen synthetase and phosphorylase in adipose tissue, and on gluconeogenesis, glycogenolysis and potassium output by the liver. Glucose transport in adipose tissue does not seem to be controlled by the cyclic AMP level and we have no evidence that the action of insulin on muscle is mediated by cyclic AMP.

Acknowledgement. We wish to thank Dr. R. W. Butcher for carrying out cyclic AMP measurements, and Mr. B. H. Empson for excellent technical help.

## REFERENCES

1. Jungas, R. L. and Ball, E. G. (1963). Biochemistry 2, 383.
2. Craig, J. W. and Larner, J. (1964). Nature, 202, 971.
3. Jungas, R. L. (1966). Proc. Natl. Acad. Sci. 56, 757.
4. Butcher, R. W., Ho, R. J., Meng, H. C. and Sutherland, E. W., (1965). J. Biol. Chem., 240, 4515.
5. Sutherland, E. W. and Rall, T. W. (1960). Pharm. Rev., 12, 265.
6. Rosell-Perez, M. and Larner, J. (1964). Biochemistry, 3, 81.
7. Crofford, O. B. and Renold, A. E. (1965). J. Biol. Chem., 240, 14.
8. Jefferson, L. S., Jr., Exton, J. H., Butcher, R. W., Sutherland, E. W. and Park, C. R. (1967). J. Biol. Chem., in press.
9. Butcher, R. W., Pike, J. E. and Sutherland, E. W. (1967). Proc. Nobel Symposium II. S. Bergstrom and B. Sammuelson, Eds., John Wiley and Sons, N. Y.

# STIMULATION OF ACTIVE SODIUM TRANSPORT BY INSULIN

Jean CRABBÉ, with the technical assistance of J. SCARLATA

Section of Endocrinology, Departments of Physiology and Medicine, University of Louvain Medical School, Louvain

Besides speeding up the transfer of glucose across cell membranes in striated muscle (Levine and Goldstein, 1955) and adipose tissue (Crofford and Renold, 1965), insulin brings about a decrease in the concentration of circulating potassium (Harrop and Benedict, 1923). Such glucose and potassium movements are not interdependant: for instance, Kestens, Haxhe, Lambotte et Lambotte (1963) have shown that isolated dog liver takes up potassium but not glucose when insulin is introduced in the perfusion circuit at 37°C.

The uptake of potassium by muscle exposed to insulin is amenable to interpretation since Zierler (1959) has demonstrated hyperpolarization of the cell membrane under such circumstances, whether or not glucose was present during incubation; the same was observed with adipocytes (Beigelman and Hollander, 1963). Hyperpolarization of cell membrane induced by insulin might in turn result from enhanced active sodium transport from the cytoplasm outward, as has been shown by Moore (1965) to occur in the case of frog sartorius.

Because of this effect of insulin on active sodium transport, it was thought worth while to evaluate the influence of the hormone on amphibian epithelia characterized by an ability to perform unidirectional sodium movement in the absence of external driving forces. A stimulation of transmembrane potential and of short-circuit current - a reflection of net, active sodium transport (Ussing and Zerahn, 1951) - was observed upon addition to Ringer's on the inner surface of incubated ventral skin of *Bufo marinus*, of 1 unit/ml ox insulin (André and Crabbé, 1966).

The significance of this stimulation of active sodium transport

Table 1. Stimulation of Active Sodium Transport by Toad Bladder, Colon and Skin after Treatment of the Animal with Insulin ($\mu A/cm^2 \pm$ S.E.)[+]

| | Untreated Toads | Toads treated with Insulin[++] |
|---|---|---|
| Bladder | 11.7 ± 1.8 | 33.6 ± 5.4 |
| Colon | 9.5 ± 1.9 | 13.6 ± 3.5 |
| Skin | 12.3 ± 1.8 | 16.7 ± 1.3 |

[+] For each preparation, 4-5 readings were obtained, at 15 min intervals, during the first hour of incubation; the latter was started as quickly as possible after dissection of the tissues.

[++] Ox Insulin, 1U/10 g body weight, injected subcutaneously 6 hours before sacrifice.

ascribed to insulin was assessed by complementary experiments which consisted in treatment of the animal with the hormone prior to dissection and incubation of their skin. Bladder and colon, also capable of efficient unidirectional active sodium transport (Leaf, Anderson and Page, 1958; Cofré and Crabbé, 1967) were examined concomitantly. When ox insulin was injected into toads 6 hours before sacrifice, an increase in short-circuit current was noted for the preparations studied (Table 1). In order to reduce variations, twelve toads were handled as pairs of which one animal received insulin while the control was treated with the excipient, and all tissues were incubated simultaneously.

An influence of insulin could be detected during the first hour of incubation only; thereafter, membrane activity was generally down to control levels. When the time interval between injection of insulin into the toad and tissue incubation was shorter than 6 hours, the changes in current and potential were less obvious; on the other hand, this insulin effect was often quite pronounced up to 16 hours after a single dose that, despite its impportance, did not kill any animal so far. Amphibians are notoriously slow to develop 'clinical' hypoglycemia (e.g. convulsions) after administration of large amounts of insulin (Hemmingsen, 1925; Houssay, 1959).

Since sodium-transporting epithelia such as toad bladder and colon, seemed to react like toad skin when insulin was injected into the animal, and since toad skin proved capable of direct stimulation, the _in-vitro_ approach was attempted with toad bladder and colon. In the case of urinary bladder of _Bufo marinus_, a stimula-

Table 2. Stimulation of Active Sodium Transport by Toad Colon incubated in the Presence of Insulin ($\mu A/cm^2 \pm$ S.E.)

| Final Concentration of Insulin | N | Activity before Insulin [+] | Activity after Insulin | Insulin Effect |
|---|---|---|---|---|
| | | A. Unstripped Colon | | |
| 1 unit/ml | 8 | 38.4 ± 6.9 | 32.7 ± 8.0 | - |
| | | B. Stripped Colon | | |
| 1 unit/ml | $15^{C}_{I}$ ++ | 15.9 ± 2.0<br>13.4 ± 0.8 | 13.9 ± 1.4<br>18.0 ± 1.3 | 6.6 ± 1.7<br>(P<0.001) |
| 0.125unit/ml | $10^{C}_{I}$ ++ | 15.1 ± 2.1<br>14.2 ± 1.9 | 11.6 ± 1.4<br>19.3 ± 1.9 | 8.6 ± 1.5<br>(P<0.001) |
| 0.025unit/ml | $10^{C}_{I}$ ++ | 21.1 ± 3.8<br>16.7 ± 3.2 | 16.0 ± 4.2<br>18.0 ± 3.6 | 6.4 ± 1.1<br>(P<0.001) |

[+] Time interval between addition of insulin and readings was 2 hours for unstripped colon, 1 hour for stripped colon.

[++] Ox insulin was added on the serosal side immediately after the first set of readings so as to take into account asymetry between matched preparations (Crabbé, 1964).

ting effect of insulin such as reported by Herrera (1965) has been seen here only occasionally. With the colon of the same animal, an increase in short-circuit current and transmembrane potential resulted regularly from introduction of insulin into the incubation fluid on the serosal side, provided the colon mucosa be dissected free from underlying muscle layers. As appears from Table 2, the sodium-transporting activity of unstripped colon declined despite insulin. When matched stripped preparations were incubated, one piece of each pair was treated with insulin, after baseline values were obtained; the other piece of the pair served as a control and, in most instances, was incubated in the presence of the excipient. Noteworthy is that the stimulating effect of insulin on sodium transport was as marked with 25 milliunits/ml as with a concentration 40 times larger.

The toad skin preparation sustains long incubations much better than colon, thereby providing a means of evaluating the repercussions of the metabolic state of the tissue on its responsiveness

Table 3. Stimulation of Active Sodium Transport by Toad Skin incubated in the presence of Insulin. ($\mu A/cm^2 \pm$ S.E.)

| Conditions of Incubation | N | | Activity during first Hour | Activity during third Hour | Insulin Effect |
|---|---|---|---|---|---|
| fresh skins glucose absent | 10 | C<br>I+ | 41.1 ± 4.7<br>42.0 ± 4.2 | 37.3 ± 3.4<br>56.9 ± 5.4 | 18.7 ± 6.9<br>(P< 0.05) |
| fresh skins glucose present++ | 10 | C<br>I+ | 45.8 ± 2.5<br>48.7 ± 2.8 | 50.5 ± 3.0<br>63.7 ± 4.4 | 10.3 ± 3.7<br>(P< 0.05) |
| skins incubated overnight; glucose added after first set of readings++ | 8 | C<br>I+++ | 12.3 ± 2.4<br>22.9 ± 3.8 | 13.2 ± 2.5<br>23.9 ± 4.3 | 10.7 ± 3.9<br>(P< 0.05) |

+ Insulin, 1 U/ml, added immediately after first set of readings; excipient added to matched preparations.

++ Final concentration : 10mM.

+++ Insulin, 1U/ml, added 2 hours before first set of readings; excipient added to matched preparations.

to insulin <u>in vitro</u>. When fragments of ventral skin were examined shortly after sacrifice, addition of glucose to incubation fluid did not modify the reaction to insulin (Table 3). Even after incubation overnight in the absence of substrate, toad skin was capable of a sizable stimulation upon addition of insulin; again, the amplitude of the response was uninfluenced by glucose. This is in contrast with what occurs with toad skin stimulated by aldosterone, as Dr. Ehrlich has shown in his laboratory.

There are arguments for interpreting the action of aldosterone on active sodium transport by toad bladder, colon and skin as due to a hormone-induced multiplication of the sites of penetration of sodium into the cell, across its apical border (Crabbé, 1967); this requires the intervention of nucleic acid metabolism and the effect of aldosterone is blocked by actinomycin D. This drug actually fails to interfere significantly with the reaction of toad skin to insulin : before paired preparations were stimulated with insulin, short-circuit current ($uA/cm^2 \pm$ S.E.) averaged 42.4 ± 2.6 for the control, 36.1 ± 1.4 for the matched fragments exposed for 2 hours to actinomycin D, $10^{-5}$M; 2 hours after addition of insulin to both sets of membranes, currents were 55.4 ± 7.1 and 46.8 ± 5.0, respectively. The small difference in the amplitude of the mean response to insulin was statistically not significant ( $P > 0.25$; N = 6). This observation suggests that, as is the case for glucose transfer across cell membrane, the influence of insulin on active sodium transport does not require the intervention of nucleic acid metabolism.

The effects of aldosterone and insulin on toad skin potentiate each other (André and Crabbé, 1966); this was taken as an argument for locating the action of insulin at, or close to, the sodium 'pump', at the basal cell membrane. Herrera (1965) had formulated the same conclusion after a study of the kinetics of sodium transport across toad bladders stimulated by insulin.

It would thus seem that the action of insulin is closely related to the operation of the sodium 'pump' itself. This might to be a reason why the acute effects of insulin on ion movement are more widespread than those of aldosterone - hence the hypokalemia developing after administration of insulin. The action of aldosterone on the other hand would be restricted to epithelia specialized in transcellular sodium transport andopposing free access of sodium to the 'pump' by interposition of a saturable permeability barrier at the apical (outer) cell membrane.

Acknowledgments : This work has been supported by the Fonds de la Recherche Scientifique Médicale, Belgium (Grants n° 688 and 840). Toads, Bufo marinus, were generously supplied by Mr. D.R. Fischer, Laboratorios Warner, Rio de Janeiro, Brazil, assisted for shipment of the animals by the Ministry of Foreign Affairs of Belgium. Recrystallized ox insulin and excipient were gifts from Novo Laboratories.

References.

ANDRÉ, R. and CRABBÉ, J., Arch. int. Physiol. Bioch. 73 : 538, 1966
BEIGELMAN, P.M. and HOLLANDER, P.B., Diabetes 12 : 262, 1963
COFRÉ G. and CRABBÉ, J., J. Physiol. (Lond.) 188 : 170, 1967
CRABBÉ, J., Acta end. 47 : 419, 1964
CRABBÉ, J., Proc. 2nd int. Cong. Horm. Steroids, 487, 1967
CROFFORD, O.B. and RENOLD, A.E., J.biol. Chem. 240 : 14, 1965
HARROP, G.A. and BENEDICT, E.M., P.S.E.B.M. 20 : 430, 1923
HEMMINGSEN, A.M., Skand. Arch. f. Physiol. 46 : 56, 1925
HERRERA, F.C., Am. J. Physiol. 209 : 819, 1965
HOUSSAY, B., in Comparative Endocrinology (A.Gorbman) A.P., 634, 1959
KESTENS, P.J., HAXHE, J.J., LAMBOTTE, L., and LAMBOTTE, C., Metab. 12, 1963
LEAF, A., ANDERSON, J. and PAGE, L.B., J.gen.Physiol. 41 : 657, 1958
LEVINE, R. and GOLDSTEIN, M.S., Rec.Prog.Horm.Res. 11 : 343, 1955
MOORE, R.D., Proc. 9th ann. Meet. Biophys. Soc. (U.S.) : 122, 1965
USSING, H.H. and ZERAHN, K., Acta Physiol. scand. 23 : 110, 1951
ZIERLER, K.L., Am. J. Physiol. 197 : 524, 1959

# Distinct antagonistic effects of glucagon and insulin in the Embden-Meyerhof-chain

H. Schimassek

J. Helms, and H.J. Mitzkat

Marburg/Germany

One of the important functions of the liver is to supply glucose for the entire organism and take up lactate from the whole body. Both these processes, i.e. the uptake of lactate and the supply of glucose, are very closely interconnected in the liver and there, coordination is obtained through hormonal control by glucagon. That means the essential effect of glucagon in the liver is not - as has been assumed so far - glycogenolysis but the reserval of the flow in the Embden-Meyerhof chain, in a word: gluconeogenesis.

Our findings, which were first reported on in 1963 are based on experiments with isolated perfused rat livers (1,2), as the method of organ perfusion is particulary suitable for the study of the effects of individual hormones. We had prepared a broad basis for our investigations enabling us to demonstrate for the first time that an isolated perfused liver is directly comparable with an in vivo organ in respect of its performance and metabolic state (3). Under the influence of glucagon the liver releases glucose into the blood and simultaneously takes up lactate (and also pyruvate) from the medium. An infusion of 0.1 mμmoles of glucagon per hour per 100 ml of medium and upwards is sufficient to release glucose with a kinetic value of 1 mmole per 10 gr liver per hour.

The concentration of hormone infused is so low that the most appropriate form of measurement is a molecular one. With an infusion of 0.1 mμmoles of glucagon per hour the liver is supplied with about $2x10^{10}$ hormone molecules per minute. Thus each liver cell is reached by only some 50 hormone molecules. - In the past, as far as concernes the liver metabolism, glucagon has always been mentioned in comparison with adrenalin. Adrenalin, however, must be administered in a quantity a 1000 times greater or even more in order to achieve an equivalent release of glucose and its effect is accompanied with a burdening increase of oxygen consumption.

The comparison of glucagon with adrenalin seemed legitimate because the release of glucose by the liver under the influence of glucagon was only explanable through the activation of phosphorylase and thus through simple glycogenolysis (4). This explanation, however, is insufficient. Perfusion experiments comparing the depletion of glycogen and the release of glucose by the liver show that when no hormone is added the values of both substrates are in balance. When adding glucagon, however, only 1/3 of the quantity of glucose released is attributable to glycogen differential. The remaining 2/3 have been produced by gluconeogenesis. Thus the essential effect of glucagon does not lie in glycogenolysis but - and here it is again in contrast to adrenalin - in the reversal of glycolysis.

According to investigations carried out by Krebs et al. (5) who adopted our experimental perfusion methods and extended them, lactate is among the most adaptable substrates for gluconeogenesis. So it is noteworthy that with glucagon absorption of L-Lactate from the medium is tripled.

Utilizing $^{14}$C-labelled lactate the experiments that at least 85% of the lactate absorbed and available under glucagon is indeed resynthesized into glucose (comp. 6). It is calculable that about 1/3 of the maximum quantity of the newly produced glucose is synthesized from lactate.

It should be mentioned in passing, that glucagon increases only the metabolism of L-lactate and not that of the isomeric D-lactate (3); this also substantiates the fact that glucagon influences only the metabolic chain localized in the cytosol. Glucagon has long been considered an antagonist of insulin because of the changes it produces in some very general parameters as, for instance, in the glucose content in the blood. We have been able - for the first time - to localize directly antagonistic effects of both hormones in a very decisive part of the

| (mμmoles/g w.w.) | in vivo | perfusion | | | |
|---|---|---|---|---|---|
| | | untreated | glucagon (3mμmoles/h) | lactate (500μmoles/h) | lactate (500μmoles/h) glucagon (3mμmoles/h) |
| G6P | 370 | 130 | 390 | 327 | 1000 |
| F6P | 74 | 26 | 78 | 65 | 197 |
| F1.6P | 22 | 31 | 19 | 90 | 9 |
| G6P/F1.6P | 17 | 4.5 | 21 | 3.6 | 110 |
| F6P/G6P | 0.2 | 0.2 | 0.2 | 0.2 | 0.2 |

Changes in the substrate content during glucagon - induced gluconeogenesis in isolated perfused rat livers.

glycolytic chain. Our extensive fundamental investigations give us an indication of this (3). - During perfusion the content of glucose-6-phosphate (G6P) falls and that of fructosediphosphate (FDP) rises. G6P and fructose-6-phosphate (F6P) are also in equilibrium in vitro so that we can simply speak of G6P and FDP and can easely detect any changes by the ratio G6P/FDP. While in vivo this ratio is 17, it is only 4 after 3 hours of perfusion. When infusing glucagon the content of G6P rises again and that of FDP falls reciprocally. Under the influence of Glucagon the ratio G6P/FDP is 21 also in the perfused liver. - By taking these measurements we investigate intracellular substrate levels in the steady state. Here the importance of the glucagon effect is not a simple increase in the substrate G6P. An increase in the intracellular substrates can - as the table shows - also be achieved by simple lactate infusion; this will, however, increase all substrates by the same quantity so that the ratio G6P/FDP remains 4. But if we infuse glucagon plus lactate simultaneously there is a sharp rise in the content of G6P and a decrease in the quantity of FDP, thereby augmenting the ratio G6P/FDP to more than 100 as a reflection of the flow reversal. - The following table shows once more the adjustment of the steady state of glucose-6-phosphate and fructosediphosphate under the effect of insulin. As is to be expected of an antagonist the effects of insulin are the exact opposite, i.e. with insulin the content of G6P falls and that of FDP rises; the ratio of G6P/FDP is only 2. The effect of insulin also becomes particulary clear by simultaneous lactate infusion; the G6P level no more rises.

| (mμmoles/gw.w.) | untreated | glucagon (3 mμmoles/h) | insulin (3 IE/h) | lactate (500 μmoles/h) | lactate (500 μ moles/h) insulin (3 IE/h) |
|---|---|---|---|---|---|
| G6P | 130 | 390 | 160 | 327 | 146 |
| F6P | 26 | 78 | 30 | 65 | 30 |
| F1.6P | 31 | 19 | 60 | 90 | 36 |
| G6P/F1.6P | 4.5 | 21 | 2.6 | 3.6 | 4 |
| F6P/G6P | 0.2 | 0.2 | 0.2 | 0.2 | 0.2 |

Antagonistic effects of insulin and glucagon on the content of G6P and F1.6P in isolated perfused rat livers.

If one representes the in vivo data on a hyperbola as function of the reciprocal values of G6P and FDP and enters on this diagram the results of the perfusion experiments we obtain the following information: The controlling of the glycolytic pathway by the hormones insulin and glucagon is reflected in the strict proportion of the substrates G6P and FDP. The data obtained in perfusions without hormone additions correspond to the values under insulin; thus the metabolism of the isolated organ remains influenced

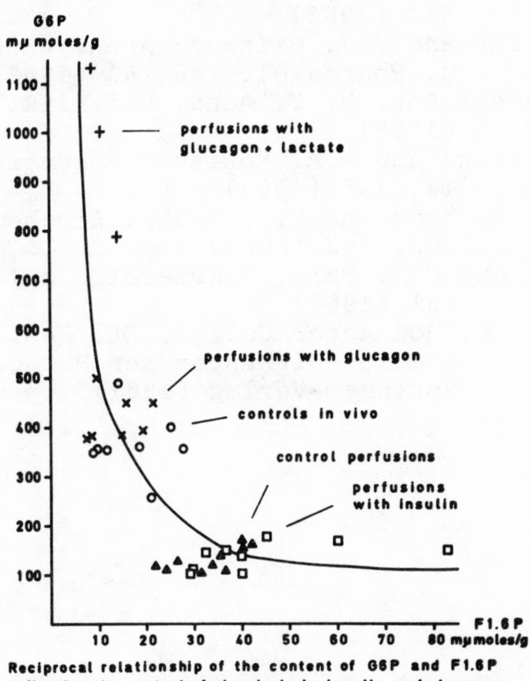


Reciprocal relationship of the content of G6P and F1.6P reflecting the control of glycolysis by insulin and glucagon.

by the insulin effects. Infusion of glucagon brings the perfused liver to the same state as the in vivo liver. This means that the Embden-Meyerhof chain in vivo is continuously controlled by glucagon. - The extreme reciprocal changes of substrate content during glucagon plus lactate infusion makes clear in the best way the glucagon-induced flow reversal in this pathway. - The control of the substrate flow takes place in the part of the glycolytic chain between G6P and FDP. The possibility of the above control in this part is readily attributable to two reactions: The ATP 'one way' directed reaction of the F6P kinase and the avoidance of this reaction by the FDPase. This section of the glycolytic chain has as the possible controller already been discussed before. By these data a specific antagonism between a hormone pair - glucagon and insulin - in a close circumscribed segment of the metabolic chain has been demonstrated for the first time.

For further data and details see (7).

References:

1. H. Schimassek and H.J. Mitzkat: Biochem. Z. 337, 510 (1963)
2. H. Schimassek and H.J. Mitzkat: Arch. exper. Path. u. Pharmakol. 246, 63 (1963)
3. H. Schimassek: Ann. N. Y. Acad. Sci. 119, 1013 (1965)
4. E.W. Sutherland and G.A. Robison: Pharmacol. Rev. 18, 145 (1966)
5. B.D. Ross, R. Hems and H.A. Krebs: Biochem. J. 102, 942 (1967)
6. J.H. Exton and C.R. Park: Pharmacol. Rev. 18, 181 (1967)
7. H. Schimassek: Mosbacher Colloq. Dt. Ges. Biol. Chemie 'Wirkungen der Hormone' Springer-Verlag (1967)

# SELECTIVE HISTOCHEMISTRY OF GLUCAGON IN THE A CELLS OF PANCREATIC ISLETS BY INDOLE METHODS*

C. CAVALLERO, E. SOLCIA, R. SAMPIETRO

Institute of Pathological Anatomy, University of Pavia, Italy (Director: Prof.C.Cavallero)

Up to now attempts to provide a selective morphological detection of glucagon in the A cells of pancreatic islets have been carried out by means of immunohistochemical techniques or indole methods. By immunohistochemistry the presence of glucagon in the cytoplasm of some islet cells has been definitively ascertained, but the exact nature of these cells has not been completely established (Baum et al., 1962; Chiappino et al., 1963). In fact, the most reliable and selective methods for islet cytology are not appliable to fresh cryostatic sections which are required by immunohistochemical techniques.

By indole methods it has been observed that secretory granules of A cells store protein(s) with reactive tryptophan, while both B and D cells do not react (Levine and Glenner, 1958; Logothetopoulos and Salter, 1960; Petersson and Hellman, 1963; Solcia and Sampietro, 1965). However, the positivity of A cells to these methods is very weak, fail to occur in some animal species and do not prove *per se* that the reactive protein is in fact glucagon. We have recon-

*This investigation was supported by grant N. 115/1139/ /991 from the Italian Consiglio Nazionale delle Ricerche.

sidered this point in an effort to obtain a sure histochemical detection of glucagon in the secretory granules (α granules) of islet A cells by increasing both the staining and the selectivity of the indole reactions.

In the first experiment samples taken from the pancreas of guinea-pigs and horses were fixed in various mixtures in order to investigate their influence on the behaviour of the dimethylaminobenzaldehyde (DMAB)-nitrite (Lison, 1960) and the xanthydrol (Lillie, 1957) methods for indoles histochemistry. In accordance with previous observations dealing with the histochemistry of another indole substance, 5-hydroxytryptamine (Solcia and Sampietro, 1967; Cavallero et al., 1967 b), it was found that protein-bound tryptophan in the secretory granules of A cells reacted far better when glutaraldehyde mixtures were employed as fixatives instead of the usual formaldehyde ones.

It was observed also that a six day fixation in 4% formaldehyde worked better that the usually recommended 24 hour fixation, provided a neutral or alkaline solution is employed. Moreover, the results of a 24 hour fixation in a formaldehyde-picric acid-sodium acetate (modified Bouin's fluid*) or in a glutaraldehyde-picric acid-sodium acetate (GPA**) mixture were better than those given respectively by formaldehyde or glutaraldehyde alone; but when acetic acid was used in these mixtures instead of sodium acetate an impairment was observed.

---

* One part of a 40% commercial solution of formaldehyde and three parts of a saturated aqueous solution of picric acid to which 1% sodium acetate is added.

** One part of a 25% commercial solution of glutaraldehyde and three parts of a saturated aqueous solution of picric acid to which 1% sodium acetate is added.

Two main factors seem to be considered in order to explain these findings: the first may be the ability of the glutaraldehyde - particularly when mixed with picric acid - to strongly fix proteins stored in the secretory granules of A cells; the second may be the tendence of the formaldehyde - at least in the presence of some acids acting as catalysts such as acetic and hydrochloric acids - to condense with protein-bound tryptophan blocking irreversibly the reactivity of their $\alpha$ (2) position, which must be free for condensation with xanthydrol or DMAB do occur.

In a second experiment most of the recommended methods for indoles histochemistry were applied to pancreatic tissue taken from several mammals and some birds and fixed in the GPA mixture. Samples from the stomach, intestine, thyroid, eye and adenohypophysis, which are known to contain tryptophan-storing structures, were also examined.

It was found that the DMAB-nitrite method and the post-coupled benzylidene reaction (Glenner and Lillie, 1957), which gave very similar results, provided the heavier staining of the $\alpha$ granules in islet A cells; the xanthydrol method gave also a sharp staining but, surprisingly, it stained these cells blue-gray, differently from all other tryptophan-containing structu-

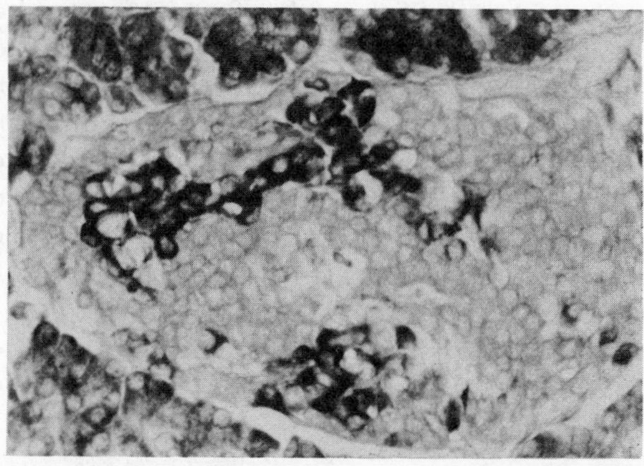

Guinea-pig pancreas fixed in the GPA mixture: groups of islet A cells heavily stained by the DMAB-nitrite method. X 350.

res, such as pancreatic and gastric zymogen, Paneth's cell granules, crystalline lens, thyroid colloid and adenohypophyseal cells.

In previous researches a blue staining by xanthydrol was also observed in the case of enterochromaffin (Ec) cells and mast cells storing 5-hydroxytryptamine (5-HT) (Solcia et al., 1966). Thus the colour reactions of α granules, zymogen granules and other structures containing tryptophan and Ec cells were systematically reinvestigated by employing both the usual methods and some aldehydes whose reactivity with indole substances was tested in previous "in vitro" experiments. The aldehydes were employed as 3% solutions in a 4:1 mixture of acetic acid and hydrochloric acid; post-treatment with sodium nitrite was performed at low concentrations in order to carefully control the staining they developed. For "in vitro" tests 1 mg of 5-HT or tryptophan and 10 mg of glucagon* or lysozyme were dissolved in one ml of the same solutions used for histochemical tests. Moreover, some crystalline proteins such as glucagon and insulin were embedded in a gelatin-sorbitol film on a slide, fixed in 6% glutaraldehyde and washed in 0,2 M phosphate buffer pH 7,2 before to be stained by indole methods. The main results are grouped in table 1. From the table it appears that α granules and glucagon give very similar reactions but, in both histochemical and "in vitro" tests, behave quite differently from tryptophan, other tryptophan-containing proteins or 5-HT. It should be added that granules did not react to any other methods detecting 5-HT, such as, diazonium, Masson-Hamperl, Schmorl, Pearse's thioindoxyl and fluorescence reactions (Vialli, 1966); conversely, they were stained by indole methods after formol-fixation also, when Ec granules and other 5-HT-storing structures failed to react.

The involvement of a tryptophan residue of both extractive glucagon and granules in the reactions

---

*Pork glucagon (sample N. 258-561b-180) kindly suported by Lilly Laboratories, Indianapolis.

TABLE I

| Method | Pancreatic zymogen | Lysozyme or tryptophan (in vitro) | Alpha granules | Glucagon (in vitro) | Ec granules | 5-HT (in vitro) |
|---|---|---|---|---|---|---|
| Xanthydrol | purple-violet | purple-violet | Blue-gray | blue-gray | blue-green | blue-green |
| DMAB | " - " | " - " | purple | purple | " - " | " - " |
| DMAB-nitrite | blue | blue | blue to violet | blue | dark-blue | dark-blue |
| Post-coupled benzyl reaction | " | " | " " | " | " - " | " - " |
| Vanilline-nitrite | violet | violet | brown to blue | brown | " - " | " - " |
| Salicylal-dehyde-nitrite | " | " | brown to violet | " | " - " | " -- " |
| Benzaldehyde-nitrite | blue | blue | blue-gray | blue-gray | green-yel-lowish | green-yellowish |

listed in the table seems to be probable on the grounds of the well known specificity of these reactions for indole substances, at least in histochemical conditions. Moreover, they were completely prevented by potassium persulphate (2% in 0,5N KOH, for 16 hours) which, according to Boyland et al. (1956), breaks the pyrrole ring of indoles, and by N-bromosuccinimide (0,1% in acetate buffer pH 4, for 5 minutes) which, according to Patchornich et al. (1958), selectively binds tryptophan at their 2 ($\alpha$) carbon and breaks the glucagon molecule. It should be added that both insulin, which notoriously does not contain tryptophan in its molecule, and insulin-storing $\beta$ granules of islet B cells failed to react to indole methods.

From the above results it seems probable that the reactive substance stored in   granules of islet A cells is in fact glucagon, or at least some compound which appears to be firmly bound to the glucagon molecule in both   granules and crystalline extractive hormone. Obviously, the disposal of synthetic or highly purified extractive glucagon could clarify the latter question. So far, we may only anticipate that the glutamine-tryptophan-leucine* sequence of glucagon molecule behaves just as tryptophan, lysozyme and other tryptophan containing proteins or structures.

Thus, at least for histochemical purposes, the relationship between $\alpha$ granules and glucagon seems to be firmly established. In this connection it should be noted that, after fixing in the GPA mixture, A cells of all species examined reacted blue-gray to the xanthydrol and blue to the DMAB-nitrite method, including A cells of man, monkey, horse, pig, dog, rabbit, guinea-pig and such species, as rat, mouse, duck and chicken, which were unreactive according to previous Authors. Then, the reactive group(s) or sequence seem to be constantly present in the hormone structure whatever species is considered. Yet, some differences in staining degrees have been observed in

---

* $NH_2$ (attached to Glu)
H-Glu-Try-Leu-$NH_2$.HCl, kindly supported by Prof. B. Camerino, Laboratori Ricerche Farmitalia, Milan, Italy.

the different species. The reactions were strong in the A cells of horse, guinea-pig and rabbit, moderate in man, monkey and pig, weak in birds, rat and mouse; in general, this behaviour parallels that of other methods detecting granules such as the dark-field microscopy (Solcia, 1962), the phosphotungstic acid hematein (PTA-Hn) and the HCl-toluidine blue method (Solcia et al., 1967), probably reflecting different concentrations of glucagon in the cytoplasm of A cells. However, birds seem to represent a conspicuous exception, because they react strongly to the conventional methods but only weakly to indole methods.

The xanthydrol and DMAB-nitrite methods were also applied to the pancreatic islets of both fetus or newborn guinea-pigs and rats. High numbers of A cells with numerous reactive granules have been observed in the 55 day old fetus of the guinea-pig as well as in the newborn guinea-pig and in a six month human fetus. Conversely no reactive cells have been observed in islets of rat fetuses and newborn rats. These findings confirm previous observations done by simple morphological methods (Cavallero and Solcia, 1964). By applying the indole methods to the pancreas of synthalin*-treated guinea-pigs, a marked decrease of reactivity was observed in the islet A cells, which, when studied by conventional morphological methods, showed a severe loss of secretory granules. Conversely, A cells of alloxan-treated rats with long-standing diabetes showed some increase as regards both secretory granules and the degree of staining by indole methods.

It should be noted that the secretory granules of islet D cells were either completely unreactive to the indole methods, or gave very weak reactions whose colours reproduced the staining of common tryptophan-containing proteins and led us to exclude the presence of glucagon in these structures. This finding, as well as the unreactivity of the same granules to the methods selectively staining insulin, support the hypothesis

---

*Synthalin A, 3 mg/Kg daily for 3 days, subcutaneously.

(Cavallero and Solcia, 1963; Epple, 1965; Cavallero et al., 1967 a) that this cell type stores a third, so far undefined, protein hormone.

## REFERENCES

BAUM J., SIMONS B.E., UNGER R.H., MADISON L.L. (1962): Localization of glucagon in the alpha cells in the pancreatic islet by immunofluorescent technics. Diabetes 11, 371.

BOYLAND E., SIMS P., WILLIAMS D.C. (1956):The oxidation of tryptophan and some related compounds with persulphate. Biochem. J. 62, 546.

CAVALLERO C., SOLCIA E. (1963): Considerazioni sulla citologia insulopancreatica; esistenza e possibile funzione di un terzo tipo cellulare. Riv.Anat.Pat. Oncol. 24/Suppl., 1000.

CAVALLERO C., SOLCIA E. (1964): Cytological and cytochemical studies on the pancreatic islets. In: Brolin S.E., Hellman B. and Knutson H. eds.:The structure and metabolism of the pancreatic islets, p. 83. Oxford, Pergamon Press.

CAVALLERO C., SOLCIA E., SAMPIETRO R. (1967 a): Cytology of islet tumours and hyperplasias associated with the Zollinger-Ellison syndrome. Gut 8, 172.

CAVALLERO C., SOLCIA E., SAMPIETRO R. (1967 b): New approach to the histochemistry of 5-hydroxytryptamine in tissue mast cells. Giornale Arterioscler. (in press).

CHIAPPINO G., CROSIGNANI P.G., SOLCIA E., POLVANI F. (1963): Dimostrazione del glucagone nelle cellule delle isole pancreatiche mediante anticorpi fluorescenti anti-glucagone. Boll.Atti Soc.It.Endocrinol. 11, 110.

EPPLE A. (1965): Weitere Untersuchungen über ein drittes Pankreashormon. Verh.Dtsch.Zool.Ges. 1965, 459.

GLENNER G.G., LILLIE R.D. (1957): The histochemical demonstration of indole derivatives by the post-coupled p-dimethylaminobenzylidene reaction. J.Histochem.Cytochem. 5, 279.

LEVINE H.J., GLENNER G.G. (1958): Observation on tryptophan staining of the pancreatic alpha cells. J.Nat.Cancer Inst. 20, 63.

LILLIE R.D. (1957): The xanthydrol reaction for pyrroles and indoles in histochemistry: zymogen granules, lens, enterochromaffin and melanins. J.Histochem.Cytochem. 5, 188.

LISON L. (1960): Histochimie et cytochimie animales. p.254 and 732. Gauthier-Villars, Paris.

LOGOTHETOPOULOS J., SALTER J.M. (1960): Morphology and cytochemistry of alpha cells of the rabbit pancreas. Diabetes 9, 31.

PETERSSON B., HELLMAN B. (1963): Effects of long term administration of glucagon on the pancreatic islet tissue of rats and guinea-pigs. Acta Endocrinol. 44, 139.

PATCHORNICK A., LAWSON W.B., WITKOP B. (1958): Selective cleavage of peptide bonds. II. The tryptophyl peptide bond and the cleavage of glucagon. Amer.J.Chem.Soc. 80, 4747.

SOLCIA E. (1962): Studio del pancreas endocrino di alcuni vertebrati mediante osservazione in campo scuro di sezioni fresche. Boll.Soc.It.Biol.Sper. 38, 1192.

SOLCIA E., SAMPIETRO R. (1965): On the nature of the metachromatic cells of pancreatic islets. Zeit.Zellforsch. 65, 131.

SOLCIA E., SAMPIETRO R. (1967): Indole nature of enterochromaffin substance. Nature (Lond.) 214, 196.

SOLCIA E., SAMPIETRO R., VASSALLO G. (1966): Indole reactions of enterochromaffin cells and mast cells. J.Histochem.Cytochem. 14, 691.

SOLCIA E., VASSALLO G., CAPELLA C. (1967): Selective staining of endocrine cells by basic dyes after acid hydrolisis. (To be published).

VIALLI M.(1966): Histology of the enterochromaffin cell system. In: Erspamer V., ed.: 5-Hydroxytryptamine and related indolealkylamines, p.1-65. Handbuch der experimentellen Pharmakologie, Bd. XIX. Springer. Berlin-Heidelberg-New York.

INTERRELATIONSHIPS GLUCAGON-ADRENERGIC SYSTEM IN MAN

INTEREST FOR THE DIAGNOSIS OF PHEOCHROMOCYTOMA

P.Lefebvre, A.Cession-Fossion, A.Luyckx
J.Lecomte, H. Van Cauwenberge
Institut de Médecine, Département de Clinique et Pathologie Médicales, et Institut Léon Fredericq, Physiologie, Université de Liège
Belgium

The existence of a stimulatory effect of glucagon on the adrenal medulla has been demonstrated on the isolated-perfused adrenal glands of the dog by SCIAN et al. (10) and on the adrenals in situ of the rat by LEFEBVRE and DRESSE (7). It has been confirmed in rats by DRESSE and LEFEBVRE (1), in dogs by SARCIONE et al. (8) and in man by SCHMID et al. (9) and von KUSCHKE et al. (12, 13) The interest of this phenomenon for the diagnosis of pheochromocytoma has been demonstrated by LAWRENCE and FORLAND (4), LAWRENCE (2, 3) and LEFEBVRE et al. (5). In the present paper we report the effects of pharmacological doses of glucagon on pulse rate, blood pressure and catecholamines blood level in normal man, in sujects suffering from hypertension of various origin and in two patients with documented pheochromocytoma. An extensive study on the mechanism of the stimulation by the glucagon of the adrenergic system in the rat will be reported elsewhere (6).

## I - MATERIAL AND METHODS

The study has been performed on 7 normal subjects, 7 patients suffering from hypertension of various origin and 2 patients with documented pheochromocytoma. Pulse, blood pressure and blood catecholamines were determined at different intervals of time before and after the intravenous injection of 1 mg of crystalline glucagon (Novo). The catecholamine content is determined on plasma according to von EULER (11). Details will

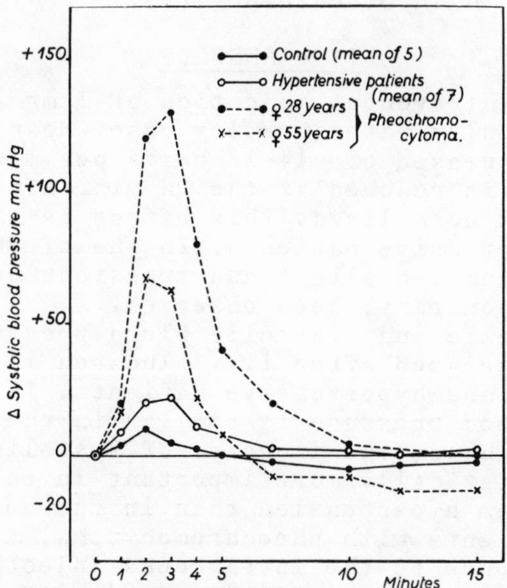


Fig. 1 : Changes in systolic blood pressure after intravenous injection of 1 mg glucagon in man.

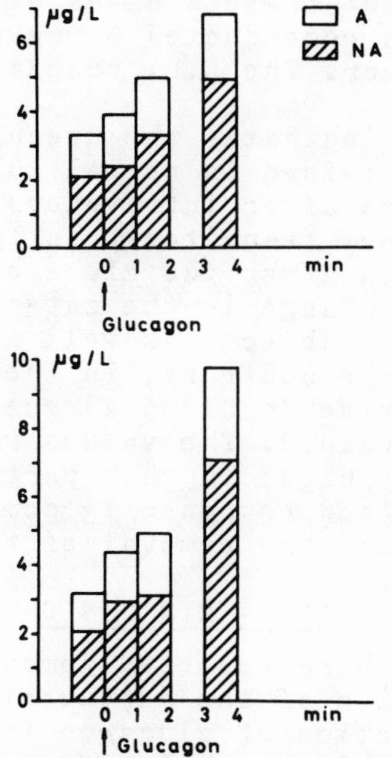


Fig. 2 : Changes in the peripheral catecholamines blood level during the first minutes after an intravenous injection of 1 mg glucagon in 2 normal subjects (A = Adrenaline; NA = Noradrenaline).

be published in an other paper (7).

## II - RESULTS

1) The intravenous injection of 1 mg glucagon increases the pulse rate in every case. Mean pulse frequency is increased of 14-18 beats per min; the maximum increase is reached at the 2d min. After 5 min, the pulse rate is normalised. This effect is similar in normal and hypertensive patients. In one of the 2 cases of pheochromocytoma, a slight and transient bradycardia has, on the contrary, been observed.

2) Systolic and diastolic blood pressure are systematically increased after i.v. glucagon injection in normal subjects and hypertensive patients. The increase of diastolic blood pressure is similar in the two groups. On the other hand, the increase of systolic blood pressure is statistically more important in the patients suffering from hypertension than in the normal subjects. In the 2 patients with pheochromocytoma, a striking pressor response to the intravenous injection of 1 mg glucagon has been obtained. Compared with the above mentioned response of the other subjects studied, the increase in B.P. is far over the 99 % confidence limits (fig. 1). In one of these 2 cases of pheochromocytoma, the test has been reconducted 6 weeks after the eradication of the tumor. The B.P. response is completely normalized.

3) Figure 2 indicates the unequivocal rise in blood catecholamines obtained in normal subjects within the very first minutes after intravenous glucagon injection. This effect is very transitory. In fact, table I shows that, 5 and 15 min after the glucagon injection, there is no detectable change in the catecholamines blood level in the normal subjects as well as in the hypertensive patients. On the contrary, in the two cases of pheochromocytoma, a rise in blood adrenaline and noradrenaline has been obtained. The values reached in case I.F. are particularely high; in this particular case, the catecholamines blood response is completely normalized six weeks after the removal of the tumor.

## III - DISCUSSION

The results here reported demonstrate the existence of a stimulation of the adrenergic system after an intravenous injection of glucagon in man. The increase in peripheral blood catecholamines, in normal men or in hypertensive subjects, is transitory, reaching its maximum during the first minutes after the injection. Afte

TABLE I

| CATECHOLAMINES (A + NA) BLOOD LEVELS (µg/L of BLOOD) | | | | |
|---|---|---|---|---|
| Time relation to glucagon injection | 15 min before | 1 min before | 5 min after | 15 min after |
| Normals | 5,61 ± 0,69 | 6,13 ± 0,77 | 5,54 ± 0,57 | 6,30 ± 1,08 |
| Hypertensive patients | 5,76 ± 0,37 | 5,96 ± 0,63 | 5,92 ± 0,65 | 6,25 ± 1,09 |
| Pheo. n°1 | | | | |
| before operation | | 6,28 | 17,43 | 18,22 |
| after operation | | 5,74 | 5,94 | 4,87 |
| Pheo n°2 | | 4,29 | 6,17 | 5,68 |

the fifth minute, the values have returned at the basal level. On the contrary, in pheochromocytoma, high blood catecholamines levels can still be observed 5 and 15 minutes after the glucagon injection. The results observed in normal men are consistent with those described by von KUSCHKE et al. (12, 13), with the difference that, in our experience, the phenomenon seems much more transitory. This stimulation of the adrenergic system is responsible for the short duration hypertensive effect of pharmacological doses of glucagon, effect which is very important in pheochromocytoma. In this respect, we agree with the proposition of LAWRENCE and FORLAND (4) to consider the glucagon-test as a new and valid test for the diagnosis of this condition. The precocity of the blood pressure response as well as the rapidity of the blood catecholamines level increase are arguments for a direct stimulation by the glucagon of the adrenergic system, including the adrenal medulla, in the human species.

## REFERENCES


1. DRESSE, A. and LEFEBVRE, P.- C.R.Soc.Biol., 1961, 155, 1168.
2. LAWRENCE, A.M.- Ann. intern. Med., 1965, 63, 905.
3. LAWRENCE, A.M.- Ann. intern. Med., 1967, 66, 1091.

4. LAWRENCE, A.M. and FORLAND, M.- J. Lab. Clin. Med., 1964, 64, 878.

5. LEFEBVRE, P., CESSION-FOSSION, A. and LUYCKX, A.- Lancet, 1966, ii, 1366.

6. LEFEBVRE, P., CESSION-FOSSION, A., LUYCKX, A., LECOMTE, J. and VAN CAUWENBERGE, H.- Arch. int. Pharmacodyn. (Submitted to).

7. LEFEBVRE, P. and DRESSE, A.- C.R.Soc.Biol., 1961, 155, 412.

8. SARCIONE, E.J., BACK, N., SOKAL, J.E., MAHLMAN, R. and KNOBLOCK, E.- Endocrinology, 1963, 72, 523.

9. SCHMID, E., ZICHA, L. and JOHN, W.- Z. Gastroenterol., 1964, 2, 214.

10. SCIAN, L.F., WESTERMANN, C.D., VERDESCA, A.S. and HILTON, J.G.- Am. J. Physiol., 1960, 199, 867.

11. von EULER, U.S.- Noradrenaline (Thomas, ed., Springfield, Illinois, USA), 1956.

12. von KUSCHKE, H.J., KLUSMANN, H. and SCHÖLKENS, B.- Arch. exp. Path. Pharmak., 1966, 253, 65.

13. von KUSCHKE, H.J., KLUSMANN, H. and SCHÖLKENS, B.- Klin. Wschr., 1966, 44, 1297.

# EVOLUTIONARY CHANGES IN ADIPOSE TISSUE PHYSIOLOGY

Daniel Rudman, Mario Di Girolamo, Luis A. Garcia

Columbia University Research Service
Goldwater Memorial Hospital, and Department of Medicine, Columbia University, College of Physicians and Surgeons, New York, New York

## I. INTRODUCTION

It is now about 10 years since the intensive investigation of adipose tissue began. Until the 1930's, this tissue was considered a metabolically inert depot of excess calories stored as triglyceride. During the next 3 decades, a series of key observations gradually kindled interest in the possibility of a more dynamic role of the adipose organ in the body's metabolism: The capacity of pituitary extracts to cause an acute mobilization of adipose tissue lipid (Best and Campbell, 1936); demonstration of the rapid turnover of adipose lipid by Schoenheimer and Rittenberg (1936); the avid uptake of glucose, and incorporation of the hexose carbons into stored lipid, by adipose tissue slices incubated *in vitro* (Shapiro and Wertheimer, 1948; Hausberger et al, 1954); discovery in 1956 of the circulating free fatty acids (FFA), recognition that this plasma lipid is secreted into the blood by the fat cells, and demonstration that its plasma concentration fluctuates continuously in rapid response to changes in carbohydrate intake and utilization (Gordon and Cherkes, 1956; Dole, 1956; Laurell, 1956). Now there could be no doubt that the fat cells were participating on a minute to minute basis in the storage and release of metabolic fuel. By 1958 a concerted effort to unravel the metabolic pathways and hormonal control of this cell had been launched in the laboratories of A.E. Renold, F.L. Engel, D. Steinberg and M. Vaughan, E.G. Ball, B.R. Landau, and many others. All of these investigators selected the epididymal adipose tissue of the young male rat, 6-9 weeks old (100-200 g) as the principal experimental tissue; the basic experimental technique was to incubate slices of this tissue in Krebs-Ringer buffer containing radioactive glucose, free fatty acids, or plasma triglycerides, and appropriate doses of hormones active on the fat cell. Within 4 years, the principal metabolic pathways of the young rat's fat cells, and their regulation by hormones of the pancreas, pituitary and adrenergic nervous system, were thoroughly elucidated. Experiments with free fat cells by Rodbell (1964), with the perfused tissue by Scow (1965), and observations in the intact rat, subsequently gave general confirmation to the concepts developed with the incubated adipose tissue slices.

Meanwhile, in several other laboratories, the adipose tissues of other species were also under study. Christophe and coworkers (1961 a and b) examined the adipose tissue of the mouse. In our own laboratory, the adipose tissues of rabbit, hamster and guinea pig were compared with that of the rat (reviewed by Rudman and Di Girolamo, 1967). Valuable comparisons of the response of rabbit and rat tissues to ACTH-type peptides of known structure were made by Raben et al (1961), Lebovitz and Engel (1965), and Tanaka and coworkers (1962). Goodridge (1964), and Goodridge and Ball (1965 and 1966) undertook studies of bird adipose tissue. The human tissue was examined by Hirsch and Goldrick (1964), Gellhorn and Marks (1961), Hamosh et al (1963), and Kahlenberg and Kalant (1964). Brilliant studies on the insect fat body were accomplished by Clegg and Evans (1961), Chino and Gilbert (1964 and 1965) and Tietz (1965).

Except for the mouse tissue, in every other case the data showed that the accepted scheme for the metabolic organization and hormonal regulation of the young rat's fat cell had to be revised in one or more major respects. Certain novel features are shared by the fat cells of insects of different species; other characteristics are exhibited by fat cells of different bird species; and even within the rodent order, taxonomically-correlated metabolic characteristics are emerging. The conclusion now seems inescapable that during animal evolution major alterations took place in the physiology of the fat cell.

In this presentation, our objectives are: (a) to review briefly the current concept of the physiology of the rat's fat cell; (b) to review the experimental data now available on the adipose tissue of other mammalian species, of birds, and of insects; (c) to attempt to

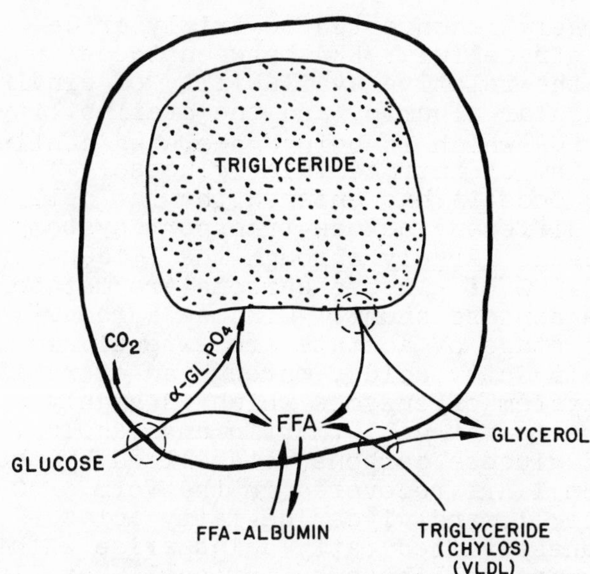


Fig. 1 Major pathways in the young rat's fat cell for uptake of extracellular FFA or triglyceride, for uptake and metabolism of glucose, and for mobilization of stored triglyceride. Circles indicate major points of hormonal control: transport of glucose across the cell membrane, activity of the hormone-sensitive lipase which hydrolyzes intracellular stored triglyceride, and activity of the lipoprotein lipase which hydrolyzes extracellular triglyceride.

arrange the species differences cited into a meaningful phylogenetic pattern; and (d) to define some of the questions which arise from these comparative observations.

## II. Physiology of the rat's adipose tissue

Fig. 1 summarizes, for the 100-200 g rat, the major metabolic pathways and their control by hormones [the reader will find a more detailed account of this subject in the following reviews: Renold et al (1965); Ball and Jungas (1964); Vaughan (1961); Rudman and Di Girolamo, (1967)]. The stored depot of intracellular triglyceride is constantly being hydrolyzed by a triglyceride, di-(mono-)glycerideease system; glycerol is not appreciably reutilized. One fraction of the FFA formed by intracellular lipolysis is released to the extracellular fluid to be transported out of adipose tissue as the albumin-FFA complex; the remaining fraction is largely reconverted to triglyceride. The partition of intracellular FFA between these two pathways depends on the relative availability of binding sites on extracellular albumin, and the availability of α glycerol $PO_4$ which is being generated continuously by the metabolism of intracellular glucose. The metabolism of glucose, once it has entered the fat cell by one of at least 3 different hexose-transport systems (Fain, 1964), proceeds largely through the Embden-Meyerhof pathway; 10-20% of glucose carbons are metabolized through the pentose shunt. Glucose carbons which arrive at the stage of acetate are extensively converted to long-chain fatty acids, through an extramitochondral lipogenic system of enzymes which is highly active in the young rat's fat cell (Martin and Vagelos, 1965). Over 70% of glucose carbons metabolized by the young rat's fat cell are recovered in the form of $CO_2$, glyceride-glycerol, and glyceride-fatty acids.

Thus unesterified fatty acids arise within the young rat's fat cell both by intracellular lipolysis of stored triglyceride, and by intracellular synthesis from glucose. A third pathway leading to this product involves the assimilation of extracellular triglyceride, circulating in the form of chylomicra and low-density lipoproteins. The first step in this assimilation is the hydrolysis of the triglyceride moiety of these fat particles by lipoprotein lipase, located apparently in the region of the fat cell membrane. The resulting glycerol is rejected by the fat cells, while the fatty acids rapidly enter the cells and are extensively incorporated into triglyceride.

Out of these interdigitating pathways emerge the 4 principal functions of the rat's fat cell: Uptake of circulating glucose and its conversion to long-chain fatty acids; uptake of fatty acids circulating as the fatty acid moiety of plasma triglycerides; storage of fatty acids thus introduced into the fat cell as intracellular triglyceride; and mobilization of stored triglyceride-fatty acids in the form of plasma FFA. Each of these functions is under hormonal control. A group of amine and peptide hormones (epinephrine and norepinephrine, glucagon, ACTH and TSH) have the property of activating the intracellular triglyceridase which catalyzes the hydrolysis of the first fatty acid of the stored triglyceride (the rate-limiting step in the lipolytic process) (Vaughan et al, 1964). These "lipolytic" hormones appear to achieve this effect by activating adenyl cyclase, with resultant increase in the intracellular concentration of 3,5 cyclic adenosine

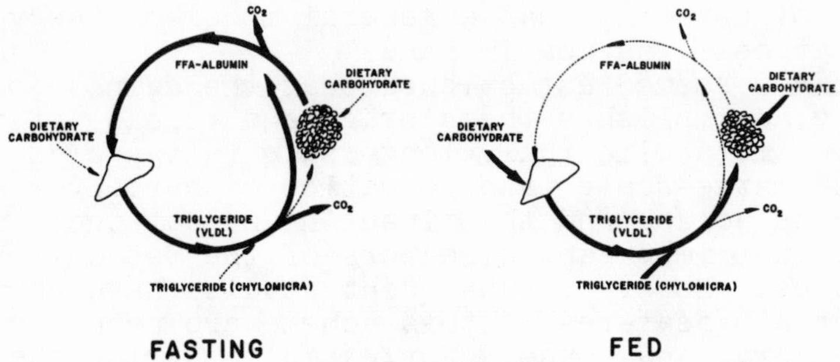


Fig. 2 Comparison of the metabolic activities of the young rat's fat cell during fasting and after eating.

monophosphate; this nucleotide then activates the triglyceridase (Butcher et al, 1965).

The opposite effect on the adenyl cyclase apparently is produced by insulin, which therefore restrains the lipolytic effect of epinephrine, norepenephrine, ACTH, etc.; this is the so-called antilipolytic action of insulin (Jungas and Ball, 1963); Butcher et al,1966). An additional effect of insulin on the fat cell is to increase the activity of one of the several hexose-transport systems, thereby accelerating the entry of glucose into the cell with resulting increased flow of glucose along all its metabolic pathways (Winegrad and Renold, 1958; Crofford and Renold, 1965). Finally, there is some evidence that the activity of lipoprotein lipase, involved in the uptake of plasma triglyceride-fatty acids, is dependent directly or indirectly on insulin (Schnatz and Williams, 1962; Kessler, 1963; Eagle and Robinson, 1964).

Thus the lipolytic hormones stimulate the immediate mobilization of stored triglyceride-fatty acids. Insulin suppresses this function, while simultaneously stimulating (a) the entry of glucose into the fat cell with subsequent extensive conversion of the hexose to stored triglyceride, and perhaps also (b) the fat cell's capacity to hydrolyze circulating plasma triglyceride with subsequent assimilation of the fatty acid moiety. The manner in which this system of hormonal signals serves the body's conservation and expenditure of metabolic fuels is shown in Fig. 2.

In 1961, Christophe et al (1961 a and b) showed that the mouse's adipose tissue conforms in all respects studied to this scheme; and although the mouse and rat are closely related phylogenetically (being members of the same rodent family) there was an understandable feeling of optimism that a general model of the vertebrate fat cell was now in hand.

It now seemed established that the adipose organ not only accomplishes the storage and release of that excellent metabolic fuel, long-chain fatty acids, but also the large-scale transformation of carbohydrate into these acids, and that insulin and catechol amines are the most important directors of the fat cell's activities. However, subsequent studies have shown that certain features of this scheme are peculiar to the young rat and mouse. For example, during the life-span of the ad-lib fed rat, two of the fat cell's properties undergo involution. After the age of 9 weeks (200 g), the conversion of glucose or acetate to long-chain fatty acids declines rapidly, and the cell no longer accelerates its rate of glucose metabolism in response to insulin (Benjamin et al, 1961; Gliemann, 1965; Di Girolamo and Rudman 1966). These changes represent some type of adaptation of the fat cell to its enlarging depot of triglyceride, since the loss of both lipogenic capacity and responsiveness to the glucose-transport effect of insulin can be completely reversed by reducing the older rat's body weight, and can be prevented by limiting the animal's food intake from an early age (Di Girolamo and Rudman, 1966).

The suitability of the young rat's fat cell as a general model of the animal fat cell has been further challenged by examination of other species, and the next section reviews the results of these studies.

## III. The adipose tissue of other mammals, birds and insects.

Fig. 3 summarizes the general course of animal evolution from the protozoan stage. In the invertebrate domain, the sizeable fat body of many insect species has been studied in considerable detail. Among the vertebrates, a small intraabodominal fat body is also present in amphibia but has not yet been studied; a massive, diffuse adipose organ first appears in the warm-blooded classes, Mammalia and Aves, and some authorities believe that the capacity of the adipose organ to store and release metabolic fuel plays an essential role in achieving homeothermy. Among the mammals, metabolic data are available on the adipose tissue of 4 rodent species, one lagomorph, one primate and one carnivore. In the bird kingdom, the adipose organ of 5 bird species has been examined. The taxo-

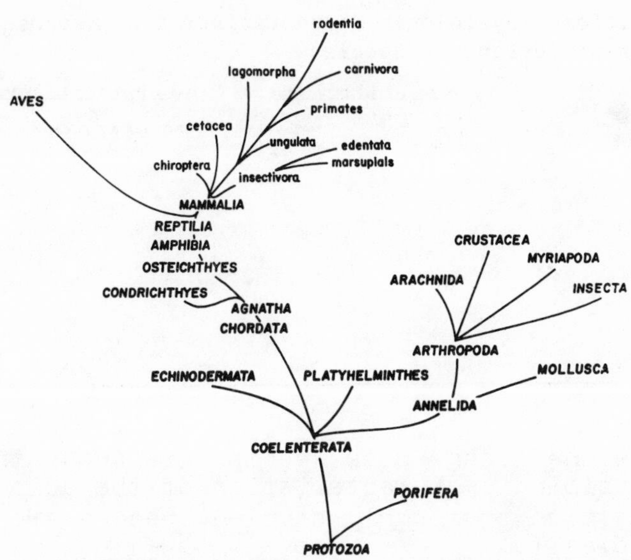


Fig. 3 Probable phylogenetic relationships within the animal kingdom (adapted from Romer, 1959).

Table I. Taxonomic Relationships of the Several Animal Species Considered in This Review. According to Young (1962), Romer (1959), Wallace (1963), and Oldroyd (1962).

| Common Name | Genus and Species | Family | Suborder | Order | Class | Phylum |
|---|---|---|---|---|---|---|
| laboratory rat | Rattus norvegicus | | | | | |
| laboratory mouse | Mus musculus | Muridae | Myomorpha | | | |
| golden hamster | Mesocricetus auratus | Cricetidae | | Rodentia | | |
| guinea pig | Cavia Porcellus | Cavia | Caviamorpha | | | |
| domestic rabbit | Oryctolagus cuniculus | Oryctolagus | Leporidae | Lagomorpha | Mammalia | |
| domestic dog | Canis familaris | Canidae | | | | |
| domestic cat | Felis catus | Felidae | Fissipedia | Carnivora | | Chordata |
| domestic pig | Sus scrofa | Suidae | Suina | Artiodactyla | | |
| man | Homo sapiens | Hominidae | Anthropoidea | Primates | | |
| domestic chicken | Gallus domesticus | Phasinnidae | | Galliformes | | |
| house sparrow | Passer domesticus | | | | | |
| white-crowned sparrow | Zonotrichia leucophrys gambelii | Fringillidae | | Passeriformes | Aves | |
| slate-colored junco | Junco hyemalis | | | | | |
| domestic pigeon | Columba livia | Columbidae | | Columbiformes | | |
| cockroach | Periplaneta americana | | | Dictyoptera | Insecta | Arthropoda |
| grasshopper | Melanoplus differentialis | | | Saltatoria | | |
| migratory locust | Locusto migratoria | | | | | |
| moth | Prodema eridamia | | | Lepidoptera | | |

Table II. Some Physiologic Characteristics of Adipose Tissue of Different Vertebrate Species.

| | | responsiveness to insulin | | responsiveness to lipolytic agents | | | |
|---|---|---|---|---|---|---|---|
| | lipogenesis | glucose-transport action | antilipolytic action | ACTH | MSH'S | glucagon | catechol amines |
| Rat | + | + | + | + | 0 | + | + |
| Mouse | + | + | + | + | 0 | + | + |
| Hamster | 0 | 0 | + | + | 0 | 0 | + |
| Guinea pig | + | ± | 0 | + | + | 0 | 0 |
| Rabbit | 0 | 0 | 0 | + | + | 0 | 0 |
| Sparrow | 0 | 0 | 0 | | | + | + |
| Junco | 0 | 0 | 0 | | | + | + |
| Pigeon | 0 | 0 | 0 | | | + | + |

nomic relations among these species are shown in Table I. The physiologic characteristics of the adipose tissues of the more thoroughly studied species are summarized in Table II.

Among the 4 rodent species, rat and mouse belong to the same suborder Myomorpha, and within the myomorphs to the same family Muridae; the hamster, although also a myomorph, represents the family Cricetidae; and the guinea pig stands apart from rat, mouse and hamster in the suborder Caviamorpha. Not surprisingly, the metabolic pathways and hormonal controls of mouse and rat adipose tissue are identical (Christophe et al, 1961 a and b). The hamster differs from rat and mouse in 2 of the 5 functions cited: the lipogenic pathway is virtually absent and there is no response to the glucose-transport action of insulin (Di Girolamo and Rudman, 1966 b). Nevertheless, the hamster's fat cell shares with those of the rat and mouse responsiveness to the antilipolytic action of insulin and to the lipolytic action of both pituitary and adrenergic agents (Rudman and Shank, 1966; Rudman and Di Girolamo, 1967). The adipose tissue of the guinea pig, from a different rodent suborder, shares with the rat and mouse tissues an active lipogenic function, exhibits a weak but detectable responsiveness to insulin's glucose-transport action, but is unresponsive to insulin's antilipolytic action or to catechol amines (Di Girolamo and Rudman, 1966 b). Pituitary peptides are the major lipolytic agents in this species. Leaving the rodent order, the lagomorph rabbit now differs in every respect from rat and mouse. Lipogenic function is minimal, the tissue is unresponsive to either action of insulin and the pituitary peptides alone regulate lipolysis. In both guinea pig and rabbit, as the sensitivity to catechol amines disappears the tissue gains responsiveness to an increasing number of pituitary peptides (see review by Rudman and Di Girolamo, 1967).

The adipose tissues of 4 different bird species, studied by Goodridge and Ball, (1964,1965 and 1966) show a distinctive profile of characteristics. As in the rabbit tissue, lipogenesis is absent and the tissue is unresponsive to either action of insulin. In responsiveness to lipolytic agents, however, the bird's tissue resembles that of the rat in possessing intense sensitivity to catechol amines and to glucagon (see also Hoak et al, 1966).

It is not unexpected that insect fat cells should differ in major respects from those of mammals and birds (see review by Tietz, 1965). Perhaps the similarities are more remarkable than the differences. The fatty acid composition of the stored triglyceride is quite similar to that found in mammals (Barlow, 1964; Van Handel and Lum, 1961). A major proportion of these fatty acids seem to be synthesized from acetate within the fat cells themselves, since a lipogenic enzyme system is present in the insect fat body closely similar to that of mammalian liver and adipose tissue (Zebe and McShan, 1959, Tietz, 1962). For the mobilization of stored triglyceride-fatty acids, the initial step is the same as in the rat tissue: hydrolysis of one ester bond of the triglyceride molecule by a triglyceridase. But while in warm-blooded vertebrates the process is then completed by a di(mono)glyceridase, in the insect fat body the diglyceride and single FFA moiety are then released into the hemolymph, where a specific lipoprotein capable of transporting diglyceride is present (Chino and Gilbert, 1964 and 1965; Tietz,1962). The insect's fat cells, like the mammal's, are efficient at transforming glucose to glycogen and the reverse; but an additional major pathway is present for breaking down glycogen to trehalose and releasing this disaccharide into the hemolymph (Candy and Kilby,1959 and 1961; Clegg and Evans,1961). The insect fat cells store considerable quantities of glycogen in addition to triglyceride. Hormones of the insect neuroendocrine system regulate various aspects of insect fat cell metabolism.

These various conclusions come largely from experiments with adipose tissue slices or free fat cells. The question is often raised,to what extent do these observations reflect the tissue's activities <u>in vivo?</u> The property of the fat cell most easily examined in the intact animal is the responsiveness to lipolytic agents; in general, the invivo and in vitro experiments in this area have yielded parallel results (Rudman and Di Girolamo,1967). The more difficult task of measuring in vivo lipogenesis within adipose tissue has been accomplished by Favarger,(1965), Goodridge and Ball (1967) in the mouse, rat and pigeon. In the pigeon, unlike the mouse and rat, lipogenesis in adipose tissue was minimal; the rate of lipogenesis in the pigeon's liver, on the other hand, was many times greater than in that of the rat. Furthermore the in vivo data of Goodridge and Ball (1967) in the pigeon, and those of Havel et al (1962) in the rabbit, have directly shown that the fat cells of these species, while lacking significant capacity to transform glucose to fatty acids, are nevertheless highly active in assimilating circulating esterified fatty acids synthesized or absorbed elsewhere in the body: a pathway for the accumulation of fatty acids in non-lipogenic fat cells is thus demonstrated.

The comparative data therefore suggest that the most versatile fat cell in the animal kingdom is probably that of the muridae rodents, rat and mouse, where all 4 metabolic functions are present: lipogenesis, uptake of circulating triglyceride-fatty acids, storage of fatty acids as triglyceride, and their mobilization as FFA; and where the fat cell is highly responsive to insulin, catechol amines and pituitary peptides. Two functions of the rat's fat cell, the storage and release of fatty acids, are seen throughout the animal kingdom, but the lipogenic function and the identity of the regulating hormone molecules undergo frequent variations.

## IV. Problems for future study

Obviously many more species must be examined before a meaningful phylogenetic pattern can be achieved, and before it will be clear at what points in animal evolution particular metabolic characteristics of the fat cell made their appearance or disappearance. In the meantime, it may not be premature to mention two lines of inquiry which arise from these evidences that the fat cell has changed during evolution. First, what may have been the survival value of these mutations? Secondly, by what cellular mechanisms were these alterations in physiologic function and control achieved?

With regard to the first question: The activities of the fat cell are intimately involved in the energy balance of all warm-blood animals and insects. During periods of positive caloric balance, excess dietary carbohydrate and protein is transformed into fatty acids somewhere in the body, and together with excess dietary fatty acids are stored in the fat cells. During periods of negative caloric balance, such as starvation, exercise, adaptation to cold, migration, and arousal from hibernation, stored triglyceride-fatty acids are released from the fat cells for utilization elsewhere in the body. Construction of eggs in oviparous animals and the processes of growth and metamorphosis also entail mobilization of adipose tissue lipid. The particular circumstances of these challenges to energy balance, survival and reproduction vary throughout the animal kingdom. Can the species differences in the site of lipogenesis (liver vs. adipose organ), and in the insulin-responsiveness of the fat cell, be correlated with the quality and timing of food intake? Can the multiplicity of naturally occurring lipolytic agents eventually be related to the differing ecologic conditions, acute and protracted, under which FFA mobilization occurs in each species? Answers to these questions may contribute to our understanding of how, at particular points in evolution, new species evolved which were capable of novel adaptations to climatic, dietary, and geographic obstacles.

The second line of inquiry concerns the chemical mechanisms within the fat cell by which these presumably useful mutations in fat cell physiology were achieved at particular points in evolution. Three types of hypothesis can be visualized. (a) Species differences in metabolic pathways presumably result from appearance or disappearance of particular enzymes. Thus in the avian fat cell, lack of lipogenic function seems to result from a low level of the system of enzymes necessary to convert acetate to fatty acids (Goodridge and Ball,1966). The mobilization of a triglyceride molecule as 1 diglyceride + 1 FFA in the insect (rather than as 1 glycerol + 3 FFA as in the mammal and bird) can logically result from presence of the triglyceridase, but absence of the di(mono)glyceridase in the insect's fat cell. (b) The species differences in hormonal responsiveness, on the other hand,might result from changes in the structure of the cellular receptor for these hormones. Thus insulin is believed in the rat to stimulate one of 3 hexose transport sites (Fain, 1964). Species differences in responsiveness to the glucose-transport action of insulin could result from appearance or disappearance of this one of the group of hexose-transport sites. If the lipolytic hormones interact primarily with adenyl cyclase in the region of the fat cell's membrane, then subtle changes in the structure of this enzyme could determine the species differences in the structural requirement for lipolytic activity. (c) Another possible mechanism for species differences in response to hormones is evolutionary changes in the fat cell's content of enzymes which alter the structure of the hormone with resulting gain or loss of biologic activity. Some evidence has been gathered that the responsiveness of the fat cells of rat, mouse and hamster to the lipolytic property of ACTH-MSH type peptides is curtailed or obliterated by the operation of peptidases within the fat cell which cleave and inactivate these hormones (reviewed by Rudman and Di Girolamo,1967). Fat cells of these 3 rodent species also contain peptidases which cleave insulin. Much further work will be needed, however, to learn which of the numerous species differences in responsiveness to hormones are determined by evolutionary changes in the peptidase content of the fat cell.

## REFERENCES

Ball, E.G., and Jungas, R.L. (1964).
Recent Progr. Hormone Res. 20, 183.

Barlow, J.S. (1964).
Canad. J. Biochem. 42, 1365.

Benjamin, W., Gellhorn, A., Wagner, M., and Kundel, H. (1961).
Am. J. Physiol. 201, 540.

Best, C.H., and Campbell, J. (1936).
J. Physiol. (Lond.) 86, 190.

Butcher, R.W., Ho, R.J., Meng, C.H., and Sutherland, E.W. (1965).
J. Biol. Chem. 240, 4515.

Butcher, R.W., Sneyd, J.G.T., Park, C.R., and Sutherland, E.W.,Jr. (1966).
J. Biol. Chem. 241, 1652.

Candy, D.J., and Kilby, B.A. (1959).
Nature 183, 1594.

Candy, D.J., and Kilby, B.A. (1961).
Biochem. J. 78, 531.

Chino, H., and Gilbert, L.I. (1964).
Science 143, 359.

Chino, H., and Gilbert, L.I. (1965).
Biochim. Biophys. Acta 98, 94.

Christophe, J., Jeanrenaud, B., Mayer, J., and Renold, A.E. (1961 a).
J. Biol. Chem. 236, 642.

Christophe, J., Jeanrenaud, B., Mayer, J., and Renold, A.E. (1961 b).
J. Biol. Chem. 236, 648.
Clegg, J.S., and Evans, D.R. (1961).
J. Exp. Biol. 38, 771.
Crofford, O.B., and Renold, A.E. (1965).
J. Biol. Chem. 240, 14.
Di Girolamo, M., and Rudman, D. (1966 a).
Fed. Proc. 25, 441.
Di Girolamo, M., and Rudman, D. (1966 b).
Am. J. Physiol. 210, 721.
Dole, V.P. (1956).
J. Clin. Invest. 35, 150.
Eagle, G.R., and Robinson, D.S. (1964).
Biochem. J. 93, 10C.
Fain, J.N. (1964).
J. Biol. Chem. 239, 958.
Favarger, P. (1965).
In "Handbook of Physiology, Section 5: Adipose Tissue", (A.E. Renold and G.F. Cahill,Jr., Eds) p. 19. Am. Physiol. Soc., Washington, D.C.
Gellhorn, A., and Marks, P.A. (1961).
J. Clin. Inv. 40, 925.
Goodridge, A.G. (1964).
Comp. Biochem. Physiol. 13, 1.
Goodridge, A.G., and Ball, E.G. (1965).
Comp. Biochem. Physiol. 16, 367.
Goodridge, A.G., and Ball, E.G. (1966).
Am. J. Physiol. 211, 803.
Goodridge, A.G. and Ball, E.G. (1967).
Biochemistry 6: 1676.
Gordon, R.S.,Jr., and Cherkes, A. (1956).
J. Clin. Invest. 35, 206.
Hamosh, M. (1963).
J. Clin. Inv. 42, 1648.
Hausberger, F.X., Milstein, S.W., and Rutman, R.J.(1954).
J. Biol. Chem. 208, 431.
Hirsch, J., and Goldrick, R.B. (1964).
J. Clin. Inv. 43, 1776).
Hoak, J.C., Connor, W.E., and Warner, E.D. (1966).
Clin. Res. 14, 440.
Jungas, R.L., and Ball, E.G. (1963).
Biochemistry 2, 383.
Kahlenberg, A., and Kalant, N. (1964).
Canad. J. Biochem. 42, 1623.
Kessler, J.I. (1963).
J. Clin. Invest. 42, 362.
Laurell, S. (1956).
Scand. J. Clin. Lab. Invest. 8, 81.
Lebovitz, H.E., and Engel, F.L. (1965).

In "Handbook of Physiology, Section 5: Adipose Tissue", (A.E. Renold and G.F. Cahill,Jr., eds.) p. 541. Am. Physiol. Soc., Washington, D.C.

Martin, D.B., and Vagelos, P.R. (1965). In "Handbook of Physiology, Section 5: Adipose Tissue", (A.E. Renold and G.F. Cahill,Jr., eds.) p. 211, Am. Physiol. Soc., Washington, D.C.

Oldroyd, H. (1962). Insects and Their World. University of Chicago Press.

Raben, M.S., Landoldt, R., Smith, F.A., Hofmann, K., and Yajima, H. (1961). Nature 189, 681.

Renold, A.E., Crofford, O.B., Stauffacher, W., and Jeanrenaud, B. (1965). Diabetologia 1, 4.

Rodbell, M. (1964). J. Biol. Chem. 239, 375.

Romer, A.S. (1959). The Vertebrate Story, 4th ed. University of Chicago Press.

Rudman, D. and M. Di Girolamo (1967). Adv. Lipid Research, 5, In Press.

Schnatz, J.D., and Williams, R.H. (1962). Diabetes 12, 174.

Schoenheimer, R., and Rittenberg, D. (1936). J. Biol. Chem. 120, 155.

Scow, R.O. (1965). In "Handbook of Physiology, Section 5: Adipose Tissue", (A.E. Renold and G.F. Cahill,Jr., Eds.) p. 437, Am. Physiol. Soc., Washington, D.C.

Shapiro, B., and Wertheimer, E. (1948). J. Biol. Chem. 173, 725.

Tanaka, A., Pickering, B.T., and Li, C.H. (1962). Arch. Biochem. 99, 294.

Tietz, A. (1962). J. Lipid Res. 3, 421.

Tietz, A. (1965). In "Handbook of Physiology, Section 5: Adipose Tissue", (A.E. Renold and G.F. Cahill,Jr., Eds.) p. 45. Am. Physiol. Soc., Washington, D.C.

Van Handel, E., and Lum, P.T.M. (1961). Science 134, 1979.

Vaughan, M. (1961). J. Lipid Research, 2, 1961.

Vaughan, M., Berger, J.E., and Steinberg, D. (1964). J. Biol. Chem. 239, 401.

Wallace, G.J. (1963). In "An Introduction to Ornithology". 2nd ed. MacMillan Company.

Winegrad, A.I. and Renold, A.E. (1958).
J. Biol. Chem. 233, 267.

Young, J.Z., (1962).
The Life of the Vertebrates, ed. 2, Oxford University Press.

Zebe, E.C. and McShan, W.H. (1959).
Biochim. Biophys. Acta 31, 513.

# ASPECTS OF GLUCAGON-INDUCED LIPOLYSIS IN ADIPOSE TISSUE.

Peter R. Bally

Department of Pharmacology, University of Bern[1] and Stoffwechsellabor Medical Department, University of Zürich, Switzerland

Glucagon is a powerful lipolytic agent for rat adipose tissue *in vitro*.

Hormone-induced lipolysis is thought to be mediated by cAMP[2] as discussed by Dr.Butcher here in Milan two years ago (1 and 2, 3). The Nashville group has observed that the tissue concentration of cAMP rapidly reaches a maximum after hormonal stimulation and then falls to fractional concentrations (s.fig.2, ref.4 - compare fig.3, ref.3). They have stated that the significance of this fall in cAMP level was not clear (3, 4). Despite the decline in cAMP concentration the rate of the final dependent reaction - FFA[2] or Glycerol-release for the present discussion - remained unchanged over the time of observation (fig.2, ref.4). It has been suggested (4) that other than cAMP-controlled steps become rate limiting after cAMP concentration reaches a certain level. However, adipose tissue content of cAMP *can* correlate with lipolytic rate over a much wider range than that reported by these authors, e.g. in a study of Hynie, Krishna and Brodie (5). And this correlation is almost linear.

A priori four other explanations are conceivable for the observed discrepancy between FFA release rate and time course of cAMP. There may be still more.

A. cAMP concentration may rise above saturation levels of a step which it activates. This has been ruled out by the data of Ho and Meng (4), as discussed below.

B. cAMP may merely be required to trigger a stable state (i.e. a non- or slowly-reversible state) of activation in one or more subsequent steps. Its concentration would be irrelevant for the activated step after that state had been achieved. This possibility is rendered unlikely by data reported in this communication.

C. cAMP may be compartmented in the adipose cell. The compartment in which lipolysis is activated may only contribute a small fraction to the total tissue-content. This concept has been discussed by Butcher, Sneyd, Park and Sutherland, but no definitive conclusion was reached (2).

D. Changing concentrations of cAMP might be required to stabilize a hormone-determined lipolytic rate against influences such as shifts in cofactor - or intermediate - levels. Such changes of concentration are known to occur as a consequence of a hormone-disturbed steady state equilibrium of metabolic flow. Cofactors are known to interact cooperatively in several cAMP sensitive reactions of hormone-influenced systems (e.g. ref.12). If such cofactor-sensitisation of a cAMP site (or the reverse) in an enzyme should take place, then a diminished cAMP concentration would be required for a given lipolytic rate. Such a diminished requirement of cAMP could conceivably be coupled with a matching regulation of cyclase. From a regulatory point of view this possibility seems attractive. But no discussion is possible on the basis of the data presented here.

A. The first possibility can be ruled out by going back to fig.2 of Butcher, Ho and Meng (4). We note that the fall of cAMP content after 40 minutes of incubation with lipolytic hormone is about 75 per cent of the full rise above control level. We assume this fall to be progressive and not abrupt (compare with fig.3, ref.3). From table III in the same article a correlation of tissue content of cAMP and FFA release obtains. This correlation extends over the range of cAMP values in the fig.2 of the same article. I have plotted the data of table III as a graph (fig.1).

The correlation is fairly linear up to a cAMP concentration which corresponds to nearly 10 μg/ml epinephrine. At 10 μg/ml epinephrine saturation of the cAMP response (fig.1, upper half) was not yet beginning and saturation of the lipolytic response was not yet achieved. In other words, to each concentration of epinephrine within the said range a single value for cAMP

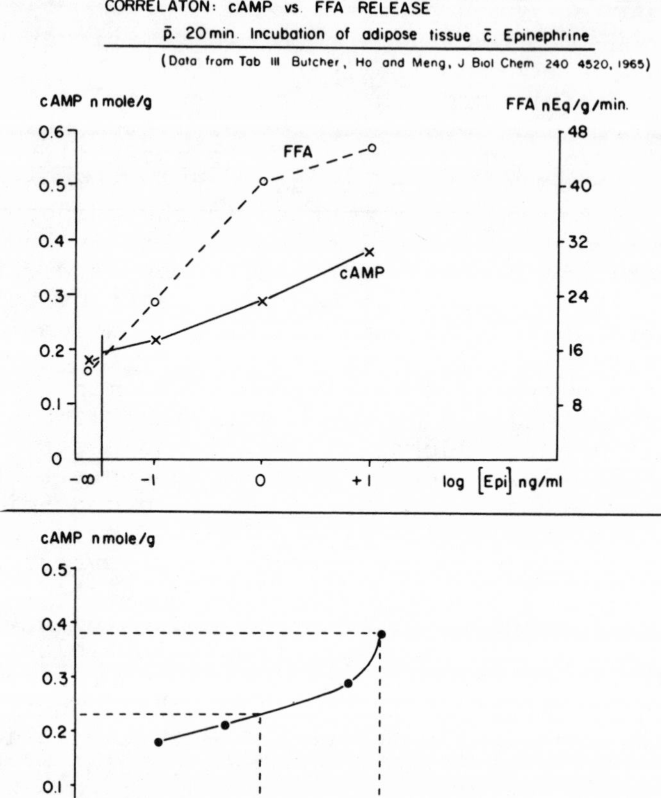


Fig.1. Replotting of data of Butcher, Ho and Meng (Table III, ref.4). See text for explanation.

tissue-content and for FFA release-rate could be assigned.

We have taken the increment of cAMP due to 10 μg/ml of epinephrine as 100 per cent (Fig.1: dashed upper coordinate in lower graph). From this value we have graphically subtracted the discussed 75 per cent fall of cAMP concentration found in fig.2 of Butcher, Ho and Meng (4) and thus obtain a cAMP value of 0.23 nmoles/g tissue. This value corresponds to a FFA release rate of 28 nEq/g/minute or 63 per cent of the FFA release rate observed with 10 μg/ml epinephrine. If the rate of lipolysis were only dependent on the simultaneous cAMP content of the tissue we would expect a corresponding fall of the lipolytic rate toward the end of the incubation shown in fig.2 of Butcher and coworkers. The lack of such a change can hence not be due to saturation with cAMP of a cAMP-sensitive step.

Such saturation is also rendered unlikely by the further increase of lipolysis observed by Hynie and collaborators upon addition of theophylline which occurred even though the tissue was maximally stimulated with a lipolytic hormone before and during theophylline addition (5). This additional increase in lipolysis was associated with cAMP values exceeding by far those shown in the graph derived from Dr. Butcher's data (fig.1). Furthermore, decrease of cAMP content after insulin addition was associated with decreased lipolytic rate even in the presence of 10 μg/ml epinephrine (2).

B. The second possibility, cAMP tiggering a stable activated state, seems unlikely because of the following observations.

Hormone induced cAMP generation in adipose tissue is very repid. One minute after addition of epinephrine to perfusate, Ho and Meng found elevated values (6). Lipolysis is stimulated to maximal rates within about 30 to 60 seconds after addition of glucagon to adipose tissue in vitro (7). This must mean that the hormone quickly equilibrates with its presumptive receptor sites.

How long will the hormone-activated state persist after the hormone has been removed? Is removal possible by washing the tissue? An answer to these questions was attempted with the following experiments. Small pieces of adipose tissue from rats were distributed in a latin-square fashion (8). Three groups of flasks were incubated - three flasks in each group - in varying sequences, with and without glucagon. Between the 15 minute incubations the tissue was quickly washed with hormone-free medium. The medium contained 4 g/100 ml human albumin in Krebs-Ringer bicarbonate buffer at pH 7.4 and 3 mM glucose (fig.2). It appears relevant in this context that cAMP concentration in a medium bathing fat cells was low and did not appreciably change with hormone addition (2) and that, generally, cAMP does not appear to penetrate the cell (2, 3). However, it seems to be able to leave certain cell types in increased amounts after hormonal stimulation (3a). cAMP concentration was not measured in this study.

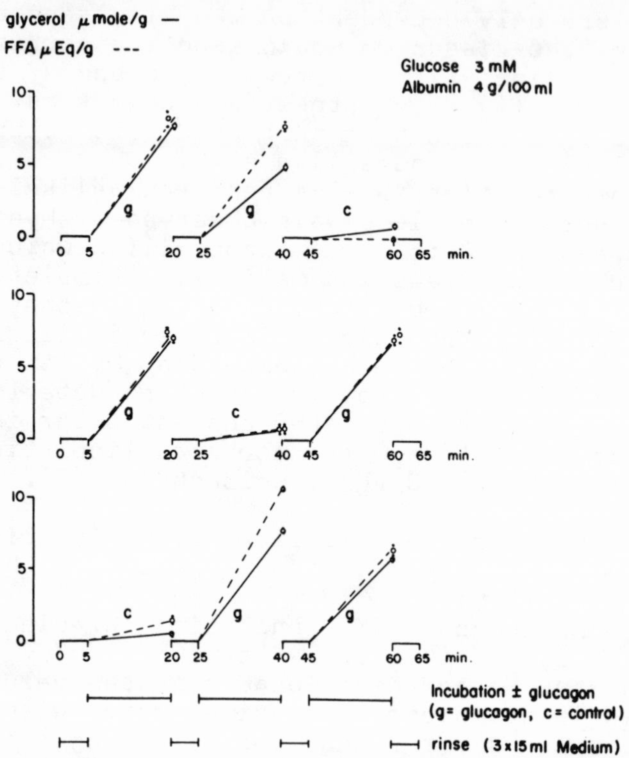


Fig.2. Wash-off experiment. 3 series of 3 consecutive 15 minute incubations with (g) or without (c) glucagon (0.3 μg/ml). Krebs-Ringer bicarbonate buffer with 4 g/100 ml albumin, 3 mM glucose and .1 ml anti-insulin-guinea pig serum/10 ml. 3.2 ml/flask. Each washing for a total of 5 minutes with 3 x 15 ml medium (without hormone). 3 adipose tissue pools in each of the three sequences. Values ± standard deviations (FFA and glycerol release, cumulative).

In order to show whether or not the washing procedure would lower the lipolytic rate, the amount of glucagon added would have to be at the upper limit of the steep slope of glycerol release plotted against hormone concentration. Fig.3 shows such plots from three experiments. Techniques were essentially as indicated before. Hormone was weighed and serially diluted with medium. It was added to each flask after 15 minutes preincubation to give the final concentration listed in the abscissa. Glycerol release (ordinate) was measured enzymatically (9). Anti-insulin-guinea pig serum was added to the medium to overcome residual insulin activity contaminating the glucagon (8b).

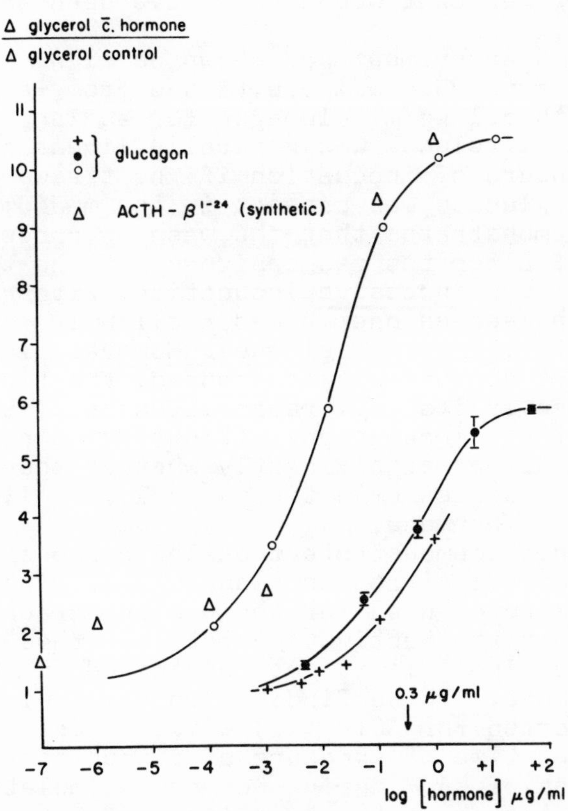


Fig.3. Hormone concentration vs. glycerol release. Experimental techniques as in Fig.2. Triangles: ACTH (synthetic: beta 1-24, courtesy Dr.R.Schwyzer) no line drawn. Three experiments with crystalline glucagon (batch 258-234-B-54-2, Lilly). Incubation time was either 2 or 3 hours. Glucose 5 mM, Albumin 6 g/100 ml in all flasks.

Maximal lipolytic rate and half-saturation concentration of glucagon with respect to lipolysis varied with different sets of rats (nutritional state? age?). Half-saturating concentration of glucagon was between $3 \times 10^{-9}$ and $2.3 \times 10^{-8}$M glucagon in these experiments. In some of them levels as low as 100 µµg/ml were detectable. This apprears to be within the physiologic range (10) in human blood. 0.3 µg/ml was used for the wash-off experiment (Fig.2). A fall of lipolytic rate of 10 or more per cent would thus have been easy to detect.

Aprevious experiment had shown no significant fall in lipolytic rate when adipose tissue from fed rats was incubated with o.1 µg/ml glucagon for as long as one hour. Glycerol release was generally linear over at least three hours of incubation if the tissue came from fed rats and glucose was present in the medium (7, 8a).

Fig.2 demonstrates that the washing procedure did not in itself alter the responsiveness of the tissue to glucagon. Of two succesive incubations with glucagon (fig.2, g) the second one showed a slightly lowered lipolytic rate (glycerol release). However, if an incubation without hormone was interposed, the tissue seemed to recover its full responsiveness. It is further apparent that the incubations without hormone (control = c) did not differ significantly whether they initiated the incubation series of a tissue pool or followed an incubation with hormone.

The results demonstrate that the hormone must have dissociated quickly from the tissue, i.e. within at least the 5 minutes used for the washing procedure. It also shows that the "activated state" of those enzymes which lead to lipolysis is not maintained in the absence of hormone. These findings contrast with results recently reported for ACTH (11) with respect to binding.

Since the time of exposure of tissue to the lower concentrations of cAMP during hormone stimulation was considerably longer in the experiment of Butcher, Ho and Meng (fig.2, ref.4) than during the wash-interval used here, it appears unlikely that persistence of a hormone-activated state could explain the apparent lack of correlation between cAMP and lipolysis with continued incubation.

Between the two remaining possibilities which were mentioned in the introduction no decision is possible on the basis of the reported data.

I would like to close by summarizing the conclusions from the reported experiments:

1. Glucagon can act as a lipolytic agent on rat adipose tissue in vitro at physiologic (10) concentrations (ca. 100 μμg/ml or $10^{-13}$M). It has a half maximal concentration for lipolysis of between 3 x $10^{-9}$ and 2.3 x $10^{-8}$M depending on the responsiveness of the adipose tissue used.

2. Glucagon can be washed off adipose tissue without leaving any residual lipolytic activity. This indicates that the glucagon-triggered activation of lipolysis is quickly reversible when the hormone is removed.

Acknowledgements:

I wish to thank Miss Susanna Diem for excellent technical assistance, Dr.E.R.Froesch for the anti-insulin serum, Dr.O.K.Behrens for the glucagon preparation and Dr.W.Wilbrandt for discussion of the manuscript. This study was supported by the Emil Barell Stiftung zur Förderung der medizinisch-wissenschaftlichen Forschung and by Grant 4047 of the Schweizerischer Nationalfonds.

Footnotes:

(1) Present address: Pharmacology Department, University of Bern, Switzerland. (2) Abbreviations used: cAMP, cyclic 3',5'-adenosine monophosphate; FFA, unesterified fatty acids.

References.

1) Butcher,R.W., 1966, Pharm.Rev. 18: 237.
2) Butcher,R.W., Sneyd,J.G.T., Park,C.R. and Sutherland,E.W., 1966, J.Biol.Chem. 241: 1651.
3) Sutherland,E.W., Øye,I. and Butcher,R.W., 1965, Rec.Progr.Hormone Res., 21: 623.
3a) Chase,L.R. and Aurbach,G.D., 1967, Proc.Nat. Acad.Sc.U.S., 58: 518.
4) Butcher,R.W., Ho,R.J., Meng,H.C. and Sutherland,E.W., 1965, J.Biol.Chem. 240: 4515.
5) Hynie,S., Krishna,G. and Brodie,B.B., 1966, J.Pharm. exp.Ther. 153: 90.
6) Ho,R.J. and Meng,C.H., 1964, J.Lipid Res. 5: 203.
7) Bally,P., Kappeler, H., Froesch,E.R. and Labhart,A., to be published.
8) a. Froesch,E.R., Bürgi,J., Ramseier,E., Bally,P. and Labhart,A., 1963, J.Clin.Invest. 42: 1816.

b. Bally, P., Kappeler, H., Froesch, E.R. and Labhart, A., 1965, Ann. N.Y. Acad. Sci. 131: 143.

9) Wieland, O. und Suyter, M., 1957, Biochem. Z. 329:320.

10) Sokal, J.E., Ezdinli, E.Z., Schiller, C. and Dobbins, A., 1967, J. Clin. Invest. 46: 778.

11) Taunton, O.D., Roth, J. and Pastan, I., 1967, J. Clin. Invest. 46: 1122 (Abstract).

12) Huijing, F. and Larner, J., 1966, Proc. Nat. Acad. Sci. U.S. 56: 647

# THE LIPOLYTIC ACTION OF CATECHOLAMINES AND ACTH AND THE INTERACTION OF PROSTAGLANDIN $E_1$

R. Paoletti, L. Puglisi and M.M. Usardi

Institute of Pharmacology, University of Milan,

20129 Milan, Italy

Rat adipose tissue is very sensitive to a variety of hormones activating the triglyceride "hormone sensitive" lipase. These factors include epinephrine (E) and norepinephrine (NE) (White and Engel, 1958b), ACTH (White and Engel, 1958a), TSH (Freinkel, 1961) and glucagon (Hagen, 1961) as recently reviewed by Renold et al. (1965).

The effect of these hormones is shown by release of glycerol (Leboeuf et al., 1959) into the medium when adipose tissue is incubated "in vitro" in Ringer solution with albumin added and by the accumulation of FFA both in the medium and in the tissue (Gordon and Cherkes, 1958). A similar effect but much more pronounced is observed when the lipolytic hormones are added to isolated adipose tissue cells, prepared according to Rodbell (1964, 1965). The effect of the hormones is extremely rapid: after an incubation as short as 3 minutes and rapid homogenization, several authors have observed a several-fold increase of the hormone sensitive lipase (Hollenberg et al. 1961; Vaughan et al. 1964). The effect on isolated cells is qualitatively comparable but several times greater, probably due to the larger surface of the cells exposed directly to the hormones and to the albumin-containing medium which facilitates the increased lipolysis and reduces the retention of FFA within the cell (Rodbell, 1965).

A common point of action in adipose tissue for the lipolytic hormones is the adenyl-cyclase system (Sutherland et al. 1962) an enzyme probably located on the cell membrane (Davoren and Sutherland, 1963) which catalyses the synthesis of adenosine 3'5' phosphate (cyclic 3'5' AMP).

The first evidence for this mechanism has been shown by Rizack (1961), who is able to induce in homogenates of adipose tissue an activation of adenyl-cyclase. Further cyclic 3'5' AMP added to a cell-free system of adipose tissue is able to activate the lipolytic action (Rizack, 1964). This effect is enhanced by methylxanthines probably through inhibition of the phosphodiesterase present in the tissue which inactivates cyclic 3'5' AMP. More recently Butcher et al. (1965) showed that an increase in the tissue level of cyclic 3'5' AMP precedes the lipolytic response when the tissue is treated with E. This effect of cyclic 3'5' AMP concentration is enhanced by caffeine and inhibited by adrenergic blocking drugs. Cyclic AMP however shows modest if any stimulating effect on lipolysis on intact cells (Butcher et al. 1965; Rizack, 1964). This is probably related to its inability to penetrate to the cellular location of the adenyl cyclase or to its excessive sensitivity to phosphodiesterase. The lipid soluble, more phosphodiesterase resistant derivative $N^6$-2'-O-dibutyryl cyclic 3'5' AMP shows to stimulate considerably lipolysis in intact adipose tissue (Butcher et al. 1965). This adenyl cyclase appears therefore to be the common final point of action for catecholamines and peptide lipolytic agents. However, it is not yet known if there is a common lipase for all the lipolytic hormones in adipose tissue or if each hormone activates a separate cyclase. Other activities are shown to be common for the lipolytic agents in adipose tissue (Lynn et al. 1960; Cahill et al. 1960):

a) Increase in glucose uptake with increased production of $CO_2$ glyceride glycerol and decreased conversion to FFA;

b) Increased conversion of the first C of glucose and decrease of the sixth C to $CO_2$;

c) Activation of phosphorylase with rapid depletion of adipose tissue glycogen. The first two effects are probably consequences of intracellular accumulation of FFA. The third effect may be a second independent effect of cyclic AMP, similar to the activation of phosphorylase which occurs in liver and adrenal cortex.

A less clear effect on the stimulation of lipolysis is shown by the adrenal cortical glucocorticoids, as shown in the classic observations by Fry (1937) and Clement and Schaeffer (1937). The response of adipose tissue from adrenalectomized rats to ACTH or E is considerably reduced (Schotz et al. 1959) while this is corrected by a pretreatment with glucocorticoids (Shafrir and Kerpel, 1964). Normal rats pretreated with cortisone become more sensitive to lipolytic agents. It may be therefore deduced that glucocorticoids play a "permissive" role in regulating the responses of adipose tissue cells to the other lipolytic hormones, while a direct lipolytic activity is not very pronounced and it can be shown only after a prolonged (2-4 hours) incubation "in vitro" (Fain et al. 1963). This effect is less constant when glucocorticoids are incubated with isolated fat cells. Glucocorticoids under such conditions appear to potentiate the lipolytic action of the growth hormone, confirming their permissive role on FFA mobilization (Fain et al. 1965).

The response of the adipose tissue preparations to the lipolytic hormones is modulated by several physiological factors: glucose metabolism, glycogen storage, presence of insulin and, probably, the presence of a newly observed factor, the prostaglandins (PG) which are the most powerful inhibitors of FFA mobilization in adipose tissue "in vitro" ( Steinberg et al. 1961) and "in vivo" (Carlson et al. 1963; Berti et al. 1964). The interaction between some of the lipolytic hormones with PG and the possible physiological significance are discussed in this presentation.

## I THE PERMISSIVE ROLE OF GLUCOCORTICOIDS ON ACTH LIPOLYSIS

The important role played by glucocorticoids as "permissive agents" in lipomobilization induced by ACTH is stressed in data obtained in our laboratory.

Rats submitted to bilateral adrenalectomy and kept on saline, maintain their ability to respond to ACTH injected "in vivo" only for about 48 hours after the operation. After that time, while controls respond with increased levels of plasma FFA to a single dose of ACTH (30 U.I./kg), the levels of FFA in adrenalectomized rats (8-11 days after the operation) are unchanged. The liver triglycerides are also unchanged in adrenalectomized rats, even for repeated administrations of ACTH (up to 240 U.I./kg in 4 hours), while in intact animals they increase four times (Table 1). If adrenalectomized rats are pretreated with corticosterone in adequate amounts, the response to ACTH is fully restored and similar to that of normal controls (Table 2).

## II THE INTERACTION OF PROSTAGLANDIN $E_1$ ($PGE_1$) WITH THE LIPOLYTIC HORMONES

Increasing concentrations of NE and of a typical methylxanthine, theophylline, induce increased release of glycerol and FFA from adipose tissue preparations through different mechanisms. The first agent, as previously described, activates the adenyl cyclase synthesizing 3'5' AMP while theophylline elicits sustained levels of cyclic 3'5' AMP in adipose tissue by inhibiting phosphodiesterase, the metabolizing enzyme. The comparative investigation of the effects of $PGE_1$ against lipolysis induced by these two agents, represents an approach to detect the point of action of this physiological inhibitor of lipolysis.

In our experimental conditions $PGE_1$ is much more active against lipolysis activated by Theophylline: at a concentration of 0.1-1.0 x $10^{-7}$ M $PGE_1$ inhibits the release of glycerol induced by Theophylline by up to

Table 1.

EFFECT OF REPEATED ADMINISTRATIONS OF ACTH ON RAT PLASMA FFA AND LIVER TRIGLYCERIDES

| GROUPS | PLASMA FFA µEg/ml | LIVER TG mg/g |
|---|---|---|
| Adrenalectomized | $0.25 \pm 0.03$ | $2.2 \pm 0.50$ |
| Adrenalectomized + ACTH | $0.23 \pm 0.005$ | $4.3 \pm 0.97$ |
| Controls | $0.23 \pm 0.01$ | $3.7 \pm 0.45$ |
| Controls + ACTH | $0.53 \pm 0.07$ | $16.0 \pm 1.90$ |

ACTH : 30 U.I./Kg for 8 times every 30 min.

Adrenalectomy performed 11 days before the experiment, controls are sham operated.

Table 2.

ACTH STIMULATION ON FFA MOBILIZATION IN ADRENALECTOMIZED

| Groups | No. of Animals | FFA uEg/ml |
|---|---|---|
| I Controls | 15 | 0.18 ± 0.008 |
| II Controls + ACTH | 12 | 0.46 ± 0.03 |
| III Adrenalectomized | 12 | 0.18 ± 0.02 |
| IV Adrenalectomized + ACTH | 9 | 0.29 ± 0.02 |
| V Adrenalectomized + Corticosterone | 5 | 0.31 ± 0.03 |
| VI Adrenalectom. + Corticost. + ACTH | 5 | 0.56 ± 0.06 |

ACTH: 30 U.I./Kg s.c.; CORTICOSTERONE 5 mg/Kg s.c. three times every 12 hours, 10 mg/Kg final dose 14 hours abefore sacrifice; rats illed 20 min. after ACTH injection, 8 days after adrenalectomy; controls are sham operated.

II IV $p < 0.001$

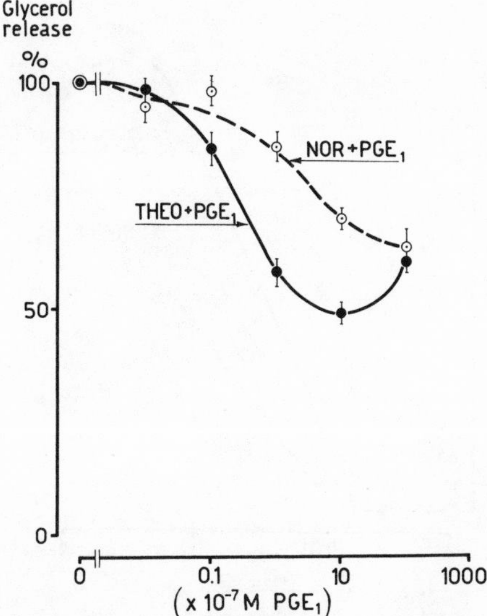


Fig. 1. Comparison of the inhibition of the lipolytic effect of norepinephrine by $PGE_1$ <u>in vitro</u>.

The interaction between the maximal lipolytic dose of norepinephrine (3.16 x $10^{-6}$M) or theophylline 1 x $10^{-2}$M) and increasing concentrations of $PGE_1$ is shown. The lipolytic effect (glycerol release) of norepinephrine and of theophylline in the absence of $PGE_1$ is taken as 100%. Each point is the mean value of 8 individual experiments ± S.E.

50%, while no significant effect can be observed against NE using the same concentration (Fig. 1).

In subsequent experiments two concentrations of $PGE_1$ ($10^{-7}$, $10^{-6}$M) are tested against increasing concentrations of NE, showing a typical dose-response curve for the release of glycerol from adipose tissue into the medium. The results (Fig. 2) showing a depression of the maximal effects of NE, are characteristic of a non-competitive inhibitor (Ariëns and Simons, 1960),

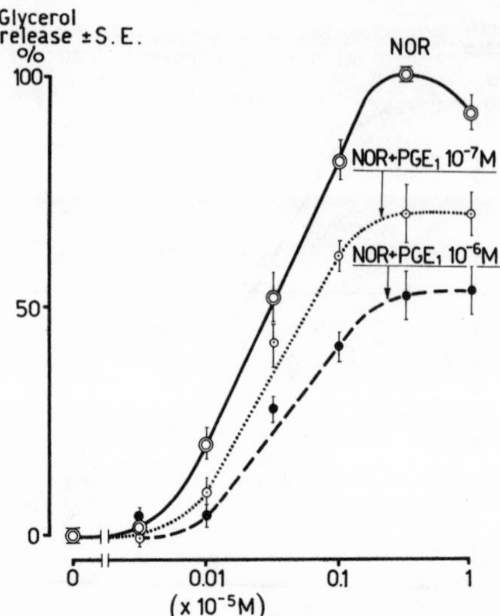


Fig. 2. The modification of the dose-response curve to norepinephrine by different concentrations of $PGE_1$.

Ordinate: lipolytic response (glycerol release) expressed as percent of the maximal lipolytic effect of norepinephrine ($3.16 \times 10^{-6}$ M). Abscissa: increasing concentrations of norepinephrine. The solid line represents the dose-response curve of norepinephrine alone. The dotted lines show the modification of this curve by the addition of $PGE_1$. Each point is the mean value of 8 individual experiments $\pm$ S.E.

the $pD_2$ are rather difficult to calculate because of the instability of $PGE_1$ in buffer solution, but they are in the range of 5.5-6.0. When the same agents are tested against Theophylline induced lipolysis, it is possible to observe (Fig. 3) that the dose-response curve to Theophylline is shifted to the right by the addition of $PGE_1$. The dose-response curve to Theophylline has been followed up to the concentration $1 \times 10^{-2}$ M which corresponds to the maximal solubility of the drug in our experimental conditions. Therefore it has not been possible to follow the upper part of

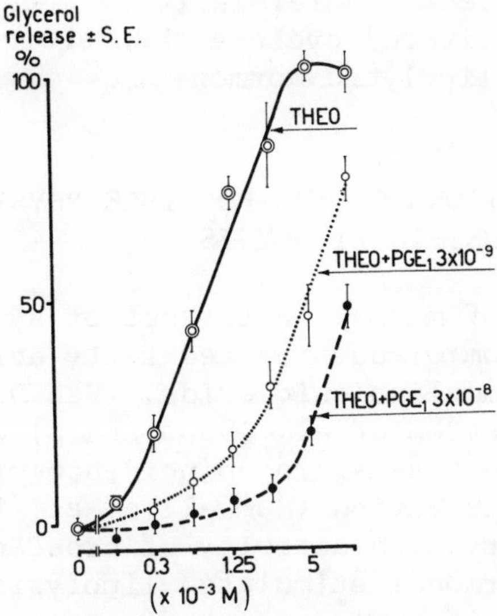


Fig. 3. Modification of the dose-response curve to theophylline by different concentrations of $PGE_1$. Each point represents the mean value of 6 experiments.

the curve. The pA2 values of $PGE_1$ have been calculated to be 10.02 at a concentration of $3 \times 10^{-9}$ M, and 9.76 at $3 \times 10^{-8}$ M. These results seem to be indicative of a functional competitive inhibition of Theophylline induced lipolysis by $PGE_1$. Our data of course do not exclude the possibility that $PGE_1$ may interfere with the formation of 3'5' AMP when catecholamines, ACTH or glucagon are added to adipose tissue preparations. However they strongly suggest that $PGE_1$ may have another point of action as well on the breakdown of 3'5' AMP. They also support previous data from this laboratory (Sólyom et al. 1967) demonstrating that $PGE_1$ acts primarily on lipolysis and not on FFA reesterification in adipose tissue. In addition to that, the high affinity parameters of $PGE_1$ as an inhibitor of lipolysis induced by NE and especially by Theophylline confirms that $PGE_1$ is one of the most potent inhibitors of lipolysis.

It may also be underlined that in adipose tissue after Theophylline treatment the adenyl-cyclase is not activated. $PGE_1$ seems therefore to be much more active against the un-activated cyclase than against cyclase activated by the lipolytic hormones.

## III INVESTIGATIONS ON THE POSSIBLE PHYSIOLOGICAL ROLE OF $PGE_1$ IN LIPOLYSIS

The ability of mammalian tissues of synthesizing prostaglandins from polyunsaturated fatty acids (arachidonic and homo-gammalinolenic acids) (Van Dorp et al. 1964a and b; Bergström et al. 1964a and b) and particularly in adipose tissue, in coincidence with the requirement for their action (Ramwell et al. 1966), combined with the very high activity of prostaglandins as antagonists of hormonal stimulated lipolysis, prompted us to investigate if these compounds may act as physiological regulators of the rate of lipolysis in the adipose tissue. A possible way to investigate this problem is to use rats kept on a diet deficient in essential fatty acids (EFA) which therefore depletes the tissues of the immediate precursors of prostaglandins. Bergström and Carlson (1965) have reported a higher basal rate of FFA and glycerol release from adipose tissue obtained from EFA deficient rats, while De Pury and Collins (1965) reported a higher concentration of plasma FFA levels and increased lipid deposition in liver of EFA deficient rats, indicating an increased mobilization of stored lipids in EFA-deficient animals and confirming the high incidence of fatty liver observed in mammals made deficient in EFA (Alfin-Slater, 1958). In our experiments a much greater release of FFA and glycerol has been observed when NE has been added to adipose tissue preparations obtained from EFA-deficient rats than in normal controls (Table 3). Under similar experimental conditions, but using the more sensitive isolated fat cells system, NE and Theophylline confirm a greater lipolytic activity when the EFA-deficient animals are used. Also "in vivo" the effect of lipolytic hormones is more evident in EFA-deficient than in normal rats. When Depot ACTH (20 U per rat) is injected subcutaneously and the plasma

Table 3.

NOREPINEPHRINE-INDUCED FFA RELEASE FROM ISOLATED FAT CELLS

| NE Concentration | μEg/mM TG/h (± S.E.) | |
|---|---|---|
| | Controls | EFA-Deficient |
| $1 \times 10^{-7}$ | 6.6 ± 3.1 | 18.8 ± 4.8 |
| $3.16 \times 10^{-7}$ | 17.3 ± 4.2 | 36.8 ± 1.2 |
| $5 \times 10^{-7}$ | 19.8 ± 3.0 | 40.1 ± 2.6 |
| $1 \times 10^{-6}$ | 21.4 ± 1.1 | 39.1 ± 2.1 |
| $3.16 \times 10^{-6}$ | 22.3 ± 1.4 | 38.8 ± 0.3 |

Experimental conditions: Atmosphere $CO_2/O_2$ (95/5)

Each figure represents the mean of 16 determinations ± S.E.

TG = Triglycerides

Basal release of FFA control and EFA-deficient fat cells negligible under experimental conditions.

Table 4.

ACTH TREATMENT IN RATS DEFICIENT IN ESSENTIAL FATTY ACIDS

| | Exper. No. | Plasma μEg/ml ± S.E. | Blood Glucose mg/ml ± S.E. |
|---|---|---|---|
| NORMAL RATS | 5 | 0.455 ± 0.021 | 0.95 ± 0.06 |
| EFA Deficients | 5 | 0.694 ± 0.055 | 0.89 ± 0.04 |
| ACTH Normal Rats | 5 | 0.903 ± 0.080 | 0.92 ± 0.03 |
| ACTH + EFA Def. | 5 | 1.234 ± 0.086 | 0.81 ± 0.02 |

FFA and glucose are measured 2 hours later there are significantly higher levels in plasma of EFA-deficient rats. Also among the untreated groups basal FFA levels in plasma of EFA-deficient rats is statistically higher than in normal controls (Table 4). Similar results have been obtained when the dibutyryl derivative of 3'5' cyclic AMP has been added to isolated fat cells (Bizzi et al., 1967). When low concentrations of prostaglandins are added to isolated fat cells from rats deficient in EFA, the inhibiting effect on lipolysis is very pronounced, of the same range or greater than in normal conditions, indicating that EFA deficient fat cells are still sensitive to exogenous prostaglandins, and suggesting that the hypersensitivity to lipolytic agents may be due to the impairment of the synthesis of endogenous prostaglandins from the unsaturated fatty acids.

Our results may be interpreted in different ways: higher levels of cyclic 3'5' AMP may be produced in EFA-deficient rats or 3'5' AMP may be slowly metabolized or eventually normal concentrations of 3'5' AMP may activate more effectively the lipolytic system. This suggests however that the 3'5' AMP system which is activated in rat adipose tissue by catecholamines, ACTH and the peptide hormones, represents a point of attack for prostaglandins as well, and that the interaction between lipolytic hormones and this physiological inhibitor of lipolysis may be critical for the understanding of normal and pathologic levels of lipolysis.

## REFERENCES

1. Alfin-Slater, R.B., and Bernick, S., Am.J.Clin. Nutr., 6, 633, 1958.
2. Ariëns, E.J., Simons, A.M., Arch.Int.Pharmacodyn., 127, 459, 1960.
3. Bergström, S., Danielsson, H., and Samuelsson, B., Biochim.Biophys.Acta, 90, 207, 1964a.
4. Bergström, S., Danielsson, H., Samuelsson, B., and Klerberg, D., J.Biol.Chem., 239, PC4006, 1964b.

5. Bergström, S., and Carlson, L.A., Acta Physiol. Scand., 64, 479, 1965.
6. Berti, F., Lentati, R., Usardi, M.M., Il Giornale dell'Arteriosclerosi, 4, 78, 1964.
7. Bizzi, A., Veneroni, E., Garattini, S., Puglisi, L., Paoletti, R., Europ.J.Pharm., 2, 1967, in press
8. Butcher, R.W., Ho, R.J., Meng, C.H., and Sutherland E.W., J.Biol.Chem., 240, 4515, 1965.
9. Cahill, G.F., Fr., Leboeuf, B., and Flinn, R.B., J.Biol.Chem., 235, 1246, 1960.
10. Carlson, L.A., and Orö, L., Metabolism, 12, 132, 1963.
11. Clement, G., and Schaeffer, G., Compt.Rend.Soc.Biol., 141, 320, 1937.
12. Davoren, P.R., and Sutherland, E.W., J.Biol.Chem., 238, 3016, 1963.
13. De Pury, G.G., and Collins, F.D., Biochim.Biophys. Acta, 106, 213, 1965.
14. Fain, J.N., Scow, R.D., and Chernick, S.S., J.Biol. Chem., 238, 54, 1963.
15. Fain, J.N., Kovacev, V.P., and Scow, R.D., J.Biol. Chem., 240, 3522, 1965.
16. Fry, E.G., Endocrinology, 21, 283, 1937.
17. Freinkel, N., J.Clin.Invest., 43, 129, 1961.
18. Gordon, R.S., Jr., and Cherkes, A., Proc.Soc.Exptl. Biol.Med., 97, 150, 1958.
19. Hagen, J.H., J.Biol.Chem., 236, 1023, 1961.
20. Hollenberg, C.H., Raben, M.S., and Astwood, A.B., Endocrinology, 68, 589, 1961.
21. Leboeuf, B., Flinn, R.B., and Cahill, G.F., Jr., Proc.Soc.Exptl.Biol.Med., 102, 527, 1959.
22. Lynn, W.S., MacLeod, R.M., and Brown, R.H., J.Biol. Chem., 235, 1904, 1960.
23. Masson, G.M.C., and Page, I.H., Proc.Soc.Exptl.Biol. Med., 101, 159, 1959.
24. Ramwell, P.W., Fed.Proc.Am.Soc.Exptl.Biol., 25, 627, 1966.
25. Renold, A.E., Crofford, O.B., Stauffacher, W., and Jeanrenaud, B., Diabetologia, 1, 4, 1965.
26. Rizack, M.A., J.Biol.Chem., 236, 657, 1961.
27. Rizack, M.A., J.Biol.Chem., 239, 392, 1964.
28. Rodbell, M., J.Biol.Chem., 239, 375, 1964.

29. Rodbell, M., Ann.N.Y.Acad.Sci., 131, 302, 1965.
30. Shafrir, E., and Kerpel, S., Arch.Biochem.Biophys., 105, 237, 1964.
31. Schotz, M.C., Masson, G.M.C., and Page, I.H., Proc. Soc.Exptl.Biol.Med., 101, 159, 1959.
32. Solyom, A., Puglisi, L., Mühlbachova, E., Biochem. Pharm., 10, 521, 1967.
33. Steinberg, D., Vaughan, M., and Margolis, S., J.Biol.Chem., 236, 1631, 1961.
34. Sutherland, E.W., Rall, T.W., Menon, T., J.Biol. Chem., 237, 1220, 1962.
35. Van Dorp, D.A., Beerthuis, R.K., Nugteren, D.H., and Vonkeman, H., Biochim.Biophys.Acta, 90, 204, 1964a.
36. Van Dorp, D.A., Beerthuis, R.K., Nugteren, D.H., and Vonkeman, H., Nature, Lond., 203, 839, 1964b.
37. Vaughan, M., Berger, J.E., and Steinberg, D., J.Biol. Chem., 239, 401, 1964.
38. White, J.E., and Engel, F.L., J.Clin.Invest., 37, 1556, 1958a.
39. White, J.E., and Engel, F.L., Proc.Soc.Exptl.Biol. Med., 99, 375, 1958b.

# LIPID MOBILIZING PEPTIDE IN HOG PITUITARY AND HUMAN URINE

Joseph Seifter, Morton Urivetsky and Eli Seifter

New York Medical College, Long Island Jewish Hospital

and Albert Einstein College of Medicine

In recent years several investigators have reported lipid mobilizing activity for low molecular weight peptides prepared from pituitary, plasma and urine of animals and humans. Seifter, Baeder and their coworkers (1956,1957) isolated their material (LM) from horse plasma and posterior pituitary of hogs. Together with Zarafonetis (1957, 1958) they found such activity in the plasma of humans during pregnancy, surgical stress and disease. Chalmers, Kekwick and Pawan (1958, 1960) isolated a peptide with LM action from the urine of fasting humans. Lelek et al. (1961) and Kadas and Nagy (1965) prepared LM from posterior pituitary, blood and urine of humans. Rudman et al. (1960, 1961) prepared a lipid mobilizing euglobulin from anterior pituitary from which they were able to detach a dialyzable peptide with LM action.

The methods of preparation and isolation as well as the bioassay for LM activity used by the various groups were not identical. Therefore the equivalence of the various peptides is uncertain. Lipovetsky (1964) concluded that Rudman's material was similar to Seifter's and reported LM activity for human blood processed by the procedure of Seifter and Baeder and assayed by the method of Rudman.

The work reported in this paper is a first step in the attempt to establish equivalence of the peptides. It was prompted by (1) Rudman's assumption that his LM is not present in posterior pituitary because whole gland contained less than anterior, (2) by reports and subsequent experience that our LM is not active in all species as originally reported, (3) the report by Chalmers et al. that more intense LM activity was present in the urine of humans in whom the fasting period was prolonged and (4) the isolation or extraction of LM by the cited Hungarian and Russian colleagues.

## MATERIALS AND METHODS

Lyophilized, anatomically intact, whole pituitary glands from hogs were kindly supplied by Canada Packers. Anterior and posterior were carefully dissected from these when necessary. Isolation of Fractions H and L was exactly as described by Rudman et al. as was the preparation of the rabbits used in the bioassay. The extracts were dissolved in 0.9% NaCl and injected into the marginal ear vein. Blood samples were drawn from the marginal ear vein (except in 10 rabbits by cardiac puncture) and analyzed for FFA (Dole), triglycerides (Van Handel), cholesterol (Abell et al;Levine and Zak), lipoproteins (Shandon), glucose, urea and uric acid (E. Seifter et al.).

Preparation of urine samples for chemical assay was as follows. Kaolin and acetone treated powder (see below) was dissolved in 0.2M sodium phosphate buffer, pH 5.8, and subjected to successive chromatographic separations on Amberlite CG-50 and DEAE-cellulose columns employing conditions identical with those for pituitary samples. The urine fractions isolated following CG-50 and DEAE-cellulose chromatography were designated as fractions UH and UL respectively.

Preparation of urine samples for LM bioassay was as follows. A 24 hour collection was adjusted with acetic acid to pH 4.5 and slurried with 10-15 gm kaolin per liter of urine. The slurry was suction-filtered and washed with one volume of acidified water in divided portions. The filter cake was suspended in 200 ml 10% ammonium hydroxide and filtered. The filtrate was centrifuged at 25000g to remove remaining debris and, after mixing with 5 volumes of cold acetone, stored for 24 hours at 4°C. The precipitate was collected by centrifugation and vacuum dried to remove ammonia. It was resuspended in 0.9% NaCl and again centrifuged. The supernatant, usually 10-12 ml, contained the peptides present in a 24 hour sample. Half this amount (5-6 ml) was injected into a rabbit.

## RESULTS

Chemical Assay of Fractions H and L in Pituitary. The posterior pituitary yielded more H and L than did anterior or whole gland. We treated commercial Vasopressin, Oxytocin and Dessicated Pituitary Powder (Parke, Davis & Co.) by the Rudman procedure and found that only the latter contained significant amounts of Fractions H and L. Before extracting large amounts of lyophilized gland, several preliminary tests of eluate from the DEAE-cellulose column subjected to paper electrophoresis consistently gave distribution patterns for H and L. Finally, several gram quantities of lyophilized pituitary were extracted on 2 trials with the following yields: posterior 9& 10 mg H and 1.8& 2.0 mg L per gm; anterior 1.6&1.8 mg H and 0.32&0.39 mg L per gm; whole gland 4.0 & 4.5 milligrams

H per gram and 0.8 & 0.9 mg L per gram.

Bioassay for LM Activity of Pituitary Fractions H and L. Rabbits were deprived of food but not water for 24 hours before receiving fractions for assay, therefore the zero time sample in the following tables reflects the changes which have occurred during the fasting period. Samples were also taken 4, 19 and 46 hours after injection and analyzed.

TABLE 1. METABOLIC RESPONSE TO INJECTIONS OF FRACTION H OF PITUITARY

| | | | Cholesterol | | Uric Acid | |
|---|---|---|---|---|---|---|
| No. | Treatment | Dose | 0 | 19 hrs | 0 | 19 hrs |
| 2 | Anterior | 2.5 mg/kg | 37 | 175 | 3.8 | 2.3 |
| 2 | Posterior | 2.5 mg/kg | 34 | 57 | 2.9 | 4.9 |
| 2 | Whole | 2.5 mg/kg | 38 | 73 | 1.7 | 1.3 |
| 2 | Anterior | 0.25 mg/kg | 52 | 69 | 3.9 | 2.5 |
| 2 | Posterior | 0.25 mg/kg | 44 | 65 | 1.1 | 1.0 |
| 2 | Whole | 0.25 mg/kg | 51 | 51 | 3.1 | 1.4 |
| 6 | 0.9% NaCl | ———— | 41 | 48 | 2.1 | 1.7 |

The doubling of mg% cholesterol in plasma (Table 1) shows that the 3 fractions were active. The values for uric acid suggest that the activity was not due to toxic tissue destruction. The apparent greater potency of anterior fraction H was not seen with the lower dose at which no extract was active.

Chemical Assay of Fractions H and L in Urine. Table 2 shows that small amounts of acetone extractable peptides were present in the daily urine specimens of healthy subjects and that 20-30% of this material was in the LM fraction. The total amount of UA was increased in several clinical states but only in pregnancy was the percent of LM fraction more than 35. Diabetic urine was not assayed.

Bioassay of Acetone Extract of Human Urine. Table 3 shows that increased amount of LM fraction was paralled by more intense LM action when UA was administered to rabbits. Lipid mobilization was striking in diabetes, hepatic cirrhosis, ingestion of hydroxyprogesterone and pregnancy. Pregnancy and diabetic urines, presumably containing larger amounts of Fractions H and L, were lethal by intravenous injection into rabbits.

Relation of Bioassay to Fasting and Starving. Demonstration of LM action requires that the recipient animal be fasted for at least 24 hours. Chalmers et al. observed that the yield of LM peptide was increased in

TABLE 2. CHEMICAL ASSAY OF ACETONE EXTRACT OF HUMAN URINES ( UA ).

| Subject | UA mg./24 Hrs. | % in UA UH | % in UA UL |
|---|---|---|---|
| C.H. Chronic Nephrosis | 615 | 16 | 8.2 |
| P.W. Chronic Nephrosis | 420 | 22 | 10.6 |
| J.H. Nephritis | 60 | 25 | 9.5 |
| D.Mc. Fam. Hyperlipemia | 21 | 25 | 2.5 |
| K.L. Pregnant 1st Trimester | 38 | 31 | 3.2 |
| J.G. Pregnant " " | 30 | 35 | 3.0 |
| T.M. Pregnant 2nd Trimester | 84 | 36 | 4.0 |
| P.F. Pregnant " " | 52 | 31 | 3.8 |
| L.F. Pregnant 3rd Trimester | 72 | 30 | 3.4 |
| P.S. Pregnant " " | 98 | 42 | 5.6 |
| R.S. 1st day Post-Partum | 192 | 32 | 8.7 |
| R.S. 2nd " " " | 122 | 40 | 6.0 |
| W.L. 1st day Post-Partum | 122 | 27 | 6.2 |
| W.L. 2nd " " " | 106 | 36 | 5.5 |
| M.G. Hyperthyroidism | 14 | 35 | 3.5 |
| B.W. Hyperthyroidism | 13 | 34 | 6.0 |
| J.P. Diabetes Insipipidus | 42 | 21 | 1.0 |
| N.S. Rheumatic Fever* | 22 | 28 | 0.8 |
| F.L. Rheumatic Fever* | 16 | 18 | 1.0 |
| H.B. Normal | 18 | 28 | 0.6 |
| W.W. Normal | 12 | 22 | 1.0 |
| F.I. Normal | 15 | 25 | 0.8 |
| B.R. Normal | 12 | 20 | 0.8 |

Yields of urinary protein (and/or peptide) fractions isolated as described in the text (see Methods). UH and UL are fractions isolated following CG-50 and DEAE-cellulose chromatography respectively.
*On 20 mg Prednisone daily.

urine by prolonging the fast to 36 and 42 hours. We therefore analyzed the blood of rabbits at frequent intervals during the 30 hours of fasting. The FFA increased to a maximum of 3-4 fold by 16 to 24 hoursand either showed no further change or decreased. The triglycerides, and usually cholesterol, progressively decreased 25-33% in 16-24 hours and then, frequently precipitously, increased. The period of triglyceride decrease was considered to be the fasting state. The increase was attributed to lipid mobilization by catabolic processes during starvation. Rabbits used in the assay are therefore in transition from fasting to starvation and should

TABLE 3. BIOASSAY OF ACETONE EXTRACT OF HUMAN URINES (UA)
(mg% Triglycerides in Plasma of Recipient Rabbits)

| Subject | 0 hr | 25 hrs |
|---|---|---|
| Control * | 64 | 76 |
| G Normal | 50 | 104 |
| K " | 55 | 143 |
| R " | 54 | 96 |
| D Porphyria | 84 | 120 |
| D " | 75 | 168 |
| T Gout | 51 | 104 |
| T " | 60 | 120 |
| B Glomerulonephritis | 79 | 169 |
| D " + | 27 | 59 |
| C Breast Tumor(androgen) | 25 | 86 |
| N Diabetes | 53 | 567 |
| C Diabetes | 81 | 318 |
| G Diabetes | 57 | 841 |
| A " - Anasarca | 51 | 218 |
| L Alcoholic (fatty liver) | 36 | 230 |
| L " | 48 | 244 |
| A Hepatic Cirrhosis | 88 | 750 |
| B Hydronephrosis | 45 | 186 |
| B " | 75 | 615 |
| A Glomerulonephritis | 45 | 218 |
| N Acute Nephritis | 44 | 77 |
| B Hyperthyroidism | 18 | 199 |
| L Hypercalcemia | 26 | 51 |
| H Dwarfism (Durabolin) | 61 | 77 |
| L Ben.Prost.Hyp.[2] | 89 | 430 |
| B " " "[2] | 33 | 376 |
| F 2nd Trimester Pregnancy | 76 | 850 |
| F 3rd " " | 153 | 1350 |

*0.9% NaCl treated as urine. Average of 15. Ranges 28-108; 34-114
+Average of 3 individual days. Ranges 24-42 and 49-77
[2] Patients receiving Hydroxyprogesterone
Extracts of urines from diabetics and pregnancies were lethal for rabbits and had to be diluted for assay.

## TABLE 4. COMPARISON OF LIPID CHANGES IN RABBITS RECEIVING ACTIVE AND INACTIVE UA

| Rabbit No. | Mg% Free Fatty Acid in Plasma, Hours After Injection of UA: 0 | 1 | 2 | 5 | 22 | 25 | 29 |
|---|---|---|---|---|---|---|---|
| 7 | 71 | 75 | 73 | 71 | 76 | 73 | 71 |
| 11 | 37 | 37 | 38 | 38 | 38 | 38 | 37 |
| 26 | 41 | 44 | 42 | 44 | 46 | 42 | 44 |
| 5 | 58 | 102 | 132 | 138 | 152 | 164 | 152 |
| 6 | 45 | 52 | 82 | 95 | 71 | 56 | 56 |
| 24 | 44 | 61 | 64 | 72 | 150 | 140 | 96 |
| 9 | 30 | 51 | 53 | 59 | 63 | 51 | 38 |

7, 11 and 26 received 0.9% NaCl processed as urine; 6, UA from patient on Testolactone for breast tumor; 5, UA from a nephritic; 24 and 9, UA from diabetics.

| Rabbit No. | Mg% Cholesterol in Plasma, Hours After Injection of UA: 0 | 1 | 2 | 5 | 22 | 25 | 29 |
|---|---|---|---|---|---|---|---|
| 7 | 73 | 73 | 77 | 77 | 83 | 73 | 77 |
| 11 | 37 | 37 | 39 | 39 | 44 | 39 | 44 |
| 26 | 51 | 55 | 55 | 58 | 62 | 55 | 60 |
| 5 | 48 | 55 | 62 | 84 | 70 | 88 | 110 |
| 6 | 44 | 48 | 53 | 55 | 61 | 70 | 62 |
| 24 | 37 | 43 | 48 | 60 | 146 | 170 | 147 |
| 9 | 44 | 54 | 54 | 61 | 93 | 86 | 78 |

| Rabbit No. | Mg% Triglycerides in Plasma, Hours After Injection of UA: 0 | 1 | 2 | 5 | 22 | 25 | 29 |
|---|---|---|---|---|---|---|---|
| 7 | 60 | 63 | 62 | 62 | 61 | 60 | 61 |
| 11 | 51 | 51 | 52 | 57 | 99 | 99 | 93 |
| 26 | 72 | 74 | 76 | 73 | 75 | 74 | 74 |
| 5 | 63 | 69 | 87 | 117 | 465 | 597 | 567 |
| 6 | 55 | 61 | 61 | 72 | 143 | 146 | 104 |
| 24 | 88 | 93 | 99 | 136 | 633 | 750 | 650 |
| 9 | 81 | 78 | 118 | 214 | 289 | 318 | 347 |

TABLE 4 -CONTINUED-

| Rabbit No. | | % α and β of Lipoproteins in Plasma, Hours after UA 0 | 1 | 2 | 5 | 22 | 25 | 29 |
|---|---|---|---|---|---|---|---|---|
| 7 | α | 21 | 20 | 20 | 16 | 16 | 14 | 17 |
| | β | 79 | 80 | 80 | 84 | 84 | 86 | 83 |
| 11 | α | 43 | 44 | 42 | 40 | 44 | 43 | 44 |
| | β | 57 | 56 | 58 | 60 | 56 | 57 | 56 |
| 26 | α | 40 | 30 | 38 | 37 | 32 | 37 | 40 |
| | β | 60 | 70 | 62 | 63 | 68 | 63 | 60 |
| 5 | α | 18 | 14 | 15 | 12 | 8 | 0 | 0 |
| | β | 82 | 86 | 85 | 88 | 92 | 100 | 100 |
| 6 | α | 27 | 24 | 18 | 12 | 4 | 0 | 0 |
| | β | 73 | 76 | 82 | 88 | 96 | 100 | 100 |
| 24 | α | 35 | 27 | 24 | 24 | 13 | 2 | 11 |
| | β | 65 | 73 | 76 | 76 | 87 | 98 | 89 |
| 9 | α | 44 | 36 | 35 | 31 | 0 | 0 | 7 |
| | β | 56 | 64 | 65 | 69 | 100 | 100 | 93 |

be sampled frequently. Typical time courses of such lipemias are shown in Table 4. Rabbits 7, 11 and 26 received only processed 0.9% NaCl. No significant changes in lipid values can be seen in 7 and 26 but 11 has an elevation of triglycerides suggestive of starvation mobilization. The rise is not as striking as that provoked by the LM containing urines administered to 5, 24 and 9.

Source of Fractions H and L During Pregnancy. Extraction of several placentas by the Rudman procedure and modifications of it failed to yield significant amounts of H and L. This and the recovery of large amounts of these from postpartum urines indicate that the placenta was not the source of the LM peptides.

## DISCUSSION

Our finding that the posterior pituitary of hogs contained more fraction H than did either anterior or whole gland is consistent with the original report of LM by Seifter and Baeder and the findings of Lelek et al. and those of Kadas and Nagy. It does not support Rudman's surmise that his LM materials were of anterior pituitary origin.

The finding that urines of healthy humans contained Fraction H in amounts that could be isolated but insufficient for bioassay and that these were markedly increased by disease and drugs confirms the observations of Zarafonetis et al. (1959). This increase in LM peptides in clinical states with severe metabolic alterations and the increase in prolonged fasting reported by Chalmers et al. suggest that LM release is a response to catabolic processes. Even bioassay for LM depends on recipient animals that have been sensitized by starvation.

The observations reported in this paper do not establish the identity of Fractions H and L or their relation to other peptide LMs. They are consistent with the suggestion that all peptides with LM activity reported since 1956 are closely related chemically.

## SUMMARY

1. Anterior, posterior and whole pituitary glands of hogs were fractionated by the Rudman procedure. Each contained Fraction H and L and each Fraction H had lipid mobilizing activity in rabbits.

2. Posterior pituitary contained more Fraction H and L than did anterior or whole pituitary.

3. Urines from healthy humans contained small amounts of H and L in a 24 hour specimen.

4. Both the absolute amount of H and L and the percent of these in extracted peptides were markedly increased in urine by diabetes and pregnancy. The increase during pregnancy was not of placental origin.

5. Transition from fasting to starvation in rabbits is associated with lipid mobilization. Injection of LM induces a similar mobilization.

6. It is suggested that LM is released from the posterior pituitary as a response to catabolic processes.

7. It is suggested that all LM peptides reported since 1956 are closely related chemically.

We are grateful to Dr. G. Rettura for valuable technical assistance. Supported by USPH Grants AM 06648 and AM 05664.

## REFERENCES

J. Seifter and D. H. Baeder, Proc. Soc. exp. Biol. Med. 91, 42 (1956).

J. Seifter and D. H. Baeder, Proc. Soc. exp. Biol. Med. 95, 318 (1957).

C. J. D. Zarafonetis, J. Seifter, D. Baeder and J. Kalas, J. Lab. Clin. Med. 50, 965 (1957). Clin. Res. 6, 265 (1958).

T. M. Chalmers, A. Kekwick and G. L. S. Pawan, Lancet 1, 866 (1958). Lancet 2, 6 (1960).

D. Rudman, F. Seidman and M. B. Reid, Proc. Soc. exp. Biol. Med. 103, 315 (1960).

D. Rudman, M. B. Reid, F. Seidman, M. Digirolamo, A. R. Wertheim and S. Burn, Endocrinology 68, 270 (1961).

V. P. Dole, J. clin. Invest. 35, 150 (1956).

E. Van Handel, Clin. Chem. 7, 249 (1961).
L. L. Abell, B. B. Levy, B. B. Brodie and F. E. Kendall, J. biol. Chem. 195, 357 (1952).
E. Seifter, D. Kambossos and G. Rettura, Symposium on Automation in Analytical Chemistry, Technicon Corp., Ardsley, N.Y. 1967
Shandon Corp., Chicago, Ill., Use of Sudan Black B, Personal Commun.

# STUDY OF THE LIPOLYTIC EFFECT OF ACTH IN ISOLATED FAT CELLS OF THE RAT: POTENTIALIZATION OF THE LIPOLYTIC EFFECT BY ADDING SMALL QUANTITIES OF SERUM OF GUINEA-PIG, RAT AND NORMAL MAN.

Dr. M. Benuzzi-Badoni, Dr. J.-P. Felber and
Prof. A. Vannotti

Département de Biochimie Clinique, Clinique Médicale
Universitaire, Lausanne, Switzerland

## MATERIAL AND METHOD

The present experiments were designed to examine the effect of ACTH on lipolysis and its modification with the addition of small quantities of guinea-pig, rat and human serum.

Female Sprague-Dawley rats (130 ± 15 g) were fed with normal diet (Altromin R, B 0100) until 18 h before the experiment when the rats were brought to the laboratory and fasted. Parametrial adipose tissue was used, approximately 1 g/experiment. Three rats were generally employed for each experiment. Isolated fat cells were prepared by the procedure of Rodbell (1). The tissue was added to 20 ml polyethylen flasks, containing 3 ml of albumine-bicarbonate buffer and 15 mg of collagenase (Worthington). Glucose was not present in the buffer solution which contained collagenase or in the solution which was used to wash the cells (2). A 4% solution of albumin (human albumin, Swiss Red Cross) was dialysed overnight against 8 volumes of the bicarbonate buffer. The solution was freshly made for each experiment. The pH was then adjusted to 7,4 in an atmosphere of 5% $CO_2$ and 95% $O_2$.

The isolated cells were incubated for 3 hours at $37^o$. A 50 mg concentration of glucose was present in the medium. The final medium contained 1 ml of the cells suspension and 1 ml of the hormone solution (ACTH Organon A' 163 U/mg, a gift of Organon, Oss, Holland).

The effect of different sera was analysed by adding varying quantities of serum directly to the medium using a micropipette (5, 10, 50, 100, 200 µl). A minimum of 5 experiments was performed for each project.

Incubations were performed in duplicate in each experiment. All statistical evaluations were based on paired comparisons. FFA were evaluated by the procedure of Dole and Meinertz and glycerol by the enzymatic method of Wieland (3, 4). The triglycerides content of the flasks, based on analysis of the free fatty acids after saponification was used for quantitative evaluation of the cells, present in each flask.

## RESULTS

The first figure shows the release of glycerol and fatty acids obtained with increasing quantities of ACTH. Doses of 2,5; 5,0; 12,5; 25,0 ng/ml of ACTH were used. The lipolysis obtained with ACTH with and without addition of 50 µl of NGPS to the incubation medium can be seen in the same figure. Different pools of serum were used in each experiment, the blood being obtained by decapitation. Statistical analysis gives $p<0.001$. The effect of 100 µl dose of NGPS on lipolysis was studied in the same group of experiments in the absence of ACTH: this dose which usually shows maximum activity has no influence on glycerol and fatty acid release (in fact, the production of glycerol and FFA ranges from 12,5 to 10,8 and from 58,1 to 67,2 (not significant)). NGPS seems therefore to be ineffective without ACTH.

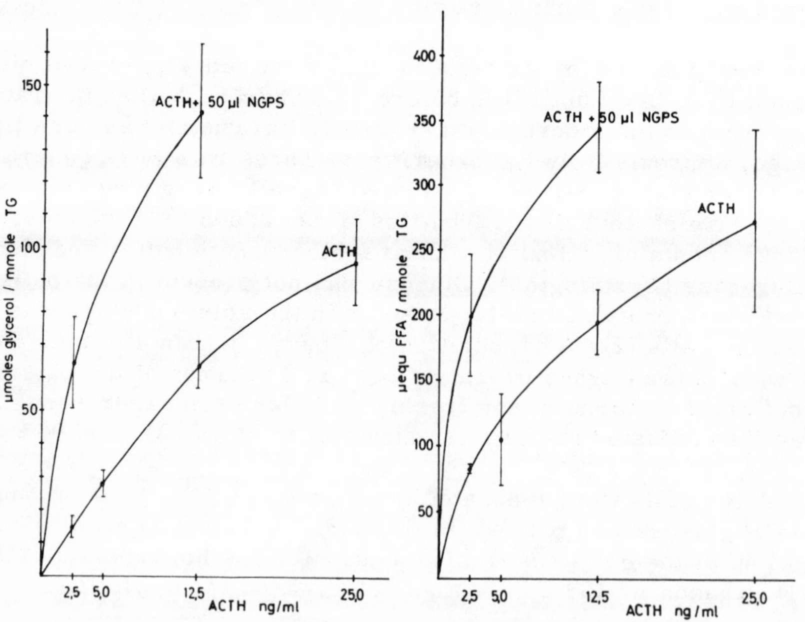


Fig. 1: Release of glycerol and FFA obtained by ACTH with or without 50 µl of NGPS. Results are presented with the standard error of the mean.

In the second group of experiments we studied the effect of increasing quantities of NGPS on a constant dose of ACTH. Results are presented in the second figure. One can see that a dose of 10 µl increases lipolysis and the plateau is attained with 100 µl (statistical analysis: $p<0.05$ for 10 µl and $p<0.001$ for 50 and 100 µl). In some experiments doses of 5 and 200 µl of NGPS were used. Stimulation was already obtained at 5 µl while a dose of 200 µl sometimes produced inhibition.

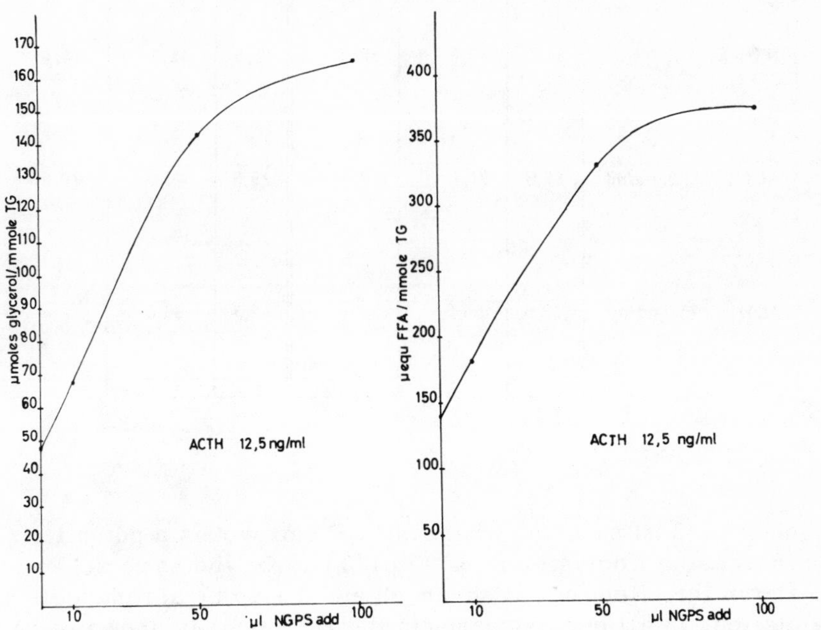


Fig. 2: Release of glycerol and FFA obtained by a same dose of ACTH to which increasing quantities of NGPS were added.

Considering the results obtained with the NGPS, one wanted to study also the effect of human serum. In table I the effect of human serum and the serum of guinea-pig are compared. Once again an absence of lipolytic effect can be seen if there is no ACTH in the medium. As to the human serum, its effect is not obvious at the lower dose of ACTH. At this level the effect of NGPS is very strong. For the higher dose of ACTH, the effect of human serum is obvious, but rather weak.

The second table shows the effect of a doubled concentration of human serum. The potentialization of the lipolysis is now obvious ($p<0.01$), though weaker than that for the guinea-pig. To better evaluate this effect we have studied the effect of increasing quantities of human and guinea-pig serum. The third schema shows the results obtained. One notes that human serum has a modest but significant effect only at a

Table 1: Release of glycerol and FFA obtained by ACTH with or without human or guinea-pig serum.

| ADDITION TO REACTION MIXTURE | FATTY ACID RELEASE µEq/mmole TG | | | GLYCEROL RELEASE µmoles / mmole TG | | |
|---|---|---|---|---|---|---|
| | without serum | serum added | | without serum | serum added | |
| | | HS 25µl | NGPS 25µl | | HS 25µl | NGPS 25µl |
| NONE | 61,7 | 60,6 | 58,5 | 12,6 | 12,2 | 14,4 |
| ACTH 12,5 ng/ml | 86,6 | 91,7 | 172,9 | 29,0 | 40,0 | 70,6 |
| ACTH 25,0 ng/ml | 127,5 | 158,6 | | 48,3 | 66,8 | |

dose of 10 µl/2 ml medium, whereas NGPS provokes a potentialization of an increasing lipolysis up to 100 µl/2 ml. On the same schema we have shown the effect of rat serum obtained also by serum pools and decapitation. In all our experiments the rat serum is shown to be more powerful than NGPS at a dose of 10 µl. However, the plateau arrives quicker. In these experiments the basal release of glycerol and FFA was 4,6 and 50.0 µmoles/m mole TG respectively and was not significantly modified by the addition of human, rat or guinea-pig serum.

In order to clear up the nature of the factor which potentiates the lipolytic effect of ACTH we have dialysed a small quantity of guinea-pig serum through membranes previously boiled 3 times in distilled water. The limit of the molecular weight which can pass this membrane is about 3000.

In the fourth figure one can see that dialysis does not modify the effect of potentialization of guinea-pig serum (Basal release of glycerol 9,6 µmoles/m mole TG. Basal content of FFA in the medium 62,4 µeq/m mole TG).

| ADDITION TO REACTION MIXTURE | FATTY ACID RELEASE µEq/m mole TG | | GLYCEROL RELEASE µ moles/m mole TG | |
|---|---|---|---|---|
| | without serum | serum added HS 50 µl | without serum | serum added HS 50 µl |
| NONE | 69,6 | 68,7 | 6,06 | 5,64 |
| ACTH 12,5 ng/ml | 103,4 | 127,6 | 32,6 | 47,08 |

Table 2: Release of glycerol and FFA obtained by ACTH with or without addition of 50 µl of human serum.

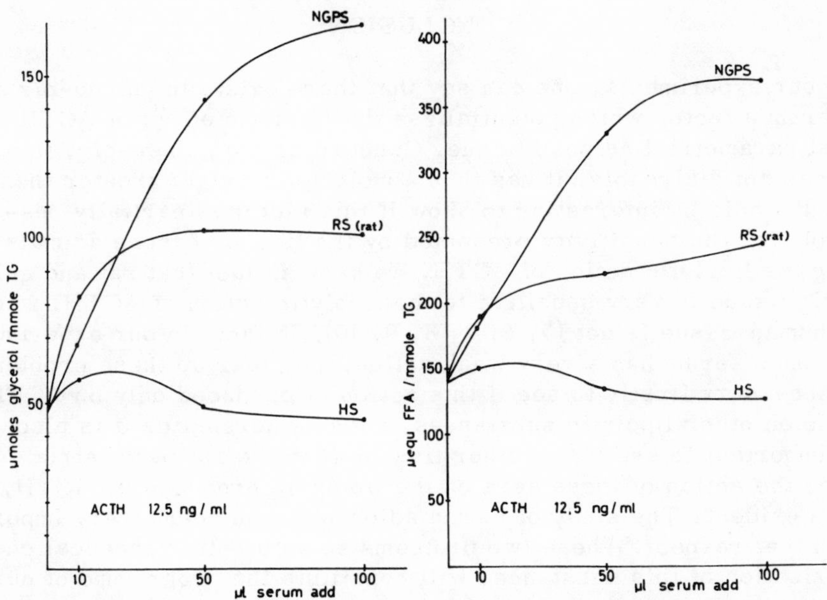


Fig. 3: Release of glycerol and FFA obtained by the same dose of ACTH with addition of increasing quantities of human, rat or guinea-pig serum.

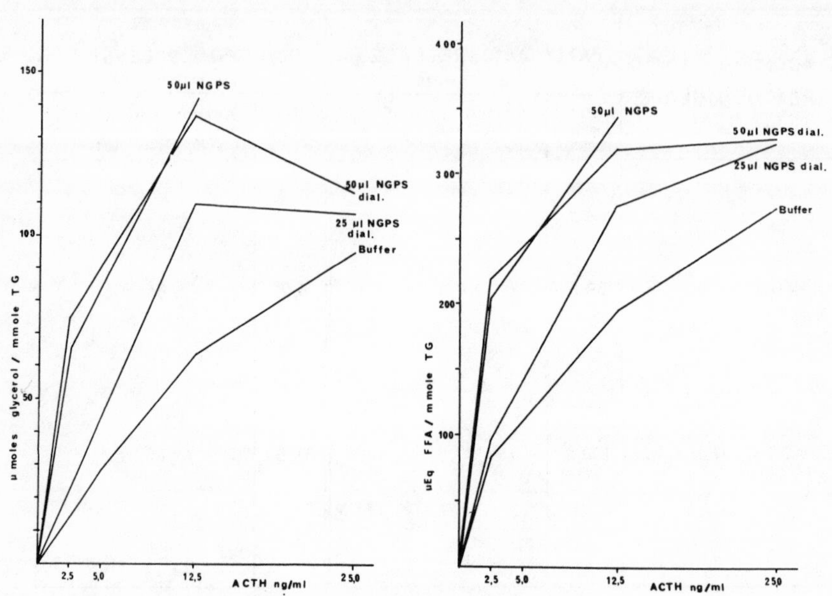


Fig. 4: Effect of the dialysis on the potentialization of the lipolytic effect of ACTH obtained by NGPS.

## CONCLUSIONS

After our experiments, one can say that there exists in guinea-pig and rat serum a factor which potentializes the lipolytic effect of ACTH on the rat parametrial adipose tissue. Concerning the guinea-pig, this factor is not dialysable, it has thus a molecular weight greater than 3000. It would be interesting to show if this factor is partially responsible for the sensitivity presented by the tissue of these animals during the lipolytic action of ACTH. We know in fact that rat and guinea-pig tissue is very sensitive to the lipolytic action of ACTH, whereas human tissue is not (5, 6, 7, 8, 9, 10). In fact, in our experiments, human serum has a very weak action. To clear up these problems, it is necessary firstly to see if this action is produced only on ACTH or also on other lipolytic substances, such as adrenalin. It is also very important to see if, on other tissues than the rat parametrial tissue, the action of these sera on the lipolysis produced by ACTH, is still evident. The study of human adipose tissue seems very important in that respect. These two problems so as a better chemical characterization of this substance, will constitute the programme of our following studies.

## BIBLIOGRAPHY

1) Rodbell M.: Metabolism of isolated fat cells. - I. Effects of hormones on glucose metabolism and lipolysis. - J. Biol. Chem. 239 No. 2, 1964, 375.
2) J.N. Fain, V.P. Kovacev, R. Scow: Effect of growth hormone and dexamethasone on lipolysis and metabolism in isolated fat cells of the rat. - J. Biol. Chem. 240 No. 9, 1964, 3522.
3) V.P. Dole and Meinertz H.: Microdetermination of long chain fatty acids in plasma and tissues. - J. Biol. Chem. 235, 1960, 2595.
4) O. Wieland: Methoden der enzymatischen Analyse. - H.U. Bergmeyer, Ed. New York, Academic Press, 1963 p. 211.
5) D. Rudman, S.Y. Brown and M.F. Malkin: Adipokinetic action of adrenocorticotropin, thyroid-stimulating hormone, vasopressin, $\alpha$ and $\beta$ melanocyte-stimulating hormones, fraction H, epinephrine and norepinephrin in the rabbit, guinea-pig, hamster, rat, pig and dog.- Endocrinology 72, 1963, 527.
6) D. Rudman: The adipokinetic action of polypeptide and amine hormones upon the adipose tissue of various animal species.- J. Lipid Research 4, 119, 1963.
7) D. Rudman: The adipokinetic property of hypophyseal peptides.- Ergebn Physiol. 56, 1965, 297-327.
8) B. Mosinger et al.: Action of adipokinetic hormones on human adipose tissue in vitro. - J. Lab. Clin. Med. 66, 1965, 380.
9) T.C.B. Stamp, A. Landon and V. Wynn: Observations on some extra-adrenal effects of corticotropin on carbohydrate and lipid metabolism in man.- Metabolism 14, No. 10, 1965, 1041-1050.
10) D.J. Galton and G. Bray: Studies on lipolysis in human adipose cells.- J. Clin. Invest. 46, 1967, 621.

This work was supported by a grant of the "Fonds National Suisse de la Recherche Scientifique" (No. 4325).

We would like to thank Miss Jocelyne Vuilleumier and Miss Franca Wyrsch for their technical assistance.

# THE ANABOLIC EFFECTS OF PITUITARY GROWTH HORMONE

Jack L. Kostyo

Duke University

Durham, North Carolina

During the past 10 years many of the details of the molecular mechanism involved in the process of protein synthesis and its control have been revealed. This advance in our understanding of the fundamental process which underwrites the growth of cells has made possible an extensive study of the mechanism of the anabolic action of pituitary growth hormone. The following will be a brief review of our present understanding of the sequence of events which occurs when target cells interact with growth hormone. In the main, the review will deal primarily with the action of growth hormone on skeletal muscle. It should be noted, however, that much recent work has been devoted to defining the mechanism of action of growth hormone on the liver. In instances where comparisons can be made, there appears to be a basic similarity in the mode of action of the hormone on both types of tissue.

One of the most rapid effects that has been observed following the injection of growth hormone into rats or following the addition of growth hormone to medium bathing isolated skeletal muscle is an increase in amino acid incorporation into total muscle protein. The *in vitro* effect of growth hormone on amino acid incorporation has been demonstrated by a number of laboratories using the isolated diaphragm of the hypophysectomized rat. Growth hormone at concentrations ranging from 1-50 μg/ml stimulates the incorporation of a variety of labeled amino acids into the total protein of the diaphragm (1-3). The stimulation of tracer incorporation into protein becomes progressively greater with time, with the effect becoming statistically significant about 20 minutes after the cells first come in contact with the hormone (4).

Recent studies (4) in our laboratory have indicated that this rapid stimulation of protein synthesis in the isolated diaphragm is clearly not an indication of the complete activation of protein anabolism. The distribution of labeled proteins in centrifugal fractions of the diaphragm was examined following incubation of the muscle with bovine growth hormone (BGH) and $^{14}C$-labeled leucine, and it was found that the hormone only increased labeling of proteins in a heavy centrifugal fraction containing nuclei, myofibrils, sarcolemma and cell debris. Like the effect of the hormone on the labeling of total muscle protein, the effect was significant only after 20 minutes of incubation had elapsed and became more pronounced with time. The hormone did not influence labeling of the proteins of mitochondria, microsomes or cell sap at any of the incubation times examined. Moreover, when nuclear proteins were examined specifically, no effect of growth hormone could be found. Muscles that were removed from hypophysectomized rats that had received an intravenous injection of BGH one hour prior to incubation also showed increased labeling only in the heavy centrifugal fraction. This suggests that the asymmetric stimulation of labeling found in response to BGH _in vitro_ is not merely an artifact of the conditions used. Rather, it would appear that the synthesis of only certain proteins is stimulated. In contrast, similar studies with insulin (5) have shown that within 5 minutes after the addition of this hormone to the incubation medium there is stimulation of labeled amino acid incorporation into all centrifugal fractions of the diaphragms. Consequently, there may be a fundamental difference in the manner in which these two agents stimulate muscle protein synthesis _in vitro_.

Further evidence that the rapid effect of growth hormone on protein synthesis in the diaphragm involves only certain proteins has been obtained in experiments (6) in which the proteins of the heavy centrifugal fraction were fractionated with water, salts and detergents. In these studies, the heavy centrifugal fraction was prepared from diaphragms that were incubated for various periods with BGH and $^{14}C$-labeled leucine. The heavy fraction was then successively treated with water, 0.1 M KCl, 0.6 M KCl and a detergent mixture containing sodium cholate, sodium deoxycholate and sodium lauryl sulfate. The proteins solubilized by these reagents and the insoluble residue were purified and counted. It was found that when diaphragms were incubated with BGH for only 20 minutes, only the residual or insoluble proteins showed a consistent and significant increase due to the hormone. A slight increase in the labeling of 0.6 M KCl soluble proteins due to BGH was also detected when the heavy centrifugal fraction was extracted with the salt solution overnight. Thus it would appear that growth hormone rapidly stimulates the synthesis of at least two types of proteins in muscle, proteins soluble at high salt concentration which may be contractile proteins, and those which are quite insoluble.

There is considerable evidence that the rapid effect of growth hormone on protein synthesis is not mediated by an action of the hormone on the transcription of new template RNA. One line of evidence comes from the work of Knobil and his co-workers (2) and Martin and Young (3). Both of these groups have demonstrated that incubation of the diaphragm of the hypophysectomized rat with a concentration of actinomycin D, which abolished precursor incorporation into RNA, had no effect on the ability of growth hormone to stimulate labeled amino acid incorporation into total diaphragm protein. Also, efforts to demonstrate _in vitro_ stimulatory effects of growth hormone on precursor incorporation into total RNA of the diaphragm have been unsuccessful (2,3,7). Even when total RNA prepared from diaphragms incubated for various periods of time with BGH and $^{3}$H-uridine was submitted to sucrose-density-gradient centrifugation, no effect of the hormone on uridine incorporation into the various RNA fractions could be detected (7). Thus it seems unlikely that the rapid stimulatory effect of the hormone on protein synthesis is related to an action on the transcription of new RNA. The effect might be accounted for by an effect of the hormone on the ability of only certain ribosomes to translate existing templates. This would explain the insensitivity of the system to actinomycin D and the finding that the synthesis of only certain proteins is stimulated. However, to the present there is no direct evidence to support this hypothesis.

Another early effect of growth hormone on skeletal muscle which can be detected following either the injection of the hormone (8,9) or its addition to medium bathing isolated muscle (2,10,11), is the stimulation of the net transport of certain amino acids into the intracellular compartment of the tissue. Recent studies (4) using the isolated rat diaphragm and the non-utilizable amino acid, α-aminoisobutyric acid (AIB), have indicated that the stimulatory effect of growth hormone on the transport mechanism is due to the enhancement of amino acid influx rather than to the retardation of amino acid efflux from the cells. The effect is not immediate but becomes evident at about the time (20 minutes of incubation) when a significant stimulation of protein synthesis has occurred (4).

The finding that there is a delay in the on-set of the transport phenomenon has suggested that certain proteins formed as a result of the action of growth hormone on protein synthesis might mediate the transport response. Experiments by Knobil and his co-workers (2) provided the first evidence in favor of this hypothesis. This group incubated diaphragms of hypophysectomized rats with BGH and $^{14}C$-AIB in the presence of the protein synthesis inhibitor, puromycin. They found a marked reduction in the effect of the hormone on the transport of AIB, but the hormonal response was not abolished completely. Their experiment was repeated in our laboratory (4), but with the modification that the muscle was preincubated with puromycin for 1 hour prior to the addition of the hormone and AIB. The amount of puromycin used was shown, in separate experiments, to abolish the incorporation of labeled leucine into the protein of the diaphragm. Using this experimental design, it was found that puromycin completely blocked the stimulatory effect of the hormone on AIB transport. Puromycin, itself, had no influence on the basal level of AIB transport, so it is unlikely that the failure of growth hormone to act was due to deterioration of the transport mechanism. That the ability of puromycin to abolish the transport effect of growth hormone is not due to the inhibition of protein synthesis but to some unknown side-effect of the poison, is, of, course, a possibility. However, cyclohexi-mide and emetine, which also block protein synthesis by a mechanism of action different than that of puromycin, have also been found in our laboratory to block the action of growth hormone on amino acid transport. Thus, the fact that three inhibitors of protein synthesis, which probably have different metabolic side-effects, all block the action of growth hormone on amino acid transport, supports the conclusion that these substances block the transport effect through their action on protein synthesis. In view of these findings it seems reasonable to conclude that protein synthesis is necessary for growth hormone to exert its stimulatory effect on amino acid transport. What relationship this increase in membrane transport has to the later actions of growth hormone on the anabolic pathway in the muscle cell has not been established.

The later effects of growth hormone on muscle appear to be directed toward expanding the protein synthetic capability of the cell. One of the next effects to appear following the occurrence of the early protein synthetic effect and the increase in membrane transport is the activation of nuclear DNA-dependent RNA polymerase. Florini and Breuer (12) have studied this phenomenon in some detail. They utilized an aggregate enzyme prepared from the leg muscles of young hypophysectomized rats. For their experiments a single injection of porcine growth hormone (PGH) was administered to rats at various times prior to the preparation of the enzyme. They found little change in the activity of the enzyme for several hours following the administration of the hormone. After about 12 hours had elapsed, there was a marked stimulation of the enzyme by the hormone, and activity reached a peak about 18 hours after its injection. In subsequent experiments (13), these investigators found that the increase in activity of the enzyme produced by growth hormone was still evident when manganese was present in the reaction mixture or when ammonium sulfate was added. In this respect, the effect of growth hormone on DNA-dependent RNA polymerase is different for muscle and liver, since in the latter case, as both Widnell and Tata (14) and Pegg and Korner (15) have shown, the effect of the hormone is not evident when the hepatic enzyme is activated with manganese or ammonium sulfate.

Breuer and Florini (13) have obtained evidence suggesting that the stimulatory effect of growth hormone on the enzyme is not due to an increase in the ability of the DNA of the muscle to serve as the template for the polymerase reaction. These investigators isolated chromatin from the nuclei of leg muscles of hypophysectomized rats that had received an injection of PGH 18 hours previously. This is the point in time at which the enzyme shows maximal activation. This chromatin was then added to a reaction mixture containing an excess of _E_. _coli_ RNA polymerase (fraction 3) prepared by the method of Chamberlain and Berg (16) and an excess of the nucleoside triphosphates with UTP being labeled with $^{14}C$. They found that the incorporation of UTP into RNA was actually somewhat less in reaction mixtures primed with chromatin prepared from rats that had received growth hormone than in those primed with chromatin from control rats. Thus, Breuer and Florini concluded (13) that growth hormone does not act on the polymerase reaction in muscle by increasing the template activity of DNA, but perhaps it affects the enzyme directly either by altering its activity or by increasing the amount of enzyme present in the nucleus.

Presumably the increase in RNA polymerase activity produced by growth hormone then leads to the enhanced synthesis of muscle RNA. Recent experiments by Earl and Korner (17) using heart muscle have shown that 12 hours after the administration of growth hormone there is an increase in the _in vivo_ incorporation of $^{32}P$-labeled inorganic phosphate into ribosomal and transfer RNA. The activation of the polymerase enzyme is also reflected in the fact that the total number of ribosomes in muscle appears to increase after the enzyme has become fully activated (12).

By itself, an increase in the number of ribosomes in a skeletal muscle cell might increase its protein synthetic capability. However, growth hormone also appears to increase the activity of the ribosomes to catalyse the incorporation of amino acids into protein. Using equal amounts of ribosomes isolated from leg muscles of rats treated for various periods of time with PGH, Florini and Breuer (12) measured the ability of these particles to incorporate $^{3}H$-leucine from leucyl-s-RNA into RNA in an _in vitro_ system in which co-factors and additives were not limiting. Ribosomes prepared from rats receiving PGH 1-3 hours previously showed only a small increase in activity, but activity was markedly increased 18 hours after hormone injection. The increase in protein synthetic activity of these ribosomes does not appear to be due primarily to an increase in the number of functional polysomal units in the ribosome preparations. Florini and Breuer (12) examined sucrose-density-gradient profiles of their ribosome preparations and found that at 18 hours after PGH injection, when protein synthetic activity was maximal, the number of polysomes in the ribosomal population was only slightly increased. Moreover, Earl and Korner (17) found no difference in the proportion of polysomes in ribosomal preparations made from heart muscle of rats that had received an injection of human growth hormone 12 hours previously. These results are in marked contrast to those of Korner obtained with the liver (18), in which the proportion of polysomes in the population of ribosomes is increased some hours following the administration of growth hormone. Thus, one is forced to conclude that the increase in protein synthetic capability of the ribosomes of muscle is due to some undefined change in their ability to translate template RNA. The combination of this increase in protein synthetic activity of the ribosomes and the number of ribosomes in the muscle cell provides a marked amplification of the protein synthetic capability of the cell, which is needed to provide the proteins necessary for the growth of the cell.

The above has been a brief summary of our present understanding of the sequence of events which occur when a muscle cell comes in contact with growth hormone. Some time, perhaps after a sequence of yet undefined events has occurred, there is an increase in the synthesis of only certain proteins which are associated with the myofibrillar-sarcolemma portion of the cell. The functions of these proteins have yet to be defined, but if their formation is prevented, the effect of growth hormone on amino acid transport will not occur. This suggests that some of the first proteins formed may in some way mediate the transport phenomenon. The indication that these proteins may actually reside in the sarcolemma favors this suggestion. Subsequent work may show that these substances also play some role in the later events which occur in response to growth hormone such as the increase in RNA polymerase activity, and the increase in ribosomal number and activity. That these later phenomena are only secondary events and not elicited by growth hormone directly seems a likely possibility. Certainly these phenomena cannot be elicited by the direct addition of growth hormone to _in vitro_ systems designed to detect them. Moreover, since these events are produced only some time after the injection of growth hormone into the animal, one cannot exclude the possibility that a humoral agent released in response to growth hormone, such as insulin, might be responsible for these events. Future work will be necessary to clarify these issues and to bring us closer to a better understanding of the anabolic action of growth hormone.

## REFERENCES


1. Kostyo, J. L. *Endocrinology 75*: 113, 1964.

2. Knobil, E. *The Physiologist 9*: 25, 1966.

3. Martin, T. E. and F. G. Young. *Nature 208*: 684, 1965.

4. Kostyo, J. L. *Ann. N. Y. Acad. Sci.* (in press) 1967.

5. Wool, I. G. *Biochim. Biophys. Acta 52*: 574, 1961.

6. Kostyo, J. L. *Program of The Endocrine Society*, 49th Meeting, 1967, p. 71.

7. Kostyo, J. L. *Biochim. Biophys. Acta 129*: 294, 1966.

8. Riggs, T. R. and L. M. Walker. *J. Biol. Chem. 235*: 3503, 1960.

9. Hjalmarson, Å. and K. Ahren. *Life Sciences 4*: 863, 1965.

10. Kostyo, J. L., J. Hotchkiss and E. Knobil. *Science 130*: 1653, 1959.

11. Reiss, E. and D. M. Kipnis. *J. Lab. Clin. Med. 54*: 937,1959.

12. Florini, J. R. and C. B. Breuer. *Biochemistry 5*: 1870, 1966.

13. Breuer, C. B. and J. R. Florini. *Biochemistry 5*: 3857, 1966.

14. Widnell, C. C. and J. R. Tata. *Biochem. J. 98*: 621, 1966.

15. Pegg, A. E. and A. Korner. *Nature* (London) *205*: 904, 1965.

16. Chamberlain, M. and P. Berg. *Proc. Nat. Acad. Sci. U.S.A. 48*: 81, 1962.

17. Earl, D. C. N. and A. Korner. *Arch. Biochem. Biophys. 115*: 445, 1966.

18. Korner, A. *Biochem. J. 92*: 449, 1964.

# THE SUBCELLULAR DISTRIBUTION OF HUMAN PLACENTAL LACTOGEN

Henry G. Friesen

Department of Medicine, McGill University and Royal Victoria Hospital, Montreal

In 1962 Josimovich and MacLaren (1) reported that the human placenta and retroplacental serum contain a lactogenic substance immunologically related to growth hormone. In the intervening years there have been many studies on the chemistry, biological effects, and secretion (2-7) of human placental lactogen (HPL) but comparatively little investigation on its intraplacental site of synthesis and storage. Sciarra (8) in an excellent review considered the evidence for the placental and cellular origin of HPL. He reported earlier that rabbit anti-HGH serum labeled with fluorescein isothiocyanate was limited to the syncytiotrophoblast layer of the villous epithelium (9). This communication describes the subcellular localization of HPL and outlines a simple method for its purification from fresh placentas.

## MATERIALS AND METHODS

Fresh term human placentas delivered per vaginam were washed in cold saline at 4°C to remove excess blood. Small pieces of placental tissue were forced through a tissue press and two volumes of 0.4 M. sucrose was added to one volume of tissue homogenate. The homogenate was centrifuged briefly at 300 to

This research was supported by the Medical Research Council of Canada Grant MA-2525 and the U.S.P.H.S., National Institute of Child Health and Human Development, Grant HD01727-02.

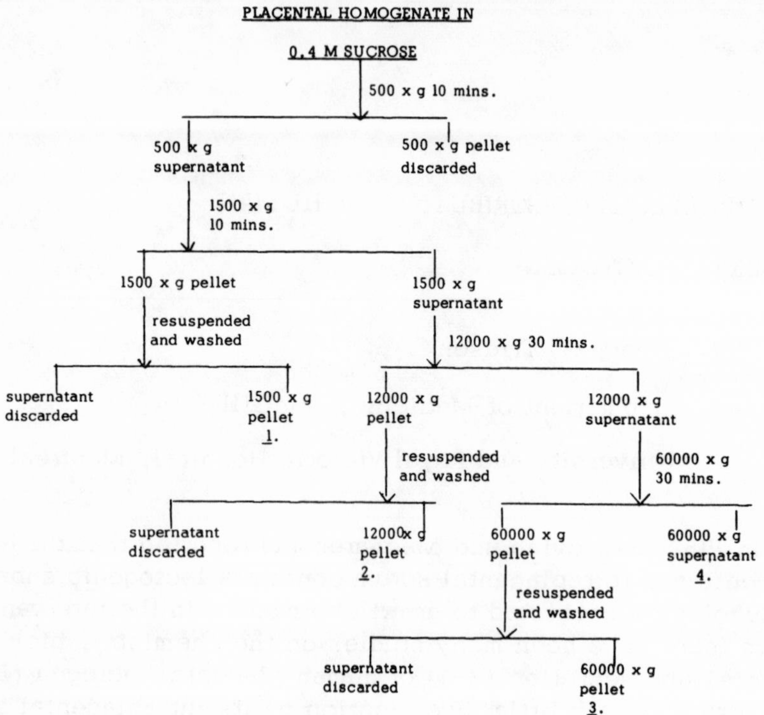


Fig. 1. Flow diagram of differential centrifugation of placental homogenate in 0.4 M. Sucrose. No. 1-4 identify the four major subfractions.

500 x g. to remove tissue fragments and unbroken cells. These and all subsequent preparative steps were performed at 4°C. The subcellular fractionation scheme applied to the 500 x g. supernatant is outlined in the flow sheet in Fig. 1.

Four principal fractions were obtained: 1) 1500 x g. pellet: 2) 12,000 x g. pellet; 3) 60,000 x g. pellet and 4) 60,000 x g. supernatant. The pellets were washed with sucrose and the HPL and proteins were extracted from each of the pellets using equal volumes of 0.1 N. ammonium hydroxide and 0.1 M. ammonium bicarbonate. In order to ensure maximum release of proteins stored in membrane bound particles, the pellets were frozen and thawed repeatedly during the alkali extraction. The HPL in each of the fractions was estimated by radial immunodiffusion with slight modifications of the method of Mancini (10), Fig. 2. The preparation of HPL used for standards was prepared according to the method published previously (11). Antiserum to HPL was obtained

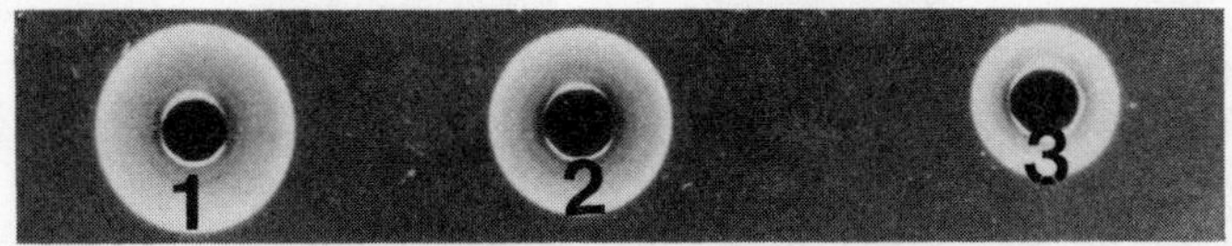


Fig. 2. Radial immunodiffusion pattern of HPL standards. Antiserum to HPL is mixed with the agar. Decreasing concentrations of HPL standards in wells 1 to 3 are reflected by smaller immunoprecipitin rings.

from rabbits which had been immunized weekly for four weeks with one milligram of hormone plus Freund's adjuvant. Protein content was determined by the Folin-Lowry method (12) and by measuring the absorbance at 280 mu. on a Bausch and Lomb Spectronic 505 recording spectrophotometer. Starch gel electrophoresis was performed according to the method of Barrett et al (13). Gel filtration studies of the extracts were carried out using Sephadex G-100 and 0.1 M. ammonium bicarbonate as buffer. Electron microscopic examination of the pellets was kindly done by Dr. H. Sheldon of the Dept. of Pathology at McGill.

## RESULTS

Table 1 summarizes the distribution of HPL and proteins in each of the four major subfractions. The percentages were calcul-

TABLE 1.

DISTRIBUTION OF PROTEINS AND PLACENTAL LACTOGEN IN FRACTIONS AFTER CENTRIFUGATION

| | Protein % | HPL % | % HPL / Protein |
|---|---|---|---|
| 1,500 x $g_1$ | 1.2 ± 0.2 | 8.2 ± 1.0 | 6.1 ± 1.3 |
| 12,000 x $g_1$ | 2.1 ± 0.4 | 16.4 ± 2.7 | 6.7 ± 1.0 |
| 60,000 x $g_1$ | 0.8 ± 0.1 | 4.3 ± 0.6 | 4.6 ± 0.6 |
| 60,000 x $g_2$ | 95.8 ± 0.4 | 71.0 ± 3.3 | 0.65 ± 0.1 |

$_1$ Pellet; $_2$ Supernatant; ± S.E. of 6 Exp'ts.

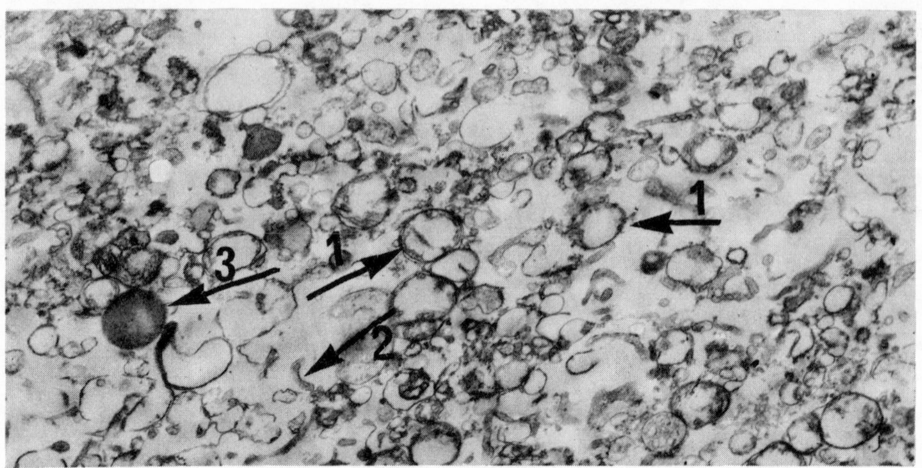


Fig. 3. Electron micrograph of 12,000 x g. pellet magnified (x 13,800). Arrows point to (1) mitochondria; (2) fragments of endoplasmic reticulum and (3) electron dense granule. The latter are relatively sparse.

ated in the following manner: the amount of HPL and proteins in the four fractions is expressed as a percentage of the total content of each in the homogenate. Almost thirty per cent of HPL is found in the three pellets sedimenting at 1500 x g. to 60,000 x g.; in the 1500 x g. and 12,000 x g. pellets, 6 per cent of the proteins in the extracts is HPL. It appears to localize preferentially in the subcellular fractions which contain predominantly nuclei and mitochondria. Fig. 3 is an electron micrograph of the 12,000 x g. pellet which morphologically appears very heterogeneous. It contains a variety of cell organelles but principally mitochondria. In addition there are fragments of endoplasmic reticulum. Very few organelles are visible that resemble secretory granules in other endocrine organs such as the pituitary. When the amount of HPL in each of the fractions was added and expressed as a percentage of the HPL content in the starting material, the recoveries varied between 85 and 90 per cent. Fig. 4 shows the distribution of HPL and proteins in particulate and supernatant fractions. The hatched bars indicate HPL and the open bars total proteins in particulate and supernatant fractions. When expressed in this way the striking difference in the ratio of HPL to proteins in the two fractions is apparent. Theoretically the subcellular distribution of HPL could be an artifact due to its non-specific adsorption to cell organelles which sediment at low speed. To exclude this possibility HPL $I^{131}$ over 90 per cent of which could be precipitated with

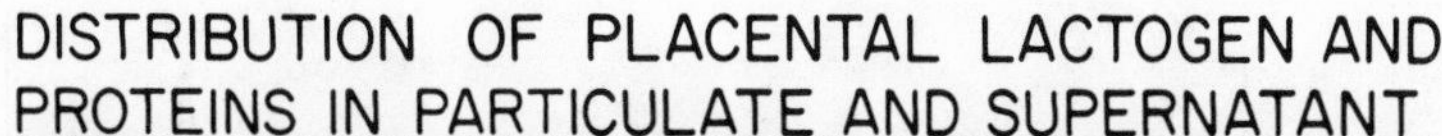


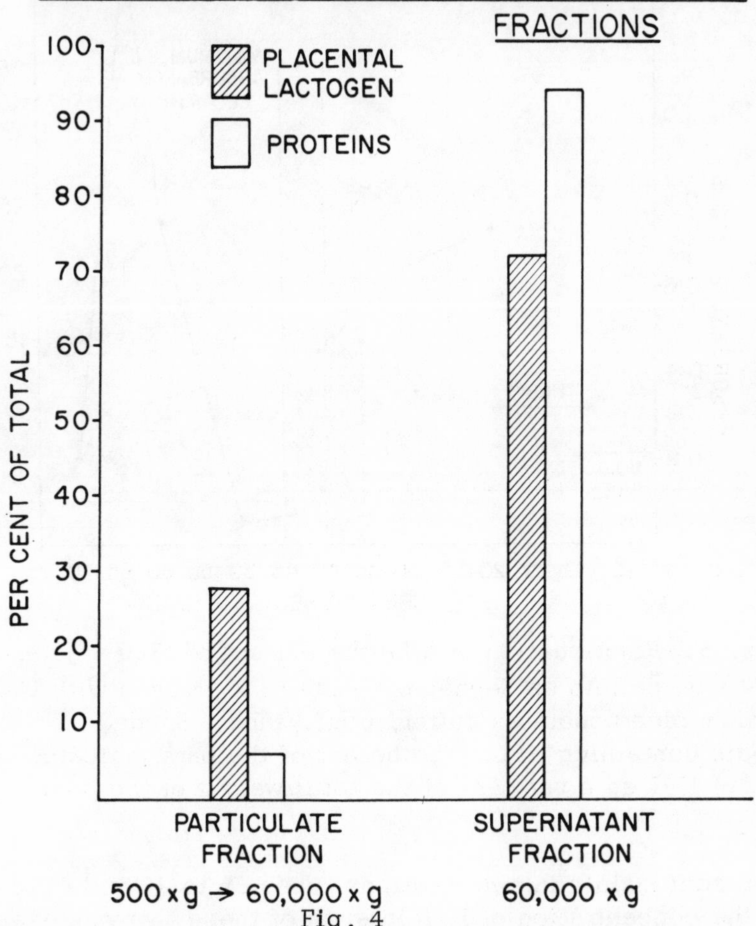


Fig. 4

antiserum was added to the 500 x g. supernatant followed by subcellular fractionation. Less than 3 per cent of the radioactivity was found in the particulate fraction compared with 30 per cent of the endogenous HPL. This finding indicates that the subcellular distribution observed was not due to non-specific adsorption.

Extracts obtained from the 1500 x g. and 12,000 x g. pellets from several placentas were pooled and examined by gel filtration on a column (3 x 45 cms) on Sephadex G-100 using 0.1 M. ammonium bicarbonate as buffer. Five ml. fractions were collected and their absorbance was measured at 280 mu. in a recording spectrophotometer. Fig. 5 shows the absorbance of eluants from the column and the vertical bars indicate the pools containing HPL.

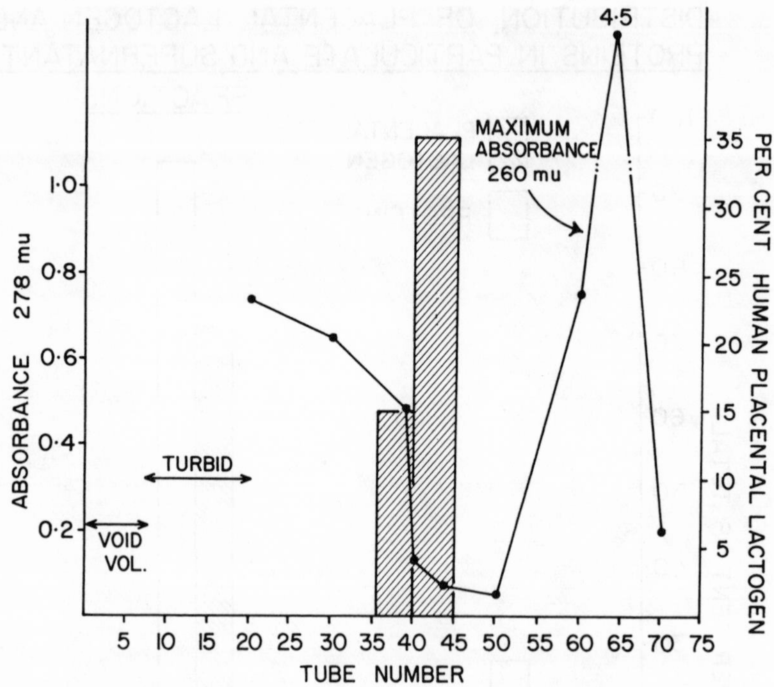


Fig. 5. Fractionation of alkaline extract of 1500 x g. and 12,000 x g. pellets on Sephadex G-100 (3 x 45 cms) with 0.1 M. ammonium bicarbonate as buffer, 6 ml./tube. Shaded bars represent pools containing HPL. The height of the bars indicates the weight of HPL as a per cent of the total weight of the pool.

It is present mainly in two fractions tubes 36 to 40 and 41 to 45. When the concentration of HPL in each of these two was examined, it was apparent as shown by the shaded bars that the concentration of HPL was greatest in pool 41 to 45; and that in this fraction HPL constituted 35 per cent of the protein content. The original homogenate contained approximately 0.6 to 0.8 per cent HPL. Using this simple two step procedure it is possible to obtain a fifty fold purification of HPL. The 60,000 x g. supernatant was examined in a similar manner. When the distribution and concentration of HPL after gel filtration was determined the maximum concentration of HPL was 8 to 10 per cent, compared with 35 per cent in similar pools from the 12,000 x g. extract.

Fractions obtained after gel filtration of the 12,000 x g. extract were lyophilized and examined upon starch gel electrophoresis and compared with a highly purified standard preparation of HPL.

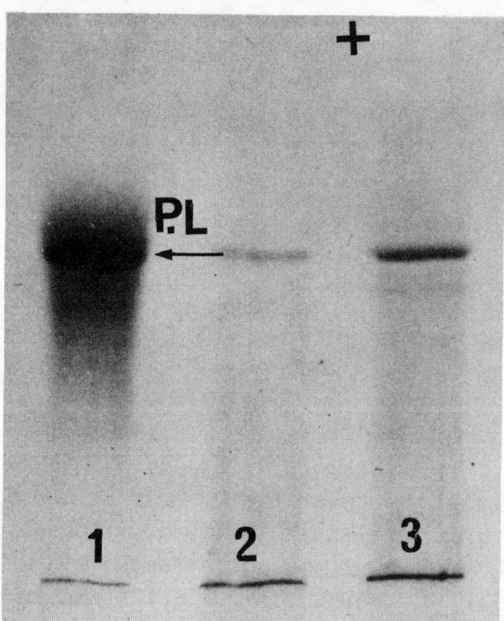


Fig. 6. Starch gel electrophoretic patterns of two pools 36 to 40, channel 2; and 41 to 45, channel 3 from the column shown in Fig. 5. Channel 1 is the pattern of the HPL standard.

As shown in Fig. 6 the principal component visible in pools 35 to 40 and 41 to 45 is HPL. It has an identical electrophoretic mobility with the highly purified standard. Final purification of the column fraction was performed using Canalco's preparative disc acrylamide gel electrophoresis system. Approximately 30 mgs. of the partially purified material was dissolved in 0.5 m. tris buffer and added to 1.0 ml. of 20 per cent Prep-cryl solution and carefully layered on top of a PD 320 separating gel with a column height of 6 cm. A current of 10 mA. was employed and the elution buffer flow was started when the ion front was several mm. from the bottom of the column and 6 ml. fractions were collected. The buffer flow rate was 1.5 ml./min. HPL in the eluant was detected immunologically on Ouchterlony plates. The appropriate fractions were dialyzed against tap water and distilled water for 24 hours each and then lyophilized. Several of these pools were reexamined upon starch gel electrophoresis against HPL standards (Fig. 7). It is apparent that one of the pools obtained in this way, channel 2, contains a highly purified preparation of HPL. One of the disadvantages of the preparative electrophoresis is that considerable losses are encountered, but attempts are underway in our laboratory to minimize these.

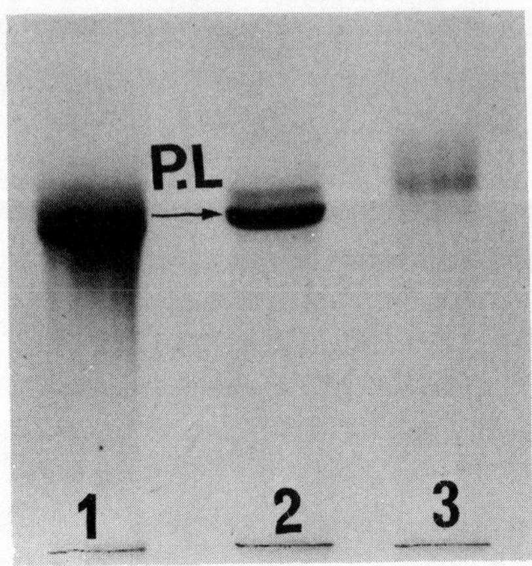


Fig. 7. Partially purified fractions from Sephadex were separated by prep disc acrylamide gel electrophoresis. Pools containing HPL were examined upon starch gel electrophoresis, channels 2 and 3. Channel 1 is the HPL standard used as a marker. The high degree of purification obtained by this method is apparent in channel 2. A faint second band is visible towards the anode.

## DISCUSSION

From electron microscopic studies of normal human placental tissue it was postulated that the electron dense granules observed were secretory granules containing HCG or steroids. Siebert and Stark (14) in studies on the subcellular distribution of HCG reported that it was localized mainly in the nuclear pellet and we have noted that a substantial proportion of HPL sediments at relatively low centrifugal force as well. Our estimates indicate that the placenta contains 8.7 mg. of HPL per gm. of pH 8.0 soluble proteins, of which 2.5 mg. is localized in the particulate fraction. Since the total protein content of the human placenta is approximately 30 gm. (15), and since half or less of this is alkali soluble, one may calculate that the placental content is approximately 120 mgs. or less. This estimate is somewhat higher than the yields obtained by others (1,16), but by their methods some losses may have occurred during sequential purification with salting out and column chromatographic procedures. One may speculate that the HPL sedimenting with the particulate fraction is a storage form, whereas the soluble portion in the supernatant is newly synthesized or secretory material which is in a soluble cytoplasmic pool. If current estimates of secretion rate which vary from 200 to 1000 mgs. per 24 hours are accurate (6,7) the intraplacental HPL pool turns over several times daily. Therefore, at any time a substantial proportion of HPL would be in transit from synthetic to storage and to secretory sites. Dynamic studies and kinetic analysis of the biosynthesis and secretion of HPL will be required to elucidate these intracellular processes.

## CONCLUSIONS

With this simple method of subcellular fractionation, gel filtration and prep disc acrylamide gel electrophoresis, it is possible to obtain highly purified HPL from fresh placentas.

## ACKNOWLEDGEMENTS

The technical assistance of Mrs. Judy Halmagyi and Mrs. Marva Greig, and the secretarial help of Miss Roseann McInroy are gratefully acknowledged.

## REFERENCES

1. Josimovich, J.B. and J.A. MacLaren. Endocrinology, 71: 209, 1962.

2. Friesen, H. Endocrinology, 76: 369, 1965.

3. Florini, J.R., G. Tonelli, C.B. Breuer, J. Coppola, I. Ringler and P.H. Bell. Endocrinology, 79: 692, 1966.

4. Catt, K.J., B. Moffat and H.D. Niall. Science, 157: 321, 1967.

5. Riggi, S.J., C.R. Boshart, P.H. Bell and I. Ringler. Endocrinology, 79: 709, 1966.

6. Grumbach, M.M. and S.L. Kaplan. Annals N.Y. Acad. of Sciences. In press.

7. Beck, P. and W.H. Daughaday. J. Clin. Invest. 46:103, 1967.

8. Sciarra, J.J. Clin. Obst. and Gynec. 10: 132, 1967.

9. Sciarra, J.J., S.L. Kaplan and M.M. Grumbach. Nature, 199: 1005, 1963.

10. Mancini, G., A.O. Carbonora and J.F. Heremans. Immunochemistry, 2: 235, 1965.

11. Friesen, H. Nature, 208: 1214, 1965.

12. Lowry, O.H., N.J. Rosebrough, A.L. Farr and R.J. Randall. J. Biol. Chem., 193: 265, 1951.

13. Barrett, R.J., H. Friesen and E.B. Astwood. J. Biol. Chem. 237: 432, 1962.

14. Siebert, G. and G. Stark. Klin. Wchsft., 32: 31/32 732, 1954.

15. Winick, M., A. Coscia and A. Noble. Pediatrics, 39: 248, 1967.

16. Cohen, H., M.M. Grumbach and S.L. Kaplan. Proc. Soc. Exp. Biol. and Med., 117: 438, 1964.

# The Biological Characterization of Purified Placental Protein (Human) [PPP(H)]

I. Ringler*

Lederle Laboratories

Pearl River, New York

Though certain of the hormonal changes occurring in the human female during pregnancy have been attributed to the placenta, the endocrine functions of this organ remain inadequately understood. Ehrhardt (1) in 1936 was first to demonstrate prolactin activity in extracts of human placenta. Subsequent evidence had been advanced by Japanese workers that the human placenta contained a protein which had both prolactin-like (2,3,4,5,6) and growth hormone-like (7,8,9) properties. The observations by Josimovich and MacLaren (10) that a partially purified substance from human placenta, designated "human placental lactogen", possessed immunological similarities to human pituitary growth hormone aroused widespread interest. Grumbach and co-workers (11,12,13) and Friesen (14,15) have confirmed the presence and immunological properties of a similar protein, the former workers designating the material "human chorionic growth hormone-prolactin". The protein isolated by Friesen has been termed "placental protein". Thus, the existence and physiochemical characteristics of material related to growth hormone in human placenta appear well established. The biological properties of the various preparations from placenta have been examined to some extent, but investigations have been limited by the relatively small quantities available. Recently, an intense effort was directed to establishing the growth-promoting efficacy of this newly isolated protein (16).

Body weight increments of hypophysectomized rats receiving

---

* The data on PPP(H) represent the collaborative effort of the following: J. R. Florini, G. Tonelli, C. B. Breuer, J. Coppola, S. J. Riggi, C. R. Boshart, and P. H. Bell.

Table 1

CUMULATIVE BODY WEIGHT GAINS OF HYPOPHYSECTOMIZED IMMATURE FEMALE RATS AFTER 7 OR 14 CONSECUTIVE DAILY SUBCUTANEOUS INJECTIONS

| TREATMENT | DOSE mg | NO. OF ANIMALS | WEIGHT GAINS (g/ANIMAL) | |
|---|---|---|---|---|
| | | | 7 DAYS | 14 DAYS |
| SALINE | - | 18 | 3.5 ± 0.4† | 8.6 ± 0.6 |
| PPP(H) | 0.25 | 8 | 4.1 ± 1.1 | 11.9 ± 2.0 |
| | 1.0 | 8 | 3.9 ± 0.6 | 11.0 ± 1.2 |
| | 4.0 | 18 | 5.9 ± 0.7 | 14.1 ± 1.0* |
| | 16.0 | 18 | 6.3 ± 0.8 | 16.1 ± 1.6* |
| | 32.0 | 10 | 8.1 ± 0.4* | 15.1 ± 2.3* |
| PORCINE GROWTH HORMONE (SIGMA) | 0.025 | 8 | 8.0 ± 1.6* | 14.3 ± 1.7* |
| | 0.050 | 8 | 9.0 ± 1.2* | 14.6 ± 2.0* |
| | 0.100 | 8 | 10.5 ± 0.7* | 16.2 ± 1.1* |
| | 0.300 | 10 | 7.8 ± 1.0* | 19.7 ± 2.8* |

Average initial body weight was 59 ± 0.9g.

* Significantly different from control, $p \leq 0.05$

† Mean ± standard error

PPP(H)[1,2] once a day for 7 or 14 consecutive days are illustrated in Table 1. Only the 32 mg dose elicited significant increases in body weight gain after 7 injections; however, the fourteen day responses to 4 and 16 mg also were statistically significant. Porcine growth hormone (PGH) enhanced weight increases after 7 and 14 injections at all doses. In comparing the effects of PPP(H) and PGH, it was apparent that the former was appreciably less efficacious in stimulating body weight gain. Similar results were obtained in adult

---

1 Thus far, to the extent that our observations include the same parameters as those published, PPP(H) appears to be identical with the "human placental lactogen", "chorionic growth hormone-prolactin" and "placental protein".

2 All biological studies were conducted with preparations containing approximately 60 to 70 percent PPP(H).

Table 2

CUMULATIVE BODY WEIGHT GAINS OF ADULT FEMALE MICE
AFTER 7 OR 14 CONSECUTIVE DAILY SUBCUTANEOUS DOSES

| TREATMENT | DOSE mg/ANIMAL | NO. OF ANIMALS | WEIGHT GAINS (g/ANIMAL) | |
|---|---|---|---|---|
| | | | 7 DAYS | 14 DAYS |
| SALINE | - | 34 | 1.1 | 1.2 |
| PPP(H) | 3.25 | 24 | 1.8 | 2.3 |
| | 7.5 | 34 | 1.5 | 1.7 |
| | 15.0 | 34 | 2.9** | 3.5** |
| | 30.0 | 34 | 3.4** | 3.7** |
| BOVINE GROWTH HORMONE (ARMOUR) | .075 | 24 | 2.2* | 2.8** |
| | .150 | 24 | 1.9 | 2.8** |
| | .300 | 24 | 1.7 | 3.0** |
| | AVERAGE STANDARD DEVIATION | | 1.3 | 1.5 |

Average initial body weight was 26 $\pm$ 1.5 g
Significantly different from control - *$p \leq 0.05$; **$p \leq 0.01$

female mice treated subcutaneously with either PPP(H) or bovine growth hormone (Table 2). At 3.25 and 7.5 mg PPP(H)/animal/day for 7 or 14 consecutive days, body weight gains did not differ significantly from those of the control animals. Significant increases in body weight gains occurred after 7 or 14 daily injections of 15 or 30 mg. Again, as was observed in the previous experiment, the body weight increments were poorly dose-related. Significant body weight increases were evinced 14 days after administration of bovine growth hormone. Substantial weight gains have been recorded also in hypophysectomized adult female rats and intact adult male and female rats receiving 1-32 mg of PPP(H) (16).

Administration of PPP(H) to hypophysectomized immature rats for periods up to 28 days resulted in a dose-related stimulation of body weight gains, with a greater rate of gain evident during the first 14 days of treatment (Figure 1). Subsequently, the rate of gain decreased tangibly. This has been a consistent observation in our studies, perhaps attributable to the formation of antibodies to PPP(H). When sera from rats receiving 16 mg PPP(H)/rat/day for 21 days were examined by micro-immunodiffusion assay against PPP(H),

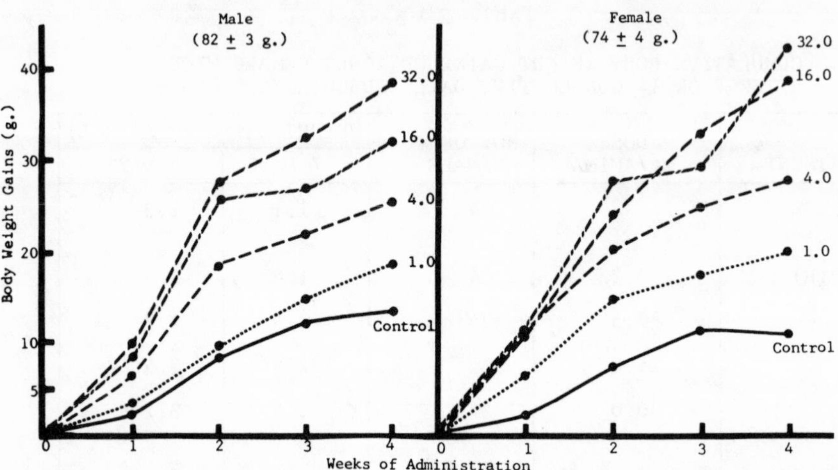


Figure 1. Cumulative Body Weight Gains of Immature Hypophysectomized Rats Following Daily Subcutaneous Administration (mg/animal/day) of PPP(H)

precipitin bands were formed in a number of samples. In addition, a dose-related increase in spleen weights of PPP(H) treated rats and mice has been observed (16). Riggi, et al. (17) also have reported splenomegaly in rabbits following administration of PPP(H).

Though it has been suggested that growth hormone affects nitrogen metabolism by decreasing protein catabolism, it appears more appropriate that the principal action is that of stimulating protein biosynthesis. To substantiate the hypothesis that the stimulation of body weight gain observed in animals treated with PPP(H) was ascribable to an increase in the biosynthesis of muscle protein, the effects of PPP(H) and PGH on the yield and activity of ribosomes isolated from muscle were studied (Table 3). Activity of ribosomes was measured as the transfer of tritiated leucine from sRNA into protein. Hypophysectomy reduced and administration of PGH or PPP(H) restored both the yield and the activity of ribosomes isolated from rat thigh muscle. The product of the yield (micrograms ribosomal RNA/rat) and the activity (dpm tritiated leucine transferred/mg ribosomal RNA incubated) of ribosomes gave an indication of the total protein synthetic capacity of the thigh muscles. This capacity was strikingly increased by the administration of either PPP(H) or PGH to hypophysectomized rats.

Various biological assays with PPP(H) have indicated, in most instances, a marked correspondence with the actions of pituitary somatotropin. In addition to stimulating protein and RNA biosyntheses in skeletal muscle, PPP(H) has been demonstrated to promote

Table 3

EFFECTS OF PPP(H) AND PGH ADMINISTRATION ON YIELD AND ACTIVITY OF RIBOSOMES ISOLATED FROM RAT SKELETAL MUSCLE

| TREATMENT (5 DAYS) | | RIBOSOMES | | TOTAL PROTEIN SYNTHETIC ACTIVITY dpm/RAT |
|---|---|---|---|---|
| mg/rat/day (NO. HYPOX. RATS) | PROTEIN | YIELD† μg RNA/RAT | ACTIVITY* dpm/mg RNA | |
| - (11) | - | 128 | $1.41 \times 10^5$ | $1.80 \times 10^4$ |
| 4 (11) | PPP(H) | 179 | 2.68 | 4.80 |
| 16 (12) | PPP(H) | 442 | 6.49 | 28.7 |
| 0.2 (12) | PGH | 504 | 7.85 | 39.6 |

* Mean of triplicate assays of the transfer of $^3$H-leucine from sRNA into protein using ribosomes prepared from pooled thigh muscle from all rats in each experimental group.

† Expressed as the RNA content of the ribosomes determined from the absorbancy of the ribosome suspension at 260 mμ using an extinction coefficient of 20 cm$^2$/mg.

the incorporation of tritiated thymidine into the DNA of rib cartilage (C. Breuer, personal communication). Costal rib cartilage segments were incubated with labeled DNA precursor according to the procedure of Daughaday and Reeder (18). Four injections of 2-8 mg of PPP(H), spaced over a period of 48 hours, in hypophysectomized immature male rats stimulated DNA synthesis markedly over that of controls. The anabolic significance of the body weight gains was further supported by observations that radioactive sulfate uptake into rib cartilage _in vitro_ (12,16) and _in vivo_ (19) was stimulated by injection of placental protein.

Moreover, PPP(H) administration to hypophysectomized immature male rats has been shown to increase the urinary excretion of hydroxyproline (C. Breuer, personal communication). Hydroxyproline was measured by a modification of the method of Neuman and Logan (20); a consistent 2-3 fold increase in 24 hr excretion over that of controls has been observed in rats receiving 5-12 mg of PPP(H)/day for periods of 4-14 days.

Increase in the width of the tibial epiphysis of the hypophysectomized rat has been found to furnish the basis of one of the better quantitative assays for growth hormone activity. Ikeda _et al._ (7) and Fukushima (8) reported the presence of a tibia test positive substance in human placenta; moreover, Otsuka and Minesita (9) obtained affirmative results in a comparable assay by administering large amounts of a placental extract. In some assays, Friesen (15) found that 1 mg or more of placental protein caused a definite but limited increase in the width of the tibial epiphyseal cartilage of hypophysectomized rats. With placental material prepared by Raben's method for the extraction of human pituitary growth hormone, Kaplan and Grumbach (12) showed that 1 mg/day elicited an equivocal response in the tibia test, whereas Josimovich and MacLaren (10, 21) concluded that human placental lactogen possessed little or no direct growth hormone-like activity in this assay. We too have been unable to demonstrate an unequivocal response in widening of the epiphyseal cartilage in hypophysectomized rats. Doses of PPP(H) from 7.5 to 30 mg/day for 14 days did not elicit a positive response in this bioassay. Moreover, animals treated for 4 days with 4-40 mg/day showed no widening of the tibial plate. Thus, doses of PPP (H) which have been demonstrated to elicit a significant increase in body weight gain of hypophysectomized rats have not, in our hands, proved effective in producing a widening of the tibial epiphyseal cartilage. This anachronism in biological responsiveness to PPP(H) is one deserving of further investigation.

The potentiation of human growth hormone (HGH) action by co-administration of PPP(H) is one of the more unique properties of the placental protein. Josimovich and associates (21,22) demonstrated that human placental lactogen, which alone had no effect, caused a striking increase in the response (tibia test) of hypophysectomized rats treated with HGH. A similar stimulation of HGH action on body weight gain in hypophysectomized immature rats by PPP(H) is apparent from the data presented in Table 4. Although the doses of HGH in this experiment were not extensive enough to allow plotting of a dose-response curve, it was notable that 0.25 mg PPP(H) given with 2.5 μg HGH caused a similar body weight gain as 10 μg HGH alone, suggesting perhaps a fourfold potentiation of HGH. A comparable degree of potentiation of HGH (tibia test) by human placental lactogen was reported by Josimovich and Atwood (22). Whereas the data in Table 4 suggest a possible potentiation between PPP(H) and bovine growth hormone, further studies are necessary to substantiate this action. Josimovich (21) was unable to observe potentiation when bovine growth hormone was substituted for HGH, or when bovine pituitary prolactin was substituted for human placental lactogen. Co-administration of bovine serum albumin with bovine growth hormone failed to potentiate the action of the latter agent.

Potentiation of other human growth hormone-stimulated systems include (a) lipolysis from isolated rat fat cells (23), and; (b) partial prevention of insulin-induced hypoglycemia in normal fasted and alloxan-diabetic rats (21). The mechanism of potentiation of growth hormone by PPP(H) cannot be determined on the basis of the data now available. Competition between PPP(H) and HGH for anti-

Table 4

SPECIFICITY OF THE POTENTIATION BY PPP(H) OF GROWTH HORMONE EFFECT ON BODY WEIGHT GAINS IN HYPOPHYSECTOMIZED IMMATURE MALE RATS*

| GROWTH HORMONE | | PPP(H) (mg/DAY) | | |
|---|---|---|---|---|
| μg/DAY | SPECIES | 0 | 0.25 | 1.0 |
| 0 | - | 9.8 ± 0.5 (10)** | 9.6 ± 0.6 (10) | 12.5 ± 1.6 (8) |
| 2.5 | HUMAN | 11.2 ± 0.5 (9) | 14.4 ± 0.7 (9) | 15.4 ± 1.5 (8) |
| 10 | HUMAN | 14.2 ± 1.4 (9) | 16.6 ± 1.0 (9) | 22.6 ± 2.0 (10) |
| 40 | HUMAN | 25.3 ± 1.4 (10) | - | - |
| 2.5 | BOVINE | 14.8 ± 1.6 (10) | 19.0 ± 1.3 (10) | 18.5 ± 1.7 (9) |
| 10 | BOVINE | 21.8 ± 1.2 (9) | 23.2 ± 1.3 (10) | 26.3 ± 1.1 (8) |

* Charles River strain, 50-70 g initial body weight

**Mean body weight gain (g/rat/14 days) ± standard error (number of rats)

body or binding sites associated with metabolic pathways or for excretion mechanisms are obvious possibilities. Josimovich and Atwood (22) also have suggested that human placental lactogen may synergize with chorionic gonadotropin in promotion and maintenance of the gestational function of the corpus luteum.

The prolactin-like activities of human placental extracts are well documented (2,3,4,5,6,10,12,15,16). The lactogenic potency of PPP(H) was assessed by the local intradermal pigeon crop assay (conducted by the Endocrine Laboratories, Madison, Wisconsin). Responses of animals treated with ovine prolactin 0.02-0.08 mg and PPP(H) are plotted in Figure 2. Approximately 35 μg of PPP(H) elicited a lactogenic response comparable to 5 μg of NIH sheep prolactin. Josimovich and Brande (24) estimated that preparations of human placental lactogen possessed 75 percent the potency of NIH sheep prolactin. They state that Raben HGH, in their assay, was only about 20 percent as potent as ovine prolactin. The lactogenic activity of preparations of Ito and Higashi (4,5) was approximately 5-10 percent that of NIH sheep prolactin. By the systemic assay in adult pigeons, the placental protein of Friesen (15) had a reported potency of 1.4 units/mg compared with 0.25-0.5 units/mg by the local intradermal method in juvenile pigeons. Milk production, induced by intraductal injection of lactogenic substances into pseudopregnant rabbit mammaries, was elicited by the placental peptide at a potency roughly 50 percent that of NIH sheep prolactin (10). Additional positive effects in a comparable test have been

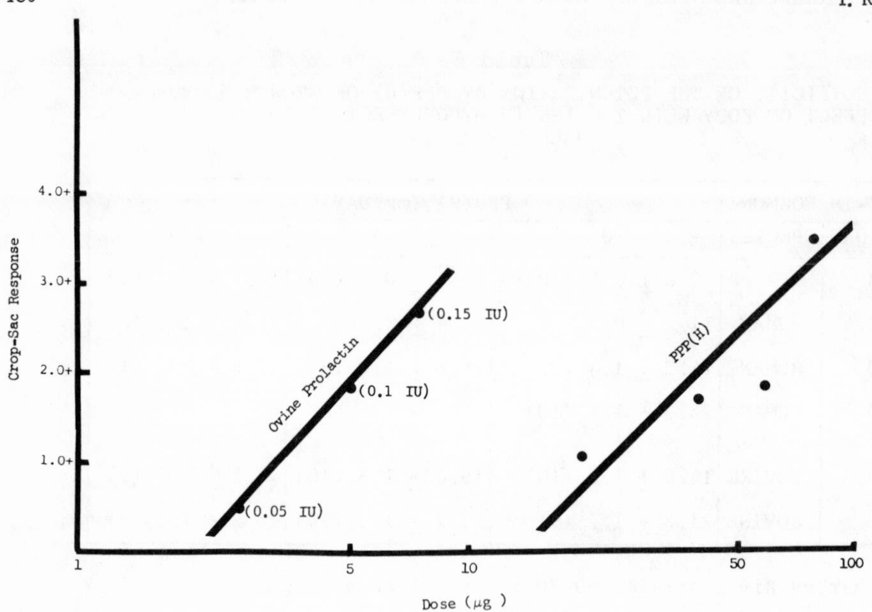


Figure 2. Lactogenic Assay of PPP(H)

documented (15). Friesen (25) also has reported that subcutaneously administered placental lactogen produced striking mammary gland development and milk secretion in pregnant rabbits.

The capacity of human placental protein, as well as human growth hormone and ovine prolactin, to evoke a luteotropic response has been demonstrated by the maintenance of the induced decidual response in experimental animals (16,19,26,27). The data in Table 5 illustrate that PPP(H) at 4.0 mg/kg was comparable to 10 IU (approximately 1.5 mg/kg of sheep LtH), but 20 mg/kg of PPP(H) caused about twice as much increase in uterine weight as 50 IU of LtH. The estimated potency of PPP(H) was roughly 30-80 percent that of LtH. The observations of Josimovich, et al. (26) and Kovacic (27) also suggest that placental peptide was less potent than LtH. The luteotropic effects of human placental lactogen could be completely inhibited by preincubation with an antiserum to human growth hormone (26).

Unfortunately, limitations of space do not allow a complete discussion of the numerous actions recorded for protein extracts of human placenta. Other biological functions responsive to placental protein are: elevation of plasma free fatty acid concentrations in fasting rabbits and monkeys (17); increased myocardial and rectus femoris glycogen concentrations in hypophysectomized rats (17); stimulated biosynthesis (in vitro) of $^{32}$P-labeled

Table 5

LUTEOTROPIC ACTIVITY OF PPP(H)

| TREATMENT | DAILY DOSE (5 RATS/GROUP) | OVARIAN WEIGHT (mg) | WEIGHT UTERINE HORNS (mg) | |
|---|---|---|---|---|
| | | | CONTROL | TRAUMATIZED |
| SHEEP LtH (NUTRITIONAL BIOCHEM. CORP. - 30 IU/mg) | 10 IU | 92 ± 13* | 153 ± 30 | 278 ± 70 |
| | 50 IU | 66 ± 7 | 135 ± 22 | 538 ± 460 |
| PPP(H) | 0.8 mg/kg | 59 ± 11 | 139 ± 6 | 174 ± 15 |
| | 4.0 mg/kg | 69 ± 8 | 189 ± 56 | 232 ± 56 |
| | 20.0 mg/kg | 56 ± 18 | 125 ± 4 | 1165 ± 94 |
| | 100 mg/kg† | 56 ± 22 | 118 ± 2 | 1253 ± 200 |

* Mean ± standard deviation

† Four rats

casein in presence of insulin and hydrocortisone (28); stimulated incorporation of tritiated thymidine into total DNA by mouse mammary explants (28); increased serum ketone bodies in fasting rats (14); stimulated uptake of U-glucose-$^{14}$C, oxidation to $CO_2$ and its conversion into fat by epididymal fat *in vitro* (14).

Though previous reports have left in doubt the growth promoting potential of the various placental extracts which react with human growth hormone antiserum, the aforementioned data affirmatively establish the anabolic actions of PPP(H); moreover, it is equally evident that the present preparations of PPP(H) are appreciably less potent than pituitary growth hormone.

## ACKNOWLEDGMENTS

I wish to express my gratitude to H. H. Bird, G. M. Campbell, E. Heyder, E. Lindemann, and J. G. Patterson for careful technical assistance, and to A. L. Jensen and F. M. Clarke for the ample supplies of PPP(H). I am indebted to the Editors of Endocrinology for permission to reproduce Figure 1 and portions of the tabular data.

## REFERENCES

1. Ehrhardt, K., Munschen. Med. Wschr. 29: 1163, 1936.
2. Ito, Y., Yakugaku Zassi (Jap.) 73: 1, 1953.
3. Kurosaki, M., Tohoku J. Exp. Med. 75: 122, 1961.
4. Ito, Y. and K. Higashi, Endocr. Jap. 8: 279, 1961.
5. Higashi, K., Endocrin. Jap. 8: 288, 1961.
6. Higashi, K., Endocrin. Jap. 9: 1, 1962.

7. Ikeda, K., A. Tanaka and T. Minesita, Shionogi Kenkyusho Nempo 7: 429, 1957.
8. Fukushima, M., Tohoku J. Exp. Med. 74: 161, 1961.
9. Otsuka, H. and T. Minesita, Folia Endocrin. Jap. 39: 897, 1964.
10. Josimovich, J. B. and J. A. MacLaren, Endocrinology 71: 209, 1962.
11. Sciarra, J. J., S. L. Kaplan and M. M. Grumbach, Nature (London) 199: 1005, 1963.
12. Kaplan, S. L. and M. M. Grumbach, J. Clin. Endocrin. 24: 80, 1964.
13. Grumbach, M. M. and S. L. Kaplan, Trans. N. Y. Acad. Sci. 27: 167, 1964.
14. Friesen, H., Endocrinology 76: 369, 1965.
15. Friesen, H., Nature (London) 208: 1214, 1965.
16. Florini, J. R., G. Tonelli, C. B. Breuer, J. Coppola, I. Ringler and P. H. Bell, Endocrinology 79: 692, 1966.
17. Riggi, S. J., C. R. Boshart, P. Bell and I. Ringler, Endocrinology 79: 709, 1966.
18. Daughaday, W. H. and C. Reeder, J. Lab. Clin. Med. 68: 357, 1966.
19. Franchimont, P., Ann. Endocrin. (Paris) 26: 346, 1965.
20. Neuman, R. E. and M. A. Logan, J. Biol. Chem. 184: 299, 1950.
21. Josimovich, J. B., Endocrinology 78: 707, 1966.
22. Josimovich, J. B. and B. L. Atwood, Amer. J. Obstet. Gynec. 88: 867, 1964.
23. Turtle, J. R., P. Beck and W. H. Daughaday, Endocrinology 79: 187, 1966.
24. Josimovich, J. B. and B. L. Brande, Trans. N. Y. Acad. Sci. 27: 161, 1964.
25. Friesen, H. B., Endocrinology 79: 212, 1966.
26. Josimovich, J. B., B. L. Atwood and D. A. Goss, Endocrinology 73: 410, 1963.
27. Kovacic, N., J. Endocrin. 35: XXV, 1966.
28. Turkington, R. W. and Y. J. Topper, Endocrinology 79: 175, 1966.

# CALCITONIN

Iain MacIntyre

Professor of Endocrine Chemistry, Joint Director, Endocrine Unit, Royal Postgraduate Medical School, Ducane Road, London, W. 12

## Existence and Origin of Calcitonin

It is now generally agreed that calcitonin is present in most mammals and that in these species it is of thyroid origin. Its identity with thyrocalcitonin is also generally accepted (Foster, Baghdiantz, Kumar, Slack and Soliman and MacIntyre, 1964; Copp, personal communication).

Pearse and colleagues (Pearse, 1966; Bussolati and Pearse, 1967; Pearse and Carvalheira, 1967) showed that the thyroid C cells secreted calcitonin and that these cells were of ultimobranchial origin. It has now been found that calcitonin can be extracted from the ultimobranchial body in chickens (Copp, 1968; Matthews, Moseley, Breed and MacIntyre, unpublished) and pigeon (Matthews et al).

## Purification of the Hormone

Calcitonin has now been obtained in very highly purified form (MacIntyre, 1967). It is a polypeptide with a molecular weight near 4,000.

## Mode of Action

Calcitonin is active in the rat in doses of less than 0.05 μg. It lowers plasma calcium by the inhibition of bone resorption (Robinson, Martin, Matthews and MacIntyre, 1967). When given for several weeks to parathyroidectomised rats it increases the radiodensity of tail vertebrae (Foster, Doyle, Bordier and Matrajt, 1966a).

## Effect in Man

Calcitonin is effective in diminishing hypercalcaemia due to secondary malignancy in bone (Foster, Joplin, MacIntyre, Melvin and Slack, 1966b). There is some evidence that it acts on the skeleton in osteoporosis. In my view it is now important that detailed clinical trials be carried out to see if it is useful in the chronic treatment of this condition.

## Conclusion

There is no doubt that former views of calcium metabolism will have to be modified. I have presented a more detailed review of the whole field elsewhere (MacIntyre, 1967; MacIntyre, 1968).

## REFERENCES

Bussolati, G. and Pearse, A.G.E. (1967). "Immunocluorescent localisation of aclcitonin in the "C" cells of pig and dog thyroid. J. Endocr., 37, 205.

Copp, D.H. (1968). "Development of the calcitonin concept - a decade in perspective." In Calcitonin : Symposium on Thyrocalcitonin and the C Cells. Edited by S. Taylor, London, Heinemann Medical Books, in press.

Foster, G.V., Baghdiantz, A., Kumar, M.A., Slack, E., Soliman, H.A. and MacIntyre, I. (1964). "Thyroid origin of calcitonin." Nature, Lond., 202, 1303.

Foster, G.V., Doyle, F.H., Bordier, P. and Matrajt, H. (1966 a) "Effect of Thyrocalcitonin on Bone". Lancet, ii, 1428.

Foster, G.V., Joplin, G.F., MacIntyre, I., Melvin, K.E.W., and Slack, E. (1966 b). "Effect of thyrocalcitonin in man." Lancet, i, 107.

MacIntyre, I. (1967). "Calcitonin : a review of its discovery and an account of purification and action." Proc. Roy. Soc. B., in press.

MacIntyre, I. (1968). "Calcitonin : An Introductory Review." In Calcitonin : Symposium on Thyrocalcitonin and the C Cells. Edited by S. Taylor, London, Heinemann Medical Books, in press.

Pearse, A.G.E. (1966). "The cytochemistry of the thyroid C cells and their relationship to calcitonin." Proc. Roy. Soc. B., 164, 478.

Pearse, A.G.E. and Carvalheira, A.F. (1967). "Cytochemical evidence for an ultimobranchial origin of rodent thyroid C cells." Nature, Lond. 214, 929.

# BIOCHEMICAL ASPECTS OF HYPOTENSIVE PEPTIDE ACTION

N. Back, and H. Wilkens

Department of Biochemical Pharmacology, School of Pharmacy, State University of New York at Buffalo

Kinins have been classified as hypotensive peptides resembling bradykinin in both structure and pharmacologic activity (1). At present three kinins have been isolated from human plasma, differing from each other chemically by an additional terminal lysine or methionine group, Fig. 1. The great general interest in the chemistry, pharmacology, and patho-physiologic significance of these peptides is reflected in recent extensive reviews (2-4) and symposia related to the topic (5-7).

The present study critically re-examines the mechanism of kinin release and explores the vascular action of kinin in the presence of other chemical mediators.

Kinins are formed by the enzymatic action of a specific protease kallikrein on a plasma substrate kininogen, Fig. 2. More

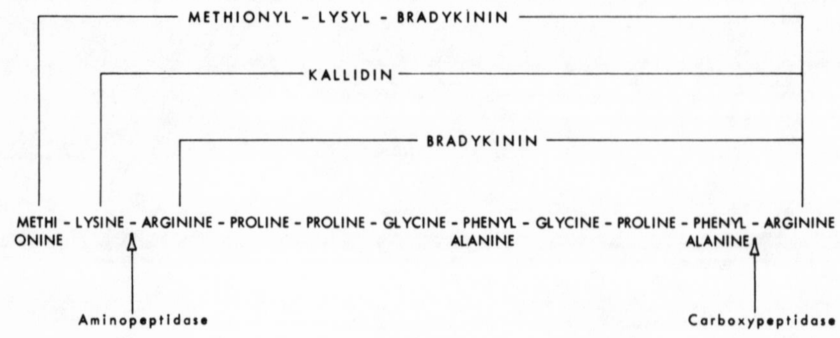


Figure 1. Chemical structure of kinins isolated from plasma.

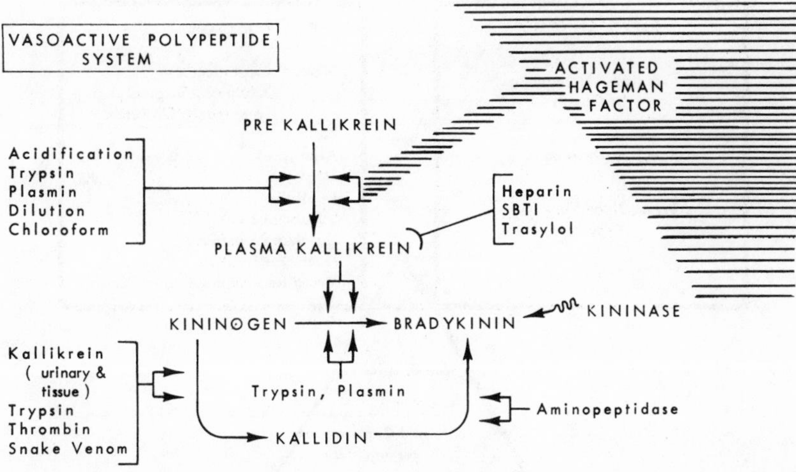


Figure 2. Mechanism of kinin formation by kallikrein and other kinin releasing enzymes.

general proteases as trypsin (8), and plasmin (9) also exhibit kinin-releasing activity. Kallikrein arises from an inactive enzyme precursor, prekallikrein, present in plasma and tissue fluids via such mechanisms as pH changes (10), enzymic proteolysis, and activated Hageman factor (11). The substrate kininogen, abundantly present in the alpha-2 globulin fraction of plasma and in tissues, yields several vasoactive kinins depending on the kinin-releasing enzyme involved. Both bradykinin and kallidin are inactivated by a kininase which splits the phenylalanine-arginine bond. A plasma aminopeptidase is capable of converting kallidin to bradykinin (12).

Nanogram concentrations of kinin cause hypotension, increase capillary permeability, contract isolated smooth muscles, and produce pain, Fig. 3. In view of these actions, kinins have been implicated as local biochemical mediators of functional vasodilation not only in such specific organs as the salivary glands, but throughout most body tissues. Immunofluorescent studies in our laboratory (13) using the indirect technique of Coons (14), Fig. 4, have demonstrated the presence and specific cellular localization of kallikrein, kininogen, and kallikrein inhibitor in such diverse tissues as the pancreas, salivary glands, small intestine, tongue, and kidney. Localization of kallikrein and kallikrein tissue inhibitor in the parotid gland of the dog are seen in microphotographs, Fig. 5a, b. Furthermore, our observation

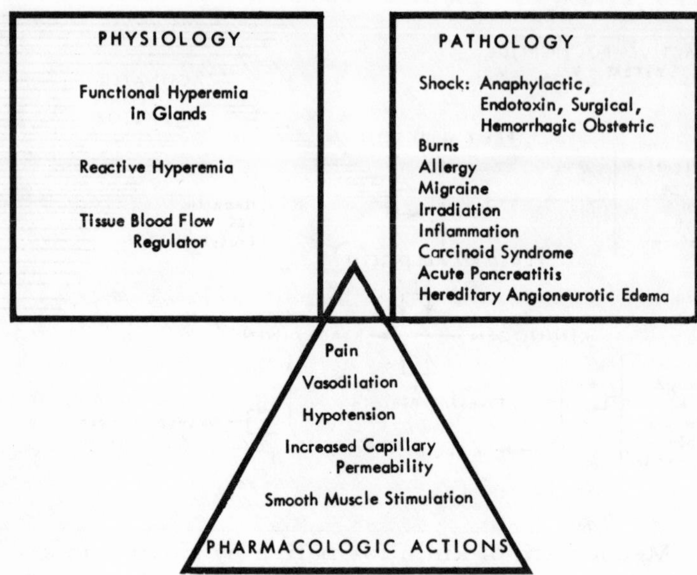


Figure 3. Pharmacologic actions and possible involvement of kinins in physiology and pathology.

on the vascular behavior of canine hind limbs made ischemic by a tourniquet supports the earlier finding of Frey, et al (10) that slight decreases in tissue pH activate kinin-releasing enzymes.

Kinins, under normal physiologic circumstances, may be characterized as local mediators since they not only are released

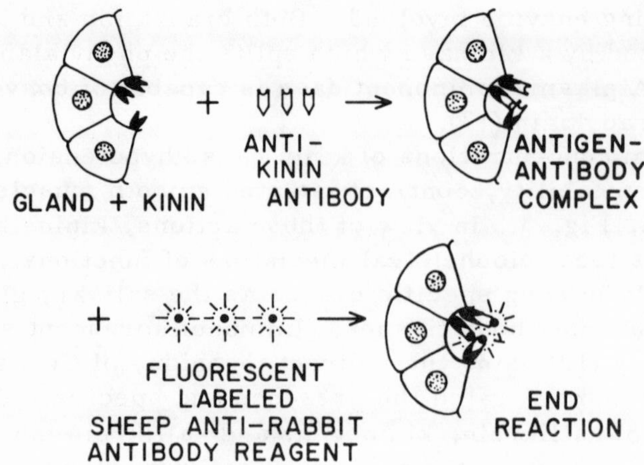


Figure 4. Indirect immuno-fluorescent technique for study of localization of components of the kallikrein-kinin system.

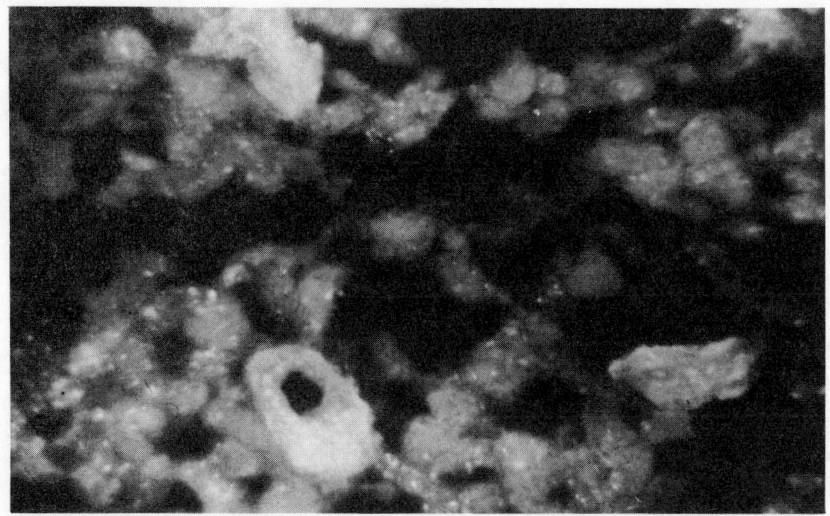

Figure 5a. Localization of plasma kallikrein in the parotid gland of the dog (450 x).

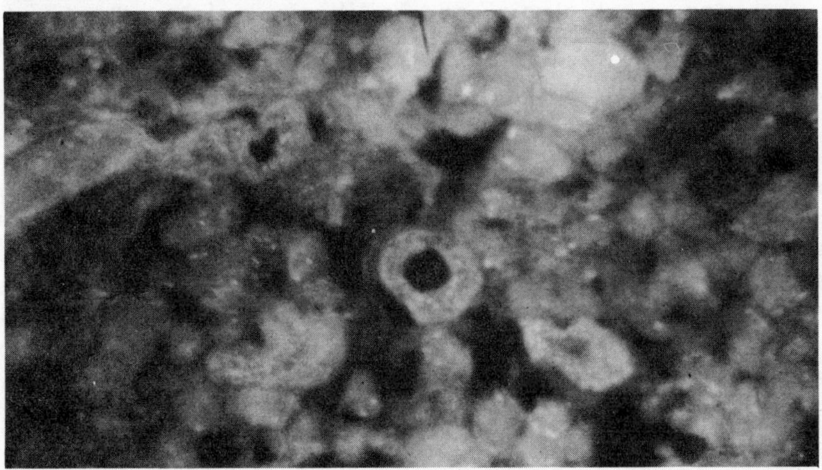

Figure 5b. Localization of tissue kallikrein inhibitor in the parotid gland of the dog (450 x).

and destroyed locally where they act but also are removed selectively from the circulation by passage through the lungs (15), thereby preventing their recirculation. However, generalized high plasma kinin levels (hyperkininemia) may occur as a result of systemic acidosis or following excessive activation of the

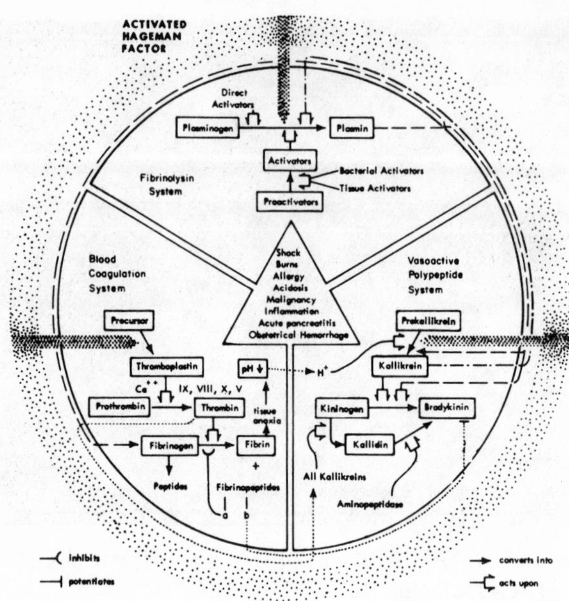


Figure 6. Proposed interrelationship amongst the blood coagulation, fibrinolysin, and kallikrein-kinin systems and pathologic states presumed to involve components of the 3 systems.

fibrinolysin and blood coagulation systems (16). A possible interrelationship amongst these three systems is proposed, Fig. 6.

Hageman factor may serve as a common activator of all three systems, involving the formation of plasmin, kallikrein and thrombin. Plasmin may release kinin directly from kininogen or indirectly by a kallikrein activation mechanism. In addition to releasing kinin, plasmin as well as thrombin act on fibrinogen to liberate fibrinopeptides that interfere with fibrinogen-fibrin conversion (17, 18). Plasmin also liberates a fibrinopeptide that potentiates kinin action on smooth muscle (19). In addition, thrombin may act as a kinin-releasing enzyme. Blood pH changes due to ischemia resulting from vascular occlusion may activate prekallikrein to kallikrein. Plasma kallikrein also is capable of activating plasminogen to plasmin, thereby establishing a cycle involving the fibrinolysin system.

Clinical conditions assumed to involve components of the three systems are listed in the triangle, Fig. 6. By virtue of their hypotensive action, the kinins have been assigned a cardinal role in some of these conditions, notably acute pancreatitis, allergy and shock due to various causes. Results obtained in our most recent studies have led to a reconsideration of some current views related to the pathological role of kinin. It will be shown that under certain circumstances, kinin raises rather than lowers the blood pressure in various vascular beds.

To study the vascular effects of kinin alone and in combination with other vasoactive mediators, notably histamine, isolated whole blood perfusions of canine hind limb, head, and kidney were established. Oxygenated heparinized (5 u/ml) blood was pumped by means of a Sigma motor peristalic pump into the cannulated femoral, carotid, or renal arteries and venous outflow collected in a specially-designed artificial oxygenator operating on the blood-over-foam principle (20). A constant pump output and a strain guage placed between the pump and the perfused organ allowed the continuous monitoring of changes in peripheral resistance, Fig. 7.

Intraarterial injection of as many as 15 consecutive doses of synthetic bradykinin, 20-200 micrograms per 600 mls. of perfusing blood, given over a period of 6 hours, caused repeatedly a fall in blood pressure, Fig. 8. The fall was immediate and dose-related. Return of blood pressure to normal pre-treatment

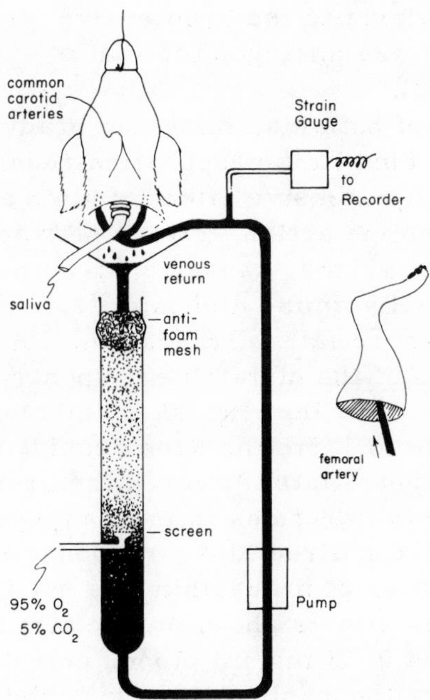


Figure 7. Pump-oxygenator apparatus used for perfusion of isolated organ.

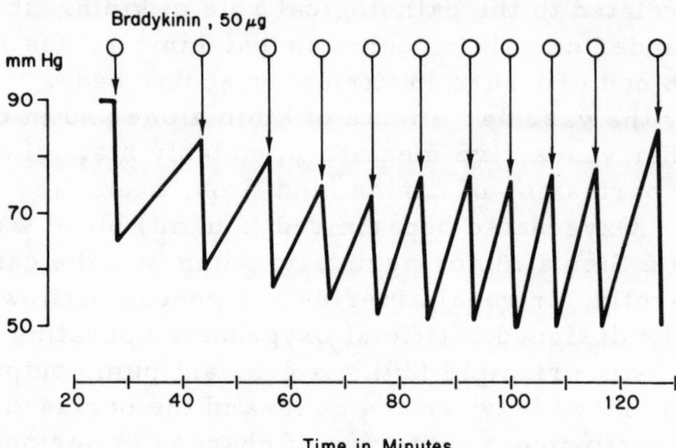


Figure 8. Effect of repeated intraarterial injections of bradykinin on the blood pressure in the perfused isolated hind limb of the dog.

levels in the hind limb occured within 10-20 minutes while return in the head preparation generally was more prolonged. Successive histamine injections produced similar responses. The absence of tachyphylaxis to the hypotensive effect of both bradykinin and histamine was analogous to that observed by us in the whole animal (21, 22).

Administration of both histamine and bradykinin successively into these perfused hind limb preparation resulted in an unexpected reversal of the hypotensive effect of kinin and a significant attentuation of histamine action, Fig. 9. Two initial doses of bradykinin, 20 micrograms, caused a 35-50 mm Hg drop in arterial pressure. Histamine diphosphate, 1 mg of the salt, caused a blood pressure fall. Bradykinin, 100 micrograms, injected at the heighth of the histamine response caused a marked blood pressure rise of 57 mm Hg. A 20 microgram dose of bradykinin gave rise to 1/5th the hypertensive effect. Histamine administered 10 minutes later lowered the pressure only 5 mm Hg. Successive kinin injections in increasing doses from 20-100 micrograms revealed a direct dose-response relationship. That the hypertensive action of bradykinin was not related to adrenergic receptor mechanisms is shown by the fact that phenoxybenzamine, in a dose of 0.25 mg/ml blood, completely abolished the epinephrine action without interfering with kinin hypertension.

Similar results were obtained in 5 other hind limb perfusions as well as in 3 head preparations, an example of which is seen in

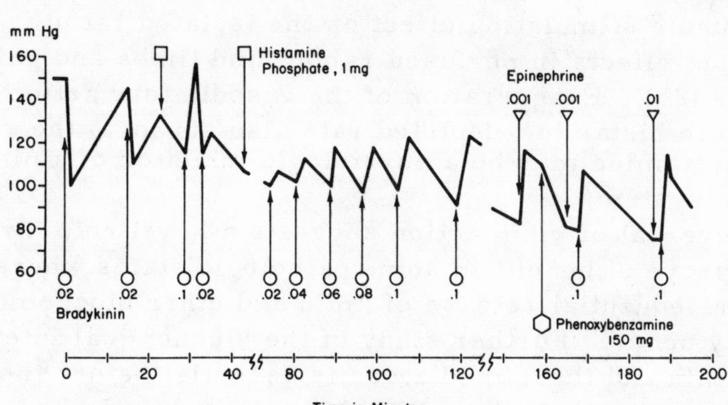


Figure 9. Reversal of bradykinin action in the perfused isolated hind limb of the dog following administration of histamine. Reactivity of the vascular system is confirmed by use of epinephrine and phenoxybenzamine.

Fig. 10. In only 1 of 4 perfused kidneys was kinin reversal after histamine noted, possibly due to the high histaminase content of the kidney (23).

The data indicate that the ordinary vascular action of both kinin and histamine can be altered or reversed when these mediators are present together. This agrees broadly with previous reports on the interaction between these two mediators relative

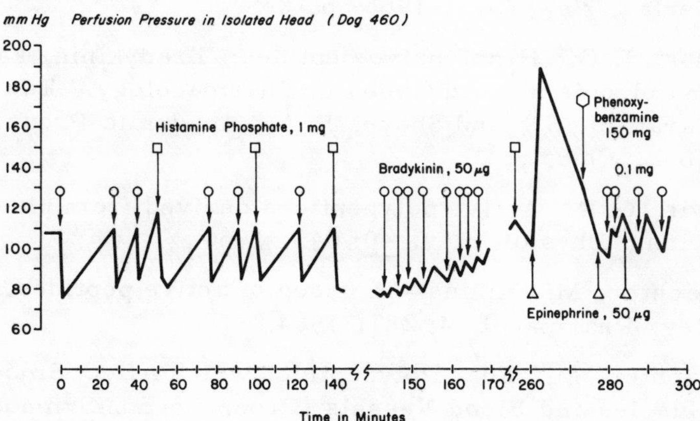


Figure 10. Reversal of bradykinin action in the perfused isolated head of the dog following administration of histamine. Reactivity of the vascular system is confirmed by use of epinephrine and phenoxybenzamine.

to their apneic and bronchoconstrictor effect in the dog (21), smooth muscle stimulating effect on the isolated rat uterus (24) and vascular effects in perfused rabbit hind limbs and guinea pig capillaries (25). Exaggeration of the vasodilatory action of bradykinin in histamine-depleted rats also supports the suggestion that histamine may be a physiologic inhibitor of kinin in certain tissue systems.

The reversal of kinin action suggests a novel role for kinin that may prove of benefit in some pathologic states where simultaneous or sequential release of kinin and other biochemical mediators may occur. Further study of the biochemical interactions between kinin and these mediators is essential before the overall physiologic and pathologic role of kinin is understood.

Acknowledgement

We are indebted to A. Weiss, La Salle College, for his assistance during the tenure of Research Participation Program Fellowship, supported by a NSF grant #Gy-2733 to the Roswell Park Memorial Institute, Buffalo, New York.

## REFERENCES

1. Webster, M.E. Report of the committee on nomenclature for hypotensive peptides in Hypotensive Peptides. Eds. Erdos, E.G., Back, N., and Sicuteri, F. Springer - Verlag, New York, 1966, p. 648.

2. Erdos, E.G. Hypotensive peptides: Bradykinin, kallidin, and eledoisin in Advances in Pharmacology. Eds. Garattini, S., and Shore, P.A. Academic Press. New York, 1966, p. 1.

3. Lewis, G.P. Active polypeptides derived from plasma proteins. Physiol. Rev. 40: 647, 1960.

4. Schachter, M. Kinins - A group of active peptides. Ann. Rev. Pharmacol. 4: 281, 1964.

5. Schachter, M. Ed. Polypeptides which Affect Smooth Muscles and Blood Vessels (Proc. London Symposium, 1959). Pergamon Press. New York, 1960.

6. Erdos, E.G. Ed. Structure and Function of Biologically Active Peptides: Bradykinin, Kallidin, and Congeners.

Ann. N. Y. Acad. Sci. 104: 1963. (Proc., New York Symposium, 1962).

7. Erdos, E. G., Back, N., and Sicuteri, F. Eds. Hypotensive Peptides (Proc. Florence Symposium, 1965) Springer-Verlag, New York, 1966.

8. Rocha e Silva, M., Beraldo, W. T., and Rosenfeld, G. Bradykinin, hypotensive and smooth muscle stimulating factor released from plasma globulin by snake venoms and by trypsin. Amer. J. Physiol. 156: 261, 1949.

9. Back, N., and Steger, R. Activation of bovine bradykininogen by human plasmin. Life Sciences, 4: 153, 1965.

10. Frey, E.K., Kraut, H., and Werle, E. Kallikrein Padutin. Enke, Stuttgart, 1950.

11. Margolis, J. The interrelationship of coagulation of plasma and release of peptides. Ann. N. Y. Acad. Sci. 104: 133, 1963.

12. Webster, M. E., and Pierce, J. V. The nature of the kallidins released from human plasma by kallikreins and other enzymes. Ann. N. Y. Acad. Sci. 104: 91, 1963.

13. Back, N., Tsukada, G. A., Castilone, B., and Aungst, C. W. Distribution of kallikrein and bradykinin in mammalian tissue. The Pharmacologist, 7: 151, 1965.

14. Coons, A. H., and Kaplan, M. H. Localization of antigen in tissue cells. II. Improvements in a method for the detection of antigen by means of fluorescent antibody. J. Exper Med. 91: 1, 1950.

15. Ferreira, S. H., and Vane, J. R. The disappearance of bradykinin and eledoisin in the circulation and vascular beds of the cat. Br. J. Pharmac. Chemotherap. 30: 417, 1967.

16. Back, N. Fibrinolysin system and vasoactive kinins. Fed. Proc. 25: 77, 1966.

17. Triantaphyllopoulos, D. C. Enzymatic effects of fibrinolysin. Fed. Proc. 24: 800, 1965.

18. Laki, K. Enzymatic effects of thrombin. Fed. Proc. 24: 794, 1965.

19. Gladner, J. A., Murtaugh, P. M., Folk, J. E., and Laki, K. Nature of peptides released by thrombin. Ann. N. Y. Acad. Sci. 104: 47, 1963.

20. Waud, R.A. A mechanical heart and lung. Canadian J. Med. Sc. 30: 130, 1952.

21. Wilkens, H., and Back, N. Bronchoconstriction and apnea in canine anaphylaxis: Role of histamine and plasma-kinins. Fed. Proc. 26: 785, 1967.

22. Back, N., Guth, P.S., and Munson, A.E. On the relationship between plasmin and kinin. Ann. N.Y. Acad. Sci. 104: 53, 1963.

23. Shayer, R.W. Catabolism of histamine in vivo. Handbook of Experimental Pharmacology, 18: 672, 1966.

24. Back, N., and Steger, R. Unpublished data.

25. Copley, A.L., and Tsuluca, V. Antagonistic action of urokinase to bradykinin and plasminogen-induced capillary permeability. Nature, 197: 294, 1963.

# PHARMACOLOGY OF ANGIOTENSIN ANTAGONISTS

T. GODFRAIND

Laboratory of General Pharmacodynamics

University of Louvain, Belgium

Angiotensin appears to be, on a molar basis, one of the most potent naturally occuring substances ; the minimal dose which e-vokes contraction of isolated smooth muscle is 100 to 1.000 times lower than that of drugs like acetylcholine or adrenaline. The formation of angiotensin results from the activation of the renin-angiotensin system occuring in physiological regulation mechanisms or in pathological states (24, 33).

## ACTIONS OF ANGIOTENSIN

Actions of angiotensin are mimetic of the motor actions of either acetylcholine or catecholamines. As discussed below, such actions are due to a direct effect of the peptide on the effector organ or to an indirect action through stimulation of the motor nerves or liberation of chemical transmitters at nerve endings. A brief description of these actions shall be given here.

Heart. On isolated preparations, the observation made in 1940 by Hill and Andrus, that natural angiotensin had a positive inotropic action on isolated cat heart, has been confirmed with synthetic peptide, not only on isolated heart (3) but also on isolated ventricular and atrial muscle (3, 21, 28). The chronotropic effect of angiotensin shows varying results depending on animal species. A positive chronotropic action has been observed with isolated rabbit, or guinea-pig heart (3, 28) but not with isolated kitten, dog or cat heart (16, 22). On whole animals, compensator mechanisms and actions of the peptide on coronary vessels leads to slowing of heart rate and decrease in cardiac output, mainly with high doses (7, 28).

Vascular smooth muscle. Angiotensin evokes a contraction of isolated vascular smooth muscle (20). This effect increases the total peripheral resistance (9, 15) and the regional circulation resistance (lungs, skin, muscle, kidney, splanchnic area and liver, cerebral circulation, for references, see 33).

Other smooth muscles. Angiotensin evokes also a contraction of several smooth muscles including guinea-pig ileum (18) longitudinal smooth muscle isolated from guinea - pig ileum (13, 14), rat duodenum (36), uterus (18).

Central nervous system. The existence of a central mechanism in angiotensin induced hypertension was first described by Bickerton and Buckley who have postulated its relation with norepinephrine action as central neurotransmitter. Recently it has been demonstrated by Palaïc and Khairallah that angiotensin reduces uptake of $^3$H-norepinephrine by rat brain.

Neuro-vegetative ganglia. The ganglion stimulating properties of angiotensin have been extensively studied by Lewis and Reit on the superior cervical ganglion of the cat. Other sympathetic ganglia are stimulated by the peptide (Turker, personnal communication). There are strong evidences to believe that the indirect action on guinea-pig ileum is due to stimulation of intramural ganglia (14).

Adrenal medulla. Angiotensin is a powerful releasing agent of catecholamines from the adrenal medulla as reported by Feldberg and Lewis and by other authors (see 33).

Nerve endings. Several reports describe the releasing action of angiotensin on catecholamine stores in various tissues (10, 30). Direct demonstration of a "cocaine-like" action of angiotensin has been recently published(30). It must nevertheless be pointed out that most of successful experiments have been done with rats and that, in similar conditions, angiotensin does not exhibit similar action in other species (39). It has recently been claimed(31) that angiotensin could liberate acetylcholine from cholinergic nerve endings, but as no increase in resting output of acetylcholine can be demonstrated, there is a need for further experimental facts.

Renal action. Angiotensin acts on renal function modifying pressure in renal artery and tubular reabsorption of sodium (26, 41). Effect on diuresis is variable in different animals. In the rabbit or rat, diuresis and increased excretion of electrolytes are the rule (34, 35). In normal man, intravenous infusion produces an antidiuresis and a reduction in excretion of electrolytes (6, 9). On the contrary, in man with hypertension, angiotensin causes diuresis with natriuresis (8). This difference suggests that in normal man, angiotensin exerts a powerful effect on renal vessels so that the direct tubular action would be masked.

Aldosterone secretion. One of the most specific actions of angiotensin is the stimulation of production of aldosterone which occurs in man (2, 23) or dog (29). As aldosterone action is delayed for half an hour, renal effects due to this action are easily dissociated from direct renal action.

## NATURE OF ANGIOTENSIN ACTION

The action of angiotensin on whole subjects appears to be the sum of a direct action of the peptide on effector organs and of the action of liberated mediators or hormones. From a pharmacological point of view, the most interesting actions to be studied are the direct ones as the indirect ones can be blocked using specific antagonists for liberated substances. It appears therefore that smooth muscle preparations are the most convenient for this study. It must here be pointed out that the action of angiotensin on vascular smooth muscle is responsible for many of the pathological, pharmacological or physiological effects of the peptide (33).

We shall therefore analyze some of the properties of angiotensin using isolated smooth muscle preparations. These preparations are the longitudinal smooth muscle of the guinea-pig ileum, the rat duodenum, the rat uterus and the rabbit aorta. On these preparations, action of the peptide is unequivocally a direct one as it is not blocked by specific antagonists for chemical transmitters (see Antagonists for Direct Actions).

Shape of smooth muscle response to angiotensin. It has been reported by Godfraind and al. that the shape of the guinea-pig ileum response to angiotensin depends on the dose of the peptide. Doses lower than $10^{-9}$ M cause a progressive increase in contraction, the maximum of which is reached after 1 min. 30 sec. With higher doses, a composite response is obtained which may be divided into two parts, an initial and fast rise with a small and transient subsequent fall called the fast component and a second and progressive increase of contraction called the slow component. The fast component has been demonstrated to be due to the stimulation of cholinergic structures and the slow component to the direct action of the peptide on smooth muscle. This observation illustrates the fact that, when compared with that of acetylcholine, the action of angiotensin on a smooth muscle needs a longer time to reach its maximum.

This slow onset of angiotensin action is associated with a slow offset of its action. When, after the full development of the contraction, the perfusion fluid containing the peptide is changed for fresh solution, relaxation of the smooth muscle occurs very slowly. We have observed that this relaxation can be accelerated by serum containing angiotensinase which, as described by Khairallah and al. (20), destroys the peptide remaining on tissue.

Auto-potentiation and auto-inhibition. The response of smooth muscle to angiotensin is influenced by a preceeding administration of the peptide. The first description of this fact is that of tachyphylaxis resulting from repeated administration of angiotensin (20, 33).

The analysis of the dose-effect curve of angiotensin on the longitudinal smooth muscle of the guinea-pig ileum has shown that maximal response produced by the peptide is always lower than that produced by histamine (14). Furthermore, this dose-effect curve is a curve with an optimum that is characteristic of partial agonists. We have also observed a similar curve with other smooth muscles.

As tachyphylaxis due to angiotensin can be prevented when the time between injections is sufficiently long in duration (14) or when the dissociation of the complex receptor - peptide is accelerated (20), it may be concluded that tachyphylaxis is due to residually bound peptide. We therefore prefer the term auto-inhibition as the reduction of response is due to angiotensin itself and not to a persistent modification of muscular tissue. It must be noted that angiotensin tachyphylaxis on guinea-pig ileum is partially specific in that angiotensin renders the smooth muscle insensitive to itself but furthermore reduces its sensitivity to bradykinin and in a lower extent to histamine ; in these conditions, action of acetylcholine is potentiated.

In some conditions, a preceeding administration of the peptide is followed by a phase of auto-potentiation. We have first observed this phenomenon on experiments performed with rat uterus. The contractile response of this tissue to angiotensin was modified after administration of a high dose such as $2 \times 10^{-6}$ M in such a way that the response to low doses was potentiated and the response to high doses was reduced. This action has been attributed to residually bound peptide as it was abolished after continous washing for 1 hour. A similar phenomenon has been observed on intestinal smooth muscle : the direct response of guinea-pig ileum to successive administrations of a low dose of angiotensin (such as $2 \times 10^{-11}$ M) is first increasing, later decreasing. This auto-potentiation causes the surprising configuration of the foot of the dose - effect curve of angiotensin on guinea-pig ileum which was described by Godfraind and al.

If we postulate that the interaction of angiotensin with its receptor is similar to that of other stimulating drugs (1), we may represent it by

$$\text{Ang} + \text{R} \underset{k_2}{\overset{k_1}{\rightleftharpoons}} \text{R Ang} \rightarrow \text{response.}$$

As angiotensin is a partial agonist, according to rate theory predictions (32), k2 of angiotensin ought to be lower than k2 of histamine or of an other agonist. The dissociation of R - Ang may be

Table I. Affinity constants and intrinsic activity constants of angiotensin.

| Preparation | $pD_2$ | Intrinsic activity (efficacity) |
|---|---|---|
| Guinea-pig ileum (direct) | 8.52 | 0.55 |
| Rat colon | 8.88 | 0.77 |
| Rat uterus | 8.70 | 1.00 |
| Rabbit thoracic aorta | 8.55 | 0.60 |
| Human pulmonary artery | 8.47 | 0.60 |

accelerated if free angiotensin concentration in the biophase is kept near zero, i.e. that reaction is becoming metastatic. Such a situation occurs in several conditions such as continous washing, presence of angiotensinase or of suspension of Dowex 50 (20, unpublished results). These factors, which accelerate peptide dissociation from its receptor, abolish auto - potentiation and auto - inhibition.

Affinity and intrinsic activity. Values of $pD_2$ and of intrinsic activity estimated according to Ariëns and Van Rossum are reported on table I. $pD_2$ constants do not differ significantly from one tissue to an other. Intrinsic activity constants are lower than 1,with exception for rat uterus. On that preparation, auto-inhibition is not so marked than on other tissues and high doses of angiotensin do not evoke a fading response. Therefore, this finding is not in contradiction with the conclusion that angiotensin is a partial agonist.

## NATURE OF ANGIOTENSIN ANTAGONISTS

If we consider now the drugs which antagonize angiotensin action we are facing with two distinct problems. The first question to be discussed is that concerning the action of antagonists for indirect actions of angiotensin, the second one, is dealing with that of antagonists for direct actions of the peptide.

### Antagonists for Indirect Actions

α-receptors blocking agents have been reported to reduce vascular action of angiotensin (4,17), but not in all conditions (37,27). An interesting observation was made by Scroop and Whelan which reported that the vasoconstrictor action of angiotensin on the hand vessels was almost abolished by phenoxybenzamine when angiotensin was administered intravenously, but was unnafected when it was given into the brachial artery. As the effect of antagonist was depending on the mode of administration of the peptide it is likely that vasoconstriction was mainly due to indirect action when angiotensin was given intravenously. This experiment shows that extreme caution must be taken when any interpretation concerning

Table II. Action of antagonists on response of guinea - pig ileum to various agonists.

| Antagonist | Agonist | $pA_2$ | $pA_h$ |
|---|---|---|---|
| Lidoflazine (1) | Angiotensin (direct) | 9.3 | 7.6 |
| | Histamine | 7.8 | 6.2 |
| | Acetylcholine | 7.44 | 7 |
| | Bradykinin | 7.14 | 6.46 |
| Atropine | Angiotensin (direct) | 9 | <4 |
| | Acetylcholine (2) | 8.8 | competitive |
| Cinnarizine | Angiotensin (direct) | 8.15 | 6.95 |
| | Histamine | 8.1 | 6.8 |
| | Acetylcholine | 6.7 | 6.1 |
| Mepyramine | Angiotensin (direct) | 5 | 4 |
| | Histamine (2) | 9.3 | competitive |

(1) : see ref. (14) (2) : see ref. (2)

the mode of angiotensin action is drawn from results obtained with antagonists.

An other important factor to be considered is the mode of recording tissue response, mainly with preparations in which both direct and indirect actions are the rule. A good illustration of this fact is given by the action of atropine on the guinea-pig ileum response to angiotensin. If experiments with atropine are performed recording mainly the indirect response, an important reduction shall be observed, but recording of the two responses shows that only indirect action is affected (14).

## Antagonists for Direct Actions

Specificity of antagonists. At present, very few drugs are known as antagonists for direct actions of angiotensin. Most of them belong to a series of piperazine derivatives originally synthetized in P.A.J. Janssen's laboratory with exception for mepyramine which was shown by Lewis and Reit to reduce the stimulating action of angiotensin on the superior cervical sympathetic ganglion of cats. Furthermore, we have also found that Pimozide, a potent neuroleptic (Janssen, personnal communication), reduces angiotensin action. Lidoflazine is a specific antagonist for angiotensin action on guinea-pig ileum smooth muscle (14). From analysis of lidoflazine action, it was concluded that although specific, this drug is a non-competitive antagonist of angiotensin. Atropine reduced also the direct action of angiotensin on guinea-pig ileum, but it behaved as an incomplete blocker : a reduction of 50 % of the maximum response to angiotensin has not been obtained with concentration such as $10^{-4}$ M. Cinnarizine, one of the first drugs reported as angiotensin antagonist (38, 43) exhibits powerful effect against angiotensin action on guinea - pig ileum,

Table III. Comparison of $pA_h$ values measured on various preparations.

| Antagonist | Agonist : Angiotensin | | | Acetylcholine | |
|---|---|---|---|---|---|
| | Rat colon | Rat uterus | Rabbit aorta | Rat colon | Rat uterus |
| Lidoflazine | 6.47 | 6.41 | inactive | 6.22 | 6.58 |
| Cinnarizine | 6.58 | 6.78 | <4.00 | 5.90 | 6.78 |
| R 7943 | 7.40 | - | inactive | 7.35 | - |
| R 7427 | 6.20 | 7.37 | 5.00 | 5.08 | 6.58 |
| R 7702 | 6.19 | - | <4.00 | 5.34 | - |
| Pimozide | 6.60 | - | 4.80 | 5.62 | - |

| | |
|---|---|
| Lidoflazine | 1-[4,4-di-(4-fluoro-phenyl)-butyl]-4-[2,6-dimethyl-anilinocarbonyl-methyl]-piperazine. |
| Cinnarizine | 1-benzhydryl-4-cinnamylpiperazine-dihydrochloride (trans). |
| R 7943 | 1-[4,4-di-(fluoro-phenyl)-butyl]-4-[2-(2,6-dimethyl-anilino)-ethyl]-piperazine trihydrochloride. |
| R 7427 | 1-[4,4-di-(4-fluoro-phenyl)-butyl]-4-[2-(N-ethyl-anilino)-ethyl]-piperazine trihydrochloride. |
| R 7702 | 1-[4,4-di-(4-fluoro-phenyl)-butyl]-4-[2-(N-propyl-anilino)-ethyl]-piperazine trihydrochloride. |
| Pimozide | 1-[4,4-di-(4-fluoro-phenyl)-butyl]-4-(2-oxo-1-benzimidazolinyl)-piperidine. |

but $pA_2$ values are similar with angiotensin or histamine as agonists (table II).

Mepyramine which is active on nervous ganglia does not exhibit any specific action on guinea-pig ileum (table II). This observation has suggested to us that angiotensin receptors could be different according to the considered effector cell. It has thus seemed desirable to design experiments in order to test the action of antagonists on various preparations. Therefore, not only guinea-pig ileum, but also rat uterus, rat aorta, rat colon , guinea-pig and rabbit aorta have been choiced for biological assay. Rat colon appeared to us as an interesting preparation because, as only sensitive to acetylcholine and angiotensin, it gave the opportunity to select drugs acting specifically on angiotensin receptors. A great number of compounds belonging to lidoflazine series have been tested in order to compare their activities as angiotensin and acetylcholine antagonists. All of them appeared to act as non-competitive antagonists for the two agonists. Therefor, their $pA_h$ values were compared, few of them are listed under table III.

If we postulate that a difference of at least 1 $pA_h$ unity (i.e. a ratio of activity of 10) indicates an action on specific receptor, it appears that R 7427 and Pimozide are the most specific antagonists of angiotensin on rat colon, and that lidoflazine has no specific affinity for angiotensin receptors of rat colon. As lidoflazine has a great affinity for angiotensin receptors of the guinea-pig ileum, we concluded that structure of receptors were different according to the two considered tissues. It was therefore desirable to test some of these compounds on vascular smooth muscle as it was postulated by Regoli and Vane that receptors are similar in rat colon and in vessels. Results of experiments performed with rabbit aorta are reported on table III. It must be first noted that phentolamine does not modify the response to angiotensin. This result disagrees with Schümann and Güther's findings but was confirmed by Khairallah (personnal communication). Lidoflazine was also inactive on angiotensin response of rabbit aorta. R 7427 and Pimozide behaved as non-competitive antagonists for angiotensin. Cinnarizine and R 7702 reduced also angiotensin response, but R 7943 was inactive on rabbit aorta. It appeared therefore that only compounds which were specific or semi-specific as angiotensin antagonists when tested on rat colon reduced angiotensin action on rabbit aorta.

On rat uterus, the specificity of these compounds is similar to that found on rat colon, but their activity is higher.

Only cinnarizine and lidoflazine have been tested as angiotensin antagonists on whole animals. Lidoflazine did not modify hypertension due to intravenous injection of the peptide. Reit (personnal communication) has also failed to demonstrate any action of lidoflazine on cervical sympathetic ganglia of cats but Cinnarizine reduced the action of angiotensin on blood pressure (38).

Receptor protection by angiotensin. During the course of our work on angiotensin antagonists which was planned in order to avoid bias due to auto-inhibition, we were impressed by a puzzling fact which was the resistance of some preparations to antagonists action, although the time of equilibration with antagonist was very long (90 min.). As measurements of time of muscle relaxation after washing out showed also variable values from one preparation to an other, the hypothesis was made that disappearance of angiotensin from the biophase - due not only to dissociation from receptor and diffusion in the bath, but also to action of tissue angiotensinase (20)-was variable from one preparation to an other. Experiments were therefore designed to test lidoflazine action in conditions in which residually bound angiotensin was present or absent. It was observed that action of lidoflazine is impaired by residually bound peptide. Protection of receptor by the peptide must therefore be taken in account when angiotensin antagonists are to be tested.

Non reversibility of lidoflazine action. In experiments designed to test the reversibility of lidoflazine action, it has been observed that its action was only partially reversible. This non-reversibility of the antagonism was obtained with angiotensin and histamine and appeared to be due to an accumulation of the drug at non-specific sites. Uptake experiments with $H_3$-lidoflazine have shown that ileal smooth muscle concentrated it approximately 200 times after an incubation of 90 min. This finding is in agreement with the very high partition coefficient organic solvent/water for lidoflazine which appears to be a limiting factor for the rediffusion of this drug in physiological solution.

## CONCLUSIONS

Experimental results here related show that in order to study angiotensin antagonists, it is necessary to take in account some particularities of angiotensin pharmacology, namely the fact that this peptide behaves as a partial agonist evoking auto-potentiation or auto-inhibition. Residually bound peptide exerts a protection of receptor against antagonist.

The analysis of action of antagonists, made on various preparations, has shown variations according to the considered effector cell. This observation suggests that angiotensin receptors could present some organ specificity.

## REFERENCES

1. Ariëns, E. J. Molecular Pharmacology, Academic Press. New-York, 1964.
2. Arunlakshana, O & Schild, H. O. Brit. J. Pharmacol. 1959, 14, 48.
3. Bianchi, A., De Schaepdryver, A. F., De Vleeschhouwer, G. R. & Preziosi, P. Arch. Int. Pharmacodyn. 1960, 124, 21.
4. Bickerton, R. K. & Buckley, J. P. Proc. Soc. exp. Biol. N. Y. 1961, 106, 834.
5. Biron, P., Chretien, M., Koiw, E. & Genest, J. Brit. Med. J. 1962, 1, 1569.
6. Bock, K. D. & Krecke, H. J. Klin. Wschr. 1958, 36, 60.
7. Bock, K. D. & Gross, F. Arch. exp. Path. Pharmak. 1961, 242, 188.
8. Brown, J. J. & Peart, W. S. Clin. Sci. 1962, 22, 1.
9. De Bono, E., Lee, G. De J. Mottram, F. R., Pickering, W., Brown, J. J., Keen, H., Peart, W. S. & Sanderson, P. H. Clin. Sci. 1963, 25, 123.
10. Distler, A., Liebau, H. & Wolff, H. P. Nature, 1965, 207, 764.
11. Feldberg, W. & Lewis, G. P. J. Physiol. 1964, 171, 98.
12. Gelfand, S. & Ganz, A. Experientia, 1966, 22, 1.
13. Godfraind, T., Kaba, A. & Polster, P. Arch. Int. Pharmacodyn. 1966, 163, 226.
14. Godfraind, T., Kaba, A. & Polster, P. Brit. J. Pharmacol. 1966, 28, 93.

15. Gross, F., Cock, K.D. & Turrian, H. Helv. Physiol. Acta. 1961, 19, 42.
16. Hill, W.H. & Andrus, E.C. J. Exp. Med. 1941, 74, 91.
17. Johnsson, G., Henning, M. & Ablad, B. Life Sci. 1965, 4, 1549.
18. Khairallah, P.A. & Page, I.H. Am. J. Physiol. 1961, 200, 51.
19. Khairallah, P.A. & Page, I.H. Ann. N.Y. Acad. Sci. 1964, 104, 212.
20. Khairallah, H.A., Page, I.H., Bumpus, M.F. & Turker, R.K. Circ. Res. 1966, 19, 245.
21. Koch-Weser, J. Circ. Res. 1964, 14, 337.
22. Koch-Weser, J. Circ. Res. 1965, 16, 230.
23. Laragh, J.H., Angers, M., Kelly, W.G. & Lieberman, S. J. Am. Med. Ass. 1960, 174, 234.
24. Laragh, J.H. Fed. Proc. 1967, 26, 39.
25. Lewis, G.P. & Reit, E. J. Physiol. 1965, 179, 538.
26. Leyssac, P.P. Fed. Proc. 1967, 26, 55.
27. McGiff, J.C. & Fasy, T.M. J. Clin. Invest. 1965, 44, 1911.
28. Meier, R., Tripod, J. & Studer, H. Arch. Int. Pharmacodyn. 1958, 117, 185.
29. Mulrow, P.J., Ganong, W.F., Cera, G. & Kuljian, E. J. Clin. Invest. 1962, 41, 505.
30. Palaïc, D. & Khairallah, P.A. Biochem. Pharmacol. in press.
31. Panisset, J.C. Can. J. Physiol. Pharmac. 1967, 45, 313.
32. Paton, W.D.M. Proc. R. Soc. 1961, 154, 21.
33. Peart, W.S. Pharmacol. Rev. 1965, 17, 143.
34. Peters, G. Proc. Soc. Exp. Biol. N.Y. 1963, 112, 771.
35. Pickering, G.W. & Prinzmetal, M. J. Physiol. 1940, 98, 314.
36. Regoli, D. & Vane, J.R. Brit. J. Pharmacol. 1964, 23, 351.
37. Renson, J., Barac, G. & Bacq, Z.M. C.R. Soc. Biol. (Paris) 1959, 153, 1621.
38. Schaper, W.K.A., Jageneau, A.H.M., Xhonneux, R. Van Nueten, J. & Janssen, P.A.J. Life Sci. 1963, 12, 963.
39. Schümann, H.J. & Güther, W.N. Schmied. Arch. Pharmak. u. exp. Path. 1967, 256, 169.
40. Scroop, G.C. & Whelan, R.F. Clin. Sci. 1966, 30, 79.
41. Tobian, L. Fed. Proc. 1967, 26, 48.
42. Turker, R.K. & Kayaalp, S.O. Experientia, 1967, 23, 647.
43. Van Nueten, J., Dresse, A., Dony, J. C.R. Soc. Biol. (Paris) 1964, 158, 1750.
44. Van Rossum, J.M. Advances in Drug Research - p. 189. Edit. N.J. Harper and A.B. Simmonds. Academic Press - New-York, 1966.

# EFFECTS OF ANGIOTENSIN ON Na AND K EXCHANGES IN BLOOD VESSELS

Sydney M. Friedman and Constance L. Friedman

Department of Anatomy, The University of British Columbia, Vancouver, Canada

It is by now reasonably well established that the distribution of sodium in blood vessels is intimately and actively related to the regulation of their tension. Our understanding of the processes involved, however, has been considerably handicapped by our lack of precise information concerning the localization of sodium and potassium in vascular tissue in general and in arteries in particular. This is especially true with regard to the role of angiotensin which appears to be complex in its actions. We have accordingly devoted a good deal of attention to characterizing the essential features of the distribution of sodium and potassium in blood vessels and will first describe some of our findings as a foundation for interpreting our observations with angiotensin.

To begin with, vascular tissue is especially rich in sodium, containing at least five times as much of this ion as does skeletal muscle. A good deal of this is contained in an especially large, free extracellular fluid volume, commonly measured with inulin or similar markers. Although there may be some minor dispute about the relative merits of one or other yardstick for measuring the exact size of this compartment, this is only a matter of degree. Regardless of the technique, all agree that the free extracellular fluid readily accounts for about half of the sodium burden of the vessel. The principal concern is with the remaining half, often loosely spoken of as "cellular" but more accurately located not only in cells but in the paracellular collagen and mucopolysaccharide ground substance as well.

In what follows, we shall show that it is possible to characterize and to distinguish the truly cellular from the paracellular components of both sodium and potassium and, further, to show how these may be related to peripheral vascular resistance. Although our findings are with inulin measurements, essentially similar results would be obtained with any of the other markers currently in use. Thus, at the start, we need make no assumptions concerning the precise localization of inulin, it being sufficient to our purposes that it defines two compartments - one which it enters, and one from which it is excluded.

Two basic paracellular components of Na may be distinguished in the phase which inulin does not enter, one temperature-sensitive the other not. These components are mobilized directly upon exposure of the tissue to a medium in which NaCl is replaced by an uncharged substitute such as lactose (Fig. 1). The actual amount mobilized is directly related to the concentration of Na in the medium and more is removed at 37°C than at 2°C; in neither case is any associated movement of potassium observed. The duration of exposure is not particularly important since the process is rapid. We classify these components as paracellular since not only are they mobilized within the short time it takes for the new medium to diffuse through the tissue but they are not affected by metabolic inhibitors. These paracellular components are thus probably trapped and bound by the abundant polyelectrolytes of the paracellular matrix. Together, the Na of the free extracellular fluid and of these paracellular components can account for as much as 95% of the total Na in the arterial wall. The remaining amount, which we shall show to be in cells, is of the same order of magnitude as that found in skeletal muscle.

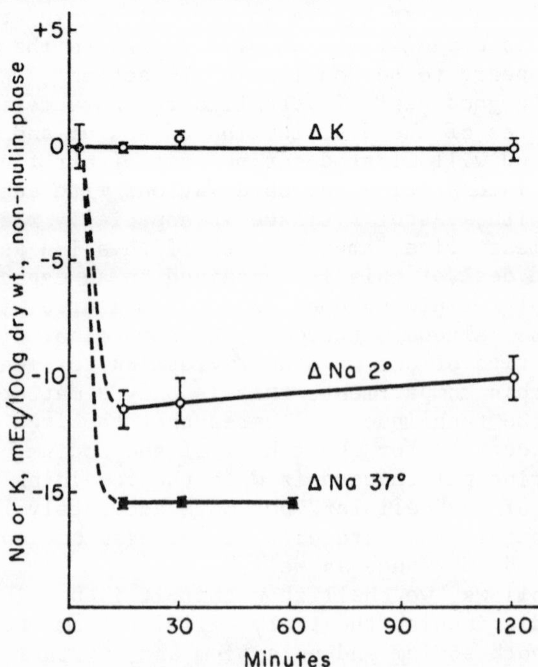


Fig. 1. Na and K content of the inulin-inaccessible phase of rat tail artery following incubation in NaCl-free medium.

We classify the bulk of the tissue potassium together with the residual (approximately 5%) sodium described above as, in all probability, truly cellular since they exchange for one another in a one-to-one coupling and are dependent upon the metabolic activity of the cell. We first observed this relation in studies of the aorta (Table 1). Here we found that the inulin-inaccessible phase of carefully excised, viable samples of rat aorta gained Na in exchange for K during cooling and reversed the process during rewarming (1). This seemingly coupled relation matched the general description for transcellular exchanges of Na and K in other tissues even to the observation that water tended to move with Na. In studies with the rat tail artery we have observed that time is basic to the process (Fig. 2). Other workers have, like us, noted that the exchange ratio often seems more like 1.2 or 1.3 to 1, in favour of Na. Our findings indicate, however, that the curve of Na efflux during rewarming can be resolved into two components, the first rapid, unrelated to K, and essentially complete within 15 minutes of rewarming; the second slow and obviously coupled in a one-to-one ratio with K. The first component is paracellular since it not only is unrelated to cell activity on the evidence of the speed of reaction but is also unaffected by metabolic inhibitors. It presumably is related to the same temperature-dependent paracellular component we have described earlier. The one-to-one coupling of the transcellular exchanges can be clearly elicited by rewarming previously cooled artery samples in the presence of metabolic inhibitors (Fig. 3).

| | $H_2O$ | Na | K |
|---|---|---|---|
| Fresh aorta | 102± 6 | 7.2±1.1 | 12.8±0.56 |
| Incubated 18 hrs 2°C + 3 hrs aeration 2°C | 130±16 | 19.0±2.4 | 1.4±0.36 |
| △ | +18 | +11.8 | -11.4 |
| Incubated 18 hrs 2°C + 3 hrs aeration 37°C | 113± 5 | 11.9±0.7 | 8.2±1.55 |
| △ | -17 | -7.1 | +6.8 |

Table 1. Exchanges of Na and K in the inulin-inaccessible phase of rat aorta during cooling and rewarming.

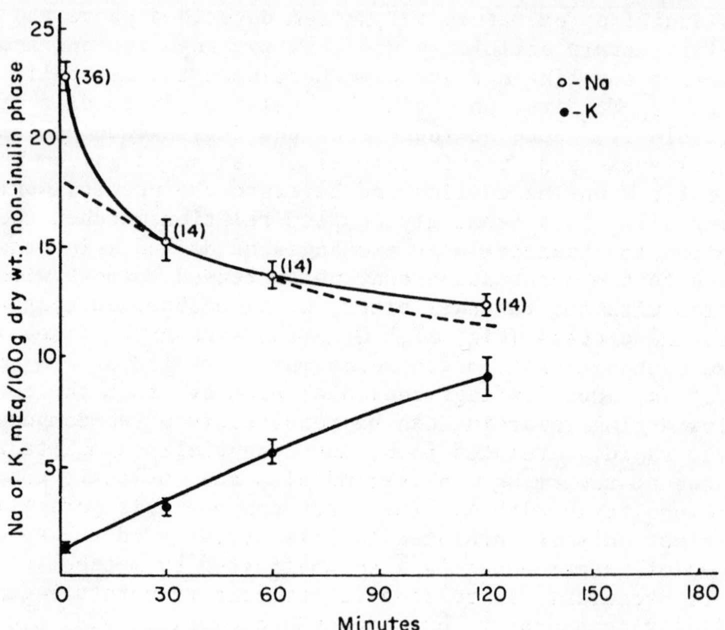


Fig. 2. Na and K content of the inulin-inaccessible phase of rat tail artery during rewarming at 37°C following overnight precooling at 2°C. Interrupted line indicates 1:1 exchange of Na and K.

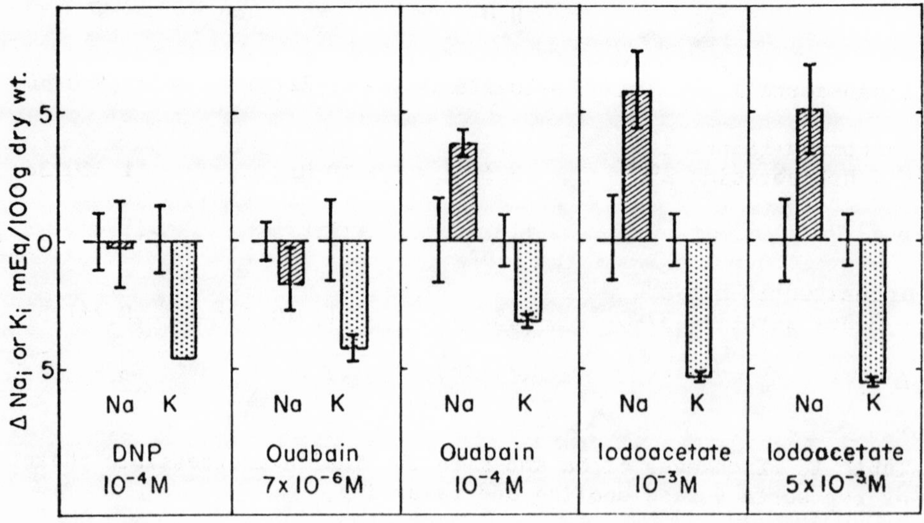


Fig. 3. Na and K content of the inulin-inaccessible phase of rat tail artery rewarmed at 37°C in the presence of inhibitors compared with rewarmed controls taken as base.

A simple experiment demonstrates the separability of the paracellular Na component from the cellular (Fig. 4). Samples of rat tail artery were pre-cooled at 2°C overnight and then transferred to fresh normal medium containing inulin for 3 hours of equilibration with continuous aeration. These control samples maintained at 2°C contain an abundance of Na and scarcely any K in the non-inulin phase. This situation is presumably static during the three-hour period of equilibration. Tissues of a second group were treated identically but 30 minutes before the end of the experiment were transferred to an isosmotic NaCl-free medium. As expected, this produced a loss of Na from the tissue without any corresponding change in K. The tissues of a third group were rewarmed at 37°C throughout the three-hour period in the expectation that they would lose Na and gain K as previously observed. Half an hour before the end of the experiment these tissues too were exposed to the NaCl-free medium. In this last group the tissue gained the expected amount of K and lost both cellular and paracellular Na.

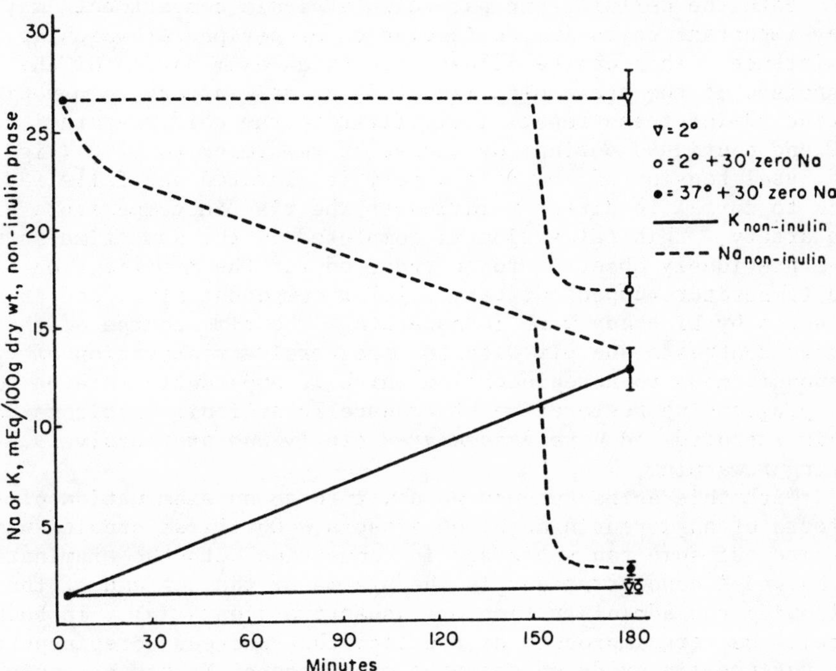


Fig. 4. Na and K content of the inulin-inaccessible phase of the pre-cooled rat tail artery exposed to NaCl-free medium at 2°C and following rewarming at 37°C.

We may note in passing that a small residual amount of Na remains in the inulin-inaccessible phase despite rewarming and exposure to the Na-free medium as in the third group. Similarly, a small residual amount of K remains in the inulin-inaccessible phase even in tissue subjected to prolonged cooling. These residual components comprise a fourth ionic phase which we will not consider further at this time.

Garrahan et al (2) have examined the exchange kinetics of Na in incubated artery samples and obtained curves similar to those of Hagemeijer et al (3) who have also studied the movements of K. Washout curves define three components for Na. The first component with a half-time of about 1 minute accounts for more than 90% of the Na of the tissue and evidently consists of that ion free in the extracellular fluid as well as that trapped (sorbed) and loosely bound in the paracellular matrix. The second component has a half-time of about 5 minutes and may well correspond to the temperature-sensitive paracellular component which we have defined. The third component is probably cellular since it has a half-time of about 70 minutes and is matched by an almost identical component in the K washout curve.

Both the cellular and paracellular ionic compartments may play important roles in the regulation of peripheral vascular resistance. This can be illustrated by an examination of the responses of the pre-cooled isolated rat tail artery to norepinephrine administered repetitively first to the cold preparation at 2°C and continued during the course of rewarming to 37°C (Fig. 5). The basal tension of the cold artery is elevated and falls rapidly back to normal in direct relation to the rise in temperature of the artery. This relaxation is completed in the same time as we have previously observed to be required for the mobilization of the temperature-dependent paracellular component of Na and is not affected by blockade with iodoacetate. The time course of this change contrasts sharply with the progressive restoration of responsiveness to norepinephrine which is apparently related to the progressive restoration of transcellular ionic gradients as Na is extruded and K re-accumulated slowly and progressively during rewarming.

With this as background we now turn to an examination of the effects of angiotensin on blood vessels. Our first studies were carried out some ten years ago in connection with the examination of Na and K concentrations in the plasma of the dog and of the rat following the administration of vasoactive agents (4). In both species we were impressed by the fact that whereas norepinephrine and vasopressin produced converse movements of Na and K, angiotensin produced a shift of Na out of the extracellular compartment not accompanied by a converse movement of K. Since angiotensin, either directly or indirectly, also excited vascular smooth muscle to constrict, this result was puzzling.

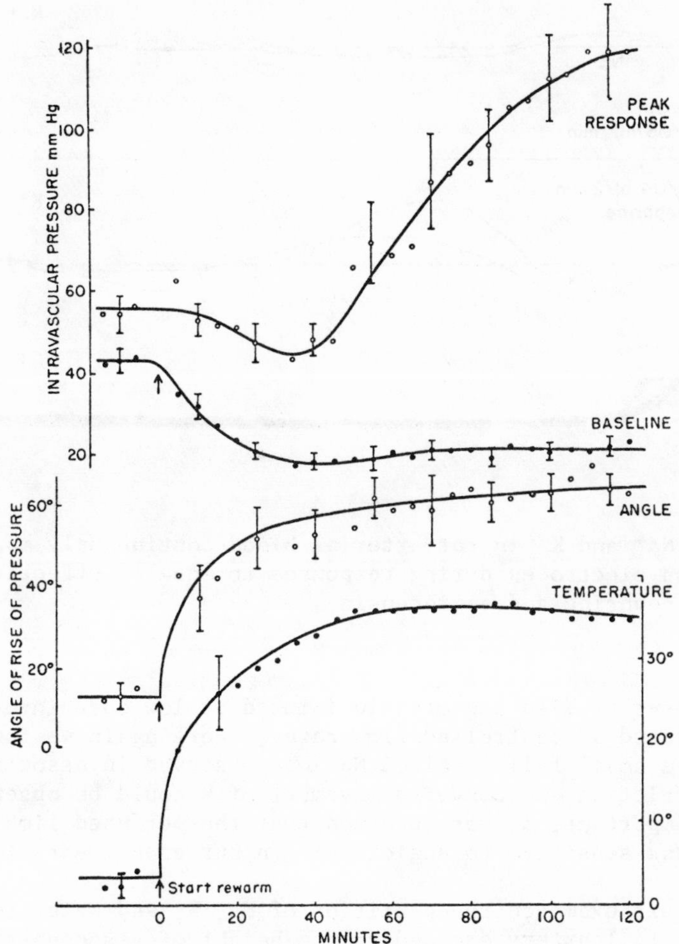


Fig. 5. Basal intravascular pressure and norepinephrine responsiveness measured during rewarming of pre-cooled rat tail artery perfused in vitro. Angle = steepness of rise of pressure.

In more recent experiments using ion-specific glass electrodes to monitor arterial Na and K in the rat we have found that the effects of norepinephrine and angiotensin are actually quite similar and differ only in degree (Fig. 6). In both cases, vasoconstriction is followed by a damped oscillation of both ions. With angiotensin, the K movement is sufficiently damped in degree as to readily escape measurement by conventional flame photometry. The findings in the whole animal thus tell us little about the special properties of angiotensin and indeed are quite compatible with the idea that the rise in blood pressure induced by the hormone is actually mediated by a catecholamine.

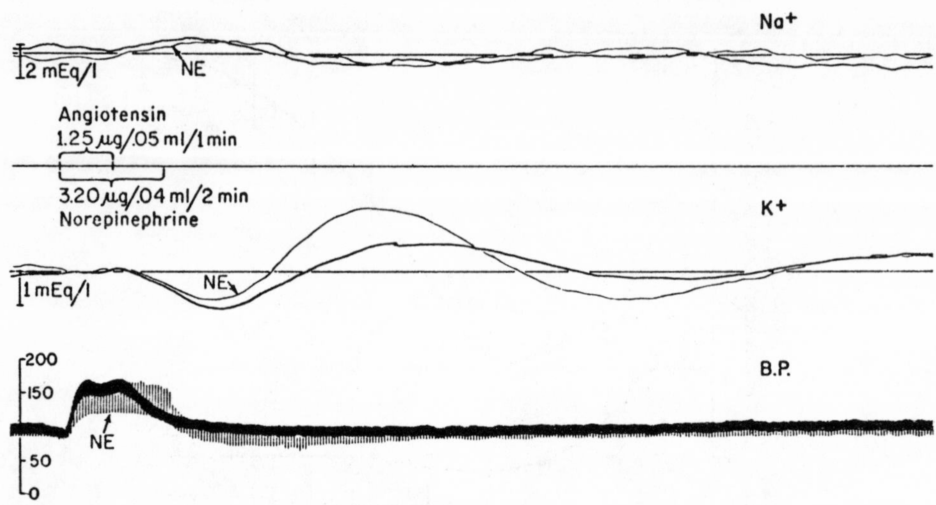


Fig. 6. $Na^+$ and $K^+$ in rat arterial blood continuously monitored with glass electrodes during responses to equally effective doses of norepinephrine and angiotensin.

We next studied angiotensin infused at low dose into the dog limb perfused at controlled flow rates. Here again we found that although a small fall in blood $Na^+$ was observed in association with vasoconstriction, no converse movement of K could be observed. Equally important, it was observed that the perfused limb was rather less sensitive to angiotensin in our experiments than anticipated.

We next examined the partition of Na, K, and water in the perfused rat tail artery excised at the height of vasoconstriction induced by a variety of agents (5). While the preparation proved to be quite sensitive to norepinephrine and to vasopressin it was exceedingly difficult to produce more than a very small response to angiotensin and correspondingly little movement of ions was measurable in this case.

Experiments such as these have led other workers (6) to conclude that angiotensin is itself not particularly vasoconstrictive but prepares the ground for other agents. If this is so, we must still inquire whether the polypeptide does affect the ionic matrix of the blood vessel. To study this we have miniaturized our electrode assembly to permit an examination of Na and K in a very small volume of medium (0.6 ml) perfusing a single rat tail artery in a closed circuit. The exchanges of Na and K between tissue and medium were examined by measuring the change in medium every 30 minutes by precise recalibration against fresh medium, a procedure capable of very close discrimination. For the most part,

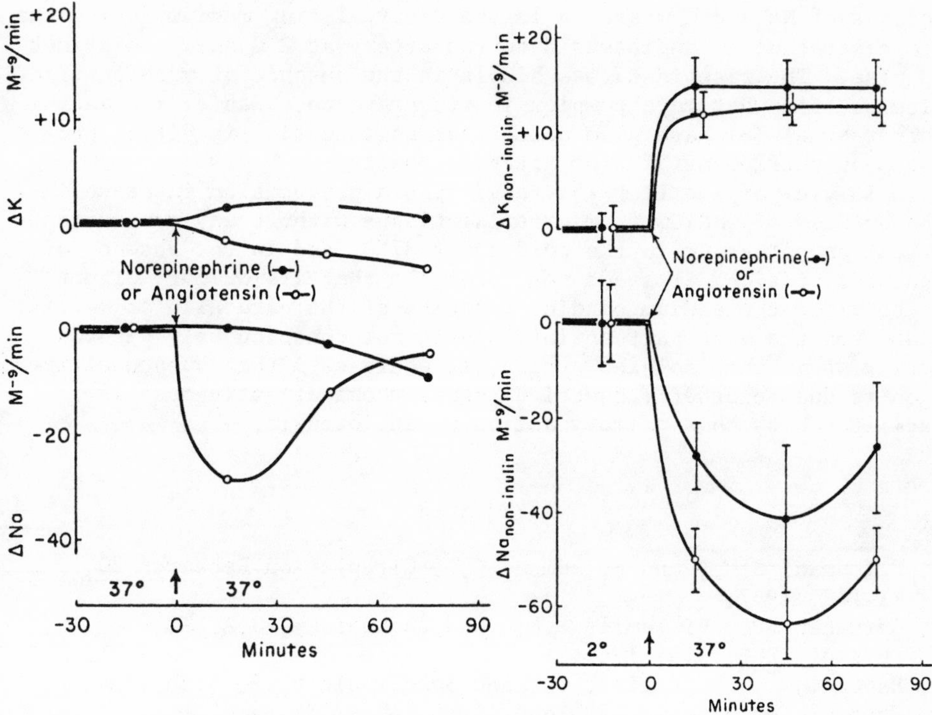


Fig. 7. Changes in tissue Na and K following single administration of 0.8 μg/0.01 ml norepinephrine or 10 μg/0.02 ml angiotensin to isolated, perfused rat tail artery at 37°C.

Fig. 8. Changes in tissue Na and K measured at 30 minute intervals as in Fig. 7. Norepinephrine or angiotensin were administered to isolated arteries pre-cooled overnight at 2°C and the subsequent exchanges during rewarming were followed.

angiotensin in this isolated system produced only small pressor responses at 37°C even though the tissue was highly responsive to norepinephrine. Following the administration of angiotensin, we observed a distinct tendency for an increased Na outflow from the tissue over the next hour despite successive washouts with fresh medium (Fig. 7). This was not in itself significant. However, when the artery was pre-cooled overnight at 2°C, although angiotensin again produced little if any pressor response, Na outflow from the tissue was distinctly increased upon rewarming without any concomitant change in K exchange (Fig. 8).

It is not possible to decide at this time with absolute certainty whether this effect of angiotensin on the outflow of tissue Na is a direct primary effect or an indirect and secondary reaction. Some of our observations are suggestive however. Thus, records of Na and K measured in the recirculating medium during the administration of angiotensin to the artery at 2°C show a distinct Na rise. The rise in Na was simple in the absence of vasoconstriction and somewhat more complex in its presence. Daniel (7) and Turker et al (8) have also noted that angiotensin, in vitro, produces an outflow of Na from artery segments.

In view of the fact that angiotensin produces an increase in the outflow of sodium from arterial tissue without any concomitant effect on potassium in the cold as at 37°C, and in the absence of vasoconstriction, it seems very probable that its primary effect is to alter the sodium binding capacity of the paracellular matrix. Since for the most part angiotensin is not conspicuously vasoconstrictive in the isolated artery, it would seem that vasoconstriction is due to other vasoactive agents whose effectiveness is facilitated by this primary action of angiotensin.

## REFERENCES

1. Friedman, S.M., and Friedman, C.L. Circulation Research 20, II-147, 1967.
2. Garrahan, P., Villamil, M.F., and Zadunaisky, J.A. Am. J. Physiol. 209, 955, 1965.
3. Hagemeijer, F., Rorive, G., and Schoffeniels, E. Arch. Intern. Physiol. et Biochim. 73, 453, 1965.
4. Friedman, S.M., and Friedman, C.L. Page 1135 in Handbook of Physiology - Circulation, Vol. 2, 1963.
5. Friedman, S.M., and Friedman, C.L. Page 323 in Electrolytes and Cardiovascular Diseases, ed. by E. Bajusz, 1965. S. Karger, Basel/New York.
6. De La Lande, S., Cannell, V.A., and Waterson, J.G. Brit. J. Pharmacol. Chemother. 28, 255, 1966.
7. Daniel, E.E. Arch. int. Pharmacodyn. 158, 131, 1965.
8. Türker, R.K., Page, I.H., and Khairallah, P.A. Arch. int. Pharmacodyn. 165, 394, 1967.

Original work described herein was carried out with the aid of grants from the Medical Research Council of Canada and the B.C. Heart Foundation.

# VARIATIONS IN THE ALDOSTERONE-STIMULATING ACTIVITY OF ANGIOTENSIN: CAUSES AND PHYSIOLOGICAL SIGNIFICANCE*

William F. Ganong, M.D.

University of California, San Francisco Medical Center

San Francisco, California, USA

Angiotensin raises blood pressure and increases aldosterone secretion. Variations in the pressor potency of angiotensin have been studied in considerable detail, but variations in its aldosterone-stimulating activity have received little attention. My associates and I have found that this latter parameter shows interesting and important fluctuations.

The effects of angiotensin and ACTH on aldosterone and 17-hydroxycorticoid secretion in a series of dogs are summarized in Table 1. The dogs had been fed a normal stock diet which on analysis was found to provide about 40 mEq of sodium per day. They were acutely hypophysectomized and nephrectomized, then tested with various doses of angiotensin and ACTH (1). It is apparent that low doses of ACTH exert their effect solely on 17-hydroxycorticoid secretion, and it is only the larger doses of ACTH that stimulate aldosterone secretion. Angiotensin stimulates aldosterone secretion at low dose levels and compared to ACTH, its effect on 17-hydroxycorticoid secretion is trifling.

It is important to remember that ACTH has two effects on the adrenal: not only does it increase glucocorticoid output acutely, but it increases the amount of glucocorticoids secreted in response to subsequent doses of ACTH. The increase in adrenal sensitivity is associated with hypertrophy of the zona fasciculata and the zona reticularis. Conversely, hypophysectomy is followed by a steady decline in the sensitivity of the adrenals to injected ACTH (2).

Does exposure of the adrenal to angiotensin increase the increment in aldosterone secretion produced by subsequent doses of angiotensin? The first evidence that this might be the case was

*Includes previously unpublished results of experiments supported by USPHS Grant AM06704.

Table 1. Adrenal responses to ACTH and angiotensin in dogs fed approximately 40 mEq sodium/day. Values are means ± standard errors and figures in parentheses are numbers of animals.

| | | Increment in output of: | |
|---|---|---|---|
| | | Aldosterone (mμg/min) | 17-hydroxycorticoids (μg/min) |
| Angiotensin (mμg/min) | 0.04 | 0.5 ± 1.3 (5) | 0.1 ± 0.4 (5) |
| | 0.17 | 11.4 ± 3.3 (8) | 0.3 ± 0.4 (8) |
| | 0.42 | 19.0 ± 5.2 (7) | 1.3 ± 0.4 (7) |
| | 1.67 | 24.0 ± 5.0 (7) | 3.9 ± 0.8 (7) |
| ACTH (mU) | 2 | 2.2 ± 3.2 (4) | 2.3 ± 0.9 (4) |
| | 5 | 3.6 ± 1.8 (7) | 3.5 ± 0.7 (8) |
| | 10 | 2.7 ± 3.4 (6) | 7.9 ± 2.0 (5) |
| | 50 | 5.4 ± 1.1 (3) | 7.0 ± 2.0 (5) |
| | 100 | 26.7 ± 1.8 (3) | 8.6 ± 1.6 (3) |
| | 1000 | 48.9 ± 9.9 (10) | 8.5 ± 0.4 (10) |

our finding that in dogs fed a low sodium diet for 14 days before being hypophysectomized and nephrectomized, the aldosterone-stimulating activity of angiotensin was increased (3). The data are summarized in Figure 1; especially at the low doses, the difference is clear cut. There was no change in the 17-hydroxycorticoid response to angiotensin. After 14 days on the low sodium diet, there was also a marked increase in the increment in aldosterone secretion produced by ACTH (Fig. 2), indeed, there was a clear-cut increase after only 5 days of restricted sodium intake (4). There were no significant changes in the increments in 17-hydroxycorticoid secretion produced by ACTH. Similar increments in the aldosterone-stimulating activity of ACTH have been reported in humans and rats fed a low sodium diet. In the humans, 17-hydroxycorticoid output remained normal (5), but in the rats, corticosterone secretion was reported to drop, possibly because the enlarged zona glomerulosa encroached on the zona fasciculata and zona reticularis (6).

Since a low sodium diet increases circulating renin and consequently angiotensin levels (7), we speculated that the increased aldosterone-stimulating activity of angiotensin in animals fed a low sodium diet was due to chronic exposure of the adrenals to angiotensin. We therefore attempted to produce the same sensitivity changes by injecting dog renin for 5 days in dogs fed a normal sodium intake (8). The dogs were then hypophysectomized and nephrectomized and their adrenal cortical sensitivity tested. The results are summarized in Figure 3. A clear-cut increase in the sensitivity of the adrenal to the aldosterone-stimulating effect of angiotensin was produced. The responses to ACTH were similarly

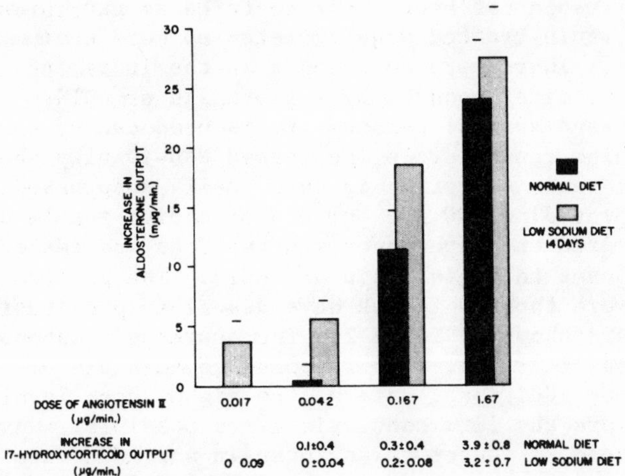


Fig. 1. Effect of a low sodium diet on adrenal responses to angiotensin. The line above each bar indicates 1 standard error. From (1).

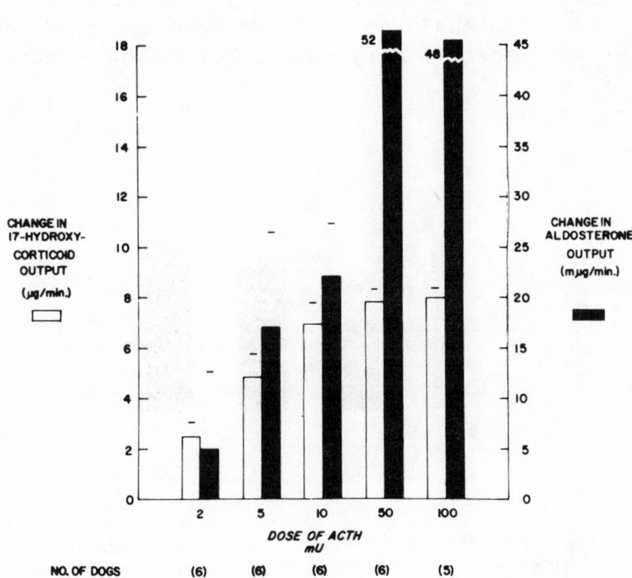


Fig. 2. Effect of a low sodium diet on adrenal responses to ACTH. From (1).

increased. The changes were not secondary to sodium loss induced by the renin; renin has been reported to cause natriuresis in rats (9), but our renin-treated dogs excreted no more sodium than control dogs (8). There were no changes in the increments in 17-hydroxycorticoids produced by angiotensin or ACTH.

If increased adrenal sensitivity is produced by chronic exposure to high renin levels, decreased sensitivity should be produced when renin secretion is chronically suppressed. We fed dogs a diet providing 100 mEq per day of sodium for 14 days, then hypophysectomized and nephrectomized the dogs and tested their adrenal responses to angiotensin and ACTH. The protocol and chemical methods were those which we have described previously (1,4,8). The results are shown in Table 2. The decreased responsiveness to angiotensin is clear cut. The response to ACTH did not seem to be suppressed, but additional data at the higher dose levels are necessary before any firm conclusions are possible. Seventeen-hydroxycorticoid responses to angiotensin and ACTH were not significantly different from those in the dogs fed the normal diet, except for a somewhat greater response to 2 mU of ACTH. This same increased 17-hydroxycorticoid response to the smallest dose of ACTH has been observed in other experiments in dogs fed the normal diet when this dose of ACTH follows infusion of a large dose of angiotensin (Ganong, unpublished data). Therefore, it is probably due to the previous angiotensin, rather than the high sodium intake. The aldosterone response to this dose of ACTH does not appear to be affected. This is the only case of short-term interaction between angiotensin and ACTH we have observed; other interactions were specifically looked for and were not found (10).

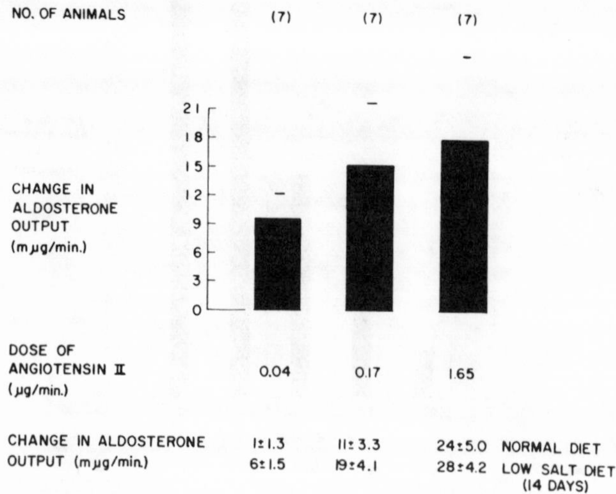


Fig. 3. Effect of dog renin on adrenal responses to angiotensin.

Table 2. Increments in aldosterone output produced by angiotensin and ACTH in dogs fed a high sodium diet. See legend for Table 1.

| | | Aldosterone output (mμg/min) | | | Aldosterone output (mμg/min) |
|---|---|---|---|---|---|
| Angiotensin (mμg/min) | 0.17 | 3.8 ± 2.7 (5) | ACTH (mU) | 2 | 4.8 ± 2.4 (4) |
| | 0.42 | 0.4 ± 2.0 (5) | | 5 | 5.5 ± 3.1 (4) |
| | 1.67 | 4.0 ± 2.6 (5) | | 10 | 0 ± 2.1 (5) |
| | 4.2 | -5.4 ± 4.2 (5) | | 50 | 11.3 ± 4.2 (5) |
| | | | | 100 | 22 (1) |
| | | | | 1000 | 28 (1) |

The decrease in adrenal sensitivity is probably due to prolonged reduction in circulating renin levels since sodium loading decreases circulating renin levels (7). A similar decreased sensitivity of the adrenal has been reported in humans fed a high sodium diet (11). In addition, there is a low circulating renin level in primary hyperaldosteronism in humans, and after removal of the aldosterone-secreting tumors in these patients, the remaining adrenal cortical tissue is insensitive to the aldosterone-stimulating effects of angiotensin and ACTH (see 1).

Thus, prolonged elevation of the circulating renin level appears to increase the sensitivity of the adrenal to angiotensin while prolonged reduction of the renin level decreases it. The effect of renin is presumably mediated via angiotensin; renin itself has no acute effect on adrenocortical secretion (12) and the only proved physiologic action of renin is catalysis of the formation of angiotensin. The long-term effects of renin on the aldosterone-stimulating activity of angiotensin and ACTH occur without any change in the 17-hydroxycorticoid-stimulating effects of these hormones. Consequently, there must be functionally separate aldosterone-secreting and 17-hydroxycorticoid-secreting mechanisms within the adrenal gland.

This functional separation is also borne out by studies of the effects of prolonged ACTH excess and deficiency on the sensitivity of the adrenal. We produced ACTH deficiency by injecting the long-acting glucocorticoid Depomedrol (the 21-acetate of 6α-methyl-11β,17α,21-trihydroxy-1,4-pregnadiene-3,20-dione) for 30 days. The animals were then hypophysectomized and nephrectomized and their adrenocortical sensitivity to angiotensin and ACTH tested (13). As expected, the increments in 17-hydroxycorticoid secretion produced by ACTH were reduced to essentially zero. The 17-hydroxycorticoid responses to angiotensin were similarly reduced. However, the aldosterone responses to angiotensin (Fig. 4) were normal; indeed, they were somewhat greater than normal, although the differences between the responses in these dogs and the controls were not statistically significant. The aldosterone responses to ACTH were also normal. Thus inhibition of ACTH secretion led to markedly diminished sensitivity of the 17-hydroxycorticoid-secreting mechanism without any inhibition of the aldosterone-secreting mechanism.

The mean weight of the right adrenal glands of the Depomedrol treated dogs was 37.4 ± 1.9 mg/kg body weight, compared to a mean value of 54.7 ± 4.8 mg/kg in 10 normal dogs. There was marked atrophy of the zona fasciculata and zona reticularis. However, the zona glomerulosa was not atrophic, and indeed was slightly but statistically significantly ($P<0.05$) wider than it was in control dogs (Fig. 5). In these and the other groups of dogs shown in Figure 5, the widths of the zona glomerulosa and the combined zona fasciculata and zona reticularis were measured with a micrometer eye piece, and the values in each dog were the mean of 3 measurements of the zones in different parts of the gland.

Dogs were also studied 1 month after surgical hypophysectomy. In these animals, the mean right adrenal weight was 35.2 ± 3.0 mg/kg and the widths of the adrenal zones were comparable to those in the Depomedrol-treated dogs (Fig. 5). There was a similar marked reduction in the 17-hydroxycorticoid increments produced by ACTH and angiotensin but the increments in aldosterone produced by ACTH and angiotensin were also diminished, in contrast to the results obtained in the dogs treated with Depomedrol (13). This suggests that some pituitary factor other than ACTH plays a permissive role in maintaining the sensitivity of the aldosterone-secreting mechanism. A similar conclusion was reached by Palmore and Mulrow on the basis of experiments in rats (14). Studies are currently underway to identify the pituitary factor. It should be noted that sensitivity decreased even though the width of the zona glomerulosa was comparable to that in Depomedrol-treated dogs.

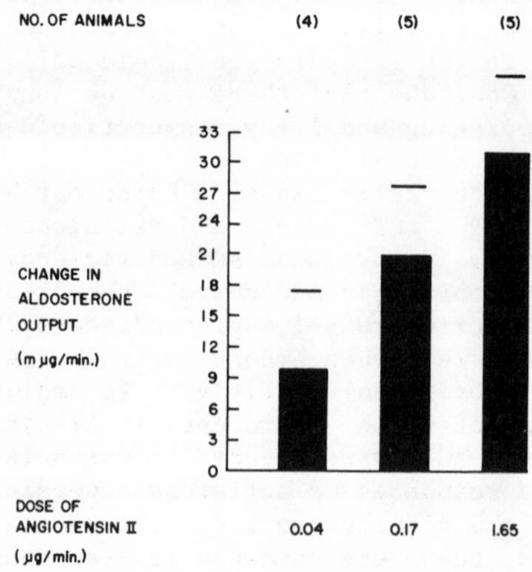


Fig. 4. Effect of ACTH deficiency on adrenal responses to angiotensin.

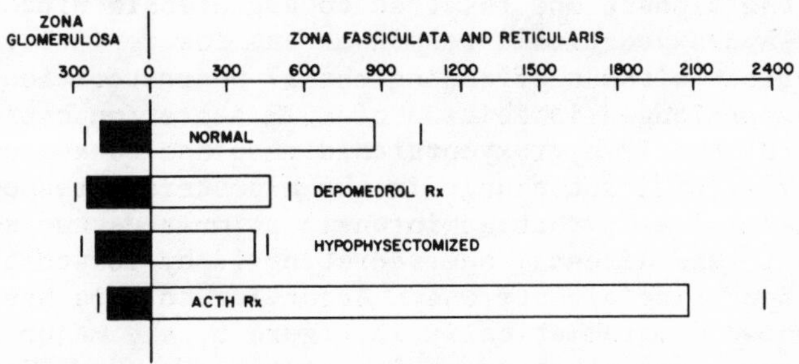


Fig. 5. Mean widths of adrenocortical zones (in μ) of 5 normal dogs, 6 dogs treated with Depomedrol for 30 days, 4 dogs hypophysectomized for 28 days and 4 dogs treated with 40 U ACTH gel twice a day for 7 days.

Data on the converse condition, chronic ACTH excess, are also available. We treated dogs with 40 U of ACTH gel twice a day for 7 days. In 2 of the dogs treated in this fashion, the width of the zona fasciculata and zona reticularis was not markedly increased, but in the remaining 4, it was 1400 μ or greater. The mean adrenal weight in these 4 dogs was 117.2 ± 19.7 mg/kg. The mean width of the zona glomerulosa was less than it was in the controls (Fig. 5), although the difference was not statistically significant, possibly due to the considerable variations from animal to animal.

In the 4 ACTH-treated dogs, 17-hydroxycorticoid responses to ACTH were increased. However, aldosterone secretion in response to angiotensin was reduced (Table 3). This reduction is statistically significant ($P<0.02$) at the 1.67 μg/min dose. The aldosterone response to ACTH did not appear to be reduced, but additional data at higher doses are needed. More data are also needed on the 17-hydroxycorticoid response to angiotensin.

These data on the variations in adrenocortical sensitivity produced by prolonged alterations in ACTH and renin secretion support a number of conclusions about adrenal function. Prolonged

Table 3. Increments in aldosterone output in 4 dogs treated with ACTH gel for 7 days.

| | | Aldosterone output (mμg/min | | | Aldosterone output (mμg/min) |
|---|---|---|---|---|---|
| Angiotensin (μg/min) | 0.04 | 0 ± 2.9 | ACTH (mU) | 2 | 2.5 ± 1.5 |
| | 0.17 | 2.8 ± 1.2 | | 5 | 4.3 ± 0.9 |
| | 1.67 | 3.3 ± 1.7 | | 10 | 3.0 ± 2.0 |
| | | | | 50 | 12.3 ± 2.2 |

exposure of the adrenal to elevated circulating renin levels increased the aldosterone response to angiotensin without affecting the 17-hydroxycorticoid response, and low renin levels had the opposite effect without affecting the 17-hydroxycorticoid response. Conversely, prolonged inhibition of ACTH secretion caused marked diminution of the 17-hydroxycorticoid response to angiotensin without any significant change in the aldosterone response. Therefore, it seems likely that angiotensin stimulates two separate mechanisms in the adrenal, one secreting 17-hydroxycorticoids and the other secreting aldosterone. According to this hypothesis, which is shown diagrammatically in Figure 6, the major stimulus to the aldosterone-secreting mechanism is angiotensin. However, the aldosterone-secreting mechanism is also stimulated acutely to a lesser degree by ACTH. The 17-hydroxycorticoid-secreting mechanism responds primarily to ACTH, but acutely, it also responds to a lesser degree to angiotensin. The sensitivity of the aldosterone-secreting mechanism is maintained by renin, presumably via angiotensin, while the sensitivity of the 17-hydroxycorticoid-secreting mechanism is maintained by ACTH.

The decreased sensitivity of the aldosterone-secreting mechanism following prolonged treatment with ACTH seem inconsistent with this hypothesis. However, the decrease may well be a secondary effect of the ACTH treatment. ACTH increases aldosterone secretion acutely, but continued treatment is associated with a return of the aldosterone secretion rate to or even below the control level (15). The initial increase leads to sodium and water retention with hypervolemia and presumably, decreased renin secretion, which may be responsible for the subsequent decline in aldosterone secretion. The low renin levels could be the cause of the

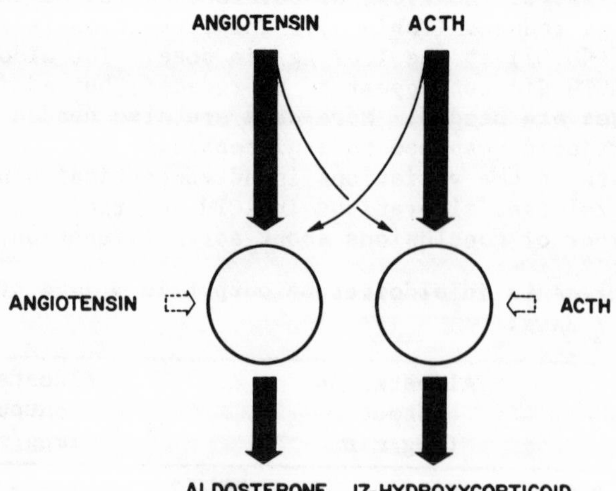


Fig. 6. Hypothetical functional organization of adrenal cortex. The circles represent independent secreting units.

decreased sensitivity of the aldosterone-secreting mechanism. A similar fall after an initial rise in aldosterone secretion has been reported to occur in humans during ACTH treatment on a low sodium diet (15); however, the subjects still gained weight, indicating that ACTH treatment caused fluid retention despite the low sodium diet. An alternative hypothesis to explain the decline in aldosterone secretion during prolonged ACTH treatment is accumulation in the steroid biosynthetic pathway of an intermediate which exerts a negative feedback effect on further aldosterone formation. There is some evidence for such feedback effects (16). Inhibition of aldosterone biosynthesis by the large quantities of glucocorticoids in the adrenal is another possibility. These hormones have been demonstrated to inhibit protein synthesis in the adrenal *in vitro* (17), and protein synthesis is claimed to be a prerequisite for the formation of some steroids (18).

It is well established that the zona glomerulosa secretes primarily aldosterone and the zona fasciculata and zona reticularis primarily glucocorticoids (19). Our data are generally consistent with the hypothesis that the aldosterone-secreting mechanism diagramed in Figure 6 is located in the zona glomerulosa and the 17-hydroxycorticoid-secreting mechanism in the zona fasciculata and zona reticularis. A low sodium diet and renin injections cause hypertrophy of the zona glomerulosa and increased sensitivity of the aldosterone-secreting mechanism while they have no effect on the inner zones of the adrenal and no effect on the sensitivity of the 17-hydroxycorticoid-secreting mechanism (4,8,19). A high sodium diet has opposite effects on the zona glomerulosa (19) and the aldosterone-secreting mechanism and a similar lack of effect on the inner 2 zones and the 17-hydroxycorticoid-secreting mechanism. Depomedrol and hypophysectomy decrease and ACTH treatment increases both the width of the inner 2 zones and the sensitivity of the 17-hydroxycorticoid-secreting mechanism. On the other hand, Depomedrol treatment causes slight hypertrophy of the zona glomerulosa and this is associated with a definite even if not statistically significant increase in the sensitivity of the aldosterone-secreting mechanism. Conversely, ACTH treatment is associated with a slight decrease in the width of the zona glomerulosa and a clear-cut decrease in the sensitivity of the aldosterone-secreting mechanism. It is true that in chronically hypophysectomized dogs, there is a wide zona glomerulosa in association with decreased sensitivity of the aldosterone-secreting mechanism, but this is probably a special case in which sensitivity is lost due to the lack of some factor other than angiotensin or ACTH.

Our experiments have several obvious physiological implications. First, the increased sensitivity of the aldosterone-secreting mechanism in animals fed a low sodium diet reduces the amount of circulating renin necessary to stimulate aldosterone secretion. This is undoubtedly why such relatively marked increases in aldosterone secretion can occur with modest elevations in circulating renin levels after several days on a low sodium diet. Surgical stress or even anesthesia alone may cause just as great an elevation in circulating renin levels with little if any increase in aldosterone secretion in dogs previously fed a normal diet (Assaykeen, Otsuka, Tu and Ganong, unpublished observations). Secondly, the inhibitory effect of prolonged hypersecretion of ACTH on the aldosterone-secreting mechanism permits prolonged elevation of 17-hydroxycorticoid secretion to occur in severe chronic stress without the development of complicating hyperaldosteronism.

In summary, the aldosterone-stimulating activity of angiotensin has been found to vary with the state of the adrenal cortex and to be independent of the 17-hydroxycorticoid-stimulating activity of the peptide. The sensitivity of the 17-hydroxycorticoid-secreting mechanism is controlled by ACTH, but the sensitivity of the aldosterone-secreting mechanism is controlled by renin.

## References

1. Ganong, W.F., E.G. Biglieri, P.J. Mulrow, Recent Progr. Hormone Res. 22:381, 1961.
2. Ganong, W.F., in Advances in Neuroendocrinology (A.V. Nalbandov, ed.), p. 92, University of Illinois Press, Urbana, 1963.
3. Ganong, W.F., A.T. Boryczka, Proc. Soc. Exper. Biol. Med. 124: 1230, 1967.
4. Ganong, W.F., A.T. Boryczka, R. Shackelford, R.M. Clark, R.P. Converse, Proc. Soc. Exper. Biol. Med. 118:792, 1965.
5. Venning, E.H., I. Dyrenfurth, J.B. Dossiter, J.C. Beck, Metabolism 11:254, 1962.
6. Eisenstein, A.B., I. Strack, Endocrinology 68:121, 1961.
7. Brown, J.J., D.L. Davis, A.F. Lever, J.I.S. Robertson, Lancet 2:278, 1963.
8. Ganong, W.F., A.T. Boryczka, R. Shackelford, Endocrinology 80:703, 1967.
9. Bartter, F.C., Recent Progr. Hormone Res. 22:415, 1966.
10. Mulrow, P.J., W.F. Ganong, Proc. Soc. Exper. Biol. Med. 118: 795, 1965.
11. Jagger, P.I., E.A. Espiner, J.R. Tucci, D.P. Lauler, G.W. Thorn, Clin. Res. 14:132, 1966.
12. Blair-West, J.R., J.P. Coghlan, D.A. Denton, J.R. Goding, M. Wintour, R.D. Wright, Recent Progr. Hormone Res. 19:311,1963.
13. Ganong, W.F., D.L. Pemberton, E.E. Van Brunt, Endocrinology (in press, 1967).
14. Palmore, W.P., P.J. Mulrow, Endocrine Society Abstracts, page 33, 1967.
15. Tucci, J.R., E.A. Espiner, P.I. Jagger, G.L. Pauk, D.P. Lauler, J. Clin. Endocrinol. 27:568, 1967.
16. Sharma, D., Recent Progr. Hormone Res. 22:425, 1966.
17. Morrow, L.B., G.N. Burrow, P.J. Mulrow, Endocrinology 80: 883, 1967.
18. Samuels, L.D., New England J. Med. 271:1252, 1964.
19. Deane, H.W., Handb. d. exp. Pharmakol. Erg. W. 14, 1-185, 1962.

# ISOLATION AND STRUCTURE OF CAERULEIN

A. Anastasi, V. Erspamer and R. Endean

Istituto Ricerche Farmitalia, Milano and Istituto di Farmacologia, Università di Parma (Italy) - Zoology Dept., University of Queensland, Brisbane (Australia)

This report summarizes the procedures of isolation and characterization of caerulein a peptide possessing a close similarity with gastrin and pancreozymin for its composition and for some of its pharmacological actions.

Caerulein is a decapeptide having the following sequence:

$$\begin{array}{c}SO_3H\\|\\ \text{Pyr-Gln-Asp-Tyr-Thr-Gly-Trp-Met-Asp-Phe-NH}_2\ \text{(I)}\end{array}$$

$$\begin{array}{c}SO_3H\\|\\ \text{Pyr-Gly-Pro-Trp-Leu-(Glu)}_5\text{-Ala-Tyr-Gly-Trp-Met-Asp-Phe-NH}_2\ \text{(II)}\end{array}$$

**FIG. 1 Structures of caerulein (I) and human gastrin-II (II)**

As can be seen it shares the whole C-terminal half of its molecule with the C-terminal end of gastrin. Furthermore it contains a tyrosine-sulphate residue and an N-terminal pyroglutamyl group also present in gastrin.

Caerulein is another peptide added to the family of active peptides found so far in that store-house of biologically interesting compounds which appears to be the skin of amphibians. It has been extracted from the skin of Hyla Caerulea, a little frog very common in Australia, after which prof. Erspamer has coined its name [1,2,3,4].

Our starting material for the isolation of caerulein was an 80% methanol extract of the dried skins. Except for a few troubles at the beginning, the work was relatively rapid and easy, since we had not to face the problem which we found to be the main problem in all the previous cases of extraction of active peptides. The extremely small peptide content of the tissues and the rarity of the specimens which had to be collected with great difficulties for instance in some tropical region of South America [1,5,6].

With caerulein we were lucky, not only for the easy availability of Hyla Caerulea but also because the skin of this frog contains a very large amount of peptide ranging from 0.8 to 1.4 mg per gram of dry skin. The peptide composition itself was another reason of help in the process of purification. It carries a strongly acidic group and lacks basic residues and this rather unique chemical character made easier its separation from the basic or amphoteric compounds more common in such a kind of tissue extracts.

At the beginning of the work, the only test system available for the assay of the caerulein containing samples, was the dog blood pressure which is not the simplest one for assaying hundred of samples. We had a change to overcome in a short time also this difficulty since we were soon able to identify the spot of caerulein on the paper electropherograms carried out to test the content of contaminants in the samples. The identification was rather easy since the caerulein spot, positive to the Ehrlich reagent for tryptophan, migrates anodically in acidic medium and therefore was separated from the bulk of the accompanying compounds migrating toward the cathode. This spot was associated with hypothensive activity which was absent from all the other points of the paper. Therefore its presence was, since then used as a test of the presence of caerulein in the samples, while bioassay was only employed to obtain quantitative data.

After a number of trial experiments it was seen that samples of caerulein pure enough to be used for degradation studies, could be obtained with a few simple steps which were applied usually on amounts of crude extract corresponding to 50 grams of dry tissue. The material previously defatted, was submitted to counter current distribution in a system of butanol and water of very mild acidity. In these solvents caerulein showed a fairly high distribution coefficient and therefore could be isolated from the bulk of contaminants remaining in the first tubes.

The second purification step was similar to that used by Gregory and Tracy[7] for the separation of gastrins I and II on anionic cellulose. We used an exchanger of stronger basicity

(DEAE-Sephadex in the $OH^-$ form) which required an higher concentration of ammonium carbonate to elute caerulein. While a concentration below 0.5M was sufficient to elute gastrin II, caerulein emerged when the effluent reached the molarity. With this procedure caerulein was rid almost completely from impurities which were retained by the gel to a much lesser extent.

Among the substances removed in this step we could identify small amounts of desulphated caerulein, which emerged in a peak well in advance from that of caerulein. Desulphation probably occurred in a previous step, in fact we had more than one occasion in the course of the work to observe the lability of the sulphate ester in acidic medium. For instance during the isolation of enzymatic fragments for the sequence studies, desulphated peptides were inevitably eluted from the electropherograms performed at acidic pH unless the paper was carefully dried in the cold. We could also find out that the sulphate radical could be completely split in 2 to 5 minutes with 0.1 N HCl and in 30 minutes with N acetic acid at 100°. After this treatment no significant hydrolysis of peptide bonds or other chemical modifications could be observed in caerulein, but the hypothensive activity was completely lost.

Caerulein was recovered from the Sephadex column in a solution containing a large amount of ammonium carbonate from the elution buffer. Desalting was made by chromatography through a carboxylic resin in the hydrogen form eluted with water.

The above procedure gave an over-all yield quite satisfactory since it ranged from 60 to 80% of the starting content. As already said it could provide samples of caerulein pure enough to be worked, as checked by the constancy of their aminoacid composition with that of purer samples, for instance those obtained by preparative electrophoresis.

The aminoacid sequence of caerulein was deduced essentially by degradation with two enzymes, chymotrypsin and subtilisin. The fragments obtained were fractionated by preparative paper electrophoresis and examined by further degradation, chemical and enzymatic.

Chymotrypsin split only one peptide bond at the carboxyl side of tryptophan thus forming two fragments. Subtilisin hydrolyzed the same tryptophyl bond and in addition, at a similar rate, the bond of tyrosine sulphate thus forming three fragments. Figure 2 shows the patterns obtained by submitting the chymotrypsin and subtilisin digests to high voltage electrophoresis in acidic and neutral medium.

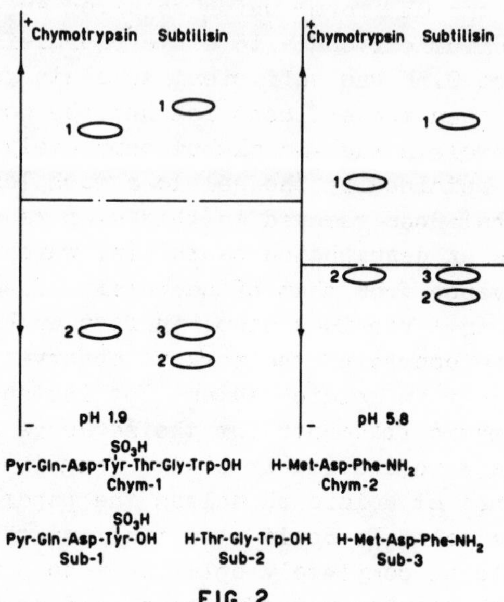

FIG. 2

As can be seen the chymotrypsin digest contains two bands. Band I, which migrates anodically at both pH's and can be stained with the reagents for tyrosine and tryptophan, corresponds to the N-terminal eptapeptide. Band 2, which can be stained with ninhydrin, marks the C-terminal tripeptide amide; according to its charge it migrates cathodically at acidic pH and stays at the origin at neutral pH.

The subtilisin digest shows three bands of which band 3 is identical with the chymotryptic 2; band 2 stained with ninhydrin and Ehrlich marks the central tripeptide and band I stained with the tyrosine reagents corresponds to the N-terminal tetrapeptide sulphate.

When caerulein is first submitted to mild acid hydrolysis as mentioned before an attack by chymotrypsin at the tyrosine bond becomes possible. Thus the N-terminal tetrapeptide can be obtained also with chymotrypsin but it carries of course a desulphated tyrosine residue. This was done to check the subtilisin results but was not convenient for preparative purposes.

The preparation of the fragments to be studied was usually made by first digesting caerulein with chymotrypsin then submitting the chymotryptic band I to digestion with subtilisin. This stepwise degradation with two enzymes permitted a better separation of the central and C-terminal tripeptides that according to their charge and molecular weight migrate too close

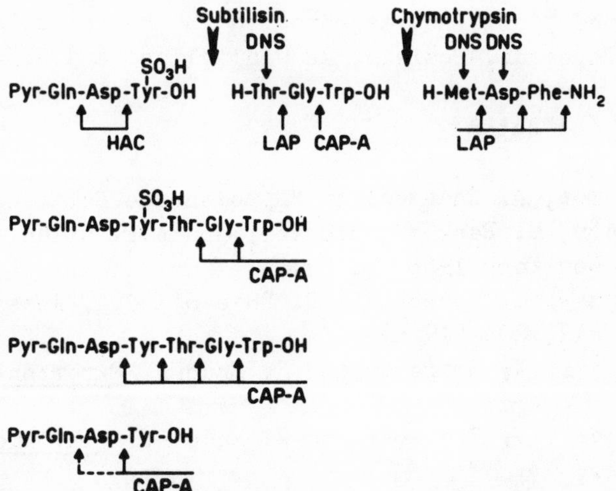


FIG. 3

to each other to be easily isolated in pure form from a single subtilisin digest.

A plot summarizing all the degradation steps is shown in fig. 3.

The sequence of the C-terminal tripeptide was readily deduced by degradation with leucinaminopeptidase and by end group determination either on the tripeptide itself or on the dipeptide obtained by controlled removal of methionine with LAP. The same approaches (end group analysis and degradation with the exopeptidases) led also to the formulation of the central peptide. The N-terminal sequence was deduced instead by partial acid hydrolysis with acetic acid which released free tyrosine, aspartic acid and the dipeptide pyroglutamylglutamine.

These results were checked in a few more ways. For instance by digesting with CAP the chymotryptic fragment I we could observe the consecutive release of triptophan and glycine followed by trace amounts of threonine. The resistance of the threonine residue to CAP attack was attributed to inhibition by the adiacent negatively charged residue of tyrosine sulphate. On the other hand when CAP treatment was applied on the previously desulphated fragment, a rapid splitting of 4 residues, from tryptophan to tyrosine, was observed. From the desulphated fragment I of subtilisin CAP removed tyrosine and only trace amounts of aspartic acid.

The decapeptide sequence of caerulein so obtained was confirmed by the identity of the natural compound with the synthetic one [8] obtained as reported in the following communication by Bernardi, Bosisio, de Castiglione and Goffredo.

## References and Notes


1) V. Erspamer, A. Anastasi in "Hypotensive Peptides", ed. by E.G. Erdös, M. Back, F. Sicuteri and A.F. Wilde. Springer Verlag. New York 1966, p. 63.
2) V. Erspamer, M. Roseghini, R. Endean and A. Anastasi, Nature, **212**, 204 (1966).
3) A. Anastasi, V. Erspamer and R. Endean, Experientia, in press.
4) A. Anastasi, V. Erspamer and R. Endean, Arch. Bioch. Bioph., in press.
5) A. Anastasi, V. Erspamer and J.M. Cei, Arch. Bioch. Bioph., **108**, 341 (1964).
6) A. Anastasi, V. Erspamer and G. Bertaccini, Camp. Bioch. Physiol., **14**, 43 (1965).
7) R.A. Gregory and J.H. Tracy, Nature, **209**, 583 (1966).
8) L. Bernardi, G. Bosisio, R. de Castiglione and O. Goffredo, Experientia, in press.

# PHARMACOLOGICAL DATA ON CAERULEIN

V. Erspamer, G. Bertaccini, R. Cheli, G. De Caro
R. Endean, M. Impicciatore and M. L. Roseghini

Institute of Pharmacology, University of Parma
Parma, Italy

Caerulein, the active decapeptide which has been detected in our Laboratory in extracts of the skin of the Australian frog Hyla caerulea, is certainly one of the most versatile polypeptides we know so far. In fact, it displays a number of striking pharmacological actions.

## Hypotensive Action

Hypotension was particularly evident in dogs and rabbits. In these species caerulein, although considerably less potent than eledoisin or physalaemin, was more potent than bradykinin, not only in regard to intensity but even more so as regards duration of action. The dog was the species in which caerulein produced the most constant and unequivocal pressure response. The threshold hypotensive dose by rapid intravenous injection ranged from 10 to 100 ng/kg, by intravenous infusion from 5 to 15 ng/kg/min and by subcutaneous injection from 5 to 10 μg/kg. Doses as high as 10,000 times the intravenous threshold dose could be tolerated with recovery of the animal.

In man, caerulein, when given by intradermal injection, displayed a wealing action approximately as intense as that produced by bradykinin. In the guinea-pig, on the contrary, the action of caerulein on capillary permeability, as assessed from leakage of circulating dye, was negligible.

The mechanism of the pressure fall is probably different in the various animal species. In the dog, hypotension is certainly due, at least to a great extent, to a direct depressor action on the vascular smooth muscle with consequent vasodilatation; in the cat, cholinergic mechanisms seem to play a major role, as shown by the reversal of the action of caerulein from hypotensive to hypertensive, produced by atropine.

## Action on Gastrointestinal Smooth Muscle

Although all examined sections of the gastrointestinal tract (stomach, small intestine, gall bladder) responded *in vivo* to very low doses of caerulein with tone increase and/or reinforcement of movements, the gall bladder seemed to occupy a unique position. In fact, this organ was tremendously stimulated by caerulein, not only when *in situ*, but also when isolated and mounted in a nutrient bath, and under no conditions was tachyphylaxis observed. Moreover, gall bladder stimulation was always atropine-resistant, whereas stimulation of the stomach or small intestine musculature was, at least to a great extent, blocked by atropine. The threshold dose of the polypeptide active on the isolated guinea-pig gall bladder was 0.5 to 2 ng/kg, the dose active on the *in situ* guinea-pig gall bladder was 0.3 to 1.5 ng/kg by quick intravenous injection, and 0.2 to 0.3 ng/kg/min by intravenous infusion. The bladder musculature of nearly all other species examined showed a similar behaviour. On the *in situ* guinea-pig gall bladder, 1 μg caerulein was equiactive to 40 to 50 Ivy dog units of cholecystokinin, i.e. to 10 to 12 μg of pure cholecystokinin-pancreozymin.

In man, cholecystographic examination showed that the threshold intravenous dose of caerulein producing an appreciable contraction of the gall bladder was 1 ng/kg. It is probable that a dose of 10 ng/kg, that is in total 0.6 to 0.8 μg, is the optimal clinical dose to be given intravenously in cholecystography (BRAIBANTI *et al.*, personal communication). For cholecystokinin JORPES, MUTT and OLBE (1959) suggested a dose of 1.5 to 2 μg in total.

The response of the smooth muscle of the gastrointestinal tract,other than that of the gall bladder was,in some respects, puzzling.In fact,whereas the *in situ* musculature of the stomach and the jejunum was,as already stated,very sensitive to caerulein,with apparently little or no tachyphylaxis,all the isolated preparations of gastrointestinal muscle were poorly sensitive to caerulein and prompt tachyphylaxis was nearly always observed.

Concerning the *in situ* musculature,the following responses to caerulein were recorded: contraction of the denervated gastric pouch of the conscious and the anaesthetized dog (threshold intravenous dose of caerulein 5 to 30 ng/kg;threshold subcutaneous dose 0.3 to 0.5 μg/kg);contraction of jejunal loops in the anaesthetized dog(threshold intravenous dose 1 to 3 ng/kg);pylorospasm in the anaesthetized rat(threshold intravenous dose 5 to 15 ng/kg);and finally vomiting and diarrhoea in the intact,conscious dog(threshold intravenous dose 0.5 μg per kg).

Research on this topic is far from complete.In order to have an adequate idea of the actions of caerulein on the gastrointestinal smooth muscle it will be necessary to study comparatively the effects of the polypeptide on the different sections of the gut of various animal species and to elucidate the mechanism which is at the root of these effects.It is,in fact, likely that some actions of caerulein are direct actions on the smooth muscle,but others are mediated through the central or peripheral nervous system.

## Action on Secretions Associated with the Gastrointestinal Tract

The secretory organ apparently most sensitive to caerulein was the pancreas.In the anaesthetized dog,the threshold intravenous dose which stimulated the pancreatic secretion was 2 to 5 ng/kg,while the threshold subcutaneous dose was 100 ng/kg. Doses as low as 0.3 to 0.6 ng/kg/min were effective by intravenous infusion.This dosage is probably of the same magnitude as that active on the gall bladder.Unlike the juice produced by secretin but like that produced by pancreozymin the juice yielded by caerulein was rich in enzymes and dry residue.

If only the increase in volume flow is considered, 1 μg caerulein was equiactive to 35 to 40 μg of human gastrin-I, 1 to 3 Jorpes clinical units of secretin or 60 Ivy dog units of cholecystokinin-pancreozymin,i.e. 15 μg of the pure polypeptide.

Stimulation of pancreatic secretion by caerulein was observed also in the cat and in the rat.

But caerulein displayed,in addition to a powerful cholecystokinin-pancreozymin activity,also a remarkable gastrin activity.

In fact,in conscious dogs provided with denervated fundic pouches,caerulein increased the volume of gastric juice and stimulated acid flow and pepsin output.The threshold subcutaneous dose was 0.2 to 0.5 μg/kg and the magnitude of the response was dose-dependent.The effect produced by a single subcutaneous dose lasted 90 to 150 minutes.Caerulein was similarly highly effective on acid and pepsin secretion when given by intravenous infusion at a rate of 0.5 to 1 μg/kg/hr. However,when administered by quick intravenous injection,the polypeptide was inactive at all examined dose levels,i.e. from 20 ng to 2 μg/kg.Pretreatment with atropine(0.2 to 0.5 mg/kg, intramuscularly)caused a strong reduction in the response to caerulein.Hexamethonium was inactive.Administration of caerulein,either subcutaneously or intravenously,during an intravenous infusion of 0.5 mg/kg/hr of histamine dihydrochloride, produced a blockade of the gastric secretion elicited by the amine.On a molar basis caerulein was 1½ times as potent as human gastrin-I in stimulating acid flow and volume of gastric juice,but 3 to 5 times as potent in stimulating pepsin output.

In the perfused stomach preparation of the rat,caerulein caused a conspicuous increase in the total acid output.The threshold dose was 15 to 25 ng/kg by the intravenous route and approximately 5 μg/kg by the subcutaneous route.On a molar basis,caerulein was 5 to 20 times as potent as human gastrin-I and 120 times as potent as carbachol.

The histamine liberator compound 48/80 strongly reduced acid flow elicited by caerulein whilst aminoguanidine,an inhibitor of diamineoxidase,increased it markedly.

Similar results were obtained when the rat stomach with ligated pylorus was used.

Preliminary experiments revealed that caerulein remarkably

increased the active transport of chloride by the isolated gastric mucosa of the frog and the tortoise. The threshold concentration was of the order of 0.003 to 0.01 ng /ml bath liquid and again caerulein was several hundred times more active than human gastrin-I(PESENTE *et al.*, to be published).

It has been previously stated that caerulein produced a powerful contraction of the *in situ* gall bladder, thus provoking an increase of bile flow. However, it is possible that the polypeptide is capable not only of producing an evacuation of the gall bladder bile, but also of displaying a true choleretic effect, i.e. of stimulating the liver cell to increase the output of bile. In the rat, for example, single intravenous doses of 2 to 5 μg/kg caerulein caused, approximately over a 2-hour period, a 20 to 30% increase in the volume of bile produced. Moreover, the dry residue content and cholesterol content of caerulein bile was as high as, or higher than the dry residue and cholesterol content of control bile.

It is obvious that before drawing definite conclusions on this topic much more experimental evidence is necessary.

We do not yet know whether caerulein stimulates the secretion of enteric juice. Some preliminary observations seem to indicate that this study might be rewarding.

## Occurrence of Caerulein or Caerulein-Like Polypeptides in Nature

Caerulein or caerulein-like polypeptides do not occur only in the skin of *Hyla caerulea* and of kindred Australian hylids, but in the skin of other amphibian species as well. Authentic caerulein seems to be present in the skin of *Leptodactylus pentadactylus labyrinthicus* from North Argentina, and a caerulein-like polypeptide (surely distinct from caerulein) in the skin of *Phyllomedusa sauvagei*, similarly from North Argentina. Isolation and elucidation of the structure of the different caerulein-like polypeptides is well advanced.

## Discussion

The pharmacological data which have been briefly summarized

in this report raise several problems,among which,first and foremost,is that of the structure/activity relationship in the field of caerulein-like polypeptides,and then the related problem of the chemical and pharmacological correlations existing between caerulein and caerulein-like polypeptides on the one hand and the gastrins and cholecystokinin-pancreozymin on the other.

As the problem concerns of the identification,in the molecule of caerulein,of the amino acid residues and sequences necessary for the appearance of the different aspects of the biological activity of the polypeptide,more than twenty caerulein-like polypeptides have been synthesized at the Farmitalia Laboratories for Basic Research,Milan,and to complete our program the synthesis of other compounds is in progress.At present it may be said that the occurrence in the caerulein molecule of the O-sulphate ester of the tyrosinyl residue is of decisive importance for pharmacological activity.Desulphated caerulein still retained about 5% of the activity of caerulein on gastric and pancreatic secretion,but was virtually inactive on dog blood pressure and on guinea-pig gall bladder.

It may be seen from the amino acid composition and sequence of caerulein discussed in a preceding communication by ANASTASI and BERNARDI(this Symposium) that the polypeptide shows a close chemical resemblance to the gastrins,especially to gastrin-II,and apparently also to cholecystokinin-pancreozymin.In fact,the C-terminal pentapeptide and the N-terminal pyroglutamyl residue are the same for caerulein and gastrin-II and similarly both peptides contain a tyrosinyl residue as sulphate ester. Quite recently MUTT and JORPES(1967) found that the C-terminal pentapeptide of cholecystokinin-pancreozymin was most probably that of gastrin and,consequently,that of caerulein.

On the basis of our preliminary results on structure/activity relationship in the field of caerulein-like polypeptides we suggest that the possibility of the occurrence of a tyrosinyl O-sulphate residue in the cholecystokinin-pancreozymin molecule should be checked.

Our pharmacological results,in their turn,raise the problem of whether cholecystokinin-pancreozymin is the actual hormone released by the duodenal wall into the blood stream upon stimulation by the acid chyme passing from the stomach into the duodenum. In fact, cholecystokinin-pancreozymin could simply be a rather large carrier-polypeptide from which a smaller peptide is split off at the moment in which a hormonal stimulation of the gall bladder and the pancreatic secretion is required.

According to available information (MUTT and JORPES, 1967 ; JORPES, this Symposium), cholecystokinin-pancreozymin is an aggregate of 33 amino acid residues with a molecular weight of approximately 4000, whereas caerulein has a molecular weight of 1350. Yet, caerulein is apparently 3 to 5 times more active than the purest available cholecystokinin-pancreozymin (= 4000 Ivy dog units), on a molar basis.

As far as the gastrin-like properties of caerulein are concerned it is worth noting that desulphation produced a 95% reduction of the activity of caerulein on gastric secretion. This is in sharp contrast to gastrins I and II, each of which produced their effects on gastric secretion and in general on alimentary tract functions, in apparently identical form and degree, as repeatedly stated by GREGORY and TRACY, 1964, 1966).

In spite of our efforts, we were unable to get a sample of gastrin-II, and hence we could not check the activity spectrum of both gastrins on our test systems. This gap is very regrettable because Prof. Gregory found that pure gastrin-II in 0.1N hydrochloric acid heated in a boiling water bath for 30 minutes showed a great decrease in its power to stimulate gastric secretion. We wonder whether by this procedure gastrin-II was simply desulphated, i.e. transformed into gastrin-I.

A question which is invariably raised is that of the possible biological significance of active amines and polypeptides in the amphibian skin. At this early stage, an answer to this question seems premature. Before having more information on the occurrence of caerulein or caerulein-like polypeptides in the skin of the different groups of amphibians any hypothesis on the function of the polypeptide would be futile.

Acknowledgement. This work was supported by grants from the Consiglio Nazionale delle Ricerche, Roma, Italy.

## References

GREGORY,R.A. and H.J.TRACY. Gut,5,103-117(1964).

GREGORY,R.A. and H.J.TRACY. In Gastrin (ed.M.I.GROSSMAN). California Press,Berkeley and Los Angeles,1966,9-26.

JORPES,E.,V.MUTT and L.OLBE. Acta physiol.scand.,47,109-114 (1959).

MUTT,V. and E.JORPES.Biochem.Biophys.Res.Communic.,26,392-397 (1967).

# "VASODILATING ACTION OF CAERULEIN AND I.C.I. 50,123 ON THE CANINE PANCREATIC VASCULAR BED"

A. H. GLÄSSER and L. DORIGOTTI

FARMITALIA - Istituto Ricerche, Milano (Italy)

Caerulein is a decapeptide isolated from the skin of the Australian frog Hyla caerulea (1). The determination of the amino-acid sequence and subsequent synthesis have been carried out in these laboratories (2). According to Erspamer et al. (3) this peptide causes a long-lasting blood pressure fall and has gastrin and pancreozymin-like activities on gall bladder motility and on gastric and pancreatic secretions. In the dog, caerulein stimulates external pancreatic secretion at doses as low as 3-6 ng/kg i. v. and decreases blood pressure at doses of 10-100 ng/kg i. v.

In this paper the action of caerulein on pancreatic and on femoral blood flow as well as on other vascular parameters are reported. A comparison with well known vasoactive peptides (angio - tensin II, eledoisin, bradykinin) and with the gastrin-like peptide I.C.I. 50,123 was made in order to elucidate the mechanism by which caerulein acts on pancreatic and on femoral blood flow. I.C.I. 50,123, like caerulein, stimulates gastric and pancreatic external secretions in dogs and in humans (4) and increases gastric blood flow in the dog (5).

Mean and phasic arterial blood flows were measured by the Nycotron electromagnetic flowmeter in dogs anesthetized with urethane-chloralose. Blood pressure was recorded by a Elema electromanometer. Pancreatic blood flow was measured with minimal manipulation of the pancreas by placing a flow probe on the caudal pancreaticoduodenal artery (6). This vessel supplies both a portion of the right pancreatic lobe and the duodenum. Therefore, in some experiments a ligature on the duodenal branch of the artery isolated the pancreatic from the duodenal vascular district.

A summary of results is given in the following table :

Comparative peack effects of peptides on blood flows (pancreaticoduodenal and femoral artery) recorded simultaneously.

| Peptides | * ng/kg i. v. | ** Blood pressure % variation | ** Pancreaticoduodenal artery % variation | ** Femoral artery % variation |
|---|---|---|---|---|
| Caerulein | 2 | 0 | + 20 | 0 |
| | 4 | 0 | + 27 | 0 |
| | 8 | 0 | + 32 | 0 |
| | 20 | - 4 | + 70 | - 22 |
| | 40 | - 6 | + 78 | - 30 |
| | 80 | - 18 | + 61 | - 45 |
| | 200 | - 35 | + 55 | - 50 |
| | 1000 | - 25 | - 36 | - 50 |
| I.C.I. 50,123 | 40 | 0 | 0 | 0 |
| | 200 | 0 | + 28 | 0 |
| | 400 | 0 | + 56 | - 14 |
| | 4000 | - 5 | + 72 | - 29 |
| Bradykinin | 50 | 0 | - 1 | + 3 |
| | 200 | - 11 | - 13 | + 66 |
| | 1000 | - 28 | - 15 | + 115 |
| Angiotensin | 4 | 0 | - 25 | + 17 |
| | 20 | + 6 | - 39 | + 33 |
| | 100 | + 16 | - 71 | + 103 |

* ng = nanograms = $10^{-9}$ g.

** Mean initial values ± S.E. : Blood pressure 161 ± 4.01 mmHg; Pancreaticoduodenal blood flow 16.6 ± 2.18 ml/min; Femoral blood flow 65 ± 5.88 ml/min.

Rapid intravenous injections of natural caerulein at doses as low as 1-4 ng/kg increase blood flow through the caudal pancreaticoduodenal artery in absence of any change of other cardiovascular parameters. Larger doses of caerulein up to 40 ng/kg produced a further marked increase of blood flow together with a decrease of total femoral flow. The vasodilating effect on pancreaticoduodenal vessels reached its maximal value rather slowly 30-50 seconds and lasted 2-15 minutes. No action on other vascular beds (common carotid artery, splenic artery, jejunal artery and duodenal branch of the caudal pancreaticoduodenal artery) were observed. Generally at higher hypotensive doses, caerulein was found to decrease blood flow in these vessels.

In accordance with Erspamer's findings (3) caerulein caused arterial hypotension at doses larger than 40 ng/kg i. v. Sometimes the hypotension was associated with bradycardia and a progressive reduction of pancreaticoduodenal blood flow.

The pentapeptide I.C.I. 50,123 behaved like caerulein being, however, about 25 times less active.

The vasodilating action of caerulein and I.C.I. 50,123 on the pancreaticoduodenal vessels seems to be peculiar to these peptides; only eledoisin, a peripheral vasodilating peptide, but not bradykinin increased temporarily for 15-30 seconds the pancreaticoduodenal blood flow. This effect was followed promptly by a decrease in flow when blood pressure dropped. Similarly the hypotension caused by bradykinin was accompained by a corresponding decrease in blood flow.

In contrast angiotensin II displayed a strong vasoconstriction effect on the same vascular district. In our experiments as in those of White and Ross (7) on cats, angiotensin reduced vascular resistance in the hind limb of the dog.

## REFERENCES

1) Erspamer V. and Anastasi Ada : In "Hypotensive Peptides", pag. 63, Springer Verlag, New York Inc., 1966.
2) Bernardi L. and Anastasi Ada : In "The Proceedings of this Symposium.
3) Erspamer V. et al. : Experientia (in press) 1967.
4) Wormsley K.G. et al. : Lancet, i, 993, 1966.
5) Haigh A.L. et al. : J. Physiol. (London), 191, 45P, 1967.
6) Miller M.E. et al. : In "Anatomy of the Dog", pag. 350, W.B. Saunders & Co., Philadelphia, 1964.
7) White F.N. and Ross G. : Am. J. Physiol., 210, 1118, 1966.

# STUDIES ON AN URINARY PROTEIN FACTOR OF GASTRIC ANTI-ULCER ACTIVITY (UROGASTRONE)

A.Corbellini, G. Lugaro, E.Crespi

Dept. Organic Chemistry, University of Milan

Italy

The literature on Urogastrone is very confused and full of gaps: many Authors have studied Urogastrone (and Enterogastrone which is a similar substance), both from the physico-chemical and the biological point of view. Yet, the results which have been obtained are only partial and often contradictory and usually rather difficult to reproduce.

Moreover, after reaching a certain stage, many people have given up researches on this subject, mainly for the difficulty to get the raw material.

The first knowledge on Urogastrone dates back to about thirty years ago, when Sandweiss (Am. J. Dig. Dis., 5, 24, 1938; ibid., 6, 6, 1939) discovered that a substance, which he called Antuitrin S, extracted from the urine of pregnant women, had a very powerful action in preventing and treating experimental ulcers in dogs (Mann-Williamson ulcers).

The prophylactic and therapeutic effects of such an extract was first attributed to the presence of chorionic gonadotropin in the urines. It was later recognised that this antiulcer factor was also present in the urine of normal women, of healthy men and in the urine of dog, horse and ox.

As we shall refer later, we have found that the anti-ulcer activity and the chorial gonadotropin activity can be clearly separated.

The likeness of effects between the urinary and the intestinal extracts on the gastric secretion inhibition could give the impression that the urinary gastric inhibitor is nothing else but Enterogastrone (probably even partly metabolized) which would have been eliminated by kidneys during the pregnancy.

The various extractive methods used, do not permit yet an exact comparison between the active factors present in the urine of several kinds of animals. Sandweiss thinks that Urogastrone is a factor having a prophylactic, therapeutic and immunizing effect against Mann-Williamson ulcers without depressing gastric secretion; however, this statement cannot be considered valid because, in all the preparations known till now, the action inhibiting the gastric secretion is always accompanied in any stage of purification by the anti-ulcer one.

However, during the physiological research, these extracts aroused such an interest as potential agents for the gastric-duodenal ulcer therapy, as to lead to a certain confusion of terminology also due to the fact that there might be a possibility that a single chemical entity can be responsible for all the biological effects.

We are defining Urogastrone as Urinary Gastric Secretory Depressant, to discriminate it from the Enterogastrone or Enteric Gastric Secretory Depressant. Two factors remain at a hypothetical state: i.e. the Gastric Motor Depressant (Urinary and Enteric respectively) and the Anthelone factor (Uroanthelone and Enteroanthelone).

As we have already said, the literature has provided us with data which are rather incomplete, fragmentary and often contrasting as the Authors who have studied the subject under this aspect have found several heterogeneous compounds having various origins. The latest researches, carried out in order to isolate in pure state the biologically active factors present especially in the pregnant woman'urines, have been carried out by Gregory in 1955 (J. Physiol., 129, 528, 1955) and by Mongar and Rosenoer in 1962 (J. Physiol., 162, 163 and 173, 1962).

Gregory tried to purify Urogastrone by using resins and calcium phosphate gel column: in this way he reached a remarkable purification.

By operating a chromatographic purification on resins, Mongar and Rosenoer could obtain an Urogastrone provided with a good antisecretory activity valued on a test of continuous secretion in the rat; however the substance was not completely pure.

Therefore even this preparate cannot yet be compared with the other ones because of its lack of definitive values of activity designed in common use units.

For some time, we have devoted ourselves to the Urogastrone study. Indeed Urogastrone is a factor provided, as we already said, with gastric anti-ulcer, antisecretory and anti-motor activity, and is extracted from the urines of three month pregnant women. The urine is passed on an active charcoal column where the active material is absorbed; it is then eluted in an acid medium and precipitated by an acetone and ether mixture. After a second precipitation by ammonium sulphate in alkaline medium, the acid solution is chromatographed on Amberlite IRC-50, then eluted at alkaline pH.

It is then passed on calcium phosphate gel column at neutral pH in phosphate buffers, in order to e-liminate several pigmented substances; then again on Amberlite IRC-50 at acid pH ($< 4$).

Elution with volatile buffers at alkaline pH and liophylization then follows. We can obtain about 10-15 mg of active substance from a hundred liters of urine. The Urogastrone prepared in that way is hydrosoluble in a slightly alkaline medium.

An excellent solubilization of the active factor can be reached by extracting it in aqueous solution at pH 10. In such a way 80% of the product and more than 95% of the biological activity is solubilized. Nevertheless we can notice that Urogastrone is very sensitive to alkaline pH. In alkaline conditions its activity is rapidly lost. On the other hand it is much more stable in a neutral or weakly acid solution.

Such a product - considered as one of the purest obtained until now - has been submitted to serial fractionations by suitable methods (electrophoresis, electrochromatography, chromatography, gel filtration, etc.) thus obtaining at least three compounds, two of which are proteic. Among these compounds, only one is active either as an antisecretory agent or as an anti-ulcer one. The best separations were obtained by gel filtration carried out in various experimental conditions.

To test the fractions obtained, we have evaluated the antisecretory and anti-ulcer activity; the anti-motor activity was studied later on the isolated stomach.

We have also examined, during the different steps of purification, the distribution of the chorial gonadotropic activity as this activity is present in our raw material at about 20 I.U./mg level. It may be added that this value is much less than that obtained by the other Authors in their preparations.

For testing the anti-ulcer activity we are using a modification of the Shay's Test: therefore, the Shay Rat Unit (S.R.U.) is defined as the amount of substance which inhibits the formation of gastric ulcers during the following 8 hours treatment. The gastric ulcers have been obtained by pyloric ligature. Due to the great variation in the sensitivity of animals which depends on the strain, season, diet, environment conditions etc., the Shay Rat Unit does not possess a constant value and can undergo even wide oscillations.

Then we are defining as Antisecretory Unit (A.S.U.) the quantity of substance which in male rats weighing 180-200 g, inhibits gastric secretion by 50% two hours after pyloric ligature.

Operating in the best conditions, the raw Urogastrone prepared by us and whose antisecretory activity is equal to 7000 A.S.U./g and anti-ulcer activity equal to 3000 S.R.U./g, can be separated by gel filtration on Sephadex G-25 in phosphate buffer 0.005 M, pH = 7.3, into five fractions conventionally named and eluted in the following order: $A_2$, $A_1$, $B_1$, B, $B_2$ (figure 1).

Fraction $A_2$ which has as we shall see the biological activity of Urogastrone, does not appear homogeneous to free electrophoresis and acrylamide gel electrophoresis.

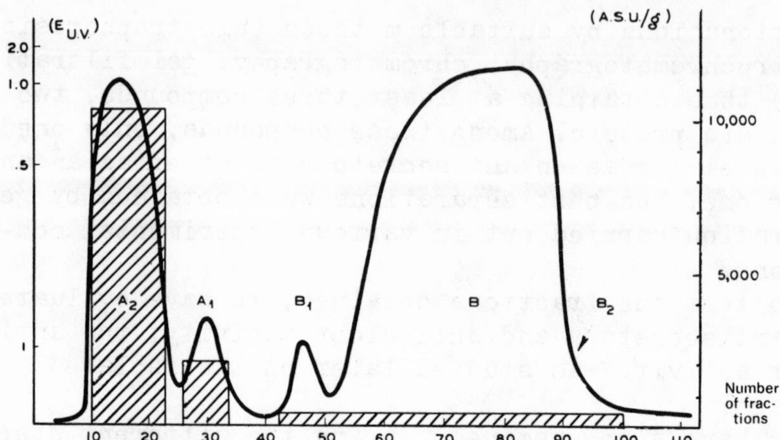


Fig. 1.- Gel filtration of raw Urogastrone in phosphate buffer pH = 7.3 on Sephadex G-25 M column 2.5 x 60 (h) cm. Elution in 4 ml portions every 6 minutes approx. (40 ml/hour). The shaded areas indicate the antisecretory activity in A.S.U./g.

Such a fraction presents a positive Schiff reaction. This would suggest that it is a glycoprotein: a merely indicative analysis shows the evidence of a glucidic compound (ca 20%). The Isoelectric Point, approximately determined, has a value of ca 4.2. We can notice that other Authors, such as Friedman (Am. J. Dig. Dis., 13, 108, 1946) and Huff (Arch. of Biochem., 25, 133, 1950), have found for their compounds, characterised by a high antisecretory activity, an I.P. included between 4.2 and 4.5.

The biological activity of fraction $A_2$ has been rising up to 5,000 S.R.U./g and 10,500-11,000 A.S.U./g; therefore 90% of the starting activity can be found again in the first case and 80% in the second one.

It can be excluded that fractions $A_1$, $B_1$, B and $B_2$ present a significant anti-ulcer and antisecretory activity.

A trace of antisecretory activity, produced by the contamination of $A_2$, can be found in the $A_1$ and $B_1$ fractions. They correspond, in terms of specific activity, to 5% from that of $A_2$.

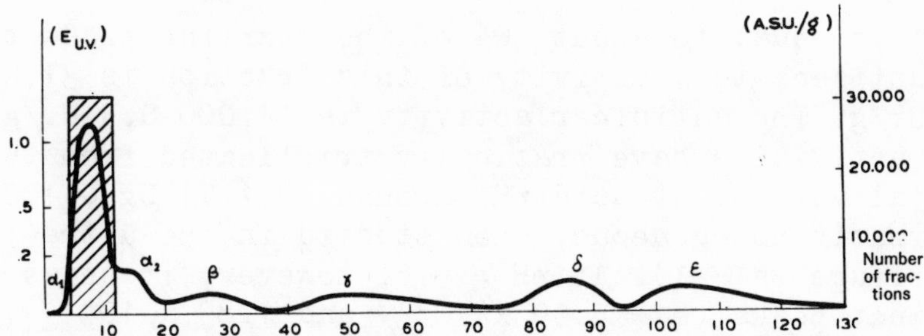


Fig. 2.- Gel filtration of fraction $A_2$ in M/1 formic acid on Sephadex G-100 column 2 x 80 (h) cm. Elution in 2 ml portions every 5 minutes approx. The shaded area indicates the antisecretory activity in A.S.U./g.

The fraction B, highly absorbing in the U.V., represents 12-15% of the whole, and is composed mainly of uric acid salts. Fraction $A_1$ would seem a degradation product of the $A_2$. This fraction is inactive and has a low molecular weight ($<$ 4,000); the $B_1$ is an urinary pigment as well as the $B_2$ which is also an urinary pigment connected with uric acid.

One can therefore observe how the fraction provided with an anti-ulcer action has also an antisecretory activity. Those fractions which are pratically inactive as anti-ulcer are equally inactive as anti-secretory.

Fraction $A_2$ has been purified by means of various methods. Ion Exchange on DEAE and on DEAE-Sephadex, followed by elution in saline concentration gradient, separate at least two pigmented and biologically inactive fractions but the biological activity is distributed on a large range of saline concentration (from 0.2 up to 0.8 M in NaCl). The yield is very low and the specific activity is much lower.

A good purification method is based on the treatment of fraction $A_2$ in 1 M formic acid and subsequently gel filtration on Sephadex G-100 column, or even in formic acid. Thus we can obtain a fraction which comes out with the "Void Volume", and which is called Ug G-100 (figure 2).

It is equal to about 30% of the starting product. The antisecretory activity of this fraction is 31,000 A.S.U./g. The antiulcer activity is 14,000 S.R.U./g. Both activities have pratically triplicated from the initial values. At acid pH (around pH 1.7) Ug G-100 is apparently homogeneous, when studied in the ultracentrifuge (s = 1). At pH = 9.6, however, it shows two distinct peaks ($s = 5.59 \times 10^{-13}$ and $1.89 \times 10^{-13}$).

The Ug G-100 is the first Urogastrone compound free from gonadotropic activity. The determination on immature male rats, immature female rats and on immature hypophysectomized rats, has not shown any evidence of gonadotropic activity even at 1 mg dose per animal - a dose which would be equivalent to a content of HCG inferior to $5.10^{-5}$ mg/mg, i.e. less than 1:20,000.

Its LH activity, valued by the ovarian ascorbic acid depletion test on pseudo-pregnant female rats is less than 0.008 Unit NIH-LH per mg, and consequently negligible. When determined by the ovarian cholesterolic depletion test, in pseudo-pregnant female rats, 1 mg of Ug G-100 corrisponds to $7.10^{-5}$ U. NIH-LH, and this value is 100 times less than the previous one.

The Ug G-100 does not possess activity of FSH type, when evaluated by the augmentation test of Steelman and Pohley.

The Ug G-100 does not inhibit an active dose of Human Chorionic Gonadotropin (HCG) on immature male and female rats.

This effect had already been noticed by various Authors on their more or less purified compounds. Therefore, the gonadotropic activity can be attributed to the presence of HCG as an impurity and not to a specific action of the Urogastrone molecule.

The HCG present in the starting material is found again quantitatively in the effluent from the column, but with a certain delay regarding the "Void Volume", and Ug G-100. This is because of its lower molecular weight. Till now the attempts to reach a subsequent purification of Ug G-100 have not yet provided valuable results; we are now trying to use block acrylamide gel electrophoresis for further fractionation.

Another way of purifying the $A_2$ fraction prepared by Sephadex G-25 consists in passing it on Sephadex G-100

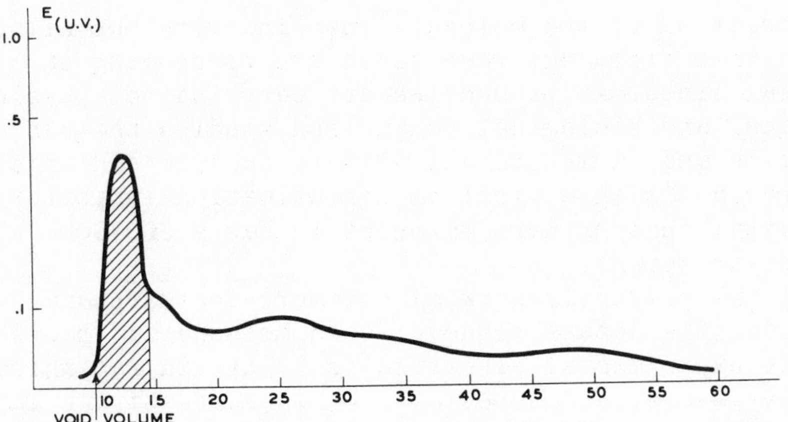


Fig. 3.- Gel filtration of fraction $A_2$ in phosphate buffer pH = 7.3 on Sephadex G-200 column 2.5 x 40 (h) cm. Elution in 3 ml portions every 10 minutes approx. The shaded area indicates the antisecretory activity.

in phosphate buffer 0.005 M at pH = 7.3 (figure 3).

Thus, we can obtain a fraction coming out with "Void Volume", whose antisecretory activity has bee rising to 50,000-60,000 A.S.U./g, giving a compound which is at least 8-10 times more active than those reported up to-day in leterature. The anti-ulcer activity is ca 25,000-28,000 S.R.U./g. Same results can be obtained when operating on Sephadex G-200, that is where the active fraction elutes with the column "Void Volume". Studies on such a fraction are presently being carried on. The ultracentrifugal results show the presence of two compounds ($S = 9.23 \times 10^{-13}$ and $2.62 \times 10^{-13}$). Besides this fact, more than one group $-NH_2$ terminal can be found.

Even in such a case, we cannot understand how $S = 2.62 \times 10^{-13}$ compound can turn out with the "Void Volume" of Sephadex G-200 column: we might think of a subsequent dissociation of the active molecule into a smaller molecule. By treating such a fraction in 0.4 per cent sodium dodecyl sulphate, this fraction looses all its antisecretory and anti-ulcer activity; with the ultracentrifuge we can obtain a single peak at $S = 2.6 \times 10^{-13}$. This would suggest the effective dissociation of the molecule into inactive subunits.

In this field our researches are proceeding on several lines. At present we are carrying out physico-chemical and biological comparison studies between this fraction and Tamm-Horsfall urinary mucoprotein which is provided with a viral emoagglutination inhibitory activity: this urinary mucoprotein has a 200,000 molecular weight.

We are presently working out more decisive and reproducible dosage methods; such methods are based on the experimental electrization ulcer in rat which is reversible, and especially on the Urogastrone inhibiting action at the level of carbonic anhydrasis of the gastric mucosa. On the other hand we are also studying the Urogastrone action on the gastric motility of the stomach isolated by nervous and chemical stimulant.

# BIOGENESIS OF ERYTHROPOIETIN*

Albert S. Gordon, Esmail D. Zanjani, Edwin A. Mirand and Nathan Back

Lab. Exp. Hematol., Dept. of Biol., Grad. Sch. of Arts and Sci., New York University, N.Y. and Roswell Park Memor. Inst., N.Y. State Dept. of Health, State University of N.Y. at Buffalo, Buffalo, N.Y.

The concept that the kidney serves as the principal site of production or activation of erythropoietin (erythropoiesis stimulating factor, ESF) has gained credence from several lines of evidence. The experiments of Jacobson *et al.* (12) established that nephrectomy inhibited the erythropoietic response of rats to hemorrhage, lowered barometric pressures and cobalt administration. Nephrectomized animals also showed a markedly diminished capacity to produce ESF when exposed to these stimuli (13). The appearance of ESF in fluids circulated through isolated and *in situ* perfused kidneys exposed to cobalt (8) and to hypoxic conditions (16,21,24) has also lent support to the renal origin of ESF. Although it has been argued that ESF is released only from injured or disintegrating renal tissue (7), other studies (9) have shown no correlation between ability to produce ESF and degenerative cellular changes in the perfused kidney preparation.

Despite the impressive evidence favoring a relation of the kidney to ESF production, attempts to extract this factor from the kidney have either proved fruitless or have met with varying degrees of partial success. In this regard, saline homogenates of kidneys from normal and anemic animals showed some erythropoietic activity and the responses obtained with anemic preparations were greater than those noted with kidney homogenates from normal animals (17,19,22). Activity has also been reported to exist in boiled and acidified (pH 5.5) homogenates of kidney tissue (22). However, the erythropoietic activity contained in supernatant fluids derived from renal homogenates, prepared at neutral pH, in general has been low (22). In addition, consideration has not been given to the possibility that the erythropoietic activity of

*Supported by Grants from the N.I.H. of the U.S.P.H.S.

these extracts was due to ESF contained in blood trapped within the kidney.

Because previous workers generally utilized isotonic media for extraction, it seemed possible that their failure to detect significant quantities of ESF in kidneys might have been due to the activity being particulate bound and not readily extractable with isotonic fluids. Accordingly, a 2-step isotonic-hypotonic extraction procedure was applied to rat kidneys (3,5,6). Adult female Long-Evans rats were rendered hypoxic by continuous exposure to 0.5 atmosphere of air for 17 hours. The animals were then immediately exsanguinated and their kidneys removed, minced and homogenized in cold isotonic saline. The homogenate was centrifuged at 37,000 g for 30 minutes and the sediment was washed in isotonic saline. The isotonic supernatants were combined. The sediment was now re-homogenized in 0.02 M phosphate buffer at pH 6.8 and the supernatant was collected by centrifugation. Erythropoietic activity of the isotonic and hypotonic supernatants was assayed in mice rendered polycythemic by discontinuous exposure to hypoxia (23). Values obtained for incorporation of radioiron into circulating red cells have more recently been converted into ESF International Units by reference to the dose-response curve for assay mice injected with the International Reference Preparation of ESF (1). This assay as now developed in our laboratory is sufficiently sensitive to detect as little as 0.02 units of ESF.

Fig. 1 indicates that the isotonic supernatant possessed small amounts of ESF. No change in activity was observed after incubation of this saline supernatant with an equal volume of normal rat serum at 37°C for 15 minutes. It will also be noted that a greater amount of erythropoietic activity was present in the hypotonic than in the isotonic supernatant. Of even greater interest was the finding that incubation of the hypotonic supernatant with normal rat serum led to the appearance of more than twice the activity in the original unincubated hypotonic extract.

We have interpreted these findings as follows. The isotonic extract contains residual ESF, that which is trapped in the renal vasculature. The hypotonic extract possesses 2 types of activity: 1. true renal ESF which is normally present on or within particles and 2. a factor which when incubated with normal serum engenders the production of greater quantities of ESF than can be extracted normally from kidneys of hypoxic animals. Kuratowska _et al_. (15) have reported a similar type of activity in Tyrode solution perfusates passed through isolated hypoxic rabbit kidneys.

In order to clarify the nature of this renal erythropoietic activity, subcellular fractionation of the kidney was undertaken (3,6). Kidneys were obtained from adult female rats rendered hypoxic by exposure to 0.4 atmosphere of air for 16 hours. The nuclear, heavy-mitochondrial, light-mitochondrial and microsomal fractions were collected. Each fraction was resuspended in 0.02 M

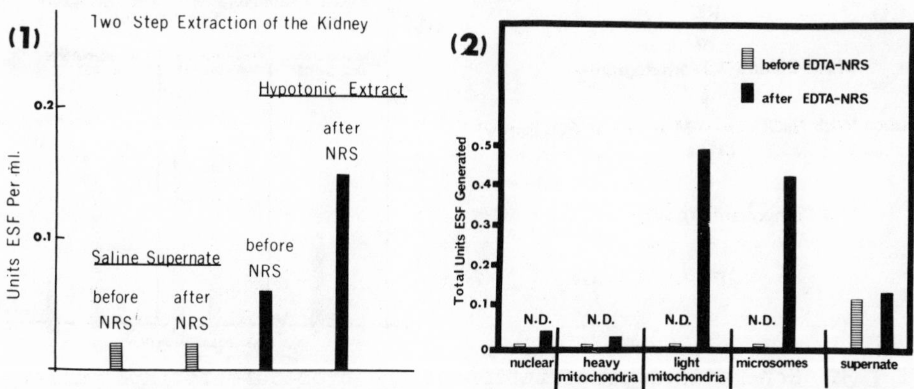


Fig. 1. Erythropoietic activity of isotonic and hypotonic renal extracts before and after incubation with normal rat serum (NRS). Data from BLOOD (6).
Fig. 2. Erythropoietic activity of subcellular fraction extracts of kidney before and after incubation with NRS. N.D.-No activity detectable. Data from BLOOD (6).

phosphate buffer (pH 6.8) and centrifuged at 37,000 g for 30 minutes. The supernatant of each fraction was then assayed for ESF activity before and after incubation with normal rat serum (NRS). No erythropoietic activity was detectable in the nuclear and heavy-mitochondrial fractions either before or after incubation with NRS (Fig. 2). It will be observed, however, from Fig. 2 that while the light-mitochondrial and microsomal fractions exhibited no erythropoietic activity when injected alone in polycythemic mice via the intraperitoneal route, considerable amounts of activity were generated after incubation with NRS. It should be emphasized that NRS showed no activity when assayed alone. The activation phenomenon is a characteristic property of the light-mitochondrial and microsomal extracts (fractions collected at 21,000-105,000 g) of renal tissue. We have designated this activity as the Renal Erythropoietic Factor (REF).

A considerable degree of purification of the REF has been achieved by utilizing DEAE-cellulose chromatographic separation procedures (4). A scheme of our recent purification procedure is shown in Fig. 3. Fraction D contains 25 times the specific activity of the relatively crude REF. It is of interest that the purified REF possessed no component which corresponded to circulating ESF in either mobility on the column or erythropoietic behavior. Moreover, the REF did not exhibit the acid and heat stability of ESF derived from hypoxic rat plasma (3,6).

Further experiments on the nature of the REF-serum interaction have indicated that when the incubations were conducted for longer than 10 minutes, gradually diminishing quantities of ESF

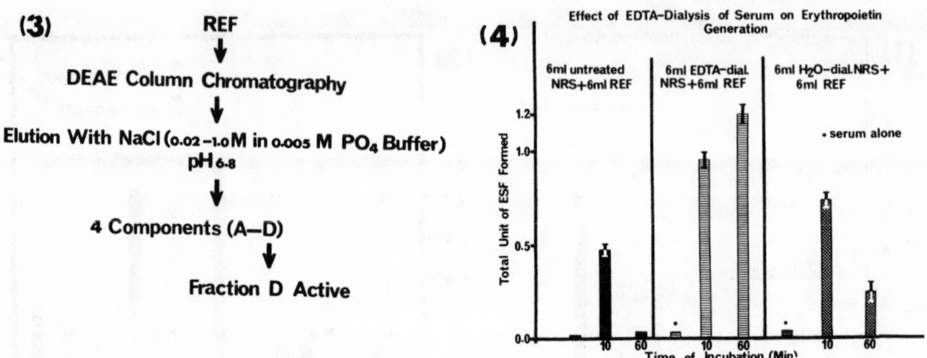


Fig. 3. Schematic representation of the procedure for REF purification. Data from Ann. N.Y. Acad. Sci. (4).
Fig. 4. Effect of dialyzing serum against EDTA on generation of ESF in incubation mixtures containing serum and REF. Data from SCIENCE (25).

appeared in the incubation mixtures. Fig. 4 illustrates a representative experiment. Here 6 ml of REF-containing fluid, extracted from 3 gm of hypoxic rat kidneys, were incubated with 6 ml of NRS. It is noted that a total of approximately 0.45 units of ESF were generated in 10 minutes. When, however, the incubation time was extended to 60 minutes, no erythropoietic activity was demonstrable in the incubation mixture. It seemed possible that the inactivation phenomenon could have been due to the presence of substances in the incubation mixture that were ion-dependent. Therefore, normal rat serum was dialyzed for 24 hours in cellulose tubing against 0.005 M EDTA at 4°C and then immediately re-dialyzed against deionized water for 24 hours at 4°C. The latter dialysis procedure removed essentially all of the excess EDTA in the serum. The dialysand was now used for the incubation experiments. It is apparent from Fig. 4 that when the incubations were carried out with EDTA-dialyzed serum, approximately twice the amount of ESF was produced at 10 minutes than when undialyzed serum was used. Whereas no detectable ESF was evident after 60 minutes when undialyzed serum was employed, incubation mixtures containing the EDTA-dialyzed serum and REF showed significantly greater erythropoiesis stimulating activity at 60 than at 10 minutes. The generation of ESF with the serum dialyzed against only deionized water (EDTA not used at all) was greater than that seen with undialyzed serum, but was significantly less, at both the 10- and 60-minute incubation periods, than that observed with the EDTA-dialyzed serum (Fig. 4).

Since the inactivation process was prevented by dialysis of serum against EDTA, the destructive factor apparently required a cation or cations present in serum. The inactivating agent was

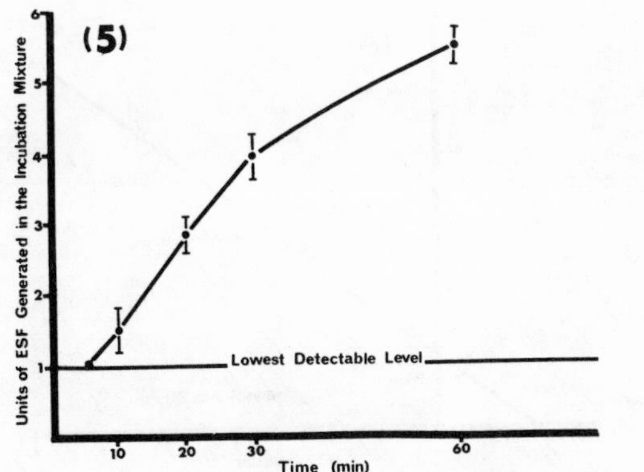


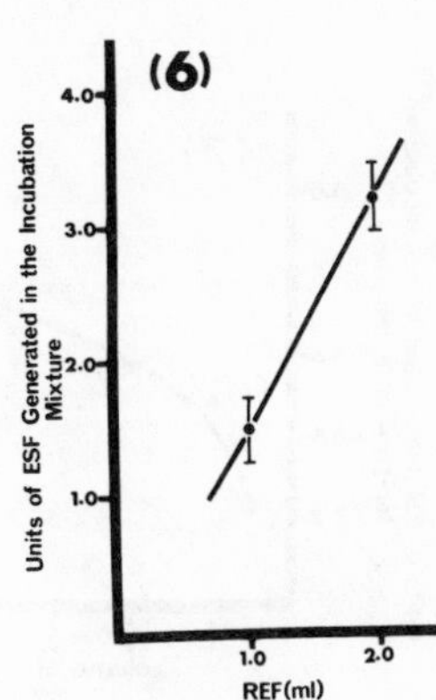


Fig. 5. Formation of ESF in the REF-serum reaction mixture as a function of time. Total volume was 54 ml (1 ml REF, 48 ml serum, 5 ml deionized water). Data from Proc. Soc. Exptl. Biol. Med. (26).
Fig. 6. Formation of ESF in the REF-serum reaction mixture as a function of REF concentration. Volume of dialyzed serum was 48 ml and incubation time 15 minutes. The volume of the mixture was adjusted to 54 ml. Data from Proc. Soc. Exptl. Biol. Med. (26).

probably not a component of serum since the incubation of relatively pure Step IV sheep plasma ESF with undialyzed serum caused either potentiation or no effect on the activity of the hormone rather than its destruction (11,18,23). The inactivating factor is probably a component of the REF-containing extract. In fact, it has been reported that kidney homogenates can inactivate exogenous ESF (10). It would thus appear that these extracts probably contain an enzyme which, in the presence of a serum-borne metal ion, destroys newly formed ESF or a precursor substance in the incubation mixture.

The elimination of the inactivation process made it possible to investigate some aspects of the kinetics of the REF-serum interaction (26). The total quantities of ESF produced in a mixture of 1 ml of REF (equivalent to 0.5 gm of hypoxic rat kidney) and 48 ml of dialyzed NRS, after varying times of incubation are shown in Fig. 5. It will be noted that a progressive increase in ESF formation occurred with a prolongation in incubation time. Keeping the amount of dialyzed serum, total volume of the mixture and incubation time constant, it was found that doubling the REF concentration resulted in approximately twice the amount of ESF produced (Fig. 6). When the volume of dialyzed serum was varied (0.5 ml to 6.0 ml), while the REF (6 ml), total volume (12 ml) and incubation time (15 minutes) were held constant, the reaction rate was seen to rise with an increase in serum content of the mixture (Fig. 7).

The finding that the production of ESF by the action of the REF on a serum substrate is proportional to the concentration of

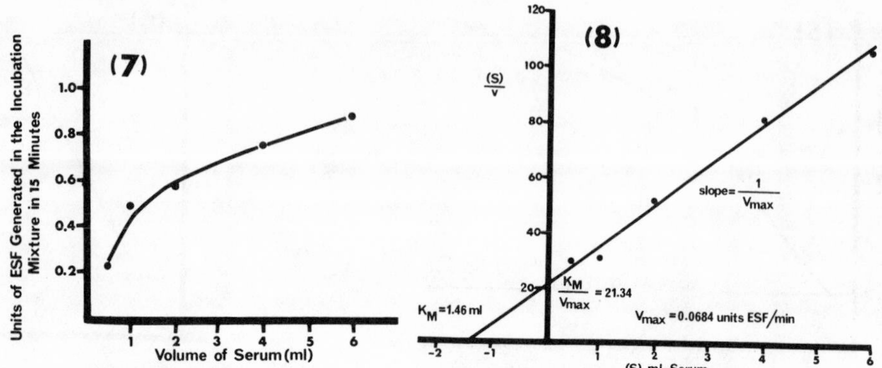


Fig. 7. Relation of volume of serum to quantity of ESF produced in 15 minutes. Volume of REF was 6 ml. Volumes of all mixtures were brought to 12 ml by addition of deionized water. Data from Proc. Soc. Exptl. Biol. Med. (26).

Fig. 8. Modified Lineweaver-Burke plot of the relation of rate of ESF production to the level of serum substrate. Data from Proc. Soc. Exptl. Biol. Med. (26).

REF (Fig. 6) suggests that we are dealing with a first- or zero-order enzymatic reaction. The reaction rate was also speeded when the substrate concentration was increased (Fig. 7). The modified Lineweaver-Burke plot of these data (Fig. 8) indicated a linear relation between substrate levels (S) and (S)/v (substrate divided by reaction rate). The reciprocal of the slope of the line gave the maximum velocity of the reaction ($V_{max}$) as 0.068U ESF/min which could occur when the enzyme was saturated with the serum substrate. This value proved to be somewhat greater than the rate of ESF production noted (0.056U ESF/min) when 6 ml of dialyzed serum was present in the reaction mixture. In these experiments the rate of ESF production was still increasing, the enzyme was not yet saturated with the substrate, and therefore a higher rate of reaction was still possible. The Michaelis constant ($K_M$) is indicated on the graph by the intercept on the abscissa, 1.46 ml. This behavior characterizes first-order enzyme-substrate reactions which supports strongly the contention that the REF-serum reaction is enzyme-catalyzed (26).

An alternative possibility concerning the *in vitro* formation of the ESF by the REF envisions the coupling or conjugation of the REF to a serum protein. Since two substrates, the REF and the protein carrier, would be involved in this type of reaction, second-order and not first-order kinetics should be demonstrable for the ESF-generating reaction. Thus it seems unlikely that ESF is formed by the coupling or activation of the REF by a serum protein moiety. In addition, if a non-enzymatic coupling of the REF to a serum protein constituted the mechanism of ESF formation, it

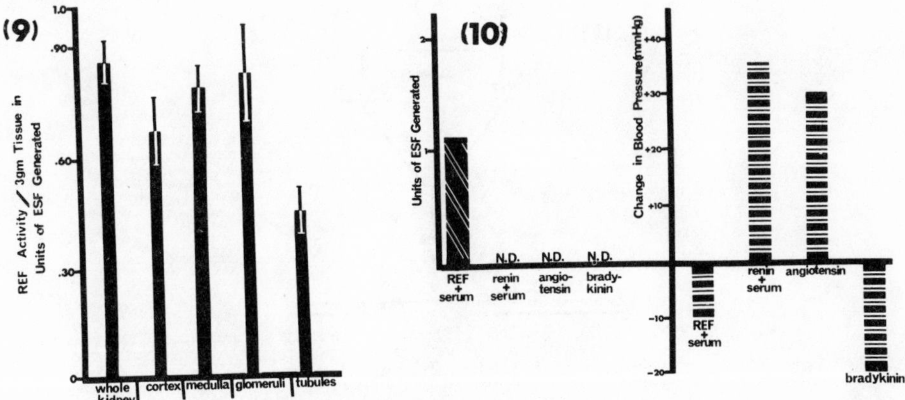


Fig. 9. Units of ESF generated by incubating REF extracted from 3 gm of various fractions of hypoxic rat kidney with 6 ml of EDTA-NRS for 60 minutes. Unpublished data.

Fig. 10. Effects of REF, renin, angiotensin II and bradykinin on erythropoiesis in polycythemic mice and blood pressure in normal rats and mice. Data from SCIENCE (25).

would be necessary for the kidney to synthesize the REF continuously during hypoxia. In this connection, Katz et al. (14) have reported that although there was significantly greater $^{14}C$-isoleucine incorporation into kidney proteins during exposure of rats to 1 hour of hypoxia, these values did not differ significantly from the controls after longer exposures to hypoxia (2-18 hours). An enzyme reaction, on the other hand, would not require continuous synthesis of the REF.

In an effort to determine the distribution of the REF, whole kidney, cortex and medullary regions were employed. In addition, renal cortical tissue was separated into glomerular and tubular regions according to the methods of Nagano and Schafer (20). Fig. 9 indicates the total quantity of ESF formed in a mixture of 6 ml EDTA-dialyzed serum and 6 ml of REF-containing fluid extracted from 3 gm of these five renal preparations from hypoxic rats. In general, the REF was approximately equally distributed throughout the kidney. Although the activity in the cortical tubular preparation appeared somewhat lower than in the glomerular fraction (Fig. 9), the difference was not significant ($P > 0.05$).

Both the REF and renin are kidney enzymes which act on a plasma protein substrate. However, their different chemical and physiologic properties militate against their being the same substance. Thus, renin is stable in acid solutions (pH 2.5), while the REF is inactivated at pH 5.0. Moreover, intraperitoneal administration to polycythemic assay mice of 3.7 Goldblatt units of purified renin, incubated for 60 minutes in saline or in dialyzed serum, did not result in stimulation of erythropoiesis (Fig. 10).

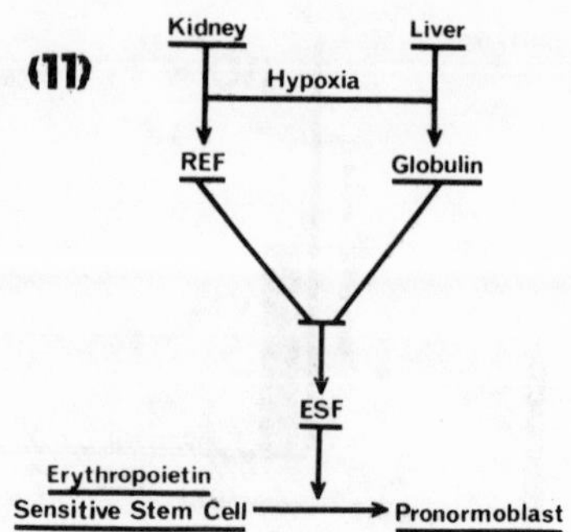


Fig. 11. Proposed scheme for the kidney-liver involvement in the regulation of erythropoiesis. Unpublished data.

Likewise angiotensin II (9 ug or 30 rat units per mouse) failed to augment erythropoiesis in polycythemic mice (25). However, similar quantities of both renin and angiotensin II increased blood pressure when administered intravenously to normal rats and mice (Fig. 10). Neither the REF nor the incubation mixture of REF and serum produced any vasopressor action. On the contrary, a slight drop in blood pressure was observed with the REF-serum mixture. This mild vaso-depressor action of the mixture of REF and serum noted in rats and the report of a kinin-forming enzyme in the light-mitochondrial fraction of rat kidneys (2) led us to test the effect of bradykinin on erythropoiesis. Intravenous injections of 2.0 and 20 ug of bradykinin triacetate evoked no demonstrable erythropoietic response in polycythemic mice (Fig. 10). This agent did, however, produce the expected drop in blood pressure of rats.

## SUMMARY

1. A renal erythropoietic factor (REF), devoid of vasopressor activity exists in light-mitochondrial and microsomal extracts of kidneys from hypoxic rats. Upon incubation of the REF with normal rat serum *in vitro*, the ESF is produced.
2. When undialyzed serum is used, loss of erythropoiesis stimulating activity occurs in the incubation mixtures, a phenomenon which appears to depend on a component associated with the REF preparations and ions existent in serum.
3. The production of ESF as a function of time of incubation of the REF with serum, conforms to a first-order reaction. The REF behaves as an enzyme acting on a substrate present in normal serum to form the ESF (Fig. 11).
4. The REF is found in approximately equal amounts in renal cortical and medullary tissues.
5. The extraction and purification of large amounts of the REF may make it possible for human ESF to be produced in quantity by

in vitro reaction of the REF with a component of normal human plasma.

REFERENCES


1. Camiscoli, J. F., Weintraub, A. H. and Gordon, A. S. (1967). Comparative assay of erythropoietin standards. Ann. N.Y. Acad. Sci. in press.
2. Carvalho, I. F. and Diniz, C. R. (1966). Kinin-forming enzyme (kininogen) in homogenates of rat kidney. Bioch. Biophys. Acta 128, 136.
3. Contrera, J. F. and Gordon, A. S. (1966). Erythropoietin: production by a particulate fraction of rat kidney. Science 152, 653.
4. Contrera, J. F. and Gordon, A. S. (1967). The renal erythropoietic factor. I. Studies on its purification and properties. Ann. N.Y. Acad. Sci. in press.
5. Contrera, J. F., Camiscoli, J. F., Weintraub, A. H. and Gordon, A. S. (1965). Extraction of erythropoietin from kidneys of hypoxic and phenylhydrazine-treated rats. Blood 25, 809.
6. Contrera, J. F., Gordon, A. S. and Weintraub, A. H. (1966). Extraction of an erythropoietin producing factor from a particulate fraction of rat kidney. Blood 28, 330.
7. Erslev, A. J., Solit, R. W., Camishion, K. C. and Ballinger, W. F., II. (1965). Perfusion of a lung-kidney preparation. Am. J. Physiol. 208, 1153.
8. Fisher, J. W. and Birdwell, B. J. (1961). The production of an erythropoietic factor by the in situ perfused kidney. Acta Hemat. 26, 224.
9. Fisher, J. W. and Langston, J. W. (1967). The influence of hypoxemia and cobalt on erythropoietin production in the isolated perfused dog kidney. Blood 29, 114.
10. Fisher, J. W., Porteous, D. D., Hiroshima, K. and Tso, S. C. (1966). Effects of cobalt on activity of sheep erythropoietin in rat kidney homogenates. Proc. Soc. Exptl. Biol. Med. 122, 1015.
11. Garcia, J. F. and Schooley, J. C. (1965). Dose-response relationships for erythropoietin given in serum or in saline. Proc. Soc. Exptl. Biol. Med. 120, 614.
12. Jacobson, L. O., Goldwasser, E., Fried, W. and Plzak, L. (1957). Role of the kidney in erythropoiesis. Nature 179, 633.
13. Jacobson, L. O., Goldwasser, E., Gurney, C. W., Fried, W. and Plzak, L. (1959). Studies on erythropoietin: the hormone regulating red cell production. Ann. N.Y. Acad. Sci. 77, 551.
14. Katz, R., Cooper, G. W., Gordon, A. S. and Zanjani, E. D. (1967). Studies on the site of production of erythropoietin. Ann. N. Y. Acad. Sci. in press.

15. Kuratowska, Z., Lewartowski, B. and Lipinski, B. (1964). Chemical and biologic properties of an erythropoietin generating substance obtained from perfusates of isolated anoxic kidneys. J. Lab. Clin. Med. 64, 226.
16. Kuratowska, Z., Lewartowski, B. and Michalak, E. (1961). Studies on the production of erythropoietin by the isolated hypoxic kidney. Bull. de l'Acad. Polonaise des Sci. 8, 77.
17. Lagrue, G., Boivin, P. and Branellec, A. (1960). Activite erythropoietique d'un extrait renal. Rev. Franc. D'Etudes Clin. Biol. 5, 816.
18. Moores, R. R., Gardner, E., Jr., Wright, C-S. and Lewis, J. P. (1966). Potentiation of purified erythropoietin with serum proteins. II. Serial dose response relationships. Proc. Soc. Exptl. Biol. Med. 123, 618.
19. Naets, J. P. (1960). Erythropoietic factor in kidney tissue of anemic dogs. Proc. Soc. Exptl. Biol. Med. 103, 129.
20. Nagano, M. and Schafer, H. E. (1963). Methode zur isolierung von glomerula und bestimmung von enzymaktivitaten im raten glomerulum. Klin. Woch. 41, 1203.
21. Pavlovic-Kentera, V., Hall, D. P., Bragassa, C. and Lange, R. D. (1965). Unilateral renal hypoxia and production of erythropoietin. J. Lab. Clin. Med. 65, 577.
22. Rambach, W. A., Alt, H. L. and Cooper, J. A. D. (1961). Erythropoietic activity of tissue homogenates. Proc. Soc. Exptl. Biol. Med. 108, 793.
23. Weintraub, A. H., Gordon, A. S. and Camiscoli, J. F. (1963). Use of hypoxia-induced polycythemic mouse in assay and standardization of erythropoietin. J. Lab. Clin. Med. 62, 743.
24. Zangheri, E. O., Campani, H., Ponce, F., Silva, J. C., Fernandez, F. O. and Suarez, J. R. (1963). Production of erythropoietin by anoxic perfusion of the isolated kidney of the dog. Nature 199, 572.
25. Zanjani, E. D., Contrera, J. F., Cooper, G. W., Gordon, A. S. and Wong, K. K. (1967). The renal erythropoietic factor (REF) II. Role of ions and vasoactive agents in the generation and activation of erythropoietin. Science 156, 1367.
26. Zanjani, E. D., Contrera, J. F., Gordon, A. S., Cooper, G. W., Wong, K. K. and Katz, R. (1967). The renal erythropoietic factor (REF). III. Enzymatic role in erythropoietin production. Proc. Soc. Exptl. Biol. Med. 125, 505.

# PHYSICO-CHEMICAL AND BIOLOGICAL CHARACTERISTICS OF ERYTHROPOIETIN

I.Baciu, V.Vasile, I.Sovrea, M.Zirbo, T.Prodan

Dept. of Physiology, Medico-Pharmaceutical Institute - Cluj - Romania

Erythropoietin, an alpha glycoprotein, appearing in the blood of mammals in conditions of hypoxia and stimulating the proliferation of red bone marrow, has not yet been obtained in pure samples.

## MATERIALS AND METHODS

### Isolation and purification of erythropoietin

Extraction of crude erythropoietin from serum of phenyl hydrazine anemic rabbits, methanol fractioning of serum, isolation of pseudoglobulins from methanolic fraction II and chromatography of pseudoglobulins on DEAE cellulose, were performed according to our previous descriptions (1,2,3).

The active fraction II, eluted from DEAE cellulose with 0.01 M acetate pH 3.6, is rechromatographed on a column of DEAE Sephadex A 50 Medium (25/1.8 cm), previously balanced at pH 4.6 with acetate buffer 0.01 M. For elution was used acetate buffer pH 3.6, with increasing molarity: 0.01 and 0.02 M. The effluent was collected in 5 ml samples, the rate of flow beeing 20 ml/h.

Chromatographic fractions were precipitated with 5 volumes acetone, dried in vacuum, solved and dialyzed against distilled water and lyophilized.

Assay of biological activity.

The active fraction orientatively identified on rat bone marrow cultures in a synthetic medium (2) and on mice "in vivo" (3) was measured by means of $Fe^{59}$ incorporated into mice erythrocytes (9).

Physico-chemical Characteristics of the Active Fraction

Paper electrophoresis was performed in veronal buffer $\eta$ = 0.075, pH 8.6 and in acetate buffer, pH 4.6 and 3.6, 0.1 M.

U.V. absorption spectrum was followed in 0.02 g% water solution with Beckman spectrophotometer.

Diffusion rate was determined by comparison with Dextran Pharmacia Uppsala Sweden (72.000 M.W.), ovalbumin (43.000 M.W.) and polyvinylpyrrolidone K 15 Fluka (10.000 M.W.). The assay was performed in Kern apparatus in 1% solutions (8).

High voltage paper electrophoresis was used in order to analyse the migration of polypeptides obtained after digestion with trypsin and of aminoacids obtained after acid hydrolysis of active fraction, in comparison with hemoglobin (4).

Sialic acid was measured spectrophotometrically (Bial reagent) using as test human orosomucoid (10).

## RESULTS

The chromatography of alpha pseudoglobulins on DEAE cellulose column leads to the separation of three distinct fractions (Fig. 1 a).

The active fraction (II) eluted with buffer 0.01 M pH 3.6 (40-60 mg proteins, from 100 ml serum) is concentrated 180-200 fold. Its rechromatography on DEAE Sephadex reveals 5 spikes (Fig. 1 b). The third spike (III), eluted at pH 3.6, 0.01 M is active. The protein content was between 2.5 and 8.5 mg%, which represents a 900-2300 fold concentration of biological activity, as compared with the serum proteins.

Erythrostimulating activity of the alpha pseudoglobulins from methanolic fraction II was found to be about 0.6-1 C.S.u/ml and for the most purified fraction by rechromatography 1.05 C.S.u./ml of anemic plasma, calculated after Magid and Hansen (9) (Table 1).

The fraction is electrophoretically homogenous (Fig. 2).

U.V. absorption is maximal at 272 mμ.

Molecular weight, estimated by diffusion, was next to that of polyvinylpyrrolidone K 15 Fluka (10,000 M.W.) (Fig. 3).

At high voltage electrophoresis it seems that erythropoietin has a lower content of peptides and aminoacids than hemoglobin (Fig. 4).

The sialic acid content was 5.5 g%.

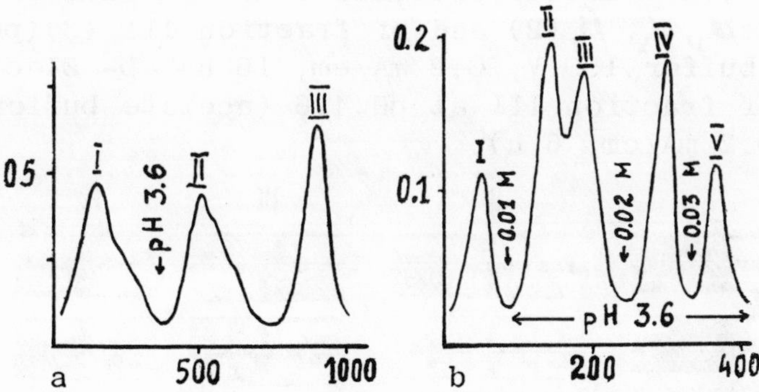


Fig. 1. Chromatography of alpha pseudoglobulins (F) on DEAE cellulose with acetate buffers (a) and rechromatography of fraction II from DEAE cellulose on DEAE Sephadex. Fraction III is active (b).

TABLE I

| Fraction: | Saline | Ps. | Fr. III | (Sephadex) |
|---|---|---|---|---|
| Quant.inj. | 0.4 ml | glob.5mg | 0.04mg | 0.06 mg |
| %$Fe^{59}$incorp. | 3.1±0.5 | 15,3±2 | 22.7±5 | 24,6±7 |

## DISCUSSION

It is still difficult to state that our erythropoietin is pure. Much high values of the specific activity were found by Goldwasser et all. (5), as compared with our own findings, amounting to 15 C.S.u./

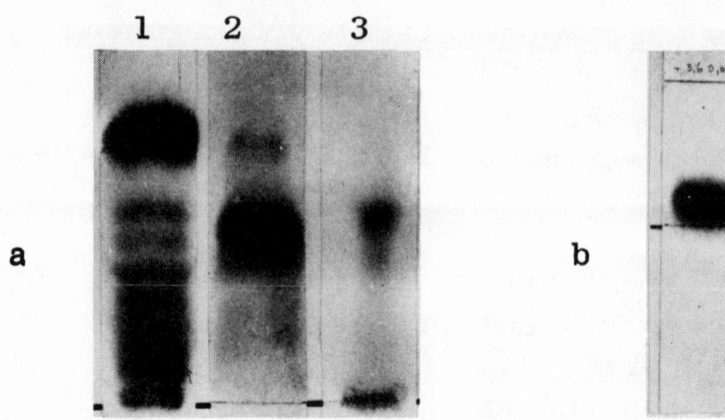


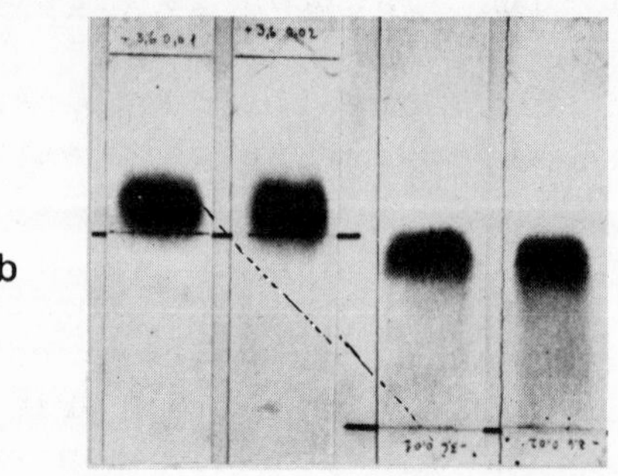


Fig. 2. a - Electrophoresis of serum (1); of pseudoglobulins $\alpha_1$, $\alpha_2$, $\beta_1$ (2) and of fraction III (3)(pH 8.6 Michaelis buffer 150 V, 0.2 mA/cm, 16 h). b- Electrophoresis of fraction III at pH 4.6 (acetate buffer 0.1 M 205 V, 0.2 mA/cm, 6 h).

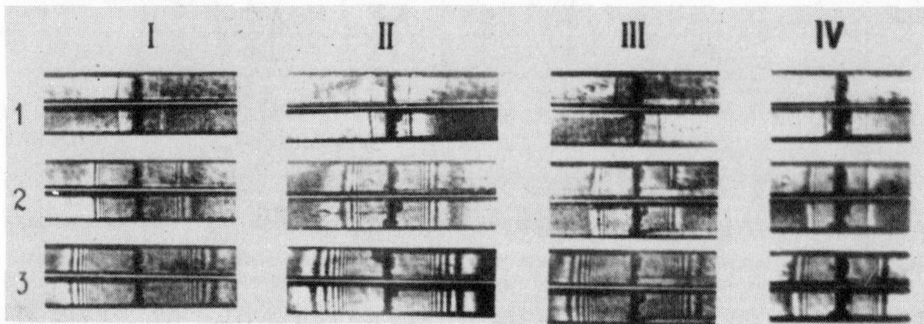


Fig. 3. Diffusion of F III (1), polyvinyl pyrrolidone K 15 Fluka (11), ovalbumin crystallized 5 x (111) and Dextran (1V) after 10 minutes (1), two hours (2) and three hours (3).

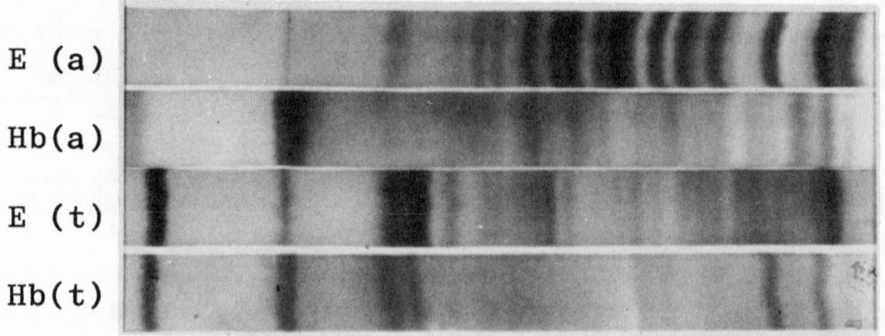


Fig.4. High voltage electrophoresis of polypeptides obtained by tryptic (t) and acid (a) hydrolysis of hemoglobin (Hb) and fraction III (E).

1 mg, but we have worked with sera from anemic rabbits and not from sheep.

The content of sialic acid was also more reduced in our extract (5.5%), pleading for a lower degree of purification, although a 3.8% content of sialic acid was found in purified kidney erythropoietin by Kuratowska (7).

The electrophoretic homogeneousness, the acid isoelectric point, the maximal ultraviolet absorption at 272 mμ however, coincide with other data (5,7).

The low molecular weight, as estimated by diffusion, agree with the data obtained by Goldwasser et all. (5), who found a value of 5,500 for 64,000 x purified erythropoietin.

The small number of polypeptides obtained by tryptic digestion and of amino acids by acid hydrolysis, also call for a low molecular weight and for a certain purification degree of erythropoietin. The low molecular weight of erythropoietin is also sustained by its equal repartition in plasma and interstitial fluid (6).

It would be premature to state that the analysed product is pure erythropoietin; there are some factors demonstrating even a less advanced degree of purification, but the technical procedure described above allows a purification with a minimal loss of biological activity up to this stage and an objective way to characterise the erythropoietin.

## CONCLUSIONS

1) Chromatography on DEAE cellulose column and rechromatography of the active fraction on DEAE Sephadex A 50 Medium, allow a 2300 fold concentration of the active fraction, with an erythropoietic activity of 15 C.S.u./mg.

2) The fraction is electrophoretically homogenous with an acid under 4.6 isoelectrical pH, and manifests a maximal absorption in U.V. at 272 mμ, and has a molecular weight under 10,000, containing a lower number of polypeptides, as well as of aminoacids.

## REFERENCES

1) BACIU I. - J. Physiol. (Paris), 1962, 54, 441.
2) BACIU I. - Rev. Roum. Physiol. 1964, 1, 149.
3) BACIU I., SECAREANU ST., VASILE V., POPA L. - Clujul Medical, 1964, 36, 118.
4) CLOTTEN R., CLOTTEN A. - Hochspannungs Elektrophorese, G. Thieme Verlag, Stuttgart 1962.
5) GOLDWASSER E., WHITE W.F., TAYLOR K.B. - Biochim. Biophys. Acta 1962, 64, 487.
6) KEIGHLEY G. - Erythropoiesis, Grune-Straton, New York 1962, 106.
7) KURATOWSKA Z. - Pat. Biol. 1967, 15, 671.
8) LONGSWORTH L.G. - J. Amer. Chem. Soc. 1953, 74, 13.
9) MAGID E., HANSEN P. - Scand. J. Lab. Invest. 1966, 18, 347.
10) WEIMER H.E., MEHL J.W., WINZLER R. - J. Biol. Chem. 1950, 185, 561.

# SECRETIN, CHOLECYSTOKININ[1]

Mutt, V. and Jorpes, J. E.

Chemistry Department II and the Wallenberg Foundation Laboratory for Physiological Chemistry
Karolinska Institutet, Stockholm, Sweden

The flow of saliva and gastric juice is stimulated by a nervous reflex elicited by visual or olfactory stimulation. The name Pavlov is intimately linked with this reflex. The pancreas gland, which takes over in the second phase of the digestion process, cannot in the same way be informed about, when it has to enter into action. For this purpose Nature has developed a system of hormonal stimulation. When the acid chyme with its digestion products enters the duodenum a series of hormones are liberated into the blood stream from the duodenal and jejunal mucosa, secretin which stimulates the secretion of water and bicarbonate from the pancreas, pancreozymin which causes enzyme secretion and cholecystokinin, which empties the gallbladder. Our knowledge of these hormones dates back to 1902 for secretin (Bayliss and Starling), 1928 for cholecystokinin (Ivy and Oldberg) and 1943 for pancreozymin (Harper and Raper).

The isolation of secretin. After the discovery of secretin eminent physiologists, Mellanby, Ivy, La Barre and others entered the field. Methods for its preparation were worked out by Mellanby (1925), by Weaver, Luckhardt and Koch (1926), Still (1930), Hammarsten, Agren and co-workers (1928, 1933a, b) and

[1]Supported by grants from U.S. Public Health Service, National Institutes of Health (No. Am 06410-01-02-03-04-05), The Squibb Institute for Medical Research, New Brunswick, N.J., U.S.A., Torsten och Ragnar Soderbergs Stiftelser, The Swedish Cancer Society, The Swedish Medical Research Council, and Magnus Bergvalls Stiftelse.

by Greengard and Ivy (1938). The earlier literature on secretin has been reviewed by Still (1931), Greengard (1948) and Grossman (1950), among others.

Every attempt at obtaining these hormones in a pure state was in these early days doomed to fail because of lack of suitable purification procedures. When we in 1952 took up work on secretin, new methods for the preparation and purification of biologically active proteins and peptides with the help of ion exchangers had been developed. By adsorbing the secretin in the first extract on alginic acid we succeeded in 1953 to concentrate twice as much secretin in a 30 times smaller amount of crude material than in the salt cake of Weaver et al. (1926). Secretin was extracted from the salt cake into methyl alcohol. It was further purified by ordinary fractionation procedures using neutral salts and alcohol. Adsorption on stearic acid in 0.1 M phosphate, pH 7.1, resulted in a product containing 1,200 clinical units of secretin/mg, in contrast to the 2-14 u./mg of the earlier secretin preparations. Electrophoresis on cellulose columns in 0.1 M ammonium bicarbonate, pH 7.7, gave preparations with up to 1,000 (4,000) u./mg (Mutt 1959 a) and chromatography on a CMC column in 0.02 M ammonium bicarbonate a preparation with up to 2,000 (7,500) u./mg (Mutt 1959 b, c, d). The final purification to 4,000-5,000 (17,500) clin. units/mg, was achieved with countercurrent distribution between 0.1 M phosphate and n-butanol, pH 7.0 (Jorpes and Mutt 1961 b).

Table I Isolation of porcine secretin

| | Weight | Clin. units/mg |
|---|---|---|
| Upper first metre of intestine from 10,000 hogs. Boiled, frozen, minced | ca. 700 kg | |
| Extracted with 0.5 N AcOH. Activity adsorbed on alginic acid, eluted with 0.2 M HCl. Precipitated with NaCl at saturation | 1 kg | 1.5 - 3.0 |
| Fractionation of aqueous solution with ethanol. Recovery in water, precipitation with NaCl and reprecipitation at pH 4 | 150 g | 5 - 15 |
| Extraction of secretin into methanol | 4 g | 150 - 300 |
| Chromatography on carboxymethyl cellulose | 100 mg | 1,000 (4,000) |
| Countercurrent distribution in 0.1 N phosphate buffer/n-butanol | 10 mg | 4,000 (20,000) |

Highly active secretin preparations were obtained in the late 1950's by Legge et al. (1957) and Newton et al. (1959). Their preparations contained about 1,500 and 900 u./mg, respectively.

The structure of secretin. The pure hormone is a low molecular polypeptide composed of 27 amino acid residues and 11 different amino acids, with cystine, methionine, tyrosine, tryptophan, proline, isoleucine and lysine lacking (Jorpes and Mutt 1961 b, Jorpes et al. 1962). The eleven amino acids are represented in the proportions:

$$\mathrm{Ala_1Arg_4Asp_2Glu_3Gly_2His_1Leu_6Phe_1Ser_4Thr_2Val_1}$$

The N-terminal sequence is histidyl-seryl-aspartyl .... Valine is carboxylterminal in the form of its amide. In two of the glutamic acids the γ-carboxyles are likewise amidated.

Complete tryptic hydrolysis with splitting of three -Arg.Leu-bonds and one -Arg.Asp-bond gave rise to five peptides, a to e. Peptide b, containing histidine, is N-terminal, No. 1, and peptide e, lacking arginine, is C-terminal, No. 5. A short time tryptic hydrolysis left the -Arg.Asp-bond unsplit, resulting in a new peptide -Leu.Arg.Asp.Ser.Ala.Arg- (d + a).

Thrombin, however, attacked only the -Arg.Asp-linkage and split secretin into two large peptides, f and g. In the thrombic peptide f, peptide d, -Leu.Arg., was found to be linked to the histidine containing N-terminal peptide b, constituting the C-terminal group. Since peptide d also is linked to peptide a, -Asp. Ser.Ala.Arg.-, as found after a short time tryptic digestion, and peptide e with valine is C-terminal, the sequence of the five tryptic peptides in secretin is given (Mutt et al. 1965; Mutt and Jorpes 1966).

Figure 1. The sequence of the tryptic peptides of secretin

The structure of the secretin molecule as presented by V. Mutt and J.E. Jorpes at the 4th International Symposium on the Chemistry of Natural Products, Stockholm, 1966, and confirmed by synthesis (Bodanszky, M. et al. 1966 a, b, Ondetti 1967, Bodanszky and Williams 1967).

His-Ser-Asp-Gly-Thr-Phe-Thr-Ser-Glu-Leu-Ser-Arg-Leu-Arg
1 2 3 4 5 6 7 8 9 10 11 12 13 14

3 4 5
Asp-Ser-Ala-Arg-Leu-Gln-Arg-Leu-Leu-Gln-Gly-Leu-Val-$NH_2$
15 16 17 18 19 20 21 22 23 24 25 26 27

Figure 2.

Secretin: His· Ser· Asp· Gly· Thr· Phe· Thr· Ser· Glu· Leu· Ser· Arg· Leu· Arg· Asp· Ser· Ala. Arg. Leu· Gln. Arg. Leu· Leu. Gln· Gly· Leu· Val· $NH_2$

Glucagon: His· Ser· Gln· Gly· Thr· Phe· Thr· Ser· Asp· Tyr· Ser· Lys· Tyr· Leu· Asp· Ser· Arg· Arg· Ala· Gln·Asp· Phe· Val. Gln· Try· Leu· Met· Asn· Thr·

Not less than 14 of the 27 amino acid units of secretin are holding the same position as in glucagon counting from the N-terminal (Fig. 2).

Action mechanism. Uvnas and associates 1966 pointed out that there is an interaction between vagal impulses and gastrin in the control of gastric acid secretion. After curtailment of the antral function the HCl-secreting glands do answer poorly or not at all to vagal impulses, and the vagus-denervated fundic mucosa shows a very low sensitivity to endogeneous or exogeneous gastrin. The nervous and humoral mechanisms are unable to operate efficiently independently of each other. After all the neural impulses are not strong enough to break, through the resistance around or within the secreting cell, without the support of a catalyser, a kind of grease which get things going. It is against this background tempting to speculate about the connection between the chemical structure and the action mechanisms of gastrin and of secretin.

Gastrin is an unusually strong acidic peptide with no basic amino acids and not less than 6 glutamic acid molecules and 1 aspartic acid amongst in total 17 amino acid units. Secretin is an extremely basic peptide with 27 amino acid units, amongst them 4 arginine units, 1 histidine unit and 2 of the 3 glutamic acid units γ-amidated. No doubt, their electrokinetical properties are essential for their function.

Whatever the real reaction mechanism may be for the accumulation of $H^+$ ions and $HCO_3^-$ ions in the gastric and pancreatic juices, respectively, the specific grease catalyser for the former reaction has a negative electric charge and that for the latter reaction a strongly positive one. One should not, however, forget that the strongest stimulant for the HCl production, histamine, has a basic charge.

Cholecystokinin. In spite of Ivy's repeated attempts during the years to call attention to this hormone (Ivy 1930, 1947, 1951,

Figure 3. The ion exchange taking place through the cell membrane during gastric and pancreatic secretion and the number of acidic and basic groups in gastrin and in secretin.

| Gastric secretion | | Pancreatic secretion | |
|---|---|---|---|
| In | Out | In | Out |
| $H^+$ | $Na^+$ | $H^+$ | $Na^+$ |
| $HCO_3^-$ | $Cl^-$ | $HCO_3^-$ | $Cl^-$ |
| Gastrin:- - - - - - - - | | Secretin: + + + + + + + (- -) | |

1955) up to 1954 the only publications dealing with cholcystokinin were one by Agren from 1939 and the writings of Harper and his group (Duncan et al. 1950, 1952, 1953, Howat 1952) about the cholecystokinin activity of their pancreozymin preparations.

Since our crude material extracted from the intestinal mucosa for the preparation of secretin contained cholecystokinin and pancreozymin as well, the methanol-insoluble material left after extracting the secretin was fractionated with ethanol and chromatographed on carboxymethyl cellulose. The material obtained could be used in man (Werner and Mutt 1954, Werner 1956), 3-4 mg of substance being injected intravenously for the cholecystographic test. The cholecystokinin activity was 22 Ivy dog units/mg and the pancreozymin activity 100-120 units/mg as defined by Crick, Harper and Raper (1949) (Jorpes and Mutt 1959, 1961 a).

Chromatography on TEAE-cellulose of this material resulted in a product with about 200-250 Ivy dog units/mg of cholecystokinin and a corresponding increase in the pancreozymin activity (Jorpes and Mutt 1961 a, 1962).

This material has been purified further to an activity of about 3,000 Ivy dog units of cholecystokinin per mg by filtration through a Sephadex G-50 column at pH 8.0 followed by chromatography on Amberlite XE-64 at pH 7.5 (Jorpes, Mutt, Toczko 1964). During these last purification steps the secretin is completely removed, but the cholecystokinin and pancreozymin activities once again go parallel.

Pancreozymin. As pointed out by the present authors in 1961 and 1962, it is quite remarkable how closely cholecystokinin and pancreozymin accompany each other during the various purification steps, whereas secretin separates very early. This applies to the chromatography on the acidic carboxymethyl cellulose and on the basic TEAE-cellulose up to a strength of 200-250 Ivy dog units of CCK per mg substance.

At this stage of purity we determined the pancreozymin activity of the cholecystokinin preparations and the ratio CCK:PZ was compared with that in Harper and Raper's original pancreozymin preparation, given to us by professor Harper in 1959, as well as in a recent sample of Pancreozymin Boots. The two last ones had about the same CCK activity, 0. 3 Ivy dog units per mg dry substance.

Evidently there is in the three different dilutions no considerable difference between the two preparations as to pancreozymin activity, and the ratio cholecystokinin: pancreozymin will be the same. The pancreozymin activity is proportional to the amount of cholecystokinin injected, irrespective of the degree of purity of the preparation of 0. 3 or 220 Ivy dog units of CCK per mg. Thus, on increasing the cholecystokinin activity of the preparation about 650 times there is an equal rise in the pancreozymin activity (Jorpes and Mutt 1966).

Table II. Purification of cholecystokinin and pancreozymin.

| | Weight | Cholecystokinin (Ivy dog units) | Pancreozymin (Crick, Harper and Raper units) |
|---|---|---|---|
| Methanol insoluble material after extracting the secretin (from 20,000 hogs) | 100 g | | |
| Adsorbed to carboxymethyl cellulose at pH 6. 5 Precipitated from eluate with NaCl | 40 g | | |
| Active material precipitated from 75% EtOH with n-BuOH | 20 g | 20 | 80 |
| Chromatography on triethylaminoethyl cellulose at pH 9. 1 | 750 mg | 250 | 1,000 |
| Filtration through Sephadex G-50 | 60 mg | 1,500 | 6,000 |
| Chromatography at pH 7. 5 on Amberlite XE-64 | 12 mg | 3,000 | 12,000 |

Table III. Pancreozymin activity of Pancreozymin Boots, (A), charge No. 23, with 0.3 Ivy dog units of cholecystokinin per mg, as compared with that of a cholecystokinin preparation with 220 Ivy dog units of cholecystokinin per mg (B).

| Sample | Dose Ivy dog units of cholecystokinin | Spectrophotometric reading (660 um) cat No. 1 | Cat No. 2 | Mean |
|---|---|---|---|---|
| - | (Secretin only) | 0.055 | 0.058 | |
| A | 1.5 | 0.155 | 0.126 | 0.140 |
| B | 1.5 | 0.140 | 0.186 | 0.163 |
| A | 3.0 | 0.227 | 0.212 | 0.220 |
| B | 3.0 | 0.245 | 0.195 | 0.220 |
| A | 6.0 | 0.292 | 0.290 | 0.291 |
| - | (Secretin only) | 0.046 | 0.034 | |
| B | 6.0 | 0.310 | 0.282 | 0.296 |

On increasing the strength of cholecystokinin from 220 to 3,000 Ivy dog units per mg a proportional rise in the pancreozymin activity was observed. Both entities have thus increased about equally, by a factor of 10,000 over the cholecystokinin and pancreozymin activities of the Pancreozymin Boots and Harper's original pancreozymin preparation.

The fact that the material in question has passed over an acidic ion exchanger, CMC-cellulose, a basic one, TEAE-cellulose, a Sephadex column and yet another acidic ion exchanger, Amberlite XE-64, resulting in a 10,000-fole increase in hormonal activity, without any detectable change in the ratio cholecystokinin: pancreozymin, speaks in favor of the assumption that both activities are exerted by one and the same substance.

Strong support for this assumption has been presented by Mutt (1964), who showed that cholecystokinin, which contains methionine, in line with the pituitary adrenocorticotropic hormone, the α- and β-melanocyte stimulating hormones and the parathyroid hormone can be oxidized by hydrogen peroxide with complete loss of activity and reactivated with cysteine to practically full strength. The pancreozymin activity simultaneously underwent the same changes. Under similar conditions the pure secretin, which does not contain methionine, did not show any change in activity.

The question as to the individuality of pancreozymin as a hormone will consequently depend upon whether the polypeptide carrying the CCK and PZ activities is a homogenous chemical entity or not.

The CCK-PZ preparation with 3000 Ivy dog units of cholecystokinin per mg seems to be a homogeneous polypeptide built up of the following 33 amino acid units, if amide groups are disregarded:

$Ala_1$, $Arg_3$, $Asp_5$, $Glu_2$, $Gly_2$, $His_1$, $Ileu_2$, $Leu_2$, $Lys_2$, $Met_3$, $Phe_1$ $Pro_2$, $Ser_4$, $Tyr_1$, $Try_1$, $Val_1$.

One of the two lysines is N-terminal. Cysteine, cystine and threonine are absent (Jorpes et al., 1964, Mutt and Jorpes 1967 a). On treatment with cyanogen bromide the dipeptide aspartyl-phenylalanine amide could be isolated from the reaction products, indicating that the polypeptide terminates in aspartyl-phenylalanine amide, and probably, because of the method used for fragmentation, in methionyl-aspartyl-phenyl-alanine amide (Mutt and Jorpes 1967 b). This C-terminal sequence is identical to that of gastrin (Gregory and Tracy 1964). It has only a trace of biological activity.

Physiological functions. Secretin stimulates the production of bicarbonate and water alone. It flushes out the enzymes already present in the ducts. It exerts only a slight stimulating influence upon the secretion of water from the liver (Jonson et al. 1964). Impure secretin preparations cause vasodilatation and facilitate the flow of blood through the pancreas. It is not certain whether pure secretin has the same effect or not.

In two concomitant publications from 1966 Murat and White, and Magee and Nakamura showed in dogs with a Heidenhain pouch that both secretin and CCK-PZ behave like gastrin, acting in small doses as stimulants and in large doses as inhibitors of the gastric secretion, the difference in dosage between that required to inhibit and that to stimulate gastric secretion being about 100 times.

Gastrointestinal hormones and the release of insulin and glucagon from the pancreas. Hyperglucemia, caused by glucose absorption from the intestine or by hormonal stimuli exerted by adrenalin and glucagon, was until recently considered to be the main or the sole stimulus to the release of insulin from the islet tissue. Now the picture has become more complicated, when irrespective of the plasma glucose level both secretin, cholecystokinin-pancreozymin as well as glucagon and probably also gastrin have been shown to be active as insulin releasing factors (Pfeiffer et al. 1965, Samols et al. 1965, Unger et al. 1966, Dupre et al.

1966). It was a fortunate coincidence that these hormones, secretin and cholecystokinin-pancreozymin as well as gastrin had been obtained in a pure state, when this question was studied by the physiologists, allowing them to draw convincing conclusions.

Secretin, CCK-PZ and the Brunner glands. In their extensive studies on the physiology of the Brunner glands Florey and Harding (1935 a, b) presented convincing evidence that the secretory activity from this area is controlled by a humoral mechanism, in which secretin seemed to be the active agent. In using dogs with innervated pouches of the Brunner gland area and pure samples of gastrin I and of secretin, Cooke and Grossman (1966) could convincingly show, that neither gastrin nor secretin exert any stimulating action on the motility of the gland area and on the secretion from it. Fundic extracts, jejunal extracts, crude gastrin, crude secretin preparations and a CCK-PZ preparation with 250 Ivy dog units per mg were potent stimulants of both motility and secretion in the Brunner gland area.

## REFERENCES

Agren, G. Skandinav. Arch. Physiol. 1939, 81, 234.

Bayliss, W.M. and Starling, E.H. J. Physiol. 1902, 28, 325.

Bodanszky, M., Levine, S.D., Narayanan, V., Ondetti, M.A., Saltza, M. von, Sheehan, J.T. and Williams, N.J. IUPAC Internat. Congress on Chemistry of Natural Products, Stockholm June 26th-July 2nd, 1966 a. Section 2C-2.

Bodanszky, M., Ondetti, M.A., Levine, S.D., Narayanan, V.L. Saltza, M. von, Sheehan, J.T., Williams, N.J. and Sabo, E. F. Chemistry and Industry, Oct. 15, 1966 b, 1757.

Bodanszky, M. and Williams, N.J. J. Amer. Chem. Soc. 1967, 89, 685.

Canfield, R.E. Biochemistry of Gastrointestinal Hormones. Rec. Progr. Hormone Research Acad. Press 1967, 23. p. 501.

Cooke, A.R., Grossman, M.I. Gastroenterology 1966, 51, 506.

Crick, J., Harper, A.A. and Raper, H.S. J. Physiol. 1949, 110, 367.

Duncan, P.R., Harper, A.A., Howat, H.T. Oleesky, S. and Valery, H. J. Physiol. 1950, 111, 63p.

Duncan, P.R., Harper, A.A., Howat, H.T., Oleesky, S. and Valery, H. Gastroenterologia 1952, 78, 349.

Duncan, P.R., Harper, A.A., Howat, H.T., Oleesky, S., Valery, H. and Scott, J.E. J. Physiol. 1953, 121, 19p.

Dupre, J., Rojas, L., White, J.J., Unger, R.H. and Beck, J.C. Lancet, 1966, II, 26.

Florey, H.W. and Harding, H.E. Proc. Roy. Soc. (Biol.) 1935 a, 117, 68-77.

Florey, H.W. and Harding, H.E. Quart, J. Exp. Physiol. 1935 b, 25, 329.

Greengard, H. and Ivy, A.C. Amer. J. Physiol. 1938, 124, 427.

Greengard, H. The Hormones. Ed. by Pincus and Thimann. Academic Press, 1948, 1, 201.

Gregory, R.A. and Tracy, H.J. Gut 1964 5, 103.

Grossman, M.I. Physiol. Rev. 1950, 30, 33.

Hammarsten, E., Wilander, O. and Agren, G. Acta Med. Scnad. 1928, 68, 239.

Hammarsten, E., Jorpes, E. and Agren, G. Biochem. Zeitschr. 1933, a, 264, 272.

Hammarsten, E., Agren, G., Hammarsten, H. and Wilander, O. Biochem. Zeitschr. 1933 b, 264, 275.

Harper, A.A. and Raper, H.S. J. Physiol. 1943, 102, 115.

Ivy, A.C. and Oldberg, E. Am. J. Physiol. 1928, 86, 599.

Ivy, A.C., Drewyer, G.E. and Orndoff, B.H. Endocrinology 1930, 14, 343.

Ivy, A.C. Am. J. Roentgenol. 1947, 57, 1.

Ivy, A.C. Oral Surgery 1951, 4, 612.

Ivy, A.C. Edit. J.H. Gaddum. Williams & Wilkins, Baltimore, 1955.

Jonson, G., Sundman, L. and Thulin, L. Acta Physiol. Scand. 1964, 62, 287.

Jorpes, J.E. and Mutt, V. Arkiv F. Kemi 1953, 6, 273.

Jorpes, J.E. and Mutt, V. Gastroenterol. 1959, 36, 377.

Jorpes, J.E. and Mutt, V. Ann. Int. Med. 1961a, 55, 395.

Jorpes, J.E. and Mutt, V. Acta Chem. Scand. 1961 b, 15, 1790.

Jorpes, J.E., Mutt, V., Magnusson, S. and Steele, B. Biochem. Biophys. Res. Com. 1962, 9, 275.

Jorpes, J.E. and Mutt, V. The exocrine pancreas, Ciba Found. Symp. Edit. Reuck and Cameron, Churchill, London, 1962 p. 150.

Jorpes J.E., Mutt, V. and Toczko, K. Acta Chem. Scand, 1964, 18, 2408.

Jorpes, J.E. and Mutt, V. Acta Physiol. Scand. 1966, 66, 196.

Legge, J.W., Morieson, A.S., Rogers, G.E. and Marginson, M.A. Austral. J. exp. Biol. 1957, 35, 569.

Magee, D.F. and Nakamura, M. Nature, 1966, 212, 1487.

Mellanby, J. J. Physiol. 1925, 60, 85.

Murat, J.E., White, T.T. Proc. Soc. Exp. Biol. Med. 1966, 123, 593.

Mutt, V. Arkiv f. Kemi 1959 a, 14, 275.

Mutt, V. Acta Chem. Scand. 1959 b, 13, 1247.

Mutt, V. Arkiv f. Kemi 1959 c, 15, 69.

Mutt, V. Arkiv f. Kemi 1959 d, 15, 75, (Thesis).

Mutt, V. Acta Chem. Scand. 1964, 18, 2185.

Mutt, V., Magnusson, S., Jorpes, J.E. and Dahl, E. Biochemistry 1965, 4, 2358.

Mutt, V. and Jorpes, J.E. IUPAC Internat. Congress on Chemistry of Natural Products, Stockholm June 26th-July 2nd, 1966. Section 2C-1.

Mutt, V. and Jorpes, E. Recent Progress in Hormone Research. Acad. Press. 1967 a, 23.

Mutt, V., Jorpes, J.E. Biochem. Biophys. Res. Com. 1967, 26, 394.

Newton, G.G.F., Love, J.W., Heatley, N.G. and Abraham, E.P. Biochem. J. 1959, 71, 6 P.

Ondetti, M.A. Recent Progress in Hormone Research, Acad. Press 1967, 23.

Pfeiffer, E. F., Telib, M., Ammon, J., Melani, F. and Ditschuneit, H. Deutsch. Med. Woschr. 1965, 90, 1633.

Samols, E., Marri, G. and Marks, V. Lancet 1965, II, 415.

Still, E. U. Am. J. Physiol. 1930, 91, 405.

Still, E. U. Physiol. Rev. 1931, 11, 328.

Unger, R. H., Ketterer, H., Eisentraut, A. and Dupre, J. Lancet, 1966, II, 24.

Uvnas, B., Emas, S., Fyro, B. and Sjodin, L. Amer. J. Digest. Dis. 1966, N.S. 11, 103.

Weaver, M. M., Luckhardt, A. B. and Koch, F. C. J. A. M. A. 1926, 87, 640.

Werner, B. and Mutt, V. Scand. J. Clin. and Lab. Invest. 1954, 6, 228.

# RELATIONSHIP BETWEEN BEHAVIOURAL EFFECT AND CIRCULATORY CHANGES PRODUCED BY INTRA-CAROTID BRADYKININ

Bertolini A, Castelli M., Mucci P., Sternieri E.

Dept. of Pharmacology, University of Modena (Italy)

Bradykinin is a typical algesic substance: it causes pain when injected by intraarterial route (1-3), applied to a blister base (4), intraperitoneally injected (5), and introduced into the cerebro-spinal fluid (6,7). However, there is some disagreement even about this generally recognised algesic property; thus, for example, by intraventricular injection of crude bradykinin into the cat Rocha e Silva got results very different from those he obtained by using the pure synthetic product (8).

We think that the injection route may be a decisive factor in determining the occurrence of the algesic action of bradykinin. As a rule we operate on the rabbit, where the intracisternal injection of small doses (3 to 10 µg per animal) of pure bradykinin elicits a syndrome characterised by extreme excitation, flight and vocalisation, which was contemporaneously reported by Sicuteri et al. (6) independently of ourselves (7). However, when we used the same doses through the intraventricular route - according to the technique by Feldberg and Sherwood - instead of the intracisternal one, we observed a quite different symptomatology: the excitation phase was much shorter or absent; a catatonic state prevailed; finally, the cry-flight response was never observed (9). Also in the dog, following intraventricular bradykinin injection, a marked catatonia was observed (7). Maybe intraventricular bradykinin is inactivated before it gets to the sites of its eventual action or, more simply, the peptide might follow different ways of distribution and thus produce different effects.

Also when small doses of bradykinin are introduced through the common carotid artery of anaesthetised rabbits motor excitation and vocalisation are induced (10). This fact might be expected because it is known that bradykinin injected into various arteries elicits a vocalisation response (1,2). However, we don't know whether the vocalisation responses to intracarotid or intracisternal route have the same meaning. We are tempted to believe that there are two different mechanisms producing vocalisation, since the two vocal responses are differently influenced by some drugs; e.g., a beta adrenergic blocking compound, 2 -isopropylamino-1-(p-nitrophenyl)ethanol or INPEA, prevents vocalisation by intracisternal bradykinin much more constantly than vocalisation by intracarotid administration (see table 1).

Moreover, while analgesic and antiinflammatory drugs, as it is known, are capable of preventing the vocalisation response to intraarterial bradykinin (11), we did not manage to anta-

TABLE 1 - Effect of ß-adrenergic blocking agents on vocalisation induced by intracisternal and intracarotid bradykinin in the rabbit.

| PRE-TREATMENT | | TIME INTERVAL in min | BRADYKININ dose and route | ANIMALS WITH VOCALISATION / TREATED ANIMALS | % INHIBITION OF VOCALISATION |
|---|---|---|---|---|---|
| ----(*) | ---------- | ---- | 5 μg/animal i.cisternal | 5/5 | — |
| ----(*) | ---------- | ---- | 10 " " | 2/2 | — |
| INPEA | 10 mg/Kg i.venous | 1 to 6 | 5 " " | 2/8 | 62 |
| INPEA | 15 " " | 1 to 7 | 10 " " | 3/5 | |
| propranolol | 4 " " | 4 to 6 | 10 " " | 2/5 | 60 |
| INPEA (**) | 10 mg/Kg i.carotid | 1 to 4 | 5 μg/Kg i.carotid | 4/4 | 22 |
| INPEA (**) | 10 " " | 1 to 4 | 10 " " | 3/5 | |

(*) control group - (**) each rabbit was utilised as control of itself: all the animals had previously responded to the same dose of bradykinin that was injected after INPEA

gonise vocalisation induced by intracisternal bradykinin by means of morphine, methadone, phenylbutazone; as for indomethacin the results were less certain (see table 2).

Concerning the mechanism of the vocalisation response to intracarotid bradykinin, a few hypotheses can be put forward: the most attractive possibility, insofar as it may refer to the origin of some kinds of headache in the human, is that pain produced by the peptide is connected with the extreme sensitivity of cortico-pial vessels to the vasodilator action of bradykinin (12-14). In an attempt to clarify this point, we carried out researches on the rabbit under moderate ethyl-urethan anaesthesia (0.75 to 1 g/Kg intraperitoneal) to which its own red cells, previously labelled with $^{51}Cr$, were transfused, like an "intravascular tracer".

On a multi-trace photograph recorder, we recorded blood pressure in a femoral artery through a Statham transducer, any occurring vocalisation with a microphone, and blood mass present in a well-defined brain area through radioactivity by means of a shielded NaJ-Tl scintillator connected to a rate me-

TABLE 2 - Effect of analgesic drugs on vocalisation induced by intracisternal bradykinin in the rabbit.

| PRE-TREATMENT | | TIME INTERVAL in min | INTRACISTERNAL BRADYKININ µg/animal | ANIMALS WITH VOCALISATION / TREATED ANIMALS |
|---|---|---|---|---|
| -----(*) | ------- | -- | 5 | 3/4 |
| morphine | 10 mg/Kg s.c. | 30 | 5 | 6/7 |
| -----(*) | ------- | -- | 10 | 6/6 |
| methadone | 5 mg/Kg i.p. | 10 | 10 | 6/6 |
| -----(*) | ------- | -- | 5 | 5/5 |
| phenylbutazone | 150 mg/Kg s.c. | 30 | 5 | 5/5 |
| -----(*) | ------- | -- | 10 | 6/6 |
| indomethacin | 10 mg/Kg i.p. | 10 | 10 | 4/6 |

(*) control group

ter. More details are reported elsewhere (10). The effects produced by intracarotid injection of synthetic bradykinin (BRS 640 Sandoz) at low doses (3 to 5 µg/Kg) are shown in fig. 1.

One can see that the peptide causes increase in radioactivity - corresponding to an increase in cerebral blood mass - , vocalisation, and a pressure alteration consisting in a transient hypotensive phase followed, as a rule, by a moderate but somewhat persistent hypertension. When hypotension occurs bradycardia sets in. The effects described above are peculiar for the intracarotid administration of bradykinin; by intravenous injection, on the other hand, marked hypotension and decreased brain blood mass occur, with no vocalisation.

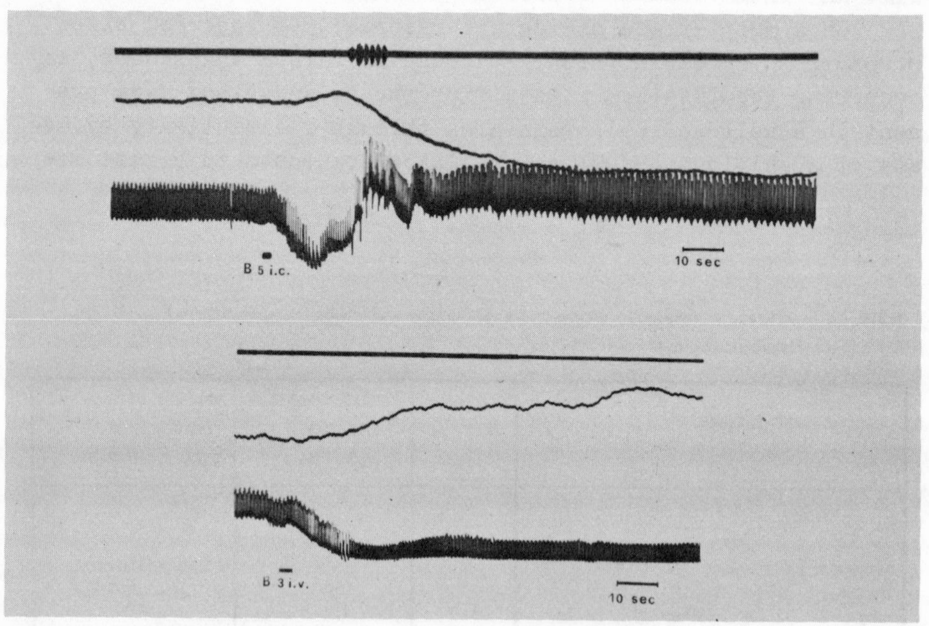


Fig. 1 - Effects of synthetic bradykinin on the anaesthetised rabbit. Upper record: 5 µg/Kg intracarotid; lower record: 3 µg/Kg intravenous. From top to bottom: phonogram, radioactivity over the brain area, femoral artery blood pressure.
Since pulses coming out from the rate meter have negative polarity, their trace lowers as radioactivity increases, and vice versa. Time constant of integration = 20 seconds.

Well now, the question arose whether vocalisation and change in cerebral blood mass might be interdependent. We approached this problem in two ways. First we investigated how other vasoactive substances might affect the brain circulation and, moreover, whether or not they could induce vocalisation. Secondly, we tried to ascertain a time relation between the two responses to intracarotid bradykinin, i.e. brain blood increase and vocalisation. In both instances we came to the conclusion that these phenomena, although associated, are not interdependent.

Effects of drugs on cerebral circulation and vocalisation. Actually, among the substances capable of affecting the cerebral circulation that we considered (histamine, serotonin, ergotamine, eledoisin, norepinephrine), only ergotamine and histamine intracarotidally injected produced vocalisation in some instances. It is of interest to point out that high doses of histamine and ergotamine injected into the cisterna magna sometimes gave a picture very close to that given by bradykinin. Table 3 shows that intracarotid ergotamine produced vocalisa-

TABLE 3 - Effect of intracarotidally and intracisternally injected histamine dihydrochloride and ergotamine methansulfonate in the rabbit.

| SUBSTANCE | DOSE AND ROUTE | ANIMALS WITH VOCALISATION / TREATED ANIMALS | CHANGES IN BRAIN RADIOACTIVITY |
|---|---|---|---|
| histamine | 50 µg/Kg i.carotid | 0/4 | decrease (in all cases) |
| " | 100 " " | 0/5 | " " |
| " | 200 " " | 3/12 | " " |
| ergotamine | 100 " " | 0/2 | no change " |
| " | 150 " " | 2/4 | increase " |
| " | 300 " " | 3/10 | " " |
| histamine | 50 µg/animal i.cisternal | 0/3 | |
| " | 100 " " | 3/3 | |
| ergotamine | 100 " " | 1/5 | |
| " | 200 " " | 2/4 | |

tion in some rabbits and, at the highest doses, an increase in brain blood mass in all cases. A typical picture of the effects exhibited by intracarotid ergotamine is shown in figure 2. The blood pressure behaviour pattern was as for bradykinin. With intracarotid histamine, on the other hand, variation in cerebral blood mass was the contrary of that obtained with ergotamine and bradykinin, even though vocalisation was produced in some cases. The effect on the blood pressure was not, however, very different. Finally, we have to take into account that the intracarotid administration of ergotamine just before that of bradykinin did not prevent vocalisation, indeed in 27 % of the experiments (four animals in 15) it favoured it.

Time relationship between brain blood increase and vocalisation by bradykinin. As to the time relations between variations in the cerebral blood mass and vocalisation, figure 3

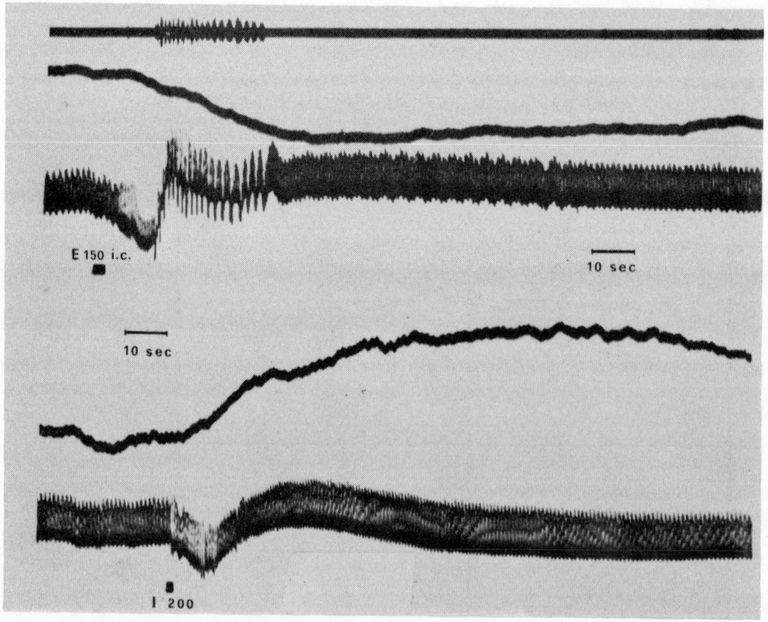


Fig. 2 - Effects of ergotamine methansulfonate 150 µg/Kg (upper record) and of histamine dihydrochloride 200 µg/Kg (lower record) intracarotidally administered to anaesthetised rabbits. The traces are arranged as in fig.1.

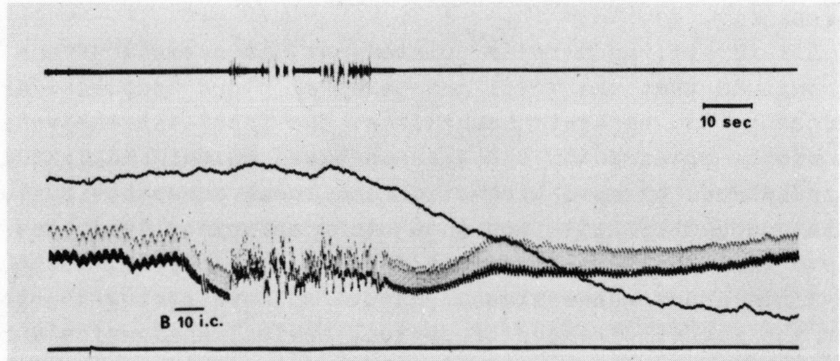


Fig. 3 - Time relation between vocalisation and cerebral blood increase induced by intracarotid injection of bradykinin.

shows that vocalisation begins with or precedes the increase of radioactivity in the brain. On the other hand, it was pos= sible to observe in few animals that the previously observed typical response to intracarotid bradykinin injection was af= fected by administration of INPEA: although vocalisation was

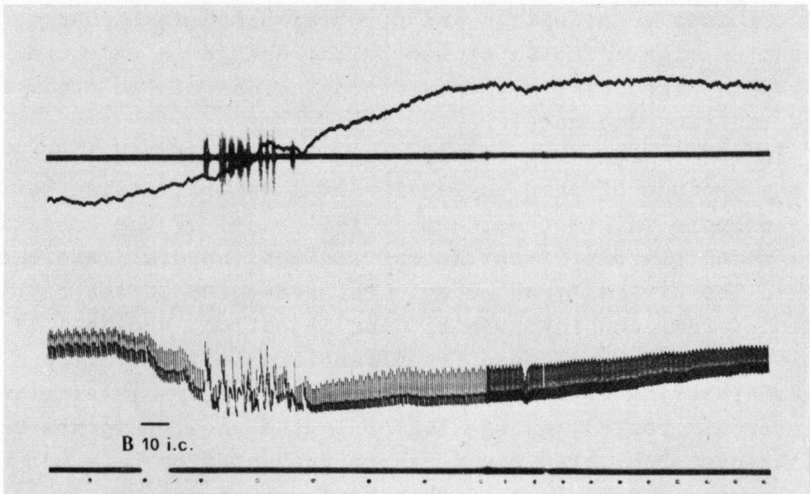


Fig. 4 - Effects of bradykinin injected to a rabbit via the carotid inflow soon after the administration of INPEA 10 mg/ Kg through the same route.

produced, the cerebral blood mass behaviour pattern was reversed (see fig. 4).

All in all, as already pointed out, it seems that one might conclude that the vocal and cerebral blood responses are produced by two separate mechanisms. The fact that the vocal response is obtained in the anaesthetised animal and that it is not influenced by antiphlogistics and local anaesthetics (unlike that caused by injection into other arteries) leads one to believe that the vocal response depends on a central action of bradykinin about whose algesic nature we have strong doubts, given the ineffectiveness of typical central analgesics such as morphine and methadon. So far, to pinpoint the sensitive sites, we have only carried out a small number of experiments. Using a stereotaxic technique, we injected bradykinin directly into predetermined areas of the brain - into the hypothalamus, the putamen, the globus pallidus - but with no vocalisation response (9).

Another problem we have not yet dealt with is that concerning the relationship between pressure and vocal responses. We have already noted that ergotamine and histamine, which are sometimes capable of inducing vocalisation, cause a change in pressure very like that caused by intracarotid bradykinin. If we examine the records of the various experiments carried out with intracarotid bradykinin we see that the variation in pressure precedes vocalisation and cerebral blood mass change. Vocalisation might also be caused by the change in pressure, or bradykinin might well act on different areas of the brain which cause the two responses.

In conclusion, our results seem to indicate that bradykinin may produce effects suggestive of a central action but that these effects may be considerably influenced by the conditions under which the experiment is carried out. Special reference is made to the diversity of behavioural responses of the rabbit to intracisternal and intraventricular injection. Besides, if we consider a common response (vocalisation) to two different ways of administering bradykinin (i.e., through intracisternal and intracarotid route), we see that only the vocal response caused by intracisternal bradykinin can be prevented by beta adrenergic blocking drugs. On the other hand, since antiphlogistics and analgesics have no effect on vocalisation by intracisternal bradykinin, one doubts whether it is true pain-vocalisation in the way that vocalisation caused by injection into other arte= ries cartainly is.

## REFERENCES

1) Guzman F. et al. - Arch. int. Pharmacodyn., 136, 353, 1962.
2) Braun C. et al. - J. Physiol., 155, 138, 1961.
3) Rocha e Silva M. et al. - J. Pharm. exp. Ther., 128, 217, 1960.
4) Armstrong D. et al. - J. Physiol., 135, 350, 1957.
5) Emele J.F. et al. - J. Pharm. exp. Ther., 134, 206, 1961.
6) Sicuteri F. et al. - Boll. Soc. it. Biol. sper., 42, 845, 1966.
7) Bertolini A. et al. - Boll. Soc. it. Biol. sper., 42, 1922, 1966.
8) Corrado A. et al. (cited by Rocha e Silva M. - Ann. N.Y. Acad. Sci., 104, 190, 1963).
9) Unpublished data.
10) Bertolini A. et al. - Boll. Soc. it. Biol. sper., 43, 794, 1967.
11) Guzman F. et al. - Arch. int. Pharmacodyn., 149, 571, 1964.
12) Sicuteri F. et al. - Sett. med., 49, 7, 1961.
13) Concioli M. et al. - Atti Accad. med. lomb., 16, 268, 1961.
14) Breda G. et al. - Atti Accad. med. lomb., 17, 235, 1962.

# THE INFLAMMATORY RESPONSE AND POLYPEPTIDES

L. M. Greenbaum, R. Freer and M. C. Carrara

Department of Pharmacology, College of Physicians and Surgeons, Columbia University, New York, N. Y.

The properties of bradykinin have given it a major place as a suspected chemical mediator in inflammation. Since several other mediators, such as histamine, serotonin and even "slow reacting" substance may be involved in the inflammatory reaction, the elucidation of the importance of each of these mediators and the role they play in inflammation is very difficult. Investigations are going on in several laboratories, including our own, to try and elucidate some of the biochemical factors in the relationship and release of these mediators during injury and inflammation.

The release of kinins into an area of injury can take place in several ways. There can be the formation of kinins by means of the kallikrein system or possibly by means of the cathepsins as demonstrated some time ago in our laboratory (1). One other source of kinins are those formed from enzymes liberated from cellular elements that are drawn from the blood to the inflammatory site.

Polymorphonuclear leucocytes are some of the earliest cells to arrive at the inflammatory site. These cells come via the blood stream through the capillaries into the inflammed area and digest or phagocytize various materials and finally break open and liberate their enzymes and other materials into the surrounding medium (2). We, therefore, felt that there was a possibility that these cells or the enzymes from these cells could possibly attack bradykininogen to form kinins. Kinins might increase the capillary permeability and more leucocytes would come to the area (3). In addition, other enzymes present in the area, including enzymes that destroy bradykinin would regulate or inactivate some of the kinin produced. In our studies, some of which have been published, we have made artificial inflammatory reactions and collected the polymorphonuclear leucocytes (PMN cells) from the inflammatory site and studied them in terms of the kinin forming and kininase enzymes present in the cells (4). We have also investigated some of the biochemical properties of these enzymes.

In order to produce polymorphonuclear leucocytes, rabbits are injected with glycogen (5) and after three hours or more the exudate which is formed is washed out with saline. The cells in the exudate are collected by centrifugation and the PMN cells are purified by various washing procedures. Finally, we can open up the PMN cells and subject their contents to differential centrifugation in order to obtain a lysosomal fraction (pellet) and what we call an extra-lysosomal fraction (supernatant). We also can analyze, in a crude fashion, the nuclei.

When we incubate human bradykininogen (a partially purified preparation that our laboratory has prepared (6) with the lysosomal fraction or the extra-lysosomal fraction we can obtain kinin activity as demonstrated by the contractions of the guinea pig ileum or rat uterus. The kinin-forming activity is thus present apparently throughout the cell and we have recently discovered (7) that it is present in the nuclei. The kinin produced is completely destroyed by carboxypeptidase B.

If we study the pH optima of the reaction between human bradykininogen and the rabbit polymorphonuclear kinin-forming enzyme, we found that the reaction takes place best under acid conditions. There is very little activity in the neutral range with this substrate. Professor Habermann was kind enough to send us some of his pure bovine bradykininogen and when we assay the activity produced by the PMN enzyme and this substrate we note that the enzyme is most active on this substrate in the neutral range although we still find a good deal of activity on the acid side. Thus, we can conclude that pH optima of the enzyme activity depends upon the substrate being used.

Mr. Freer of our laboratory has begun to analyze the kinin that is formed from human bradykininogen and the kinin-forming enzymes of the polymorphonuclear leucocytes of the rabbit. He finds that if he chromatographs the material produced by the PMN cell and human bradykininogen on Sephadex G-25 he obtains a material which is eluted much earlier than bradykinin and which has a greater ratio of activity between the rat uterus activity and the ileum than does bradykinin. Thus, it would appear from the information we have so far that the peptide that is being produced from human bradykininogen by the PMN cell is not bradykinin but probably something related to methionyl-lysyl-bradykinin or kallidin. Our studies are continuing so as to identify the peptide chemically.

Dr. Carrara in our laboratory has begun investigation of the purification of the kinin-forming enzyme or enzymes in the polymorphonuclear leucocytes. She finds that if she chromatographs the lysate of the white cells on Sephadex G-200 she obtains at least three peaks of activity. Since this is Sephadex, we do not know whether this may be breakdown products of an original enzyme or may represent several different enzymes. We are studying this interesting observation further.

We have also investigated the time course of the inflammation exudate with relation to the formation of kinin-forming activity with time. Rabbits are injected with glycogen and then at various times exudates are washed out and analyzed for kinin-forming activity as well as kininase activity. We find that over a period of time after glycogen injection there is a continuing increase in kinin-forming activity per volume of exudate up to 24 hours after the glycogen injection. However, if one measures the number of polymorphonuclear leucocytes in the area, one sees that there is a simultaneous increase in the number of leucocytes in the area. Thus, this increase in kinin-forming activity in the exudate is a function of the number of polymorphonuclear leucocytes in the area. We also find an increase in kininase activity with time after glycogen injection. In collaboration with Dr. Yamafuji in Japan, we have been studying this activity. We have found that most of the kininase activity resides in the extralysosomal fraction. This is in contrast to the kinin-forming activity which is present throughout the cell.

Analysis of synthetic substrates hydrolyzed by the supernatant; one sees in these cells, a good deal of dipeptidase and perhaps even tripeptidase activity, indicating a varied group of peptidases in these cells. In contrast to the blood enzyme (carboxypeptidase-N), enzymes in the white cells do not hydrolyze hippuryl-L-arginine or hippuryl-L-lysine. Therefore, from this evidence and other evidence we have, we have concluded that the type of attack by the kininase in the white cell is not the same as that of the plasma carboxypeptidase. The latter attacks at the carboxyl-terminal arginine of bradykinin. The attack by leucocyte kininases is probably somewhere in the middle of the molecule.

Summary. I can summarize then by saying that our laboratory has been investigating one of the cellular elements of inflammation, the polymorphonuclear leucocyte, for enzymes that form and destroy kinins. Our investigations have shown that there are several different species of kinin-forming enzymes (at least by

Sephadex) and that the polypeptide produced by the PMN cell may be related to methionyl-lysyl-bradykinin rather than to bradykinin. The kininase activity in these cells is different than the blood carboxypeptidase since the activity of the PMN kininase(s) appears to involve the center of the bradykinin molecule rather than the terminal carboxyl position. The question remains as to the role of cells like these in producing chemical mediators such as kinins in inflammation. Our system is a slow one compared to the system of Cline and Melmon who claim a rapid or explosive formation of kinins from granulocytes under specific conditions (8). On the other hand, the inflammatory response is known to consist of various phases with the later phase being most consistent with a slow build up of a chemical mediator. This would also be consistent with our findings on the rate of kinin formation by the enzymes in the rabbit cells. It should be noted that Jasani, Katori, and Lewis (9) have found little contribution of white cells in producing or destroying kinins in synovial fluid of rheumatoid patients while Movat (10) feels that polymorphonuclear leucocytes play an important role in releasing kinins during anaphylactic reactions. Austin and his group have recently shown that PMN cells may also release slow reacting substance (11).

As stated at the outset, the situation is complex. One thing is clear, PMN cells isolated from inflammatory reactions do contain enzymes that can form and destroy kinins. The relative importance of this reaction needs further clarification.

## References

1. Greenbaum, L. M. and Yamafuji, K., Br. J. Pharmac. Chemother., 27, 230-238, 1966.
2. Rogers, D. E., In: International Symposium on Injury, Inflammation and Immunity, ed. Thomas L., UHR, J. W. & Grant, L. H. Williams and Wilkins, Baltimore, 1964.
3. Greenbaum, L. M. and Yamafuji, K., In: Proceedings of the 3rd International Symposium on Hypotensive Peptides, ed. Erdös, E. G., Back, N. and Sicuteri, F., held in Florence, Italy, Springer Verlag, New York, 1965.
4. Greenbaum, L. M. and Kim, K. S., Brit. J. Pharmac. Chemother., 29, 238-247, 1967.
5. Cohn, Z. A. and Hirsch, J. G., J. exp. med., 112, 983-1004, 1960.
6. Greenbaum, L. M. and Hosoda, T., Biochem. Pharmac., 12, 325-330, 1963.
7. Greenbaum, L. M. and Freer, R., Unpublished observations.
8. Cline, M. J. and Melmon, K. L., Nature, 213, 90, 1967.
9. Jasani, M. K., Katori, M. and Lewis, G. P., J. Physiol., 190, 28P, 1967.
10. Uriuhara, T. and Movat, H. Z., Proc. Soc. Exp. Biol. Med.,

124, 279, 1967.

11. Orange, R. P., Valentine, M. D. and Austen, K. F., Science, 157, 318, 1967.

This investigation was supported by grants from the Life Insurance Medical Research Fund and the General Medical Research Fund of the U.S.P.H.S. L. M. Greenbaum is a Career Scientist of the Health Research Council of the City of New York.

# KININS IN CHRONIC INFLAMMATION

G.P. Lewis

CIBA Laboratories, Horsham, Sussex, England

There are several factors which indicate that kinins play a role in inflammatory reactions. Firstly they possess the pharmacological actions which characterise the inflammatory response. For example kinins give rise to changes in the blood vessels and in the extravascular space similar to those which occur during the early stages of inflammation. Blood flow through the inflamed tissue is increased as a result of vasodilatation; fluid containing a high concentration of plasma protein escapes from the vessels into the interstitial space; and the flow of lymph leaving the area and its protein content are both increased. In addition kinins produce pain and although there is tachyphylaxis the pain sensation is like that produced during tissue injury. Finally in higher concentrations kinins cause accumulation of polymorphonuclear leucocytes, although it is difficult to ascertain whether this is a direct action of the peptides or the result of other substances which have escaped from the plasma after the kinins had increased vascular permeability.

Another characteristic which is consistent with the view that kinins play a role in such a general reaction as inflammation is the ubiquity of the whole system. Both the enzymes, or at least their inactive precursors, which are responsible for the formation of kinins and the protein substrate from which kinins are derived are present in all tissues, in the blood and tissue fluids.

The questions that we are led to ask are, how does this system produce the kinins? and how can we estimate the extent of kinin formation?

There appear to be two possible answers to the first question. Firstly the kinin-forming enzymes could be activated by local changes in the tissues such as denaturation of protein or direct cellular damage. Secondly the forming enzymes could be brought in an active form to the site of injury and released there from specific cells. In support of the latter view several investigators have recently suggested that polymorphonuclear leucocytes contain kinin-forming enzymes and that when these cells accumulate at the site of inflammation they release their enzymes flooding the area with kinins (Eisen 1966; Greenbaum & Kim 1967; Melmon & Cline 1967).

Jasani, Katori and Lewis (1967) have examined the origin of intracellular enzymes and the enzymes of the kinin system in the synovial fluid collected from patients with rheumatoid arthritis and related diseases. The results of this investigation have a direct bearing on the question of whether the kinin system is activated in situ or whether the kinin-forming enzymes are brought from another site inside migratory cells during the inflammatory response.

We determined the concentrations of kinin-forming, kinin-destroying and kininogen activities as well as lysosomal and cytoplasmic enzymes and protein in centrifuged cell-free synovial fluid, in non-centrifuged synovial fluid which still contained cells and in the plasma of the same patients. The cell-free samples were centrifuged before freezing immediately after collection. All samples were frozen and thawed several times and stored at -20°C so that the cells in the non-centrifuged samples were completely disrupted before the enzymes assays were made.

As shown in Table I the concentrations of acid phosphatase (acid phos.), β-glucuronidase (β-gluc.), lactic dehydrogenase (LDH) and glutamic oxalacetic transaminase (GOT) were higher in the synovial fluids which had not been centrifuged and contained cells than in the centrifuged cell-free samples. Further the concentrations of the enzymes in the cell-free fluid were higher than those in plasma. The concentration of protein which we know is derived mainly from plasma is the same whether the fluid was centrifuged or not. It therefore appears that these enzymes, like potassium were derived from the cells of the synovial fluid and not from the plasma.

However when we consider the kinin-forming, and kinin-destroying enzymes, kininogen and protein the distribution pattern is different. Their concentrations in cell-containing synovial fluid were equal to or lower than the concentrations in either cell-free fluid or in plasma. It would therefore appear that the enzymes of the kinin system were derived not from the cells but from the plasma. This view was confirmed by the finding that the pellet obtained after centrifugation of synovial fluid contained high concentrations of the intracellular enzymes, with the exception of LDH, but only very small amounts of the kinin enzymes.

Free kinins were not measured in this investigation since the available methods of collection and estimation seem to be inadequate. For instance the kinin-forming system is readily activated and even with the most stringent precautions there is a danger that the small amount of kinin that might be detected in a sample might represent the formation during collection rather than formation at the site of inflammation. Webster and Gilmore (1965) suggested that for the estimation of free kinins in body fluids the fluid should be collected into a syringe containing acid and soya bean trypsin inhibitor. However Greenbaum and Kim (1967) have shown that certain catheptic enzymes which act at an acid pH optimium are not inhibited by this trypsin inhibitor. In addition a

TABLE I

Comparison of Enzyme Activities in Synovial Fluid Before and After Centrifugation with Those in Plasma

| | Acid phos. (u./100 ml.) | β-gluc. (u./ml.) | LDH (u./ml.) | GOT (u/l.) |
|---|---|---|---|---|
| Cell-containing synovial fluid | 16.0 ± 2.6 | 22 ± 3.4 | 1.29 ± 0.23 | 29.1 ± 3.5 |
| Cell-free synovial fluid | 5.4 ± 1.2 | 7.2 ± 1.2 | 0.59 ± 0.13 | 14.7 ± 1.7 |
| Plasma | 1.6 ± 0.4 | 4.7 ± 1.0 | 0.11 ± 0.02 | 10.8 ± 1.6 |

| | Protein (mg/ml.) | Kinin-forming activity (ng/ml.) | Kinin-destroying activity (ng/min/ml.) |
|---|---|---|---|
| Cell-containing synovial fluid | 51.9 ± 3.4 | 359 ± 92 | 168 ± 11 |
| Cell-free synovial fluid | 50.5 ± 3.7 | 568 ± 94 | 161 ± 18 |
| Plasma | 65.1 ± 3.5 | 708 ± 109 | 208 ± 29 |

Mean value ± S.E. of mean.

From Jasani, Katori and Lewis, J. Physiol. (1967), 190, 26-28P.

small formation might occur in the hypodermic needle or in the syringe before the sample is properly mixed with the inhibitors. Measurements of amounts of kinin as small as ng/ml, appear therefore to be unreliable using present methods. Even if there were large amounts of kinins present there would arise the problem of preventing their destruction as in most body fluidsthere is a high kininase activity.

Although these artefacts could make important differences to the small amounts of kinins usually observed, the amounts of enzyme or substrate used up would not significantly affect their assay. The answer to the question of how to estimate the extent of kinin formation at a site of injury would therefore be to examine the enzyme and substrate concentrations rather than those of the peptides themselves.

It was possible to show that the kinin-forming and kinin-destroying enzyme systems in synovial fluid behaved similarly to those in plasma. The forming enzymes could be activated by treatment with acid, by dilution, or by contact activation with glass and could be inhibited by soya bean trypsin inhibitor. Greenbaum and Yamafuji (1966) have reported that leucocytes contain an enzyme which is capable of forming kinins at pH 5. In this investigation synovial fluid containing its cells was acidified to activate the kinin-forming enzymes and instead of neutralising to pH 7.4 for incubation with the substrate the pH was brought to 5.0. Incubation now failed to produce kinin formation. However when this acid incubation mixture was subsequently neutralised and incubated at pH 7.4, kinin formation commenced. Thus the kinin-forming activity of synovial fluid resembles that of plasma in its pH optimium.

The kinin-destroying enzyme or kininase activity in synovial fluid is also similar to that in plasma and different to that found in leucocytes (Zachariae, Malmquist and Oates, 1966; Greenbaum and Kim 1967). The kininase activity of synovial fluid is inhibited by EDTA but not by DFP like that of plasma, whereas the kininase activity of leucocytes is inhibited by

DFP but not by EDTA.

These experiments do not themselves show that kinins play a role in rheumatoid arthritis and related chronic inflammatory conditions. But they show that the components of the kinin system which are present in the synovial fluid of chronically inflamed joints are derived from the plasma and not the invading leucocytes.

## REFERENCES

Eisen, V. (1966). Urates and kinin formation in synovial fluid. Proc.roy.Soc.med., 59 (4), 302-305.

Greenbaum, L.M. and Kim, K.S. (1967). The kinin-forming and kininase activities of rabbit polymorphonuclear leucocytes. Brit.J.Pharmacol., 29, 238-247.

Greenbaum, L.M. and Yamafuji, K. (1966). The in vitro inactivation and formation of plasma kinins by spleen cathepsins. Brit.J.Pharmacol., 27, 230-238.

Jasani, M.K., Katori, M. and Lewis, G.P. (1967). Origin of kinin enzymes and intracellular enzymes in the synovial fluid of rheumatoid patients. J.Physiol., 190, 26-28P.

Melmon, K.L. and Cline, M.J. (1967). Interaction of plasma kinins and granulocytes. Nature, 213, 90-92.

Webster, M.E. and Gilmore, J.P. (1965). The estimation of the kallidins in blood and urine. Biochem.Pharmacol., 14, 1161-1163.

Zachariae, H., Malmquist, J. and Oates, J.A. (1966). Kininase in human polymorphonuclear leukocytes. Life Sci., 5, 2347-2355.

# ADRENERGIC RECEPTORS AND KININS

Sicuteri F., Franchi G., Fanciullacci M., Del Bianco P.L. and (°) Siddel N.

Dep. of Medicine, Univ. of Florence
(°) Dep. of Medicine Univ. of Illinois,Chicago

Epinephrine, as it is known, exhibits a diphasic effect on the blood pressure. In hypertensive patients, if small amounts of epinephrine are used, the hypotension is more evident than hypertension (Greppi 1932, Goldenberger et al. 1948). The effect of nor-adrenaline is simply monophasic. The first phase by epinephrine, namely slight hypertension , is due to the alpha-receptor stimulation; second phase, namely hypotension, is due to a beta-receptor stimulation (Alhquist 1948).

At this point one may put a question: is hypotension a pure vascular effect or does an other mechanism, humoral for instance, also contribute? The fact that, after administration of a beta-blocking drug hypotension disappears, while only hypertension remains (Fig. 1), does not resolve the problem. In fact a hypothetic humoral mechanism may be inhibited by beta-blockers, similarly to other adrenergic metabolic effects.

Now the kininogen-kinins system, when activated, strongly affects the blood pressure. Epinephrine for example could activate this metabolic system (Back 1966, Sicuteri et al. 1963, Periti et al. 1963).

The measurable elements of this system are principally the kinins and the kininogen. Kinins are released enzymatically from the plasmatic precursor, in the alpha-2-globulines area. An excess of released kinins di-

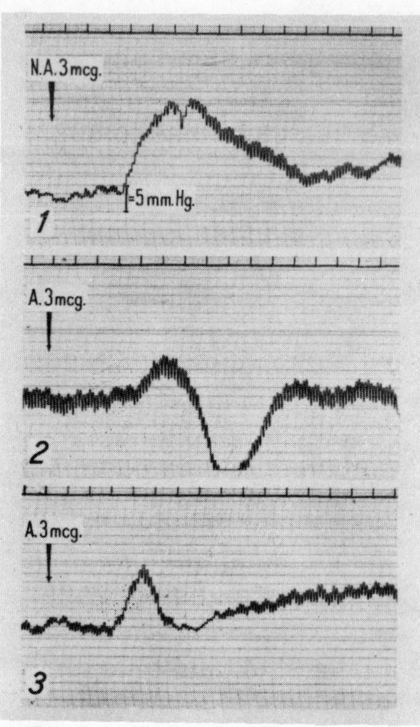


Fig.1 - 1) Arterial hypertension after nor-epinephrine - i.v.
2) Diphasic effects of epinephrine - i.v.
3) Hypertensive effect of epinephrine i.v. after beta-blocker INPEA Metronome: 5 sec.

sappears very quickly from the blood stream. This important "clearance" of kinins depends principally from microvessels which have the capacity to promptly inactivate or fix free kinins (Sicuteri et al. 1963). Plasma also exhibits a kinin inactivating power, due to the carboxipeptidase B enzyme (Erdős and Yang 1966). In this way the measure of plasmatic kinins in aims to focuse a small kinin release, as in clinico-pharmacological conditions, does not appears too suitable. Small kininogenolysis is better evaluated by measuring the change of the plasmatic precursor, namely kininogen.

Some years ago we observed that epinephrine provokes an important fall in kininogen, while nor-epinephrine does not (Sicuteri et al. 1962; Periti et al 1963).

We hypothesized then that this kininogenolysis may be considered as one of the metabolic phenomena induced by epinephrine. Before our studies, Hilton and Lewis (1958) demonstrated in animals an increase in kinin while perfusing the salivary gland with adrenalin.

This problem then seemed to merit more extensive research to confirm the probable kininogenolytic effect of epinephrine and to try understand, with the help of beta-blocking agents, if this effect may be coupled really to beta-receptor stimulation.

In the second group of these studies we have placed the accent on yet another pharmaco-clinical effect of adrenalin. "Metabolic" doses of adrenalin, that is not affecting cardiac or circulatory function, given by vein, are able to provoke angina pectoris and headache rispectively in chronic coronary insufficiency and migraine. This question then follows: is this effect due to a adrenergic kinin release or does epinephrine have also a direct pain facilitating activity?

## METHODS AND RESULTS

Kininogen: the kininogen extraction was carried out according to technique of Diniz and Carvalho (1961), with biological assay on isolated guinea-pig ileum.
Kinins: the extraction was carried out with cold ethanol and with addition of tioglicolic acid. The biological assay was carried out on isolated rat uterus.
Kinin inactivating power of the plasma: 500 ng. of synthetic bradykinin were incubated with 1 ml. of plasma at 37°C. Every 10 min. 1/10 ml. of incubated plasma was diluted with 9/10 ml. of phosphates buffer solution (pH 7,4) and quickly boiled. The residual bradykinin was assayed on isolated rat uterus.
Epinephrine: was injected by intramuscular way (1 mg) or by venous slow infusion (5 mcg/min. within 50 min.).
Nor-epinephrine: was injected by venous slow infusion (5 mcg/min. within 50 min.).
Isoproterenol: was injected by venous infusion (3 mcg/min. within 50 min.).
The blood samples were taken before and every 5 min. after the start of the administration of the drugs.

The study in 27 voluntary subjects has produced the following results.

Adrenalin: a statistically significant lowering in kininogen. This fall occurs chronologically with "beta" adrenergic phenomena, that is tachycardia, tremor, excitability and concomitantly with the augmentation of fibrinolysis. The acute dosis of 1 mg intramuscularly, in comparison to the slow infusion of 5 mcg/min. induces the most intense reaction and the greatest decline in kininogen.

Nor-epinephrine: the same dose and the same route of administration as with adrenaline, were used. A marked hypertension with bradycardia and unchanged kininogen levels are noted.

Isoproterenol: during the slow infusion of 3 mcg per min. cutaneous vasodilation, hypotension and tachycardia are observed. Kininogen tends to diminish in few cases, but is not statistically significant.

Beta-blocker "INPEA": given by rapid venous injection or slow infusion does not provoke lowering of kininogen. If given previously to adrenaline, the vasomotor effect of hormone is diminished or eliminated: the fall of kininogen, at least as measured statistically, does not appear inhibited (fig. 2).

In previous researches, we have noted non significant variation in kinin after epinephrine and nor-epinephrine. In few cases we have observed an increase in the kinin inactivating power of the plasma. After adrenalin in other words the normal kininolytic activity of the plasma appears enhanced.

In a series of experiments using 20 normal rabbits, we attempted to demonstrate the effect of beta-blockade on the kininogenolytic effect of adrenalin. In order to better evaluate the effects of adrenalin which has both alpha and beta actions, we chose also to use a beta stimulator, isoproterenol and a mainly alpha stimulator nor-epinephrine. The femoral artery and peripheral ear vein of unanesthetized animals was cannulated and 2.000 U. heparin were injected; 20 minutes were allowed for recovery. 10 mcg of isoproterenol (IPA) were then administered in the ear vein and blood pressure recorded simultaneously from the femoral artery.

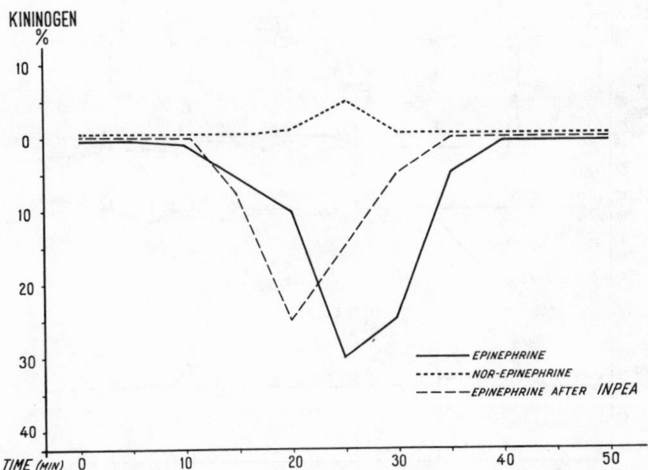


Fig. 2 - Per cent rate of kininogen lowering after i.v. administration of epinephrine, nor-epinephrine and epinephrine after INPEA in man.

At 5, 10 and 20 min. after drug administration 1 ml. blood samples were withdrawn and prepared for kininogen evaluation. In an orderly sequence 25 mcg animal of adrenalin followed 20 min. later by 25 mcg animal of nor-epinephrine were given, with blood samples and pressure recordings as previously described. INPEA in doses of 12,5,25 or 50 mg. animal was then given followed by a repetition of the administration of the IPA, adrenalin and nor-epinephrine sequence 20 min. later. Isoproterenol, at the same time of lowering of blood pressure, induces constantly a fall of kininogen. Epinephrine also lowers the kininogen of the time of blood pressure decreases. Nor-epinephrine, provokes a slight rise in kininogen or no change and as usually increases blood pressure. INPEA by self does not affect blood pressure and kininogen. After beta-blockade the hypotension of isoproterenol is inhibited but little change occurs in the lowering of kininogen. The most striking effect is observable on epinephrine effect. The lowering of kininogen, wich constantly happens within 5 min. is blocked by INPEA. A rise of kininogen may be observed with beginning of a slight fall within 20 minu tes. The adrenaline hypotension is not inhibited even after the highest doses of INPEA. (fig. 3).

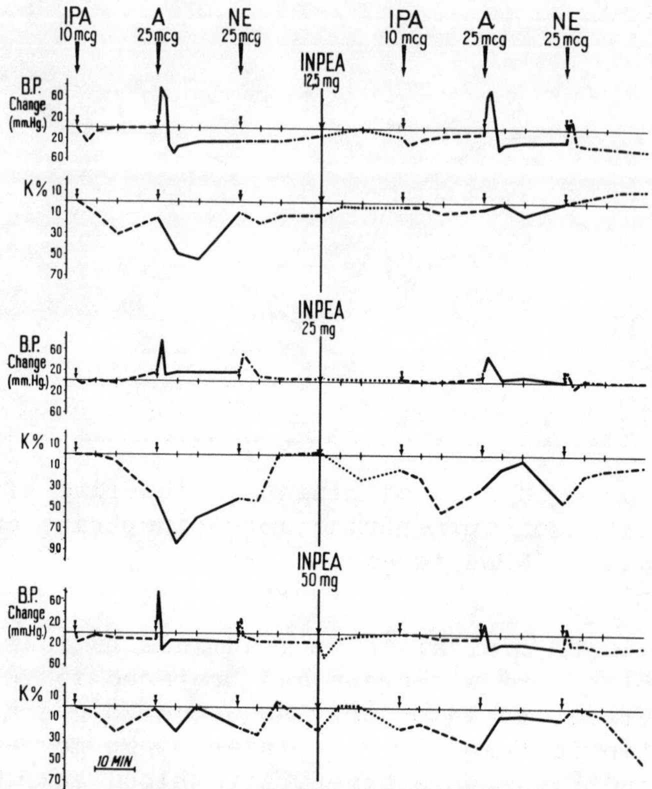


Fig. 3 - Effects of INPEA on blood pressure and kininogen changes induces by isoproterenol, adrenaline and nor-epinephrine in rabbits.

Although this is a preliminary study, it can be noted that beta-blockade affects the kininogenolysis by epinephrine but not blood hypotension. Singularly the beta-blocker, while inhibits the isoproterenol hypotension does not effect isoproterenol kininogenolysis.

## COMMENT

By discussing this group of experiments, it seems possible to deduce as follows:

1) In man and in rabbit epinephrine exhibits a kininogenolytic activity.

2) Our data do not permit us to know whether this metabolic effect of adrenalin is direct or mediated, for example by an induction of hyperfibrinolysis.

3) In rabbit kininogenolysis seems coupled with beta-receptor stimulation as it is inhibited by beta-blockers. In man the kininogenolysis, as other metabolic effects, such for instance as glycogenolysis, seems due to a sub-type of beta-receptor, in accordance with the slight inhibition obtainable with beta-blocking drugs.

4) In man the beta-blockade inhibits the circulatory beta effects of adrenaline but not the kininogenolysis; the opposite happens in rabbit.

In conclusion, kininogenolysis probably because of its slowness, would not seem to contribute to the vasodilatory effect of adrenaline, at least in the clinical field.

The second clinico-pharmacological aspect of adrenaline remains to be considered. This hormone, also in non vasoactive, but only "metabolic" doses evokes clinical pains as those of angina pectoris and migraine. The kinin release may be implicated: in fact the injection of bradykinin or kallikrein is able to induce identical effects.

Furthermore, we would like to point out a particular observation, concerning the correlation between adrenaline, kinin and pain. Without going into detail, the local contact of pain-receptors with adrenaline enhances strongly the sensitivity to bradykinin, in the sense that this peptide evokes pain when administred in lower doses than usual.

In identical conditions the contact of pain receptors with nor-adrenaline, does not induce this sensitization to bradykinin.

The pain-facilitation by adrenaline may be simply due to a summation phenomenon, when considering the kininogenolytic ability of this amine. It is more reliable to think that beta receptors have something to do with pain receptors. In conclusion, when beta-receptors are stimulated by epinephrine and in minor extent by isoproterenol, the pain-producing property of bradykinin seems enhanced.

Adrenalin is able to induce pain of angina pectoris and headache when injected not only in non vasoactive, but also in "only metabolic" doses. These clinical pains may be provoked by particular emotional stresses; the same type of emotion, that is acute an-

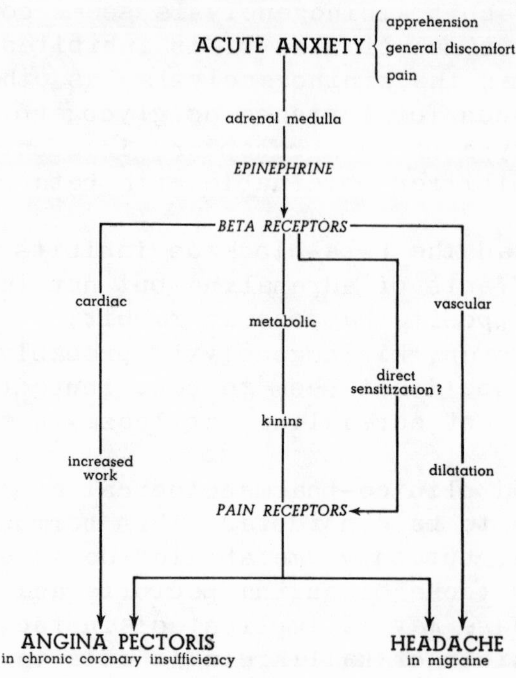


Fig. 4

xiety, provokes a marked release of adrenaline but not of nor-adrenaline.

We want to propose a possible sequence of events in which emotional stresses release adrenaline which, in turn, releases kinin and evokes pain (fig. 4). Apprehension, general discomfort and pain are emotions able to release epinephrine from the adrenal medulla, which on one hand affects, the vascular and heart receptors and on the other releases kinins and perhaps sensitizes pain receptors.

In angina pectoris as in headache the attacks induced by emotional stress or by injection of vasoactive doses of adrenaline may be mediated partially by the kinin mechanism, however the increase of heart work and respectively the dilatation of head vessels remain the primary factors. On the other hand, when angina pectoris is induced by slow infusion of non vasoactive but only metabolic amounts of adrenaline the most reliable mechanism appears that of the kinins. In this case at least apparently, the heart and vascular motor beta-receptors are not implicated.

This point seems to be most stimulating aspect of the problem, especially because of the practical impor tance of angina pectoris; in this field investigations are in progress.

## SUMMARY

Epinephrine administered intramuscularly and by venous infusion lowers plasmatic kininogen in man and in rabbit, whilst nor-epinephrine is inactive in this sense. Decrease of kininogen (that is, release of ki-nins) is perhaps another metabolic feature of epineph-rine, and is appears to be correlated to stimulation of adrenergic beta-receptors. As an indirect evidence of this, we can consider the results of researches by means of isoproterenol and beta-adrenergic blocking agents.

Epinephrine sensitizes the nociceptors to the pain producing properties of bradykinin in man. Some clini-cal-pharmacological implications are considered.

## REFERENCES

Ahlquist P.R.: "A study of the adrenotropic receptors" Am. J. Physiol. 153, 580, 1948.

Back N.: "Int. Symp. on Hypotensive Polypeptides" Fi-renze, ottobre 1965; Springer Verlag, Ed. N.Y., pag. 578 (Discussion).

Diniz C.R., Carvalho L.F., Ryan J. and Rocha E. Silva M.: "A micro metod for the determination of brady-kininogen in the blood plasma" Nature 192,1194,1961.

Erdös E.G. and Yang H.T.: "Inactivation and potentia-tion of the effects of bradykinin" Relaz. Int. Symp. on Hypotensive Polypeptides Firenze, ottobre 1965, Springer Verlag Ed. N.Y. 1966, pag. 235.

Greppi E.: "L'ipertensione arteriosa come automa dis-funzione e malattia" Ed. Pozzi, Roma, 1932, Relaz. 38° Congr. Soc. It. Med. Interna.

Goldenberger M., Pineas F.L., Greene D. and Roh G.E.: "The hemodinamic response of man to nor-epinephrine and its relation to the problem of hypertension"

Amer. J. Med. 6, 792, 1948.

Hilton S.M. and Lewis G.P.: "Vasodilatation in the tongue and its relationship to plasma kinin formation" J. Physiol. L44, 532, 1958.

Periti P., Sicuteri F., Gasparri F. and Franchi G.: "Plasmachinine, bradichininogeno ed attività chininasica del sangue umano dopo somministrazione parenterale di adrenalina e noradrenalina" Boll. Soc. Ital. Biol. Sper. 39, 318, 1963.

Sicuteri F., Periti P., Gasparri F., Franchi G. e Fanciullacci M.: "Diminuzione del tasso ematico di bradichininogeno nell'uomo dopo somministrazione di adrenalina e callicreina" Boll. Soc. Ital. Biol. Sper. 38, 1033, 1962.

Sicuteri F., Periti P., Anselmi B. e Fanciullacci M.: "Livelli plasmachininici nell'uomo dopo somministrazione endovenosa ed endoarteriosa di bradichinina sintetica e naturale" Boll. Soc. Ital. Biol. Sper. 39, 314, 1963.

# THE FORMATION, ACTIONS AND PROPERTIES OF ANAPHYLATOXIN

W. Vogt

Department of Pharmacology, Max-Planck-Institut

für experimentelle Medizin, Göttingen, Germany

It is not the intention of this paper to collect the many details about anaphylatoxin (AT) which have accumulated during nearly 6o years but to present an up-to-date picture of what the nature of this powerfully toxic compound is, what its actions are and how it is formed. This in order to draw the attention of biochemists, pharmacologists and physiologists interested in peptides, to this compound. As a tradition, the exploration of AT forming has so far been largely a domaene of scientists engaged mainly in immunological or allergological problems. A comprehensive review has been given by Giertz and Hahn (1966). More recent results are also collected (Vogt 1967 a).

AT was detected after it had been predicted on the base of a theory, which later proved to be wrong. Friedberger (191o) found that it formed in guinea-pig serum when the serum was incubated in the presence of immune precipitates at 37°. He recognized it by the toxicity and even lethal effects which the incubates produced on injection into normal guinea pigs. According to the theory the toxic principle was a cleavage product of the antigen split from it by guinea-pig serum complement in the presence of antibody. It is now certain, that AT is a product formed exclusively from plasma constituents. The addition of antigen-antibody complexes serves merely to initiate its formation. The formation can be induced in plasma or serum of rat, guinea pig or hog, not only by contact activation with immune precipitates but also with various polymer substances, mainly undissolved neutral polysaccharides. Under the same conditions AT activity does not develop in human, bovine or dog plasma.

## PREPARATION AND PROPERTIES

AT has been purified from activated rat and hog plasma (Stegemann, Vogt and Friedberg, 1964; Vogt 1967a; 1967 b). The procedure used more recently has been adapted to large scale preparation. As shown in fig. 1, it contains 5 steps of purification.

AT is a large basic peptide which does not dialyse in aqueous solution, probably because of aggregation. According to gel chromatography in phenol-acetic acid solution the molecular weight of hog AT appears to be 76oo. The compound contains 15% N, and essential disulphide bridges (Vogt, 1967 a,b). Analytical data on the amino-acid composition have been obtained by Stegemann, Bernhard and O'Neill (1964) and Stegemann et al. (1965). The results may, however, have to be reexamined with preparations now available as they were obtained with very small samples of material of low nitrogen content.

In acid solution AT is surprisingly stable resisting even boiling for several minutes. At alkaline pH it is destroyed quickly at moderately elevated temperatures.

Hog serum
|
activation with yeast
|
batch adsorption on CM-cellulose
|
Amberlite XAD-2
|
Sephadex G 100
|
CM-Sephadex C 50
|
Sephadex G 25
|
lyophilization

Fig. 1. Purification of Anaphylatoxin.

Unlike bradykinin AT is stable in plasma once formed. Trypsin inactivates it (Stegemann, Vogt and Friedberg, 1964).

## ACTIONS

The effect which originally defined the compound is the "shock" which AT produces on i.v. injections into guinea pigs. Of activated, AT-containing rat plasma 3 ml/kg may be lethal. Purified hog AT is lethal at doses of 25 - 5o µg/kg (Bodammer, 1967). It is thus more toxic than histamine, even on a weight basis. The main adverse effect is bronchospasm which leads to asphyxia. When an animal has recovered from a sublethal dose a second injection of AT will have a much less severe or no bronchoconstrictor effect, i.e. the action on the bronchi is tachyphylactic. The constrictor action is preceded by a short period of respiratory stimulation which effect is not tachyphylactic (Bodammer and Vogt, 1967).

The blood pressure of guinea pigs shows a biphasic response after injection of AT. Initially, it falls and then it rises above normal. The latter effect but not the former is subject to tachyphylaxis and can be blocked by sympatholytics. It is probably due to catecholamine liberation.

Various smooth muscle organs are contracted by AT (Rothschild and Rocha e Silva, 1954; Friedberg, Engelhardt and Meineke, 1964; Vogt and Zeman, 1964); all these effects are subject to more or less pronounced tachyphylaxis. Isolated guinea-pig ilea contract maximally to 0.1 µg/ml AT. They are now mostly used for bioassay. The phenomenon of tachyphylaxis facilitates the identification of AT but it is an obstacle to its quantitative estimation, at the same time. By suitable arrangements a reasonably quantitative assay is, nevertheless, possible (Rocha e Silva and Rothschild, 1956; Randall et al., 1961; Vogt, 1967 a).

Guinea pigs can be protected from lethal doses of AT by antihistaminics (Hahn and Oberdorf, 195o). Also the action on the guinea-pig ileum is blocked by this group of drugs. This finding led to the hypothesis that AT acts by liberating histamine, and, in fact, a release of histamine from perfused guinea-pig organs and mast cells has been demonstrated (Rocha e Silva, Rothschild and Aronson, 1952; Mota, 1957). The phenomenon of tachyphylaxis could then be easily explained by assuming an exhaustion of histamine stores. However, several facts point at direct effects of AT. 1) Tachyphylactic isolated organs become responsive again to AT when a rest of 1o - 3o min is allowed (Randall et al., 1961; Friedberg et al., 1964; Vogt and Zeman, 1964). 2) Antihistaminics do not block the action of AT on other isolated organs but the guinea-pig ileum (Vogt and Zeman, 1964). 3) AT contracts isolated organs under conditions under which histamine is inactive (Vogt and Zeman, 1964). 4) When the dose of AT is increased sufficiently (5 times) it may be lethal to guinea pigs despite pretreatment with high doses of antihistaminics (Bodammer and Vogt, 1967). 5) Some effects of AT (see above) are not subject to tachyphylaxis (Bodammer and Vogt, 1967).

Bodammer (1967) has recently found that the bronchial action of extremely high doses of AT in guinea-pigs can be entirely suppressed by a combined treatment with tripelennamine and atropine. This seems to indicate that the antihistamine-resistant part of AT-action is mediated by a cholinergic stimulus. AT then appears to be a general liberator of local hormones: histamine, catecholamines and acetylcholine.

In contrast to the powerful effects of AT on guinea pigs the compound does not induce serious reactions in other animals, as far as is known. Likewise the mast cell disrupting action seems to be restricted to the guinea pig. In anaesthetized cats, bronchospasm and a hypotensive effect have been observed (Marquardt and Hedler, 1965; Bodammer, to be published).

## FORMATION OF AT

As mentioned already, AT is formed from plasma constituents when the plasma is incubated with suitable contact agents. The contact leads to the activation of an AT-forming enzyme which then acts on a specific substrate, anaphylatoxinogen (AT-ogen). Evidence for an enzymic process was obtained by the following facts (Vogt, 1963, 1964; Vogt and Schmidt, 1966): 1) The formation of AT can be separated into two steps: activation of the enzyme by a contact agent followed by its action on the substrate. The latter step can proceed in the absence of the contact factor. 2) Two serum fractions have been separated one of which acts catalytically (enzyme) whereas the other one (substrate) limits the yield of product. (For older views on the mechanism of contact activation and AT formation see the review of Giertz and Hahn, 1966.

The enzyme has been found in the plasma of all species investigated (Schwoerer, cited from Vogt, 1967a). Only the AT-forming enzyme of rat plasma has been investigated in more detail. It is a thermolabile globulin which contains an essential metal group. The latter is evident from the fact that the enzyme is inactive inthe presence of EDTA but can be reactivated simply by dialysis. Further, carbonyl groups are contained in the enzyme and are involved in its action. Ammonia,hydroxylamine and hydrazine inactivate it, the interaction of $NH_3$ being reversible (Vogt, 1963; Vogt and Schmidt, 1966).

During contact activation with sephadex or zymosan the enzyme is (partly) adsorbed. For this to occur, preparatory biochemical steps are apparently necessary, for the adsorption onto and activation by sephadex does not take place at $0^{\circ}$. Efforts to elute the active enzyme from the contact agent have so far not met with success. By gel chromatography purified enzyme preparations have been obtained but the results have been varying and the yields were poor.

More successful was the purification and characterization of another AT-forming enzyme which is present in cobra venom (Vogt and Schmidt, 1964). It has been separated by gel and ion exchange chromatographies. It is also thermolabile but less so than the plasma enzyme, inhibited by EDTA and inactivated by hydroxylamine. The cobra enzyme acts on the same anaphylatoxinogen which is also utilized by the rat plasma enzyme (Vogt and Schmidt, 1966). Probably the product is also the same (Stegemann, Vogt and Friedberg, 1964) although some quantitative differences in activity have been found (Poppe and Vogt, cited from Vogt, 1967 a). For the action on anaphylatoxinogen the cobra enzyme needs a further serum protein which can be separated from the substrate proper (Jensen, 1967; Vogt and Schmidt, 1967). It has been assumed that the cofactor is a pre-enzyme itself activated by the cobra enzyme and then inducing the formation of AT (Jensen, 1967). However, quantitative experiments have shown that it does not produce AT activity from AT-ogen after incubation with the cobra enzyme unless the latter is still present (Vogt and Schmidt, 1967). This indicates a joined action of the two compounds, cobra enzyme and serum co-factor. They may form a complex. Whether the AT-forming enzyme of plasma also needs this co-factor to be active, cannot be stated yet.

The substrate, anaphylatoxinogen, has been purified from rat and hog serum by gel and ion exchange chromatographies (Stegemann, Vogt and Friedberg, 1964; Vogt and Schmidt, 1966). It is a thermolabile globulin.

Anaphylatoxinogen is present also in guinea-pig serum, but not in human, bovine or dog blood (Schwoerer, cited from Vogt, 1967 a). This is the reason why no AT is detected when sera of these species are incubated with contact agents or cobra enzyme.

The nature of the reaction, that leads to AT formation, is unknown. As anaphylatoxinogen has a larger molecular size than AT, a proteolytic cleavage is probable. However, neither plasma nor cobra enzyme are inhibited by the esterase inhibitor DFP nor by soybean trypsin inhibitor, Trasylol$^{(R)}$ or ε-aminocaproic acid. The cobra enzyme does not split casein nor TAME. Apparently the AT-forming enzymes are different from the group of proteases, trypsin, chymotrypsin, plasmin and kallikreins, which have so many properties in common. On the other hand, they show properties of pyridoxal-phosphate enzymes, inasmuch as they contain essential carbonyl groups as well as a metal. Further, pyridoxal-phosphate at high doses specifically inhibits the reaction which effect might be understood as competition (Vogt and Schmidt, 1966).

The idea that the AT-forming enzyme of plasma may be identical with complement or part of it, goes back to Friedberger himself. In recent years it became doubtful whether complement is involved since Rothschild and Rocha e Silva (1954) were able to generate AT activity in decalcified serum and Hahn (196o) purified AT-forming factors which had a rather low content of C'2 and C'4. On the other hand Osler et al. (1959) showed that various treatments known to destroy one or the other component of complement also deprived rat serum of its AT-forming capacity. These experiments suggestive as they are do, however, not prove the involvement of complement, as the reactions carried out are not specific. Vogt (1967 a; further publication in preparation) has demonstrated that under conditions under which C'1, C'2 or C'4 are inactive or eliminated, AT formation is still possible in rat and guinea-pig plasma.

Results obtained by Jensen (1967) suggest that anaphylatoxinogen may be identical with C'3c. If this is so, C'3c can be acted upon by proteins other than complement factors.

Dias da Silva and Lepow (1967) have recently produced AT-like activity from purified complement factors of human serum. The product showed, however, differences to AT formed by contact in rat or guinea-pig plasma. This and the fact that generation of AT activity in humanplasma by contact agents or cobra venom has been unsuccessful raise doubt whether the product of this complement reaction is AT. Further results have to be waited for to solve this problem.

## FUNCTIONS OF AT

As stated already, Friedberger believed that AT was the toxic principle causing the signs of anaphylactic shock. In the light of later experiments this seems rather unlikely. Not only can true anaphylactic shock be elicited after desensitization to AT but experiments aimed at demonstrating the appearance of AT during anaphylactic shock have failed (Giertz et al., 1958). A role in inflammation or allergy has been presumed by authors in favour of the complement theory of its formation (Osler et al., 1959; Dias da Silva and Lepow, 1967; Jensen, 1967). However, evidence for this is lacking, as yet. Further work is needed to clarify the role of AT. More will have to be known about its actions. The conditions of its formation in vivo will have to be studied, and independently, the nature of the process that leads to its formation. The latter the more, as many species do not form AT in detectable amounts but have the AT-forming enzyme (system) in their serum. This raises the question what kind of reactions are catalyzed by that enzyme.

## REFERENCES

Bodammer, G.: In preparation (1967).

Bodammer, G., and W. Vogt: Actions of anaphylatoxin on circulation and respiration of the guinea pig. Int. Arch. Allergy (1967, in print).

Friedberg, K.D., G. Engelhardt und F. Meineke: Untersuchungen über die Anaphylatoxin-Tachyphylaxie und über ihre Bedeutung für den Ablauf echter anaphylaktischer Reaktionen. Int. Arch. Allergy 25, 154 - 181 (1964).

Friedberger, E.: Weitere Untersuchungen über Eiweißanaphylaxie. IV. Mitt. Z. Immun.-Forsch. 4, 636 - 689 (1910).

Giertz, H. und F. Hahn: Makromolekulare Histaminliberatoren. C. Das Anaphylatoxin. In: Heffters Handbuch der Pharmakologie, Erg.-Band 18/1, p.517 - 545. Springer: Berlin-Heidelberg-New York (1966).

Giertz, H., F. Hahn, I. Jurna und A. Lange: Zur Frage der Beteiligung des Anaphylatoxins im anaphylaktischen Schock. Int. Arch. Allergy 13, 201 - 212 (1958).

Hahn, F.: Anaphylatoxin: Formation, actions and role in anaphylaxis. In: Polypeptides which affect smooth muscles and blood vessels, p. 275 - 292. Pergamon Press: Oxford, London, New York, Paris (1960).

Hahn, F. und A. Oberdorf: Antihistaminica und anaphylaktoide Reaktionen. Z.Immun.-Forsch. 107, 528 - 538 (1950).

Jensen, J.: Anaphylatoxin in its relation to the complement system. Science 155, 1122 - 1123 (1967).

Marquardt, P., und L. Hedler: Anaphylatoxin und DAS ("Frühgift"). Arzneimittel-Forsch. 15, 1261 - 1265 (1965).

Mota, I.: Action of anaphylactic shock and anaphylatoxin on mast cells and histamine in rats. Brit.J. Pharmacol. 12, 453 - 456 (1957).

Osler, A.G., H.G.Randall, B.M.Hill and Z.Ovary: Studies on the mechanism of hypersensitivity phenomena. III. The participation of complement in the formation of anaphylatoxin. J.exp.Med. 110, 311 - 339 (1959).

Randall, H.G., S.L.Talbot, H.C.Neu and A.G.Osler: Studies on the mechanism of hypersensitivity phenomena. IV. An isometric smooth muscle assay system. Immunology 4, 388 - 400 (1961).

Rocha E Silva, M., and M. Aronson: Histamine release from the perfused lung of the guinea pig by serotoxin (anaphylatoxin). Brit.J.exp.Path. 33, 577 - 586 (1952).

Rocha E Silva, M., and A.M.Rothschild: Experimental design for bioassay of a material inducing strong tachyphylactic effect (anaphylatoxin). Brit. J. Pharmacol. 11, 252 - 262 (1956).

Rothschild, A.M., and M.Rocha E Silva: Activation of a histamine-releasing agent (anaphylatoxin) in normal rat plasma. Brit. J. exp. Path. 35, 507 - 518 (1954).

Stegemann, H., G. Bernhard und J.A. O'Neil: Endgruppen von Anaphylatoxin. Einfache Sequenzanalyse von carboxyl-endständigen Aminosäuren. Hoppe-Seylers Z. physiol. Chem. 339, 9 - 13 (1964).

Stegemann, H., R.Hillebrecht und W.Rien: Zur Chemie des Anaphylatoxins. Hoppe-Seylers Z. physiol. Chem. 340, 11 - 17 (1965).

Stegemann, H., W. Vogt und K.D.Friedberg: Über die Natur des Anaphylatoxins. Hoppe-Seylers Z. physiol. Chem. 337, 269 - 276 (1964).

Vogt, W.: Anaphylatoxinbildung durch ein Metallferment. Naunyn-Schmiedebergs Arch. exp. Path. Pharmak. 246, 31 - 32 (1963).

Vogt, W.: Weitere Untersuchungen zur Fermentnatur der Anaphylatoxinentstehung. Naunyn-Schmiedebergs Arch. exp. Path. Pharmak. 247, 327 (1964).

Vogt, W.: The anaphylatoxin-forming system. Ergebn. Physiol. 59, 160 - 184 (1967 a).

Vogt, W.: Preparation of anaphylatoxin (in preparation, 1967 b).

Vogt, W., und G.Schmidt: Abtrennung des anaphylatoxinbildenden Prinzips aus Cobragift von anderen Giftkomponenten. Experientia (Basel) 20, 207 - 208 (1964).

Vogt, W., and G. Schmidt: Formation of anaphylatoxin in rat plasma, a specific enzymic process. Biochem. Pharmacol. 15, 905 - 914 (1966).

Vogt, W. and G. Schmidt: Separation, purification and properties of two proteins from rat serum which are essential for anaphylatoxin formation. Biochem. Pharmacol. (in print, 1967).

Vogt, W., und N. Zeman: Analyse der erregenden Wirkung von Anaphylatoxin auf glatte Muskulatur. Naunyn-Schmiedebergs Arch. exp. Path. Pharmak. 247, 328 - 329 (1964).

# CHOLINERGIC EFFECTS OF ANGIOTENSIN AND BRADYKININ

Jean-Claude Panisset

Department of Pharmacology, Faculty of Medicine, University of Montreal, Montreal, Canada

Investigations of the stimulating effect of angiotensin on isolated intestinal preparations have suggested that part of its action on the smooth muscle is mediated through the release of acetylcholine (1, 2). In a recent report on the effects of angiotensin on the superior cervical ganglion of the cat, we postulated that the stimulating effect of small doses of this polypeptide on the nictitating membrane contraction and on ganglionic transmission was mediated through a cholinergic pathway affecting preganglionic nerve endings (3). In order to test this hypothesis, we first studied the effect of angiotensin on the release of acetylcholine using the isolated and perfused cat's superior cervical ganglion preparation.

Cats of 2-4 kg of both sexes were used. After inducing anesthesia with a trichlorethylene and ether, chloralose was injected intravenously. The right superior cervical ganglion was prepared for perfusion by the method of Kibjakow (4) with the modification made by Feldberg and Gaddum (5). Injections were made through the central end of the lingual artery into the common carotid artery, while the external carotid artery was occluded. The injected substance was thus diverted towards the superior cervical ganglion. Effects of preganglionic sympathetic nerve stimulation were estimated by recording nictitating membrane contractions with an isotonic myograph transducer. For the measurement of acetylcholine release, the ganglia were perfused with fresh and filtered heparinized cat plasma to which eserine sulfate ($10^{-5}$ g/ml) was added. The preganglionic cervical sympathetic trunk was stimulated with square waves of 0.8 msec duration, the stimulus strength being supramaximal for the contraction of the nictitating membrane. The frequency of stimulation was 20 shocks/sec. The content of ace-

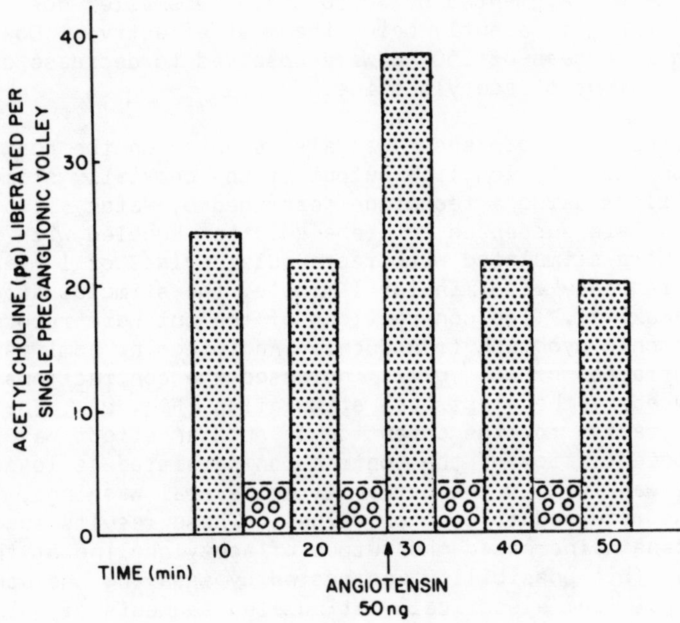


Figure 1. Histograms of output of acetylcholine from perfused eserinized superior cervical ganglia showing the effect of 50 ng of angiotensin. Dotted bars represent acetylcholine released by electrical stimulation. Bars with circles indicate that the acetylcholine content of the control samples was below the threshold for detection.

tylcholine in perfusate samples was measured by comparison with standard acetylcholine chloride solutions on the blood pressure of eviscerated cats, as described by MacIntosh and Perry (6). For these experiments, the following procedure was adopted. The perfusion having been established, samples of the effluent were collected over 5 minute periods. Every 10 minutes, preganglionic stimulation was applied for 3 minutes and 2 further minutes were allowed to insure that all the acetylcholine liberated had been washed out. Each such period was preceeded and followed by a control period during which no stimulation was applied.

Figure 1 shows histograms of output of acetylcholine from a perfused ganglion. Grey bars represent acetylcholine released by electrical stimulation. Dotted bars indicate that acetylcholine content of the control samples was below the threshold for detection. Angiotensin (50 ng) was injected 30 seconds prior to the electrical stimulation. It caused in this instance a 50% increase in the output of acetylcholine. In 12 other preparations with doses of angiotensin ranging from 1 ng to 100 ng, the output of acetylcholine was augmented by 30 to 50%, the smaller doses (between 5 and 20 ng) apparently being the most effective. Doses of angiotensin in excess of 150 ng were observed to decrease or to inhibit the output of acetylcholine.

The effect of angiotensin was also studied on the contractions and on the acetylcholine output of the coaxially stimulated guinea-pig ileum using a technique described by Paton (7). The preparations were suspended in Krebs solution bubbled with 95% $O_2$-5% $CO_2$ and were stimulated with rectangular pulses of 1 msec duration at a frequency of 6/min for 1 minute, the stimulus strength being supramaximal. The contractions of the gut were recorded with an isotonic myograph transducer. Angiotensin, administered at a concentration of $10^{-9}$ g/ml, increased the contractions of the ileum produced by the electrical stimulation (Fig. 2). The threshold concentration for the obtention of such an effect was $10^{-11}$g/ml. This potentiation of the contraction persisted as long as the polypeptide was left in the bath. After several washings, the contractions returned to control values. These results suggested that angiotensin increased the output of acetylcholine at the nerve endings. This possibility was tested by assaying the acetylcholine output from electrically stimulated segments of guinea-pig ileum in the presence of neostigmine ($10^{-5}$ g/ml). These results are shown in Table 1. Assays of acetylcholine output were conducted on strips of guinea-pig ileum treated with neostigmine sulfate (5 mg/l) and morphine sulfate (10 mg/l). In each of 5 experiments, the output of acetylcholine into the bath fluid was measured before and after the administration of angiotensin. The output of acetylcholine due to stimulation was obtained by subtracting the spontaneous output. Angiotensin, at a concentration of

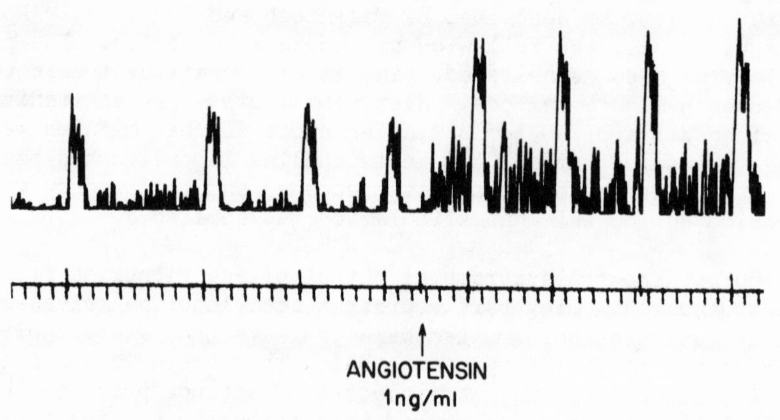


Figure 2. Effect of angiotensin, $10^{-9}$ g/ml, on contractions of coaxially stimulated guinea pig ileum; time, 30 sec.

TABLE 1.

Effect of angiotensin on output of acetylcholine from stimulated guinea pig ileum (ng/shock, corrected from resting output; mean ± S.D.)

| Experiment | Without angiotensin | With angiotensin |
|---|---|---|
| 1 | 0.3 ± 0.2 | 1.5 ± 0.6 |
| 2 | 0.3 ± 0.3 | 1.9 ± 0.4 |
| 3 | 0.6 ± 0.2 | 2.2 ± 0.6 |
| 4 | 0.4 ± 0.1 | 1.9 ± 0.4 |
| 5 | 0.1 ± 0.6 | 1.5 ± 0.1 |

$10^{-9}$ g/ml increased the output of acetylcholine induced by electrical stimulation. The mean output of acetylcholine was .34 ± .28 ng/shock in the control periods and, 1.8 ± .34 ng/shock in the presence of angiotensin, this difference being statistically significant. The effect of angiotensin on acetylcholine output in the absence of stimulation was variable. In some experiments, there was a small increase which neither exceeded 20%.

The effect of angiotensin and bradykinin were finally studied on the output of acetylcholine released from parasympathetic ganglia in the perfused submaxillary gland of the cat. We used the technique described by Emmelin and Muren (8): the submaxillary gland was perfused with Ringer Solution containing eserine sulfate ($10^{-5}$ g/ml). In order to avoid liberation of acetylcholine from the postganglionic nerve endings on stimulation of preganglionic fibres of the chorda tympani, d-tubocurarine ($10^{-5}$ g/ml) was added to the perfusion fluid in order to paralyse the ganglion cells to preganglionic impulses. The gland was perfused through the external carotid artery. All branches except the artery of the submaxillary gland were tied. Perfusion samples were collected from the external jugular vein, all of its branches except the vein draining the gland were previously ligated. For these experiments, the usual procedure was the following: each sample was collected during 3 minutes. After two control periods, the chorda tympani was stimulated during the first two minutes of a 3 minute period. Two control samples were then collected whereupon another stimulation period followed. The whole procedure was repeated after the close arterial injection of angiotensin or bradykinin. The content of acetylcholine in perfusate samples was measured on the blood pressure of eviscerated cats. Figure 3 shows histograms of output of acetylcholine from a perfused gland. Grey bars represent acetylcholine released by electrical stimulation. Dotted bars

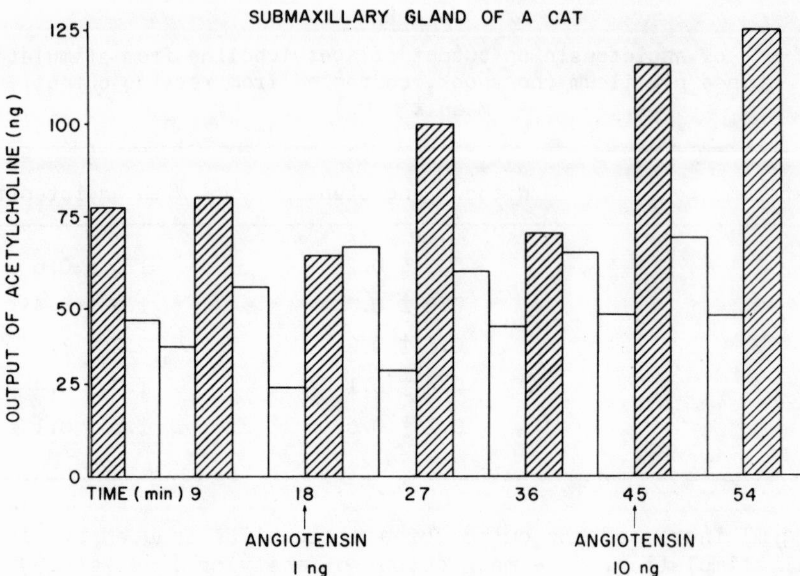


Figure 3. Histograms of output of acetylcholine from perfused cat submaxillary gland showing the effect of 1 ng and 10 ng of angiotensin. Hatched bars represent acetylcholine released by electrical stimulation. White bars indicate the acetylcholine content of the control samples.

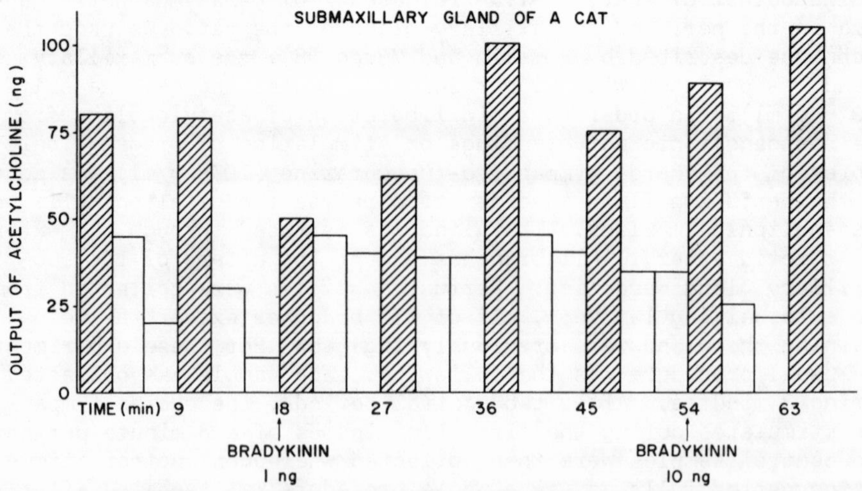


Figure 4. Histograms of output of acetylcholine from perfused cat submaxillary gland showing the effect of 1 ng and 10 ng of bradykinin. Hatched bars represent acetylcholine released by electrical stimulation. White bars indicate the acetylcholine content of the control samples.

represent the control samples. Angiotensin was injected 15 seconds before the stimulation. In 10 such preparations, the output of acetylcholine stimulation period ranged from 65-80 ng before the administration of angiotensin. The initial effect of angiotensin (1 ng) was an apparent reduction of acetylcholine output (15-30%) which was always followed by an enhancement of 25-40% over the control periods during the second period of stimulation. The second administration of angiotensin (10 ng) was always followed by an immediate enhancement of acetylcholine output (40-55%). Angiotensin did not show any effect on the spontaneous output of acetylcholine.

Figure 4 shows that identical results were obtained with bradykinin administered at doses of 1 ng and 10 ng. An initial reduction of acetylcholine output was followed by an enhancement of acetylcholine release.

## SUMMARY

It has been shown:

1) in the isolated and perfused superior cervical ganglion of the cat, that small doses of angiotensin increase the output of acetylcholine liberated by electrical stimulation of preganglionic fibres.

2) in the coaxially stimulated guinea-pig ileum, that angiotensin enhances the release of acetylcholine from postganglionic nerve endings.

3) in the parasympathetic ganglia of the cat submaxillary gland, that angiotensin and bradykinin enhance the output of acetylcholine released by the electrical stimulation of the chorda tympani.

## REFERENCES

1. P.A. ROBERTSON and D. RUBIN. Brit. J. Pharmacol. 19, 5 (1962)
2. P.A. KHAIRALLAH and I.H. PAGE. Am. J. Physiol. 200, 51 (1961)
3. J.C. PANISSET, P. BIRON and A. BEAULNES. Experientia 22, 394 (1966)
4. A.W. KIBJAKOW. Pflüger's Arch. Ges. Physiol. 232, 432 (1933)
5. W. FELDBERG and J.H. GADDUM. J. Physiol., London 81, 305 (1934)
6. F.C. MACINTOSH and W.L.M. PERRY. Methods Med. Res. 3, 78 (1950)
7. W.D.M. PATON. Brit. J. Pharmacol. 11, 119 (1957)
8. N. EMMELIN and A. MUREN. Acta physiol. scand. 20, 13 (1950)

This work was supported by grant MA-1860 from the Medical Research Council of Canada and by grant DA-159 from the National Research Council of Canada.

# ON THE MECHANISM OF BRADYKININ POTENTIATION

U. Hamberg[1], P. Elg, and P. Stelwagen

Department of Biochemistry, University of Helsinki

Finland

Bradykinin potentiating effects have been demonstrated with various compounds, some of them due to a sensitization during bioassay with proteolytic enzymes (1), specific fibrinopeptides (2,3) or peptides isolated from snake venom (4). In earlier reported work we found that the bradykinin-induced contraction of isolated guinea pig ileum was influenced by tryptic peptides split from plasma proteins or obtained by a spontaneous activation of an endogenous plasma protease (5). Since kininase inhibitors have been shown to produce similar effects (6, 7) we have explored the possibility that a competitive enzyme inhibition may regulate the biological activity of bradykinin. Several kininases may have to be accounted for (6). The importance of the carboxyl end arginine for the activity of bradykinin (8) appears first to suggest an enzyme mechanism correlated to a carboxypeptidase-like action (9). As shown below a competitive inhibition of carboxypeptidase B action is obtained with the tryptic peptides isolated from human plasma, which may be released simultaneously with bradykinin.

Material and methods. Human plasma was acid-heated at pH 2, neutralized and incubated with 200 ug of trypsin (Worthington x 2 crystallized, salt free) per ml for 40 minutes at +37°C.

---

[1] Part of this work was carried out during a stay as visiting scientist to the Department of Physiology I, Karolinska Institutet, Stockholm, Sweden.

Released bradykinin was removed by stirring the incubation mixture with IRC-50 (XE-64) resin preequilibrated with 0.5M ammonium acetate pH 5.0. The supernatant was dialyzed for 48 hours against 0.001 N HCl (1:10) at +5°C with stirring, with one change of the outside fluid. The dialysate was evaporated to dryness and redissolved to a concentration of 20 ml of the originally used plasma per ml. Salt was removed by gel filtration on Sephadex G-10 (5) of G-25. Fig. 1 illustrated a recycling experiment with

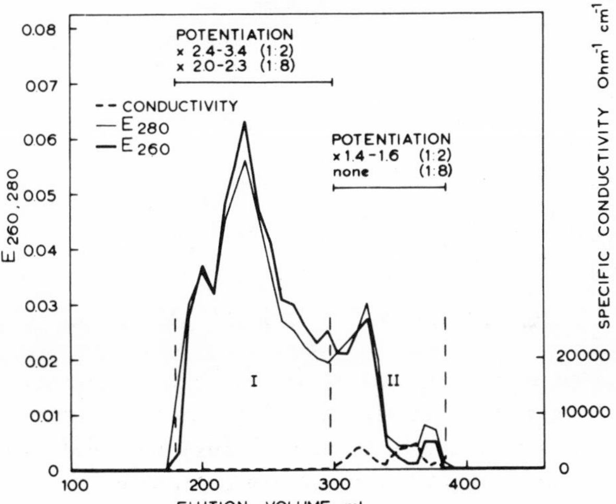


Fig. 1. Recycling on Sephadex G-25 of tryptic peptides obtained from acid-heat denatured human plasma.

fractions containing the potentiating peptides. Further purification was performed by ion exchange chromatography on a Dowex-50 W X-2 column (200-400 mesh) preequilibrated to pH 3.1 with 0.1 M ammonium formate buffer (10). Elution was performed by stepwise change of buffer between pH 3-8; the potentiating peptides were obtained between pH 5-8.

Carboxypeptidase B (Worthington COB-DFP, 3.4.2.2) contained 2.4 mg enzyme protein per ml. Activity was measured according to Folk et al. (11) using hippuryl-L-arginine (HLA, Sigma) as a substrate. The initial reaction velocities were measured at 254 mu on a Beckman DU spectrophotometer. The enzyme was tested for lack of chymotryptic activity on acetyl-L-tyrosine ethyl ester (obtained through the courtesy of Dr. S. Iwanaga) (12).

Potentiation of bradykinin was measured in bioassay on the isolated guinea pig ileum suspended in Tyrode solution at +37°C and aerated or oxygenated with carbogen. The experiments were performed by first determining the linear dose response range with synthetic bradykinin (BRS-640, Sandoz Ltd.). The standard bradykinin (0.625-2.5 ug per ml) was compared at three dose levels with the peptide solution mixed with an equal volume of the same standard solution (Fig. 2, 3).

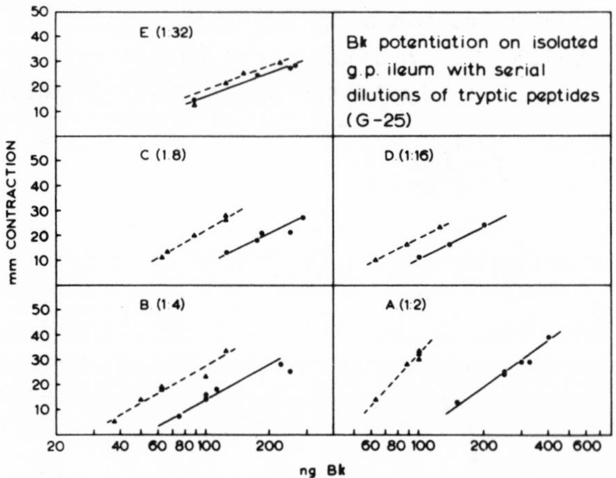


Fig. 2. Effect of serial dilutions (A to E) upon the potentiation of bradykinin. Dose response lines at (minimum) three dose levels with (o) standard synthetic bradykinin (BK) and test solution (4); see methods. The contractions (isotonic-isometric) were registered with a Grass Polygraph. The sensitivity of the ileum was 20-100 ng BK per ml bath volume (2.5 ml); the threshold dose was 25 ng BK. The doses were given at three minutes intervals. Times potentiation obtained: 2.4-3.4 (A), 2.0-2.5 (B), 2.0-2.3 (C), 1.6 (D) and 1.0-1.3 (E).

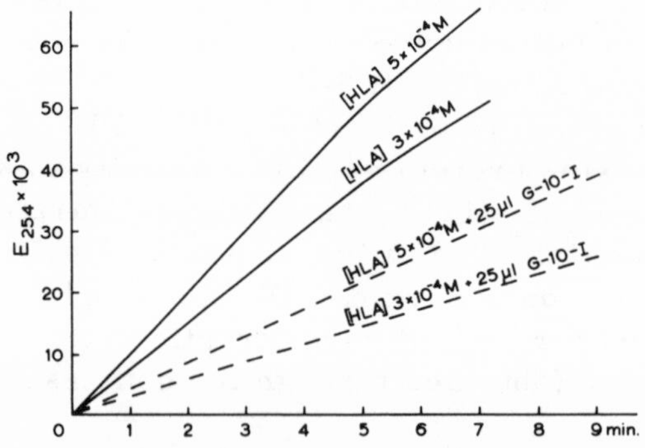


Fig. 3. Inhibition (dotted line) of carboxypeptidase B action (full line: controls) on HLA with the tryptic peptide preparation G-10 at two different substrate concentrations. Enzyme concentration 0.29 ug per ml.

Salt concentration was estimated by determination of specific conductivity (Philips conductivity measurement bridge PR 9500). The potentiation of bradykinin with doses giving equal contractions was calculated as follows:

$$\text{times potentiation} = \frac{\text{ng bradykinin in standard dose}}{\text{ng bradykinin in test dose}}$$

Results and discussion. The potentiating effect obtained with a peptide preparation recycled on Sephadex G-25 is shown in Fig. 1. The fraction pools I and II were evaporated to dryness and redissolved in 5.0 ml redistilled water. As appears from the elution profile the different concentration of peptides in the respective pools presumably accounts for the different potentiation effects obtained. This further indicates that several peptides may cause potentiation. Fig. 2 illustrates the effect of a serial dilution of the pool I material upon the potentiation. With a 1:32 dilution (E) of this peptide preparation the potentiation of bradykinin almost disappeared. No contraction of the ileum was obtained with the tryptic peptide solutions alone.

The inhibition of the carboxypeptidase B action on HLA is demonstrated in Fig. 3 using a crude preparation obtained after gel filtration on Sephadex G-10. In this type of gel filtration the potentiating peptides were obtained in the first fraction pool (G-10-I) (5). The amount of peptide preparation used (25 ul) equals 0.5 ml of the original plasma applied to the experiment. The competitive nature of the inhibition is demonstrated in Fig. 4

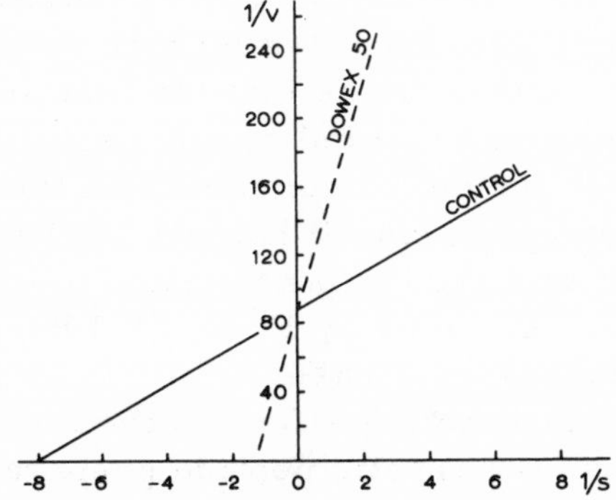


Fig. 4. Competitive inhibition of carboxypeptidase B action on HLA caused by a tryptic peptide preparation from human plasma (G-10-I/Dowex 50-purified) plotted according to Lineweaver and Burke. Assay conditions: 0.31 ug enzyme protein per ml: HLA concentrations 0.1 to 1 mM: 10 ul peptide preparation (dotted line).

with a peptide preparation further purified on Dowex-50 W X-2 resin. The similar result is shown with the synthetic salt free bradykinin triacetate (Fig. 5).

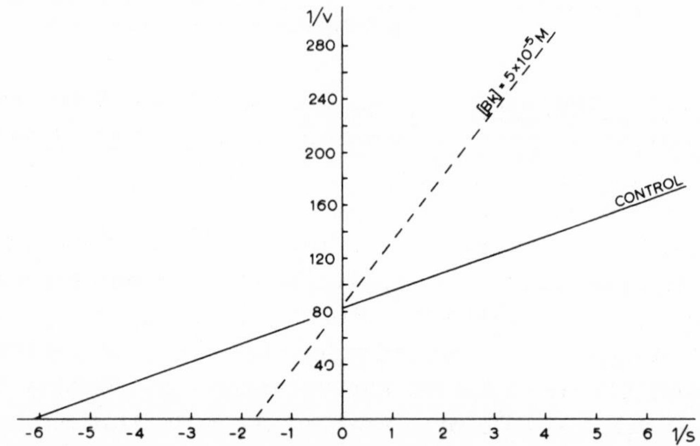


Fig. 5. A Lineweaver-Burke reciprocal plot showing the competitive inhibition of carboxypeptidase B action on HLA by synthetic bradykinin triacetate (Bk; kindly supplied by Sandoz Ltd. Basle). Assay conditions comparable to those in Fig. 4.

This indicated that the *in vitro* mechanism of inhibition obtained with the tryptic peptides may depend on the presence in the carboxyl end of arginine or/and lysine.

The presence of arginine in the potentiating Dowex-50-purified peptide fraction was sustained by Sakaguchi reaction. After incubation of the same preparation with carboxypeptidase B, free arginine could also be demonstrated by high voltage paper electrophoresis. These results indicate that the inactivation of bradykinin presumably may be retarded by any peptide having a C-terminal arginine (or lysine) to compete with the carboxypeptidase action. Since carboxypeptidase B has not yet been found in blood the question as to if this type of potentiation mechanism occurs *in vivo* is still open. As found by Erdos (13) human blood contains another type N of carboxypeptidase.

While having no contracting effect on the isolated guinea pig ileum by themselves, the tryptic peptide preparations strongly potentiated the smooth muscle contraction of bradykinin, when assayed in a mixture. To avoid the interference of these peptides in a bioassay of bradykinin separation is easily achieved by ion exchange chromatography on IRC 50 (XE-64) resin by a method described in detail before (14). The broader aspects concerning the possible physiological or pathophysiological role of this phenomenon in blood clotting and fibrinolysis have been discussed elsewhere (5).

## REFERENCES

1. Edery, H. Brit. J. Pharmacol. 22: 371, 1964.

2. Gladner, J.A., Murtaugh, P.M., Folk, J.E. and Laki, K. Ann. N.Y. Acad. Sci. 104: 47, 1963.

3. Gladner, J.A. Hypotensive Peptides, Ed. Erdos, E.G., Back, N. and Sicuteri, F. Springer-Verlag, New York Inc. p. 344, 1966.

4. Ferreira, S.H. Ibid. p. 356.

5. Hamberg, U. Ann. N.Y. Acad. Sci. 1967. In Press.

6. Erdos, E.G. and Wohler, J.R. Biochem. Pharmacol. 12: 1193, 1963.

7. Lewis, G.P. Physiol. Rev. 40: 647, 1960.

8. Stewart, J.M. and Woolley, D.W. Nature 207: 1160, 1965.

9. Erdos, E.G. and Yang, H.Y.T. Hypotensive Peptides, Ed. Erdos, E.G., Back, N. and Sicuteri, F. Springer-Verlag, New York Inc. p. 235, 1966.

10. Blomback, B., Blomback, M., Edman, P. and Hessel, B. Biochim. Biophys. Acta 115: 371, 1966.

11. Folk, J.E., Piez, K.A., Carroll, W.R. and Gladner, J.A. J. Biol. Chem. 235: 2272, 1960.

12. Schwert, G.W. and Takenaka, Y. Biochim. Biophys. Acta 26: 570, 1955.

13. Erdos, E.G. and Sloane, E.M. Biochem. Pharmacol. 11: 585, 1962.

14. Hamberg, U. Ann. Acad. Sci. Fenn. A 113, p. 33, 1962.

# MECHANISM OF ACTION OF RELAXIN

H. Struck und N. Bhargava

Biochemistry Department of the II$^{nd}$ Surgical Clinic of Cologne University
Cologne-Merheim, West Germany

The findings of HISAW on pubic symphysis lead to the discovery of relaxin in 1926. The peptide hormone relaxin is produced in ovaries, specially of pregnant animals and in the placenta (1,2). During pregnancy relaxin content in the blood increases and decreases at once after the birth (3). Therefore besides progesterone one calls it hormone of pregnancy. To display its full activity one has to give a pre-treatment of estrogen. The relaxin is not available in homogeneous form up till now. The purest preparations obtained so far had a molecular weight of 8 000 - 10 000 (4,5,6). Recently we were successful in obtaining active relaxin fractions having a molecular wt of 4 000 - 5 000 (15) as determined on sephadex.

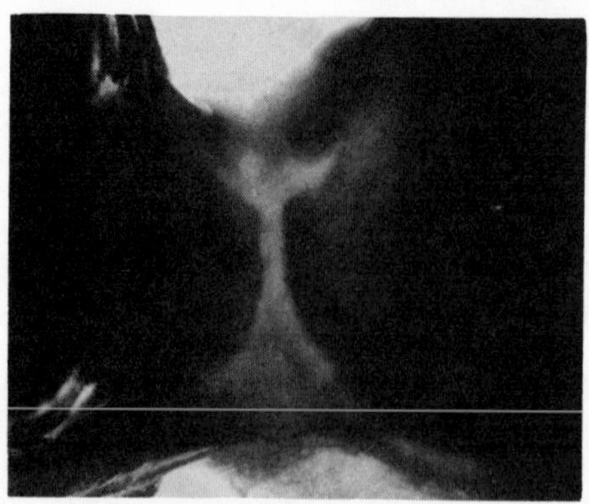

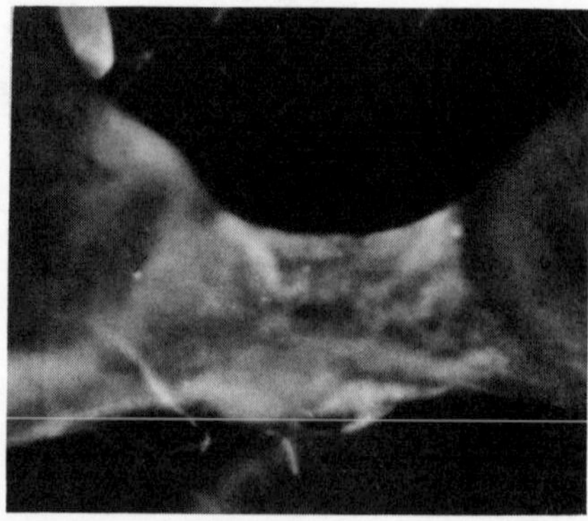

Figure 1. The effect of relaxin on pubic symphysis
a) without relaxin b) with relaxin

The determination of biological activity is based on the opening of pubic symphysis. The figure 1 shows the effect of relaxin on pubic symphysis.

The table 1 shows important effects of relaxin on endocrine organs. After relaxin administration histological examination of pubic symphysis showed depolymerisation of the ground substances, loosening and swelling of the collagen fibers. Chemically an increase in water content and soluble collagen has been noticed. The total collagen decreases. The relaxin administration causes also an inhibition in contraction of uterus. After treatment with relaxin, uterus shows similar histological changes as in the case of pubic symphysis. Collagen fibers loosen and swell and vacuoles appear. Chemically an increase in water content and a decrease in total collagen was found.

Dilatation of cervix caused by relaxin yields same histological and chemical findings as in the case of uterus. The keratinization of vaginal epithelium in mice by estrogen (as examined by Allen-Doisy-Test) is potentiated by relaxin. Histological and chemical investigations were not carried out in this direction.

The lobules of the mammary glands of mice, rats and guinea pigs showed an increased development after the treatment of relaxin. Here also histological and

Table 1. ***Effects of RELAXIN on endocrine organs***

| Organ | Species | Effects | Histological findings | Chemical findings | Investigators |
|---|---|---|---|---|---|
| pubic Symphysis | monkeys, guinea pigs, rats, mice | opening and softening | depolymerisation of ground substances, loosening and swelling of collagen fibers | increase of $H_2O$ content, decrease in total collagen content, increase of soluble collagen | Catchpole, Kroc, Crelin, Steinetz, Frieden, Struck, Hall, Talmage, Hisaw, Zarrow |
| Uterus | human, rats, monkeys, mice, guinea pigs | inhibition of contraction | loosening and swelling of collagen fibers, formation of vacuoles | increase of $H_2O$ content, decrease in total collagen | Folsome, Steinetz, Harkness, Wiqvist, Kelly jr., Zarrow |
| Cervix | human, rats, cows, pigs, monkeys | dilatation | depolymerisation of ground substances, loosening and swelling of collagen fibers | increase of $H_2O$ content, decrease in total collagen | Eichner, Hall, Hisaw, Zarrow |
| Vagina | mice | potentiation of keratinization of vaginal epithelium induced by estrogen | | | Dewar, Struck |
| Mammary glands | rabbits, mice, guinea pigs, rats | increase development of lobules | | | Hamolsky, Smith, Wada |

chemical investigations are not available in literature.

Inhibition of uterus contraction as well as softening of the cervix were also found in human beings. These findings lead to clinical application of the relaxin.

Further it was of interest to evaluate whether the effect of relaxin is limited to endocrine organs only. The first investigations on this problem were carried out by BOUCEK et al (11) and SOBEL et al (12). We have also carried investigations on this subject in our laboratories. The results are compiled in table 2. The general effect of relaxin were mainly investigated on rats. In particular they were concerned with metabolism of collagen and cholesterol. Although these investigations were carried out on rats, it seems that similar collagen metabolism also occurs in human beings as relaxin has successfully been used in scleroderma.

An inhibition of cholesterol biosynthesis (13) and increase in acid soluble collagen (11) also increase in hydrolysis of collagen fibers by trypsin was noticed in the skin of female rats by STRUCK (8).

In the hypercholesterinemic rats the relaxin depresses the cholesterol level in the liver and serum as observed by STRUCK (8). The effect of collagen metabolism by relaxin is similarly shown by an increase in urinary hydroxyproline after treatment with relaxin (8).

Table 2. *General Effects of RELAXIN*

| Organ | Species | Effects | Methods | Investigators |
|---|---|---|---|---|
| Skin | Rats ♀ | Inhibition of cholesterol biosynthesis<br>Increase in acid soluble collagen<br>Increased hydrolysis of collagen fibers by Trypsin | chemical | Noble and Boucek<br>Sobel<br>Struck |
| Liver | Rats | Decrease in cholesterol level | chemical and histological | Struck |
| Serum | Rats ♀ | Decrease in cholesterol level | chemical | Struck |
| Urine | Rats ♀ | Increase in hydroxyproline level | chemical | Struck |
| Pituitary body | Rats ♀ | Increase in TSH synthesis | chemical | Plunkett |

Relaxin causes an increase in the biosynthesis of thyrotropic hormone as PLUNKETT (14) could confirm by radiological methods. Mechanism of action of relaxin was up till now not very clear. There is a reciprocal action between the other hormones and relaxin as pretreatment of estrogen is necessary for achievement of maximum activity of relaxin. STEINETZ (7) has published a hypothesis for the mode of action of relaxin on pubic symphysis, which is not applicable for other endocrine organs and particularly for general effects of relaxin.

In the figure 2 I would like to show you a scheme of action of relaxin in endocrine organs which is in good agreement with the results known so far.

The investigations of B.G. STEINETZ (9) and others have shown that the somatotropic hormone, as well as the estrogen are necessary for the endocrine action of relaxin. Inhibition of activity of relaxin by androgens and corticosterone can be partially reversed by a high dose of estradiol, so that one may assume that there is a regulation mechanism. Depending on concentration, progesterone can inhibit or activate

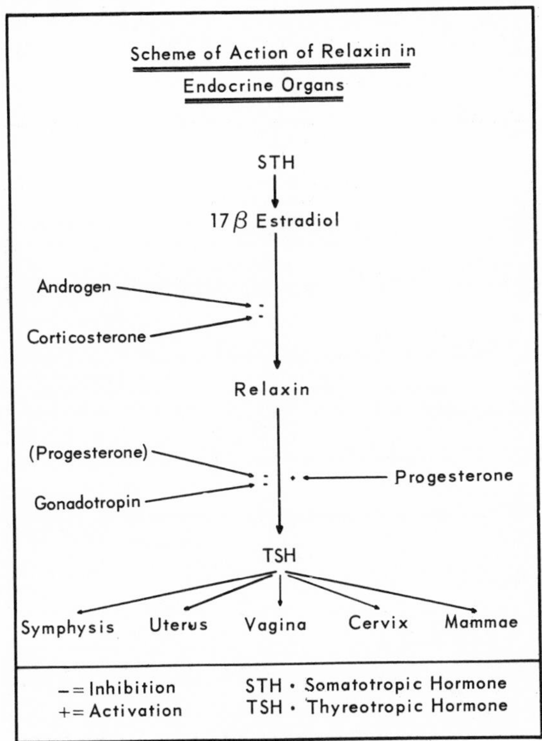


Figure 2.

the effect of relaxin. ZARROW (10) also found that the small concentration of progesterone activates the biosynthesis of relaxin. Gonadotropic hormone increases the endogenic synthesis of estrogen. The investigations of STEINETZ (9) and coworkers have shown that the effect of the relaxin on pubic symphysis of mice can be inhibited by high concentration of 17ß-estradiol. The scheme shown in figure 2 is not as such applicable for general effects of relaxin. In this case reciprocal effect of estrogen and relaxin is taken into consideration. Such reactions are known in which for example estradiol and relaxin do not behave synergistic but as antagonist (e.g. depression of serum cholesterol level). Taking these results into consideration a second scheme for general effects of relaxin has been out lined. This scheme is shown in the figure 3.

The general effects of relaxin on collagen and cholesterol metabolism show a fundamental difference with endocrine activities: pre-treatment of estradiol is not absolutely necessary for general effects of relaxin. All other effects are similar e.g. inhibition by androgens and progesterone and the mode of action of relaxin along with the TSH of pituitary body. In this case role of somatotropic hormone has still to be investigated.

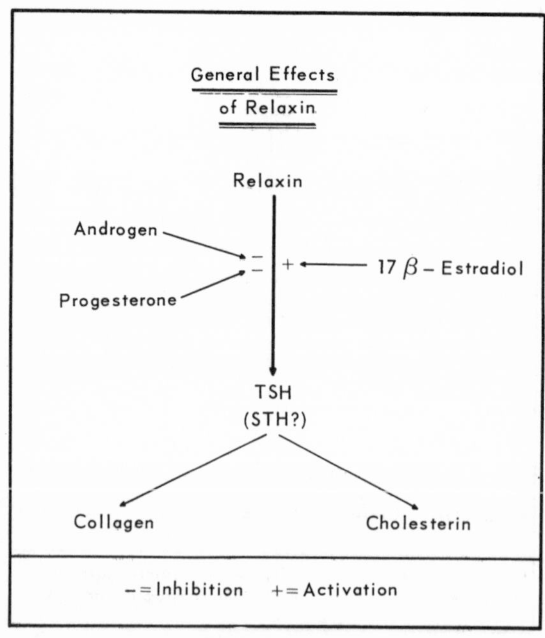


Figure 3.

In accordance with known results and the mechanism of action proposed by us, it seems that the action of relaxin is mainly dependent on thyrotropic hormone of pituitary body.

To know exact mechanism of action of relaxin still extensive experimental work is necessary. Further investigations on the influence of relaxin on enzyme systems and purification of relaxin would give important informations on this interesting hormone and its properties. Experiments in this direction are in progress.

We thank Dr. STEINETZ of Warner-Lambert Research Institute, Morris Plains, N.J., research affiliate of Warner Chilcott Laboratories for providing us W 1164 A, Lot 8, Releasin (R).

Literature.

1) F.L. Hisaw, M.X. Zarrow — Proc. Soc. Exptl. Biol. Med. 69 395 (1948)

2) F.D. Dallenbach, G. Dallenbach-Hellweg — Virchows Arch. path. Anat. 337 301 (1964)

3) M. X. Zarrow, E.G. Holmstrom, H.A. Salhanick — J. Clin. Endocrinol.and Metab. 15 22 (1955)

4) E.H. Frieden, N.R. Stone, N.W. Layman — J. biol. Chem. 235 2267 (1960)

5) H. Cohen — Transact. N.Y. Acad. Sci. 25 313 (1963)

6) H. Struck — Vitamin-, Hormon- und Fermentforsch. 14 370 (1967)

7) B.G. Steinetz, J.P. Manning, M. Butler, V.L. Beach — Endocrinology 76 876 (1965)

8) H. Struck — Thesis for habilitation submitted to the faculty of medicine of Cologne University 1967

9) B.G. Steinetz, V.L. Beach, R.L. Kroc — Methods in Hormone Res. Vol. II S. 559 Academic Press New York-London 1962

10) F.L. Hisaw, M.X. Zarrow, W.L. Money, R.V.N. Talmage, A.A. Abramowitz — Endocrinology 34 122 (1944)

11) H.R. Elden, R.J. Sever, N.L. Noble, R.J. Boucek — Fed. Proc. 16 293 (1957)

12) H. Sobel, H.A. Zutrauen, J. Marmorston — Arch. Biochem. 46 221 (1953)

13) N.L. Noble, R.J. Boucek — Circulation Research 5 573 (1957)

14) E.R. Plunkett, B.P. Squires, F.C. Heagy — J. Endocrin. 26 331 (1963)

15) H. Struck, N. Bhargava — Unpublished results 1966

# COMMENTS ON THE ROLE OF PLASMA KININS IN INFLAMMATION.

Peter Hebborn

Department of Biochemical Pharmacology, School of Pharmacy State University of New York at Buffalo

This symposium contains a series of papers on the production, destruction and possible function of vasoactive polypeptides. Dr. Lewis and Dr. Greenbaum in their presentations were cautious about the role of kinins in inflammation and summarize the evidence which leads to the conclusion that bradykinin and related polypeptides may be involved as mediators of the inflammatory response. A more definitive statement cannot be made at this time because, although all tissues have the capacity for kinin production, the response of tissues to exogenously administered bradykinin does not correlate completely with the microvascular changes observed during an inflammatory response (see Cotran and Majno, Ann. N.Y. Acad. Sci. 116, 750, 1964). In addition, model systems used to investigate the mechanism of action of anti-inflammatory agents, particularly their modification of kinin formation and/or kinin action, have generally failed to give convincing evidence for or against the involvement of kinins as mediators of inflammation. This is the consequence of a poor correlation between anti-inflammatory properties and antagonistic effects in the model systems used.

Data presented in this symposium indicates that the response to bradykinin may vary depending on the events which preceded the application of the kinin. Thus, Dr. Sicuteri has reported that agents with affinity for β-adrenergic receptors modify the response to bradykinin in his vein model. Of great significance is the report by Drs. Back and Wilkens that bradykinin will cause vasoconstriction after pretreatment with substances including histamine. Dr. Hamburg has presented data on polypeptides which are in themselves inert, but which potentiate the effects produced by bradykinin.

The above data are important because inflammation is a sequence of events, starting in many model systems with the transient release of histamine and possibly other vasoactive agents. The subsequent production of bradykinin may produce effects in this "prepared" environment which differ from those produced by bradykinin in "non-prepared" tissues.

It seems likely that further advances in the elucidation of the mechanisms of action of anti-inflammatory agents may follow the development of new model systems of inflammation. Perhaps the approaches in the past have been too analytical, i. e. the effects of one mediator acting alone have been stressed. The indications from this symposium are that one should study the overall effects of potential mediators acting in combination or in sequence and determine the ability of anti-inflammatory agents to modify the response to such combinations.

## AUTHOR INDEX

**(Underscored numbers indicate complete papers in this volume.)**

# SUBJECT INDEX